NUMERICAL COMPUTING AND MATHEMATICAL ANALYSIS

NUMERICAL COMPUTING AND MATHEMATICAL ANALYSIS

Stephen M. Pizer

Department of Computer Science
University of North Carolina
Chapel Hill, N.C.

S R A

SCIENCE RESEARCH ASSOCIATES, INC.
Chicago, Palo Alto, Toronto, Henley-on-Thames, Sydney, Paris
A Subsidiary of IBM

Library of Congress Catalog Card Number: 74–78163
ISBN 0–574–19155–0

Acknowledgments: For problems on pages 99, 123, 169: J. N. Franklin, MATRIX THEORY, © 1968. Reprinted by permission of the author and Prentice-Hall, Inc., Englewood Cliffs, New Jersey. For table 4-11 on page 323: Richard W. Hamming, *Numerical Methods for Scientists and Engineers,* Copyright © 1962 by the McGraw-Hill Book Company, Inc. Reprinted by permission of the McGraw-Hill Book Company, Inc.

to Lynn

TABLE OF CONTENTS

2 SYSTEMS OF LINEAR EQUATIONS

LIST OF PROGRAMS

PREFACE

Numerical computing is a major part of all computing. The characteristics of the high-speed digital computer greatly affect the science of efficiently producing accurate numerical answers to problems. It is therefore necessary for the computer scientist, as well as for all scientists doing numerical computing above the "cookbook" level, to know the basic concepts supporting the numerical use of computers.

To understand these concepts, the computer scientist must use the branch of mathematics called *analysis*: calculus and real functional analysis. This part of mathematics is useful to him (or her) not only in understanding the theory of numerical computation (called *numerical analysis*), but also much more broadly in developing and understanding the many mathematical models used both in computer science and in the process of the application of computers. The ability to apply this mathematics to problems is one of the skills that the computer scientist must develop.

This book then, while focused at the computer scientist, is appropriate for anyone who wants to learn the basic concepts of numerical computing, some important concepts and tools from mathematical analysis, and the application of mathematical analysis to the problems of numerical computing. Thus it is suitable for numerical analysis courses oriented toward computer scientists, applied mathematicians, and/or users of numerical computing who desire more than minimum understanding of these subjects. The book differs from other numerical analysis or numerical methods books in that they tend to focus at either the pure mathematician or the unsophisticated user of numerical methods. The computer scientist or the sophisticated user needs deeper and more careful knowledge of the basis for the methods than the latter group, and more of a feel for application of the methods and their relations to the computer than the former group. He needs to develop mathematical tools and insights related to numerical analysis, but with a regard to the application of these tools and insights in the real world of computers. He needs to understand how and when to use the available numerical methods, but with a depth sufficient to identify the numerical analytic issues relevant to a problem and to deal with those issues so that he can choose from among the methods and decide when to modify them.

Not all computer scientists and sophisticated users of numerical methods have to be numerical analysts. Nor is this text basically meant to produce numerical analysts. It is meant to produce computer scientists, applied mathematicians, and users who can do basic numerical analysis and who can converse at a deep level with colleagues who are numerical analysts. It is meant as a core computer science course for students who may but probably will not become numerical analysts but rather will become scientists who understand the various modes of information processing (including numerical information processing) and computers, and their relationships.

Based on the above goals, this text emphasizes mathematical and intuitive understanding of the problems of numerical computing and their solutions, and of many common numerical methods. Though the text surveys a large number of these methods, it does not cover the method of choice for all cases. Rather, the numerical methods covered were chosen because (1) their development involves common mathematical tools, (2) they bring out essential mathematical insights related to numerical analysis, or (3) they teach important lessons in the application of numerical methods on computers. I believe that such an approach will allow the student to understand, as needed, specific methods described in the present or future literature, whether or not they are described in this book.

Unlike many numerical analysis or numerical methods texts, this text gives significant attention to mathematical tools needed for computer science. It contains surveys of probability, linear algebra, and Fourier analysis, as well as applications of these areas to numerical analysis. The instructor of a course for which one or more of these areas are prerequisites can pass lightly over this material and plunge directly into the applications of the mathematics.

The text is based on a one-year course taught at the University of North Carolina at Chapel Hill. The students in the course are largely senior-year undergraduates and first-year graduate students. The course is required of candidates for the Ph.D. in Computer Science and the B.S. in Mathematical Sciences with an Option in Applied Mathematics. It is recommended for candidates for the M.S. in Computer Science and the B.S. in Mathematical Sciences with an Option in Computer Science who have a relatively strong mathematical background. It is open to seniors and graduate students in all areas who have done well in mathematics and have had at least two semesters of calculus (three or four semesters of calculus are desirable), a basic knowledge of the definitions related to matrices, and an ability to write computer programs in a problem-oriented language. Some algorithms discussed in the book are shown not only as mathematical formulas or general procedures but also in PL/I. However, it is not intended that every student using the text be fluent in PL/I.

Assuming the instructor omits 10% of the material in the book, the text can be covered in one academic year in the order presented. Since this class-tested one-year sequence of presentation does not subdivide easily into separable semester courses, an alternative order is indicated in the table on pages xx through xxvi. In this alternative two-semester sequence, also class-tested, the first semester is self-contained and can be offered by itself if desired. This first semester-course surveys many numerical methods and develops basic numerical analytic and mathematical concepts. The second semester-course develops three major mathematical areas and applies them to increase considerably the student's understanding of numerical computing. It also presents a few numerical methods not covered in the first semester.

To allow instructors to skip material or cover material in an order differing from either of those presented in the table, the prerequisites for each chapter are listed below:

Chapter 1: Single- and multiple-variable calculus; familiarity with floating-point arithmetic operations.

Chapter 2: Basic definitions of matrix and determinant; chapter 1 excluding 1.4.

Chapter 3: Chapter 1 excluding 1.4; single-variable calculus; convergence of sequences.
Sections 3.5 and 3.6: Multiple-variable calculus.

Chapter 4: Chapter 1 excluding 1.4; single-variable calculus.
Section 4.2.1: Definition of $\binom{n}{m}$, $e^{i\theta}$.
Section 4.2.4: Inner product and orthogonality.
Section 4.3: Linear algebra including vector space, basis, dimension, norm, inner product, orthogonality, matrix multiplication, and inverse; Gaussian elimination (chapter 2); probability (section 1.4); elementary concepts of complex analysis, including e^{x+iy}.
Section 4.3.7b,c: Simple eigentheory (section 2.6.1).

Chapter 5: Chapter 1 excluding 1.4; single- and multiple-variable calculus; definition of differential equation; polynomial interpolation and integration (sections 4.2.1 and 4.2.4).
Section 5.4: Iterative solution of nonlinear equations (chapter 3).

Subject	Numerical Methods Included	Lecture Time (in weeks)	Book Sections Covered (not necessarily by lectures)
First Semester			
I. Intuitive Survey of Numerical Methods (simple methods without "bells and whistles," giving handwaving motivation for the methods; homework should involve carrying out methods on sample problems)		4.0	No text reading other than programs:
	Linear equations: Gaussian elimination (without pivoting), Jacobi and Gauss-Seidel iteration;		Programs 2-1, 2-4, 2-5
	Nonlinear equations: methods of Picard, Newton, secant, constant slope, false position, bisection, multidimensional extensions;		Programs 3-1, 3-2, 3-3, 3-4, 3-5, 3-6
	Approximation: Lagrange form of polynomial exact matching, least squares with sinusoids (Fourier), cubic splines;		Programs 4-1, 4-4, 4-10, 4-11
	Integration: Trapezoidal rule, Simpson's rule, Gaussian;		Programs 4-5, 4-6, 4-8

Subject	Numerical Methods Included	Lecture Time (in weeks)	Book Sections Covered (not necessarily by lectures)
	Differential equations: Euler's method, Euler-Heun predictor-corrector method.		Programs 5-1, 5-2
II. Computer Representation of Mathematical Objects (theory development using above methods as motivation and examples)			
A. Numerical methods as operators		0.5	1.1, 1.2, 1.3.1
1. Properties of operators			
2. Repetitive application of operators			
3. Error			
B. Vector Algebra		2.0	2.1, 2.2, 2.4.1, 2.5
1. Vector spaces through basis and dimension			
2. Linear transformations—rank, null space, properties of solutions to linear equations			

Subject	Numerical Methods Included	Lecture Time (in weeks)	Book Sections Covered (not necessarily by lectures)
3. Inner products, norms, orthogonality, determinant			
C. Error generation and propagation			
1. Computational error (with bound as measure)	Linearly converging iterative methods, Newton-Cotes integration, pivoting and scaling in Gaussian elimination.	1.0	1.3, 1.5, 2.3.1, 2.3.2, 2.3.4, 2.3.5
2. Conditioning	Linear equations.	0.5	2.4.2 – 2.4.4
3. Approximational error; polynomial exact matching, numerical integration	Newton divided-difference method, unaveraged equal-interval difference methods, Newton-Cotes integration methods. numerical differentiation methods.	2.0	4.1, 4.2.1 a, b, d, g (through truncation error analysis), 4.2.2, 4.2.3, 4.2.4a
III. Repetitive Application of Operators			
A. Direct methods			

Subject	Numerical Methods Included	Lecture Time (in weeks)	Book Sections Covered (not necessarily by lectures)
1. Averaging independent estimates	Iterated linear interpolation, averaged polynomial exact-matching formulas, Richardson extrapolation-Romberg integration, mop-up in Euler-Heun method, overrelaxation.	1.0	4.2.1e, remainder of 4.2.1g, 4.2.4b
2. Stability; look-ahead and look-back in the solution of differential equations	Euler's method, modified Euler-Heun method, Adams method, Runge-Kutta method (fourth-order).	1.0	Appendix A, chapter 5, 2.8.5
B. Iterative methods 1. Convergence and its properties	Picard method, Newton's method (as overrelaxation), secant method, Aitken's δ^2 acceleration.	1.0	3.1 – 3.5
IV. Choice of Methods	All above plus sketches of multi-dimensional δ^2 acceleration, projection method for linear equations, rational function approximation, Clenshaw-Curtis integration, differentiation using least-squares approximation.	0.5	4.2.1h, 4.4, 2.8.6
		Total = 13.5	

Subject	Numerical Methods Included	Lecture Time (in weeks)	Book Sections Covered (not necessarily by lectures)
Second Semester			
I. Probabilistic Analysis of Error			1.4, 4.3 through 4.3.4
A. Introduction		3.0	
B. Survey of probability theory		↓	
C. Probabilistic treatment of error generation and propagation	Linearly converging single-variable iterative methods.	0.5	
D. Introduction to estimation theory		0.5	
E. Least-squares approximation		1.0	
1. Justification and development			
2. Error analysis for linear case			
II. Orthogonal Functions			
A. In linear least-squares approximation			4.3.5 – 4.3.7, 2.5.2

Subject	Numerical Methods Included	Lecture Time (in weeks)	Book Sections Covered (not necessarily by lectures)
1. Importance		1.0	
2. Orthogonalization		↓	
3. Polynomial least-squares approximation	Polynomial least-squares approximation using recurrence relation.	↓	
4. Properties of orthogonal functions and least-squares approximations using them		0.5	
5. Fourier approximation and analysis	Fourier approximation, including odd reflection with linear tendency subtraction, Fast Fourier transform, discrete filtering.	3.5	
B. Other uses	Gaussian integration; weighted integration; choosing argument points for polynomial approximation.	0.0	4.2.4c, d
1. Classical orthogonal polynomials; Gaussian integration			
III. Eigentheory			
A. Properties of eigenvalues and eigenvectors		1.5 ↓	2.6.1 – 2.6.4 ↓

Subject	Numerical Methods Included	Lecture Time (in weeks)	Book Sections Covered (not necessarily by lectures)
B. Applications		↓	↓
1. Geometric interpretation of mappings			
2. Conditioning			2.6.5
3. Convergence of iterative methods	Jacobi iteration, Gauss-Seidel iteration, projection method, multidimensional δ^2 acceleration applied to these.	0.5	2.8.1 – 2.8.3
C. Numerical methods to compute eigenvalues and eigenvectors	Full pivoting in Gaussian elimination to compute an eigenvector given an eigenvalue, power method including reduction, methods to find eigenvalues using similarity transformations.	1.0	2.7.1, 2.7.2, 2.7.4
		Total = 13.0	

I wish to acknowledge the following textbooks, which have served in some degree as models: *Matrix Theory* by J. N. Franklin† and *Numerical Methods for Scientists and Engineers* by R. W. Hamming.‡

I wish to acknowledge gratefully the important help of the following people in the construction of this text: Dr. Victor L. Wallace for using the manuscript in a one-year course, for his many helpful suggestions, and for being the prime mover in the development of the alternative order presented above; Dr. Frederick P. Brooks, Jr., Dr. Peter Calingaert, Dr. Martin Dillon, Dr. Lester M. Levy, Dr. Charles E. Metz, Dr. David F. McAllister, Dr. Donald F. Stanat, and Dr. Stephen F. Weiss for their important comments; Miss Marilyn Bohl for her careful and comprehensive editing; Mr. Fuad Abukhayr and Mr. Thomas H. Dunigan, Jr., for writing many of the problem solutions; Dr. Lois Mansfield and Dr. Guiliano Gnugnoli for their careful and helpful review of the manuscript; the many U.N.C. students in COMP 151, 152 who found errors, gave constructive criticism, and put up with using the manuscript as a text; Mrs. Linda Vann, Mrs. Roberta Williams, Mrs. Judith Todd, Miss Sarah Jo Wood, and Mrs. Mary Kirkland for typing the difficult manuscript; and my wife Lynn for interminable proofreading help and, particularly, for her encouragement and patience during my many hours of work on this book.

Stephen M. Pizer
Chapel Hill, N.C.

† See the list of references in chapter 2 of this book for details of publication.
‡ See the list of references in chapter 4 of this book for details of publication.

1

NUMERICAL COMPUTING

ACCURACY AND EFFICIENCY

1.1 INTRODUCTION

This book deals with the question of how to take a problem stated in mathematical terms and arrive at numerical answers to the problem. Mathematically stated problems are produced in many areas: the physical sciences, the biological sciences, the social sciences, and business, to name a few. Stating the problem mathematically requires understanding of the area in which the problem has come up. Describing how this step is accomplished is beyond the scope of this book. Rather, this book will assume that the problem is stated in mathematical terms.

Given a problem stated in mathematical terms, the desirable next step is to solve the problem analytically. We cannot overemphasize the degree to which analytic solutions are superior to numerical solutions, which are approximations at best. A search for an analytic solution should be carefully carried out before the necessity for a numerical solution is considered. Usually, at least part of the solution can be obtained analytically. As little as possible should be left to the approximate solution.

Unfortunately, the majority of mathematically stated real-world problems do not have a full analytic solution. This is particularly true of problems that are brought to the attention of the computer scientist. With such problems, proper methods for determining answers must be found or created. Since the methods will produce and work with numbers, they are called *numerical methods*. As indicated above, the problem solution often involves the combination of a numerical method with an analytic method. Furthermore, the basis for many numerical methods is in analytic methods. Therefore, we will spend some of our time discussing some analytical tools for problem solving in applied mathematics and another part of our time discussing the numerical methods themselves.

After a proper numerical method has been found or created by the computer scientist, the method is normally programmed for execution on a computer. It is very important to take this fact into account when creating the numerical method. Furthermore, the problem-solution process does not end with computer output. Rather, the output must be carefully analyzed for its correctness with regard to both numerical accuracy and reasonability. Unreasonable answers can be caused by a poorly stated problem or by blunders in programming as well as by the use of a numerical method that causes large errors. Techniques for stating the problem carefully and for avoiding blunders are beyond the scope of this book, but the analysis of methods and of their output is its principal subject.

1.2 QUALITY OF NUMERICAL METHODS

There are two properties by which a numerical method is evaluated. The first is *accuracy,* the second *efficiency.* Efficiency is normally measured in terms of the amount of computer time used in the execution of a particular method. This time is often estimated on the basis of the number of additions, subtractions, multiplications, and divisions required. Since multiplications and divisions are significantly more time-consuming than additions and subtractions in many computers, the number of multiplications and divisions required by the method may suffice. In this text, a reference to the number of multiplications required by a method should be interpreted as meaning the number of multiplications and divisions required.

When we speak of the accuracy of a particular answer to a problem, we mean the closeness of that answer to the correct answer. Accuracy should be distinguished from precision, which is the number of significant digits in the answer. For example, the number 42.3968127 is a more precise value for π than the number 3, but the number 3 is more accurate.

1.3 ERRORS IN NUMERICAL METHODS

1.3.1 Definitions and Context

There are many sources of inaccuracy in the answers produced by a numerical method. First, there is error in the input to the method. Such errors often occur when the inputs come from measurements of real-world entities. Second, there are errors which are inherent in the method. Numerical methods by their very nature are approximative. For example, they may involve approximating an infinite series solution by a finite series solution. Or they may find an answer by making a guess, improving a guess, improving the improved guess, and so on, but never fully reaching the correct answer.

A third source of error comes from the fact that digital computers can represent values only to a fixed number of significant digits. Any real number must be rounded off when it is represented in a digital computer, and the imprecision caused by this roundoff is a source of error. The roundoff may be executed by *truncating* a series of digits at some point or by increasing the last retained digit of a series by one unit if the next rightmost digit (closest dropped digit) was greater than one half a unit. The latter process is called *symmetric rounding.*

All of these sources of error fall into the category of *generated error.* Another important source of error is *propagated error*—that error which is propagated through a numerical method after it has been generated. This error is defined as that which appears in output values of a step of a numerical method in which calculations are assumed to be error-free but for which input values are in error. Errors can be increased significantly by propagation. Thus we must study both the question of how much error is generated at any step in a numerical method and the question of the degree to which later steps propagate this error.

In this text, we will let Roman letters such as x, y, and z stand for variables in which there is no error; they stand for the correct values for the solution of the problem in which the variables arise. We will let the same letters with asterisks as superscripts stand for the corresponding variables with error, that is, the ones we are actually dealing with. We will let ε subscripted by a variable name stand for the error in that variable, the difference between the variable with error and the variable without error. Thus

$$x^* = x + \varepsilon_x. \tag{1}$$

ε_x in equation 1 is often called *absolute error* to distinguish it from *relative error,* which is defined as ε_x/x^*. The relative error tells the fraction of the computed value by which that value is off. The relative error is often a more useful index for error than the absolute error. Being off by .01 in a number as large as 1,000,000 may be a matter which can be neglected, but being off by the same amount in a number whose value is .001 is a disaster. The relative error, 10^{-8} in the first case but 10^1 in the second case, is a good indicator of the negligibility of the first error and the seriousness of the second one.

In some of the literature, relative error is defined as ε_x/x. We should note that if $\varepsilon_x \ll x$, where $\ll$ means "is very much less than," then

$$\frac{\varepsilon_x}{x^*} = \frac{\varepsilon_x}{x + \varepsilon_x} = \frac{\varepsilon_x}{x}\left(\frac{x}{x + \varepsilon_x}\right) = \frac{\varepsilon_x}{x}\left(1 - \frac{\varepsilon_x}{x} + \left(\frac{\varepsilon_x}{x}\right)^2 - \cdots\right) = \frac{\varepsilon_x}{x} + \mathrm{O}\left(\frac{\varepsilon_x}{x}\right)^2. \tag{2}$$

In equation 2, the "O" is read "order of" and denotes a sum of terms with $(\varepsilon_x/x)^2$ and higher powers of ε_x/x. Since we have assumed that ε_x/x is $\ll 1$, $(\varepsilon_x/x)^2$ is negligible compared to $1 - \varepsilon_x/x$, and thus we have

$$\varepsilon_x/x^* \approx \varepsilon_x/x, \tag{3}$$

where $\approx$ means "approximately equal to." In such circumstances, it matters little which definition for relative error is used. In this book, although ε_x/x^* is the primary definition of relative error, ε_x/x is used where it is more convenient.

In practical situations, we do not know what the error is (if we did, we would simply subtract it from the inaccurate value to obtain the correct value). Rather, we have only some bounds on the error. Furthermore, commonly, wc do not know the sign (direction) of the error, and must concern ourselves with bounds on the magnitude of the error, be it absolute or relative. Bounds on errors of measurement in values which are input to numerical methods can usually be produced empirically by the measurer. Bounds on errors in computer-generated numbers must be made by the computer scientist.

When a number is represented in fixed-point form in a computer, it is normally thought of as an integer. The error due to the representation is at most one unit. Thus, we have a bound on the absolute error due to representation in fixed-point numbers. The relative error in these numbers, on the other hand, can vary from infinite in case the number is 0 to one part in the largest integer representable by the computer, so the relative error of a fixed-point number is not easy to bound in a useful way.

The representation of a floating-point number consists of an exponent and a fraction with a fixed number of significant digits. The error due to representation in a computer is incurred by truncating the fraction. Hence, the absolute error depends strongly on the exponent. The relative error depends mainly on the (constant) precision of the fraction, so it is a less variable, more convenient indicator than the absolute error.

1.3.2 Propagation of Errors

We will now discuss the effect of using inaccurate numbers in computations. More precisely, we will discuss how to find an error bound on the result of a computation, given error bounds on the inputs.

First, let us consider binary operations such as addition, subtraction, multiplication, and division. Say we wish to execute an operation represented by ω (for example, $\omega = +$). That is, we wish to compute $z = x\omega y$. In fact, what we compute is $z^* = x^*\omega^*y^*$; we have errors both in the inputs and in the operation itself (normally caused by roundoff).

$$\varepsilon_z = z^* - z = x^*\omega^*y^* - x\omega y = (x^*\omega^*y^* - x^*\omega y^*) + (x^*\omega y^* - x\omega y), \quad (4)$$

where the first term on the right side of equation 4 is ε_z^{gen}, the error generated at this step, that is, the error resulting from the inaccuracy in the operation ω (so ε_z^{gen} may also be written ε_ω), and the second term is ε_z^{prop}, the error propagated by this step, that is, the error resulting from the fact that the inputs are in error. By the triangle inequality ($|a + b| \le |a| + |b|$), we obtain

$$|\varepsilon_z| \le |x^*\omega^*y^* - x^*\omega y^*| + |x^*\omega y^* - x\omega y| = |\varepsilon_z^{gen}| + |\varepsilon_z^{prop}|. \quad (5)$$

Equation 5 states that the magnitude of the overall absolute error is bounded by the sum of the magnitude of the generated error and the magnitude of the propagated error.

As an example, take the error in addition: $z = x + y$. By definition

$$z^* = x^* +^* y^* = (x^* + y^*) + \varepsilon_+, \quad (6)$$

where ε_+ is the error generated by $+^*$. Since $x^* = x + \varepsilon_x$ and $y^* = y + \varepsilon_y$,

$$z^* = x + \varepsilon_x + y + \varepsilon_y + \varepsilon_+ = (x + y) + (\varepsilon_x + \varepsilon_y) + \varepsilon_+$$
$$= z + (\varepsilon_x + \varepsilon_y) + \varepsilon_+, \quad (7)$$

or

$$\varepsilon_z = z^* - z = (\varepsilon_x + \varepsilon_y) + \varepsilon_+. \quad (8)$$

Since ε_+ is the generated error, from equations 4 and 8 it follows that the propagated error is $\varepsilon_x + \varepsilon_y$.

Equation 8 implies that

$$|\varepsilon_z| \le |\varepsilon_x| + |\varepsilon_y| + |\varepsilon_+|. \quad (9)$$

If we are given bounds b_x, b_y, and b_+ such that $|\varepsilon_x| \le b_x$, $|\varepsilon_y| \le b_y$, $|\varepsilon_+| \le b_+$, then

$$|\varepsilon_z| \le b_x + b_y + b_+, \quad (10)$$

with addition (and subtraction; see exercise 1.3.1), where the equality can hold if ε_x, ε_y, and ε_+ have like signs and the magnitudes b_x, b_y, and b_+, respectively. Thus

$$b_z = b_x + b_y + b_+; \quad (11)$$

the bound on the propagated error in the sum is the sum of the bounds on the input errors, and the bound on the overall error is the sum of the propagated and generated error bounds, that is, *absolute error bounds add*. Thus, for example, if z^* is the computed sum of x^* and y^*, where $x = 10 \pm .5$, $y = 4 \pm .1$, and $|\varepsilon_+| \le .04$, then $z = 14 \pm (.5 + .1 + .04) = 14 \pm .64$.

As another example of error propagation, consider multiplication: $\omega = \times$. Then

$$z^* = x^* \times^* y^* = x^*y^* + \varepsilon_\times, \tag{12}$$

where $\varepsilon_\times$ is the error generated by $\times^*$.

$$\begin{aligned} x^*y^* &= (x + \varepsilon_x)(y + \varepsilon_y) = xy + \varepsilon_x(y + \varepsilon_y) + x\varepsilon_y \\ &= xy + y^*\varepsilon_x + (x^* - \varepsilon_x)\varepsilon_y = xy + y^*\varepsilon_x + x^*\varepsilon_y - \varepsilon_x\varepsilon_y. \end{aligned} \tag{13}$$

Thus
$$\varepsilon_z = z^* - z = y^*\varepsilon_x + x^*\varepsilon_y - \varepsilon_x\varepsilon_y + \varepsilon_\times, \tag{14}$$

and
$$\frac{\varepsilon_z}{z^*} \approx \frac{\varepsilon_z}{x^*y^*} = \frac{\varepsilon_x}{x^*} + \frac{\varepsilon_y}{y^*} - \frac{\varepsilon_x}{x^*}\frac{\varepsilon_y}{y^*} + \frac{\varepsilon_\times}{z^*}. \tag{15}$$

Assuming $|\varepsilon_x| \ll |x^*|$ and $|\varepsilon_y| \ll |y^*|$, the term $(\varepsilon_x/x^*)(\varepsilon_y/y^*)$ is negligible compared to the other terms, and we have

$$\frac{\varepsilon_z}{z^*} \approx \frac{\varepsilon_x}{x^*} + \frac{\varepsilon_y}{y^*} + \frac{\varepsilon_\times}{z^*}. \tag{16}$$

Thus $|\varepsilon_z/z^*| \lesssim |\varepsilon_x/x^*| + |\varepsilon_y/y^*| + |\varepsilon_\times/z^*|$, where the symbol $\lesssim$ means "less than a number approximately equal to." If we are given bounds r_x, r_y, and $r_\times$ on the respective magnitudes of the relative errors in x^* and y^* and that generated by multiplication, we find

$$r_z \approx r_x + r_y + r_\times. \tag{17}$$

With multiplication (and division; see exercise 1.3.2), *relative error bounds add*. For example, if z^* is the computed product of x^* and y^*, where $x = 10 \pm .5$, $y = 4 \pm .1$, and $|\varepsilon_\times| \leq .04$, then $z^* = 40$ and $|\varepsilon_z/z^*| \lesssim \frac{.5}{10} + \frac{.1}{4} + \frac{.04}{40} = .076$, so $z = 40 \pm .076\,(40) = 40 \pm 3.04$. Note that in this example we use the general relationship

$$b_z = (r_z)(z^*), \tag{18}$$

which follows from
$$\varepsilon_z = (\varepsilon_z/z^*)(z^*). \tag{19}$$

In the next section, we shall consider how to determine the absolute error b_+ generated in addition and the relative error $r_\times$ generated in multiplication.

Now let us consider unary operations, $z = f(x)$, for example, $z = sin(x)$. As with binary operations, we find

$$\begin{aligned} \varepsilon_z = z^* - z = f^*(x^*) - f(x) &= (f^*(x^*) - f(x^*)) + (f(x^*) - f(x)) \\ &= \varepsilon_f + (f(x^*) - f(x)). \end{aligned} \tag{20}$$

Here ε_f, the error generated in computing f, is due either to roundoff or to using an approximate numerical method to compute f, for example, a truncated infinite series for $sin(x)$. The error propagated by the operator, given by $f(x^*) - f(x)$, can be computed in terms of ε_x using a *Taylor series*, a mathematical tool that we shall apply repeatedly to evaluate a function at a small offset from a point at which it is known.

$$f(x) = f(x^* - \varepsilon_x) = f(x^*) - \varepsilon_x f'(x^*) + \varepsilon_x^2 f''(\xi)/2!, \tag{21}$$

where ξ is in the interval bounded by, and including, x and x^*, a relation we write $\xi \in [x, x^*]$. If $|\varepsilon_x| \ll 1$ and $f''(\xi)$ is not large $(|f''(\xi)|/|f'(x^*)| \ll 1/\varepsilon_x)$, then

$$f(x^*) - f(x) \approx \varepsilon_x f'(x^*). \tag{22}$$

This equation is interpreted graphically in figure 1-1.

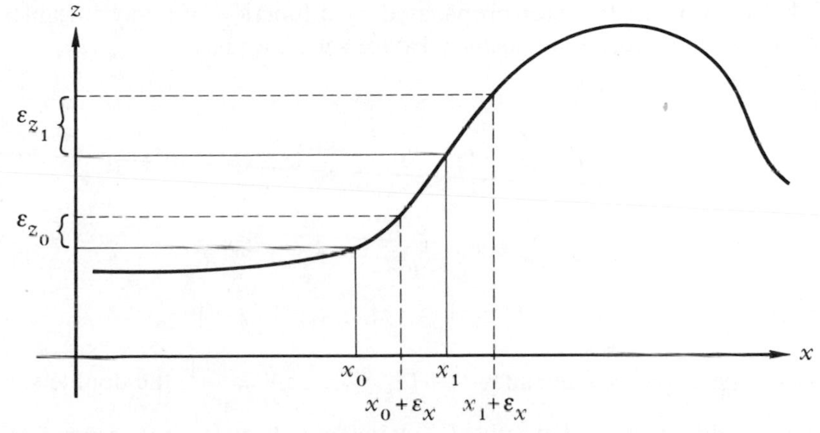

FIG. 1-1 Error propagation by function evaluation

In the region near x_0 where $|f'(x^*)|$ is small, the propagation of a given ε_x into $z^* = f(x^*)$ does not result in a larger output (z) error than input (x) error. In fact, just the opposite is true. In contrast, in the region near x_1 where $|f'(x^*)|$ is large, a small ε_x becomes a large ε_z.

As an example, consider the function $z = f(x) = ax$, where a is an error-free constant. Then $\varepsilon_z^{prop} \approx a\varepsilon_x$ by equation 22, and ε_z increases as a

increases. However, notice that the propagated relative error,

$$\frac{\varepsilon_z}{z^*} \approx \frac{a\varepsilon_x}{ax^*} = \frac{\varepsilon_x}{x^*}, \tag{23}$$

is independent of the constant a.

Similarly, if $z = f(x) = a/x$,

$$\varepsilon_z^{prop} \approx -\frac{a}{x^2}\varepsilon_x, \tag{24}$$

which increases in magnitude with a. But the propagated relative error,

$$\frac{\varepsilon_z}{z^*} \approx \frac{(-a/x^{*2})\varepsilon_x}{a/x^*} = \frac{-\varepsilon_x}{x^*}, \tag{25}$$

is independent of the constant a. So multiplication by a constant or division of a constant can change the magnitude of the absolute error but not of the relative error.

Let us consider the error propagated by a function of many variables, all with some error. Again using a Taylor series, we have

$$\begin{aligned} f(x_1,x_2,\ldots,x_n) &= f(x_1^* - \varepsilon_{x_1}, x_2^* - \varepsilon_{x_2}, \ldots, x_n^* - \varepsilon_{x_n}) \\ &= f(x_1^*,x_2^*,\ldots,x_n^*) - \sum_{i=1}^{n} \varepsilon_{x_i} \frac{\partial f}{\partial x_i}(x_1^*,x_2^*,\ldots,x_n^*) \\ &\quad + \sum_{i=1}^{n}\sum_{j=1}^{n} \varepsilon_{x_i}\varepsilon_{x_j} \frac{\partial^2 f}{\partial x_i \partial x_j}(\xi_1,\xi_2,\ldots,\xi_n), \end{aligned} \tag{26}$$

where $\quad \xi_1 \in [x_1,x_1^*], \xi_2 \in [x_2,x_2^*], \ldots, \xi_n \in [x_n,x_n^*].$

Assuming all $|\varepsilon_{x_i}| \ll 1$ and all $\left|\frac{\partial^2 f}{\partial x_i \partial x_j}(\xi_1,\xi_2,\ldots,\xi_n)\right| \ll \left|\frac{1}{\varepsilon_{x_i}}\right|$, the double sum is negligible and we have the following equation for the propagated error:

$$f(x_1^*,x_2^*,\ldots,x_n^*) - f(x_1,x_2,\ldots,x_n) \approx \sum_{i=1}^{n} \varepsilon_{x_i} \frac{\partial f}{\partial x_i}(x_1^*,x_2^*,\ldots,x_n^*). \tag{27}$$

Thus the magnitude of the propagated error is bounded by

$$\sum_{i=1}^{n} |\varepsilon_{x_i}| \left|\frac{\partial f}{\partial x_i}(x_1^*,x_2^*,\ldots,x_n^*)\right|.$$

Most numerical methods consist of the repetitive application of a relatively simple operation. If error propagation in each execution of the operation results in even a small error increase, the fact that the operation is repeated many times can result in a large propagation of the initial error into the final result. Thus, we must be able to analyze the error propagation, not only in a single operation but also in the repeated application of the operation. We must deal with the interaction, in one application of the operation, of errors propagated in previous applications of the operation.

Let
$$z_{i+1} = f(z_i, z_{i-1}, \ldots, z_{i-m}), \tag{28}$$
where the final answer is $z = z_n$, and $z_0^*, z_1^*, z_2^*, \ldots, z_m^*$ are inputs to the computation. Applying the error propagation analysis methods developed above to equation 28, we can derive an equation for ε_{i+1}, the error in z_{i+1}, in terms of $\varepsilon_i, \varepsilon_{i-1}, \ldots, \varepsilon_{i-m}$, and $z_i, z_{i-1}, \ldots, z_{i-m}$:

$$\varepsilon_{i+1} = g(\varepsilon_i, \varepsilon_{i-1}, \ldots, \varepsilon_{i-m}, z_i, z_{i-1}, \ldots, z_{i-m}). \tag{29}$$

In certain cases, we can analytically solve the recurrence relation 29 for ε_j as a function of j, where $z_0, z_1, \ldots, z_m$ and $\varepsilon_0, \varepsilon_1, \ldots, \varepsilon_m$ are parameters of the problem. In particular, such an analytic solution is possible when g is a linear function of the ε_k and not a function of the z_k:

$$\varepsilon_{i+1} = \sum_{k=0}^{m} a_k \varepsilon_{i-k}. \tag{30}$$

The solution method of such so-called linear difference equations is covered in appendix A.

The solution of equation 29 can be used to bound the propagated error in z_n, given bounds on the errors in the z_k, $0 \le k \le m$. More realistically, the common question is whether for large n, ε_n becomes small or large. This question is approximated by asking for the value of $\lim_{n\to\infty} |\varepsilon_n|$. Usually the value is either zero or infinite. In the former case, we say that the numerical method is *absolutely stable*; in the latter, we say that it is *absolutely unstable*.

We have argued that relative error is often a more useful measure of accuracy than absolute error is. For this reason, we often compute, not $\lim_{n\to\infty} |\varepsilon_n|$, but $\lim_{n\to\infty} |\varepsilon_n/z_n|$, using a solution of equation 28 to obtain z_n. If this limit is zero, we say the method is *relatively stable*; if infinite, *relatively unstable*.

For example, suppose we are computing z_i from

$$z_{i+1} = z_i + 2z_{i-1}. \tag{31}$$

Analyzing the error propagation in a single application of the above formula, we find

$$\varepsilon_{i+1} = \varepsilon_i + 2\varepsilon_{i-1}, \tag{32}$$

that is, the error follows the same relation as the z_i, except that the initial values, ε_0 and ε_1, are not the same as z_0 and z_1. By the methods described in appendix A, we can solve the difference equation 32 to produce

$$\varepsilon_n = \frac{\varepsilon_0 + \varepsilon_1}{3} 2^n + \frac{2\varepsilon_0 - \varepsilon_1}{3} (-1)^n. \tag{33}$$

As $n \to \infty$, $|\varepsilon_n| \to \infty$, so the above computation is absolutely unstable. However, since z_i satisfies the same difference equation as ε_i,

$$z_n = \frac{z_0 + z_1}{3} 2^n + \frac{2z_0 - z_1}{3} (-1)^n, \tag{34}$$

so

$$\lim_{n \to \infty} \left| \frac{\varepsilon_n}{z_n} \right| = \frac{\varepsilon_0 + \varepsilon_1}{z_0 + z_1} \quad \text{if} \quad z_0 + z_1 \neq 0. \tag{35}$$

Thus the relative error does not become infinite as $n \to \infty$. Rather, it approaches a nonzero constant. The computation is not relatively unstable, though neither is it relatively stable.

The stability of a method pertains only to its error propagation properties. Even in a stable method, the combined effect of error generation and error propagation may produce large errors. Nevertheless, stable methods usually produce accurate answers, and stability analysis often is tractable mathematically, at least in restricted cases. Stability should be an issue in all numerical methods consisting of a repetitive application of an operation (a class which includes most numerical methods). Stability analysis is particularly emphasized and amplified in chapter 5, since the problems caused by repetition are particularly virulent in methods for the solution of differential equations.

1.3.3 Generation of Errors in Arithmetic Operations

The subject of the errors generated by addition ($+*$), subtraction ($-*$), multiplication ($\times *$), and division ($/*$) is an important one. We will discuss two of these operations, $+*$ and $\times *$, here and leave the other two to exercises (see exercises 1.3.5 and 1.3.6).

With fixed-point addition or multiplication, if there is no overflow, no error is generated. With fixed-point addition, there will be no overflow unless the summands are of the same sign, at least one has a nonzero high-order digit, and the sum produces a 1 in the next, higher order digit. This situation seldom occurs. With fixed-point multiplication, however, there often is overflow, because multiplication of a p-digit number by a q-digit number, each with a nonzero high-order digit, results in a $(p+q)$-digit or $(p+q-1)$-digit number. Thus, some scaling must be done (truncating low-order digits, and storing only the high-order part of the result, but remembering how many digits were truncated). Doing this scaling automatically is a primary reason for floating-point arithmetic. But the scaling produces error. If the scaling is done by a computer program, the amount of error depends on the number of digits lost in the scaling. Since the number of digits lost depends on the algorithm used for scaling, we cannot further discuss the matter generally here.

Moving on to floating-point arithmetic, let us assume we have a base-b floating-point computer with an n-digit normalized fraction (high-order digit $\neq 0$). With floating-point addition, the addition process involves shifting the fraction with the smaller exponent to the right, with the loss of the digits shifted out, and correspondingly increasing the exponent until it agrees with that of the summand of greater magnitude. With computers such as System/370, one guard digit (hexadecimal) to the right of the shifted fraction is saved for further use, together with the sign of the fraction from which it came. The summands are then added as fixed-point numbers.

Case 1: If the addition overflows, the exponent is increased by one and the sum is shifted one digit position to the right. The rightmost (low-order) digit is lost, and a 1 is placed in the leftmost (high-order) digit position, thus completing the operation. For example, for $n = 3$ and $b = 10$, $.990 \times 10^0 + .109 \times 10^{-1} = .100 \times 10^1$ with an error of $-.0009 \times 10^0$.

Case 2: If the addition does not overflow but its result has a nonzero high-order digit, the guard digit, if saved, is ignored, and the operation is complete. For example, for $n = 3$ and $b = 10$, $.890 \times 10^0 + .109 \times 10^{-1} = .900 \times 10^0$ with an error of $-.0009 \times 10^0$.

Case 3: If the high-order digit of the initial addition result is zero due to the addition of operands of opposite signs, the result is shifted one digit position to the left, the signed guard digit is added in the low-order position, and the exponent is decreased by one. If the high-order digit of the result of this operation is zero but the result is nonzero, the result is shifted to the left until the high-order digit is nonzero and the exponent is decreased accordingly. For example, for $n = 3$ and $b = 10$, $-1.00 \times 10^0 + .999 \times 10^{-2} = -.901 \times 10^{-1}$ with an error of $-.00009 \times 10^0$ if a guard digit is kept, and $-.100 \times 10^0 + .999 \times 10^{-2} = -.910 \times 10^{-1}$ with an error of $-.00099 \times 10^0$ if no guard digit is kept.

In these cases, the error generated by the addition is due to the truncation of the digits shifted out either before the summation or, in the case of overflow, after the summation. In the worst case, each of the lost digits is $b-1$. In cases 1 and 2, the result of the addition has n correct digits, and the error is at worst k digits, each $b-1$, for some $k < n$. That is, the error is at worst approximately 1 in the nth place of the sum. Thus, in cases 1 and 2, the magnitude of the relative error is less than $1/v$, where v is the magnitude of the integer represented by the n-place sum, ignoring the exponent and radix point of the floating-point number. But since the sum has a nonzero high-order digit, $v \geq b^{n-1}$, so the magnitude of the relative error in cases 1 and 2 is less than $1/b^{n-1}$.

Case 3 can occur only when adding numbers of opposite signs. The error results from the loss of digits truncated from the summand of smaller magnitude during shifting and not saved as guard digits to be recovered after the sum is computed and the result shifted left. The largest relative error results from having the largest digit $(b-1)$ in all truncated and untruncated places of the summand of smaller magnitude and a summand of larger magnitude which consists of a high-order digit of 1 followed by all zeros. If the exponent of the summand of smaller magnitude is less than the exponent of the summand of larger magnitude by more than 1, the magnitude of the relative error in the worst case (assuming the sum $\neq 0$) is $1/b^n$ if the floating-point addition keeps a guard digit and $1/b^{n-1}$ if it does not. However, if the difference in the exponents of the summands is 1, in the worst case the relative error is 0 if a guard digit is kept but $(b-1)/b(\approx 1)$ if no guard digit is kept. For example, for $n = 3$ and $b = 10$, $.100 \times 10^0 - .999 \times 10^{-1} = .100 \times 10^{-3}$ with no error if a guard digit is kept, and $.100 \times 10^0 - .999 \times 10^{-1} = .100 \times 10^{-2}$ with an error of $.090 \times 10^{-2}$ if no guard digit is kept. It is to avoid this disastrous error that a guard digit is kept.

To summarize, assuming a guard digit is kept, the absolute error generated by a floating-point addition depends on the exponent of the sum, and a bound on that error depends upon the exponent. In contrast, the relative error in a nonzero sum generated by a floating-point addition with a guard digit is bounded in magnitude by $1/b^{n-1}$, where b is the floating-point base and n is the number of digits in the fraction representation.

Now consider floating-point multiplication. With most computers, the fraction part of the product is produced by fully multiplying the fraction parts of the factors, normalizing the result, and then truncating the result to the n digits of the fraction representation.† Thus, as with addition, the

† This is true of the short floating-point multiply of System/370 but not of the long floating-point multiply.

absolute error is due to truncation of the normalized product and is dependent on the value of the exponent of the product. The magnitude of the generated relative error, in contrast, is bounded (assuming neither factor is 0) by $1/b^{n-1}$, the quotient of the largest possible truncation error (the fraction made by n zeros followed by a number of digits with values of $b-1$) divided by the smallest normalized fraction ($1/b$).

1.3.4 Error Estimates vs. Error Bounds

Error bounds produced using methods discussed in the previous section give the error generated and propagated by arithmetic operations (*computational error*) for the worst case that can be encountered. Especially when the number of operations is large, such bounds are very conservative. The bounds may be many orders of magnitude larger than the actual errors likely to be encountered. Some method for getting more realistic estimates of error must be developed.

At the simplest level we would like to know the character of the error propagation in a particular method as a function of the number of arithmetic operations in the method and the number of applications of the method necessary to get an answer. For example, we are interested in whether the propagated error increases linearly or exponentially with the number of operations.

For any character of the error propagation, generally the computational error in the final result is a monotonically increasing function of the number of arithmetic operations, beginning with the last point in the computation where all values used have only generated error, up to the end of the computation. With so-called direct methods, this is the total number of arithmetic operations required by the method. With convergent iterative methods, the computational error estimate is a constant multiple of the error estimate for the number of arithmetic operations required by the last iteration (see exercise 1.3.9). Since the final error increases with the number of arithmetic operations required, we see that, everything else being equal, an efficient method (that is, one requiring a minimal number of computations) is likely to be an accurate one.

As shown in figure 1-2, while the computational error rises with an increase in the number of arithmetic operations in a method (or in its last iteration), the approximational error of the method generally decreases. That is, better approximations can be achieved with the cost of additional operations. Everything else being equal, our objective is to minimize the overall error, which is the sum of the computational error and the approximational error. This minimization determines an optimum number of operations for a given method.

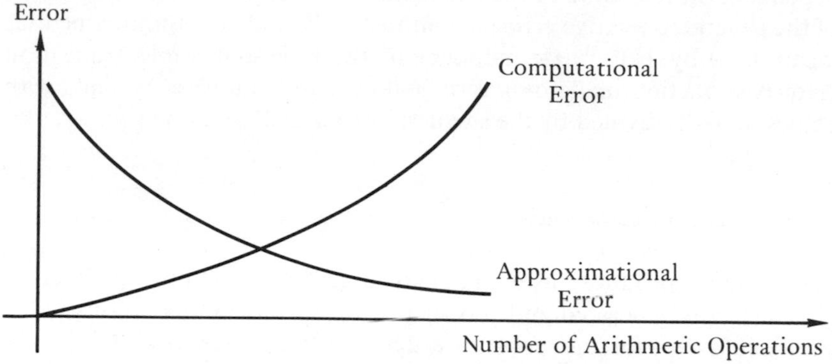

FIG. 1-2 Behavior of error as a function of number of computations

Techniques for obtaining actual numerical estimates of computational errors to be encountered have been developed. One is to apply the method to be tested in a small number of cases using double-precision arithmetic and to compare the double-precision results (which are assumed to be correct) to the single-precision results to obtain an estimate of the error. Another technique is to perturb the set of input values by a slight amount for a number of cases and check the change in the answer due to the perturbation, having computed with both the unperturbed and perturbed values in single-precision arithmetic. The difficulty with both of these approaches is that it is hard to know whether or not the few cases tested are characteristic of all inputs to which the numerical method being tested is likely to be applied. Furthermore, with neither technique do we compare the computed answer to the correct answer. In the first case, double-precision computing does not necessarily give a correct answer. In the perturbation method, we are comparing two inaccurate answers rather than an inaccurate and a correct answer.

There are techniques that give realistic error estimates. One is referred to as *interval arithmetic.* It involves calculating, for each arithmetic operation, not only the output value as a function of the input values but also (or instead) the largest and smallest answers which could result as a function of the largest and smallest possible values for the inputs in question. That is to say, for every application of the method, not just for some test cases, one enters the method with inputs designated by their maximum and minimum possible values, and one produces for every executed arithmetic operation, a maximum and minimum possible output value from that operation. When one finishes, one has a range within which the answer can lie as well as (or instead of) a single answer produced by using the best estimate of the input values in the calculations. This technique takes three times as many calculations as the technique of simply calculating a single answer because each calculation has to be carried out for a maximum value, a minimum value, and the normally computed value. Its advantage is that the range is specifically related to the inputs in question rather than to any possible set of inputs, as may be the case when error bounds are calculated on a theoretical basis. However, the calculated range still may be quite conservative compared to the actual error incurred.

To obtain a more realistic estimate of the error, one may treat the problem probabilistically. We assume that the input error is a sample from a distribution of errors with known probability. We want to know the probability distribution of the output error, given the probability distribution of the input errors (or, more simply, the average error magnitude in the output, given the average error magnitudes in the inputs). Before treating this area, we need to briefly survey the theory of probability (see section 1.4).

PROBLEMS

1.3.1. Show that the bound on the propagated error in the difference of two variables is the sum of the bounds on the input errors.

1.3.2. Show that the bound on the propagated relative error in the quotient of two variables is approximately the sum of the bounds on the relative errors of the inputs, if these input relative errors are much less than 1 in magnitude.

1.3.3. Bound the propagated error in the following functions, given that the error bounds in x and y are $b_x = 10^{-6}$ and $b_y = 10^{-5}$.

a) $f(x) = log(x^2 + 5)$ near $x = 2$

b) $f(x,y) = e^{x^3y} cos(y)$ near $x = 2,\ y = \pi/4$

1.3.4. Give the bounds for the propagated absolute error and for the propagated relative error in each of the following computations.

a) $(3 \pm 10^{-2}) + (3.1 \pm 10^{-3})$

b) $(3 \pm 10^{-2}) - (3.1 \pm 10^{-3})$

c) $(3 \pm 10^{-2})(3.1 \pm 10^{-3})$

d) $(3.1 \pm 10^{-3})/(3 \pm 10^{-2})$

1.3.5. Compare the error generation in floating-point subtraction to that in floating-point addition.

1.3.6. Discuss the error generation in floating-point division in a base-b floating-point computer with an n-digit fraction. Assume n digits are produced in the quotient.

1.3.7. a) Consider multiplying the two matrices A and B where each element of A is positive and has a maximum relative error magnitude r_a and each element of B is positive and has a maximum relative error magnitude r_b. What is the maximum propagated relative error magnitude in each of the elements of C, where $C = AB$? (Assume all errors can be positive or negative.)

b) Consider computing $\|Ax\|^2 = (Ax)^T Ax$, where A is a matrix with positive elements and x is a vector with positive elements. What is the maximum relative error magnitude in $\|Ax\|^2$ if the elements of A have maximum relative error magnitude r_a and the elements of x have maximum relative error magnitude r_x?

1.3.8. The following exact numbers are given in a decimal floating-point computer with a two-digit normalized fraction:

$a = 9.9$
$b = 9.8$
$c = .098$

Perform the following operations, assuming a single guard digit is maintained during each arithmetic operation but all results including intermediate results are truncated to two digits. Assume all expressions are evaluated left to right as in PL/I. Give the generated absolute and relative error in each case.

1) $a + b$	6) $a \cdot b$
2) $a - b$	7) $a \cdot c$
3) $a + c$	8) a/b
4) $a - b - c$	9) b/a
5) $a - (b + c)$	10) a/c
	11) $(a + b)/c$
	12) $a + (b/c)$

1.3.9. Assume we have an iterative method in which an estimate, x_i, of a desired value is improved by applying a function g. That is, let $x_{i+1} = g(x_i)$, and assume $\lim_{i\to\infty} x_i = x$, the desired value. Assume we have a method which has the property called simple linear convergence: assuming no generated error in computing $x_{i+1} = g(x_i)$, $\lim_{i\to\infty}((x_{i+1} - x)/(x_i - x)) = k$, where k is a constant with magnitude less than 1. Let $\varepsilon_i \equiv x_i - x$, where $\equiv$ means "is defined as" or "is equivalent to." Assume that, at each iterative step, the propagated error is given by $\varepsilon_{i+1}^{prop} = k\varepsilon_i$ and an error bounded in magnitude by b is generated. Show that the error in $\lim_{i\to\infty} x_i^*$ is bounded by $b/(1 - |k|)$.

1.4 SURVEY OF PROBABILITY

1.4.1 Basic Concepts

Probability is the mathematical theory for analyzing situations in which chance is assumed to play a part. That is, the outcome of a particular action or occurrence of interest is assumed not to be fully predictable, as with the flipping of a coin. In many cases where probability is applied, the outcome would be predictable if more information were available. For example, if one knew the weight and the initial position, speed of rotation, and velocity of the coin, together with the amount and direction of wind and the characteristics of the surface on which the coin would fall, one could predict the outcome of the coin flip. However, when one uses probability theory, such information is assumed not to be available or if it is available, we choose not to use it because the information or its use is too complicated to be of interest. In such cases we use a probabilistic model to understand the process being studied. Such a model is useful in analyzing errors in numerical computation.

In probability theory, we study situations, called *experiments,* which are assumed to have one of a number of mutually exclusive possible outcomes, called *events.* The set of possible outcomes is called the *sample space.* Thus the experiment of a coin flip may produce an event, heads, from the sample space {heads, tails}. We often associate with each experiment a variable which takes on a value as a result of the experiment. Thus, for example, if x is the variable associated with our coin flip, $x = 1$ could mean that the flip resulted in heads and $x = 2$ that the flip resulted in tails. The variable in question is called the *random variable* of the experiment.

If the sample space is discrete (that is, if the number of possible outcomes is countable), we commonly associate an integer with each outcome; the result of the experiment is for the random variable x to take on the integer value i, for some i. The sample space may not be discrete, as in the case when the random variable is the height of the next person to enter some office. In such a case, the random variable x takes on other than integer values (say, 65.3947 inches).

Let us consider the finite discrete case. That is, let the random variable be x which can take on values i from the sample space $\{1,2,\ldots,N\}$. With each event we associate a number, called its *probability.* The probability associated with event i, written $P_x(i)$, is the probability that the random variable x takes on the value i. The set of probabilities associated with the respective events is called the *probability distribution* of the experiment. These probabilities are chosen to have the properties required to model the following interpretation of probabilistic processes: Given the situation defining the experiment, consider in the abstract the existence of a large number of such situations, and assume each of these experiments is run. Then the probability of event i is the fraction of the number of times the experiment was run that event i occurred. Note that it is not necessary for the event to be logically repeatable for the notion of probability to make sense.

The above implies that $0 \leq P_x(i) \leq 1$. If event i never occurs, $P_x(i) = 0$; if event i occurs every time the experiment is run, $P_x(i) = 1$.

We have defined events i and j for $i \neq j$ to be mutually exclusive. They cannot occur simultaneously in a single experiment (as is true of 1 and 2 in a die throw). However, we can define a combined event, the occurrence of either event i or event j, with probability $P_x(i \text{ or } j)$. Since the fraction of the experiments in which either event i or event j occurs is the sum of the fraction in which event i occurs and that in which event j occurs,

$$P_x(i \text{ or } j) = P_x(i) + P_x(j). \tag{36}$$

Since every time the experiment is run, either event 1, or event 2, or event i, for some i between 3 and N, occurs,

$$\sum_{i=1}^{N} P_x(i) = 1. \tag{37}$$

We can also define the event that occurs whenever event i does not occur. We call it the event *not i*. Either the event i or the event *not i* must occur when an experiment is run, so

$$1 = P_x(i \text{ or } \mathit{not}\ i) = P_x(i) + P_x(\mathit{not}\ i), \tag{38}$$

so

$$P_x(\mathit{not}\ i) = 1 - P_x(i). \tag{39}$$

A sample space may be continuous rather than discrete, as is the case with numerical error. Here we wish to speak of the probability that a number x is computed as $x^* = x_1$ for a particular x_1 among a continuous set of possible values for the random variable x^*. In this case, the number of events with nonzero probability is continuously infinite. It is not meaningful to ask the fraction of this number for which any event, $x^* = x_1$, occurs. But it is meaningful to ask what fraction of the time x^* is in the interval $[x_1, x_2)$, where $x_1 < x_2$ (x_1 is in the interval but x_2 is not). We express this fraction as $Pr(x_1 \leq x^* < x_2)$. If we let $x_2 \rightarrow x_1$, we can define the *probability density* as

$$p_{x^*}(x_1) \equiv \lim_{x_2 \to x_1} \frac{Pr(x_1 \le x^* < x_2)}{x_2 - x_1}.\dagger \tag{40}$$

If we speak of the probability that x^* is in the infinitesimal interval $[x_1, x_1+dx)$ and the probability density is given by the probability density function $p_{x^*}(x_1) \ge 0$, then the probability that $x^* \in [x_1, x_1+dx)$ (that is, $Pr(x_1 \le x^* < x_1 + dx)$) is $p_{x^*}(x_1)dx$; the probability that x^* is in the interval is the probability density in the interval times the width of the interval. The probability that $x^* \in [x_1, x_2)$ is the sum of all of the probabilities of it being in the infinitesimal intervals which make up $[x_1, x_2)$:

$$Pr(x_1 \le x^* < x_2) = \int_{x_1}^{x_2} p_{x^*}(u)du. \tag{41}$$

Since each event is defined by "x^* is in some particular infinitesimal interval" and we assume some event occurs,

$$\int_{-\infty}^{\infty} p_{x^*}(u)du = 1. \tag{42}$$

For example, consider the probability density function

$$p_x(u) = \begin{cases} \frac{1}{2}, & -1 \le u \le 1 \\ 0, & |u| > 1. \end{cases}$$

The function $p_x(u)$ satisfies the requirement that $\int_{-\infty}^{\infty} p_x(u)du = 1$, and states that the probability that the random variable x will take on a value other than one between -1 and 1 is zero, and the probability that x will take on a value between a and b where $-1 \le a \le b \le 1$ is $(b-a)/2$.

It is important to reemphasize here that a probability density function does not give a probability; rather, it gives the probability per unit of the random variable. Probability is given by probability density multiplied by the width of the infinitesimal interval in question, or by sums (integrals) of these products.

We noted that with a discrete sample space we can talk about the probability that $x = i$, whereas in the continuous case we must talk about

---

† Since $Pr(x_1 \le x^* < x_2) = Pr(x^* < x_2) - Pr(x^* < x_1)$, probability density can alternatively be defined as

$$\lim_{x_2 \to x_1} \frac{Pr(x^* < x_2) - Pr(x^* < x_1)}{x_2 - x_1} = \frac{d}{dz} Pr(x^* < z)\bigg|_{z=x_1}.$$

This definition has some mathematical advantages we will see later.

the probability density at $x = i$. We ask what the probability density looks like for a discrete probability distribution with random variable x which takes on integer values. We require that for any $0 < \varepsilon < 1$, $Pr(i+\varepsilon \leq x < i+1-\varepsilon) = 0$. But

$$Pr(i + \varepsilon \leq x < i + 1 - \varepsilon) = \int_{i+\varepsilon}^{i+1-\varepsilon} p_x(u)du. \tag{43}$$

Since $p_x(u) \geq 0$ for all u, the integral being zero implies that $p_x(u) = 0$ for $u \in [i + \varepsilon, i + 1 - \varepsilon)$ for any ε, no matter how small. Furthermore,

$$Pr(i - \varepsilon \leq x < i + \varepsilon) = P_x(i), \tag{44}$$

so

$$\int_{i-\varepsilon}^{i+\varepsilon} p_x(u)du = P_x(i). \tag{45}$$

As $\varepsilon \to 0$, the integral approaches $2\varepsilon p_x(i)$, so $p_x(i) \to P_x(i)/2\varepsilon$. Thus $p_x(u)$ in $i-1 < x < i+1$ is zero at all points except at $x = i$ where it is the limit as $\varepsilon \to 0$ of a rectangle 2ε wide and $P_x(i)/2\varepsilon$ high (that is, a rectangle of infinitesimal width and infinite height proportional to $P_x(i)$, but whose area is constant and equal to $P_x(i)$). This shape is called a *spike* or *impulse* of area $P_x(i)$ at $x = i$. The spike of area 1 at $x = 0$ is called the *Dirac* δ *function* and is written $\delta(x)$. Thus $\delta(x - a)$ is the spike of area 1 at $x = a$, and it has the following properties:

$$\delta(x - a) = \begin{cases} 0, & x \neq a \\ \infty, & x = a. \end{cases} \tag{46}$$

For any $\varepsilon > 0$,

$$\int_{a-\varepsilon}^{a+\varepsilon} \delta(x - a)dx = 1. \tag{47}$$

Note that $\delta(x)$ is not truly a function but rather a limit of a sequence of functions. However, it can usefully be treated as a function.†

Thus if $P_x(i)$ is nonzero for $i = 1, 2, \ldots, N$, the probability density corresponding to this distribution is

$$p_x(u) = \sum_{i=1}^{N} P_x(i)\delta(u - i), \tag{48}$$

a series of spikes at the integers of respective heights $P_x(i)$.

† To be rigorous, one must deal with the distribution function $Pr(x < z)$ mentioned earlier and not probability density. A discrete probability $P_x(i)$ at $x = i$, which has density there, $p_x(i) = P_x(i)\delta(x - i)$, corresponds to a discontinuity of amount $P_x(i)$ in $Pr(x < z)$. Since $p_x(u) = d/dx\ Pr(x < z)_{|z=u}$, $\delta(x)$ has an alternative definition as the derivative at a unit discontinuity.

Let us now consider the case of a compound experiment which consists of a number of component experiments. For example, let the compound experiment be the flipping of two coins and the casting of four dice. We need a concept of *joint probability* where we define $P_{x_1,x_2,\ldots,x_n}(i_1,i_2,\ldots,i_n)$ as the probability that simultaneously the random variable x_1 will have the value i_1, the variable x_2 will have the value i_2, and so on. For example, let heads be specified as 1, tails as 2, and dice results as 1, 2, 3, 4, 5, or 6. Let x_1 and x_2 respectively refer to the results of flipping the two coins and x_3, x_4, x_5, and x_6 respectively refer to the results of casting the four dice. Then we might specify that $P_{x_1,x_2,x_3,x_4,x_5,x_6}(2,1,4,6,1,5) = \frac{1}{5184}$. Similar probabilities would hold for all other possible outcomes of the compound experiment. As with the single-variable case,

$$\sum_{i_1=1}^{N_1} \sum_{i_2=1}^{N_2} \cdots \sum_{i_n=1}^{N_n} P_{x_1,x_2,\ldots,x_n}(i_1,i_2,\ldots,i_n) = 1. \tag{49}$$

The joint probability distribution for a subset of the variables $x_1,\ldots,x_n$ is the sum, over all possible values of the variables not in the subset, of the overall joint probability distribution for all of the variables. For example,

$$P_x(i) = \sum_{j=1}^{N_2} P_{x,y}(i,j). \tag{50}$$

The fraction of the compound experiments in which $x = i$ is the fraction of the compound experiments in which $x = i$ and $y = 1$, plus the fraction of the compound experiments in which $x = i$ and $y = 2$, and so on, through the compound experiments in which $x = i$ and $y = N_2$.

If the value for one random variable does not affect the value for another random variable, as in the case of the respective results of throwing two dice, we say that the variables are *independent*. Mathematically, independence is defined by the relationship

$$\forall i \forall j \; P_{x,y}(i,j) = P_x(i)P_y(j). \tag{51}$$

Interpreting this equation, we see that the fraction of the compound experiments in which $x = i$ and $y = j$ is simply the fraction of the compound experiments in which $x = i$ times the fraction of the compound experiments in which $y = j$. For any i, the fact that $x = i$ does not change the fraction of the compound experiments in which $y = j$, and vice versa.

Analogous to discrete multiple-variable probability, we have continuous multiple-variable probability. For example, we can specify $Pr(a \leq x < b,\, c \leq y < d)$, where x may be the weight and y the height of the next person to enter some office. As in the single-variable continuous case, we define a probability density function $p_{x1,x2,\ldots,x_n}(u_1,u_2,\ldots,u_n)$ such that

$$Pr(u_1 \leq x_1 < u_1+du_1,\ u_2 \leq x_2 < u_2+du_2,\ldots,u_n \leq x_n < u_n+du_n)$$
$$= p_{x1,x2,\ldots,x_n}(u_1,u_2,\ldots,u_n)du_1du_2\ldots du_n, \quad (52)$$

and

$$Pr(x_1 \in [v_1,w_1),\ x_2 \in [v_2,w_2),\ldots,x_n \in [v_n,w_n))$$
$$= \int_{v_1}^{w_1}\int_{v_2}^{w_2}\cdots\int_{v_n}^{w_n} p_{x1,x2,\ldots,x_n}(u_1,u_2,\ldots,u_n)du_1du_2\ldots du_n. \quad (53)$$

The probability distribution for a subset of $\{x_1,x_2,\ldots,x_n\}$ is given by integrating the overall joint probability density function over all possible values (over $[-\infty,\infty]$) of each variable not in the subset. For example,

$$p_x(u)du = Pr(u \leq x < u+du)$$
$$= \int_{v=-\infty}^{\infty} Pr(u \leq x < u+du,\ v \leq y < v+dv) = \int_{-\infty}^{\infty} p_{x,y}(u,v)dv\ du. \quad (54)$$

Since du on each side of equation 54 refers to the same interval width, the density function $p_x(u)$ is given by

$$p_x(u) = \int_{-\infty}^{\infty} p_{x,y}(u,v)dv. \quad (55)$$

Similarly, for independence we require

$$p_{x,y}(u,v)du\ dv = (p_x(u)du)(p_y(v)dv), \quad (56)$$

which leads to the requirement on the density functions:

$$p_{x,y}(u,v) = p_x(u)p_y(v). \quad (57)$$

The next concept we must develop in the context of compound experiments is that of the probability of one event in one experiment given that another event has occurred in another experiment of the compound experiment. This concept is called *conditional probability.* For example, we may be interested in the probability distribution of the height of people who weigh 170 pounds. Consider the compound experiment which gives values to two random variables, x and y. We write $P_{x|y}(i|j)$ to mean the probability that $x=i$ given $y=j$, and similarly $p_{x|y}(u|v)$ to mean the probability density at $x=u$ given $y=v$. Note that

$$\sum_{i=1}^{N_1} P_{x|y}(i|j) = \int_{-\infty}^{\infty} p_{x|y}(u|v)du = 1. \quad (58)$$

That is, $P_{x|y}(i|j)$ gives the fraction of the compound experiments in which

$y = j$ that $x = i$. Conceptually, only x varies in this conditional probability; y is fixed.

If x and y are independent, giving y does not affect the probability of each value of x:

$$P_{x|y}(i|j) = P_x(i), \tag{59}$$

and similarly

$$p_{x|y}(u|v)du = p_x(u)du, \tag{60}$$

which implies

$$p_{x|y}(u|v) = p_x(u). \tag{61}$$

Weight and height, for example, do not satisfy equation 61, so they are not independent random variables.

Now consider the fraction of the compound experiments in which $x = i$ and $y = j$ together. This value is the product of the fraction of the experiments in which $y = j$ and the fraction of the experiments in which $x = i$ given $y = j$. That is,

$$P_{x,y}(i,j) = P_y(j)P_{x|y}(i|j). \tag{62}$$

Similarly,

$$p_{x,y}(u,v)du\,dv = (p_y(v)dv)(p_{x|y}(u|v)du), \tag{63}$$

which implies

$$p_{x,y}(u,v) = p_y(v)p_{x|y}(u|v). \tag{64}$$

Assuming we know the value of x first, we obtain

$$P_{x,y}(i,j) = P_x(i)P_{y|x}(j|i) \tag{65}$$

and

$$p_{x,y}(u,v) = p_x(u)p_{y|x}(v|u).\dagger \tag{66}$$

Using equation 65 together with 62, and 66 with 64, we obtain

$$P_y(j)P_{x|y}(i|j) = P_x(i)P_{y|x}(j|i) \tag{67}$$

and

$$p_y(v)p_{x|y}(u|v) = p_x(u)p_{y|x}(v|u), \tag{68}$$

or

$$P_{x|y}(i|j) = P_{y|x}(j|i)P_x(i)/P_y(j) \tag{69}$$

and

$$p_{x|y}(u|v) = p_{y|x}(v|u)p_x(u)/p_y(v). \tag{70}$$

Equation 69 is called *Bayes' Rule*. It relates the conditional probability of x given y with the conditional probability of y given x, using $P_x(i)$ and $P_y(i)$, the so-called "*a priori* probabilities" (that is, before knowing the value of the other variable). Bayes' rule is very useful because we often know from theory *Pr*(*computed value* | *correct value*) and we wish to know *Pr*(*correct value* | *computed value*), because the computed value is what we are given.

---

† This relationship can also be obtained by noting that $P_{x,y}(i, j) = P_{y,x}(j,i)$ and then using equation 62.

1.4.2 Probability Distribution Measures

Often, either we find it unnecessary to obtain the probability distribution of a random variable (such as the computed value of some function) or we are not given enough information to determine this probability distribution. In such cases, various measures which partially describe the distribution may be computable and of interest. The two most important features of the distribution given by such measures are *centrality* and *width*. A measure of centrality specifies the middle value of the probability distribution; that is, it gives an indication of the value near which the random variable is likely to be. A measure of width specifies how near the central value the variable is likely to be.

Three common measures of centrality are the mode, the median, and the mean. In considering these, we will discuss only continuous random variables. Corresponding relations for discrete random variables are easily developed by using the probability densities corresponding to discrete variables (see equation 48), thereby replacing "∫" by "Σ."

The *mode* of a probability distribution gives the most likely, or most common, value. This is the value for which the probability density is maximum.

$$mode(x) = u_0 \quad \ni: p_x(u_0) = \max_{u \in [-\infty, \infty]} p_x(u), \tag{71}$$

where "∍:" means "such that."

The *median* gives the value for which, half of the time, the random variable takes on a larger value and, half of the time, takes on a smaller value.

$$median(x) = u_0 \quad \ni: \int_{-\infty}^{u_0} p_x(u)du = \tfrac{1}{2}. \tag{72}$$

The *mean* gives the weighted average of the random variable where each value u of the random variable x is weighted by the probability $p_x(u)du$ that x is in the infinitesimal interval du about u. A *weighted average* of $f(x)$ with weight density function $w(x)$ is defined as $\int_{-\infty}^{\infty} f(u)w(u)du / \int_{-\infty}^{\infty} w(u)du$. Thus,

$$mean(x) \equiv \mu_x = \frac{\displaystyle\int_{-\infty}^{\infty} up_x(u)du}{\displaystyle\int_{-\infty}^{\infty} p_x(u)du} = \int_{-\infty}^{\infty} up_x(u)du. \tag{73}$$

All three of the above measures are reasonable, though the mode has the drawback that it depends on a local property (the maximum of $p_x(u)$ may occur at a high, very narrow peak so that x is near the maximum in only a very small fraction of the experiments). The mean seems superior to the median in that the position of the mean is sensitive to the value of the position of an increase in probability density rather than just to which side of the center the increase is on (see figure 1-3). A further important advantage of the mean, not related to its conceptual basis, is that its definition is much more mathematically tractable than that of the median or the mode.

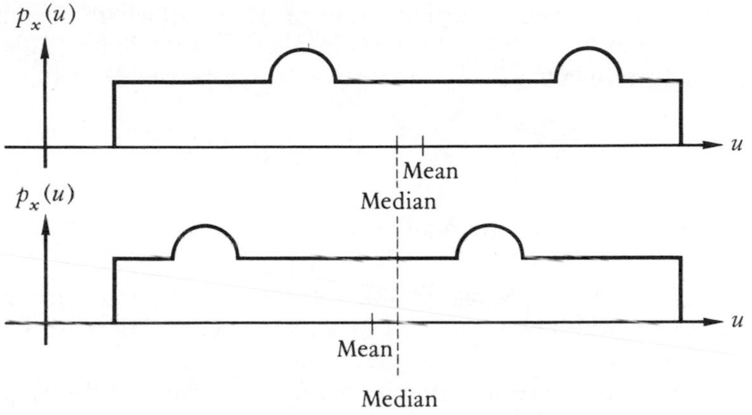

FIG. 1-3 The mean is sensitive to the positions of increase in $p_x(u)$ whereas the median is not.

A measure of the width of a probability distribution should indicate the variability of the random variable from the central value. The first candidate that presents itself is the average (weighted by its probability) deviation from the mean. Note that we must speak of the average deviation magnitude, $\int_{-\infty}^{\infty} |u - \mu_x| p_x(u) du$, because the average deviation about the mean, $\int_{-\infty}^{\infty} (u - \mu_x) p_x(u) du$, is zero (see exercise 1.4.4); the positive and negative deviations cancel each other. Other possibilities for a width measure involve the weighted average of a positive integer power of the deviation magnitude: $[\int_{-\infty}^{\infty} |u - \mu_x|^n p_x(u) du]^{1/n}$. Large deviations have more effect on the measure in the latter case with $n > 1$ than they do when the average deviation magnitude ($n = 1$) is used. The difficulty with the average deviation magnitude and any other width measure with an odd value of n is lack of mathematical tractability due to the discontinuity in the derivatives of the absolute value function. Because higher powers of n give too much weight to the larger deviations, $n = 2$ is commonly chosen. The measure of width for $n = 2$ is called the *standard deviation*, denoted

$$\sigma_x = \left[\int_{-\infty}^{\infty} (u - \mu_x)^2 p_x(u) du \right]^{1/2}. \tag{74}$$

The square of the standard deviation is called the *variance* and is written σ_x^2 or $var(x)$. The standard deviation is closely related to the mean in that the mean can be defined as that value of M for which the standard deviation about M, $s = [\int_{-\infty}^{\infty} (u - M)^2 p_x(u) du]^{1/2}$ is minimum. This can be shown by setting to zero the derivative with respect to M of s^2:

$$0 = \frac{ds^2}{dM} = \int_{-\infty}^{\infty} -2(u - M) p_x(u) du, \tag{75}$$

leading to

$$\int_{-\infty}^{\infty} u p_x(u) du = \int_{-\infty}^{\infty} M p_x(u) du, \tag{76}$$

which implies $M = \mu_x$. Similarly, if $n = 1$, the centrality measure which minimizes the width of $p_x(u)$ about M is the median of the distribution (see exercise 1.4.5).

Another important relationship is obtained by expanding the equation

$$\sigma_x^2 = \int_{-\infty}^{\infty} (u - \mu_x)^2 p_x(u) du \tag{77}$$

to obtain

$$\begin{aligned} \sigma_x^2 &= \int_{-\infty}^{\infty} (u^2 - 2\mu_x u + \mu_x^2) p_x(u) du \\ &= \int_{-\infty}^{\infty} u^2 p_x(u) du - 2\mu_x \int_{-\infty}^{\infty} u p_x(u) du + \mu_x^2 \int_{-\infty}^{\infty} p_x(u) du \\ &= \int_{-\infty}^{\infty} u^2 p_x(u) du - 2\mu_x^2 + \mu_x^2 = \int_{-\infty}^{\infty} u^2 p_x(u) du - \mu_x^2. \end{aligned} \tag{78}$$

Thus

$$\int_{-\infty}^{\infty} u^2 p_x(u) du = \sigma_x^2 + \mu_x^2. \tag{79}$$

We have repeatedly encountered the form $\int_{-\infty}^{\infty} f(u)p_x(u)du$ when seeking a measure for the average value for $f(x)$. This form has the special name, the *expected value* of $f(x)$, and is written

$$E(f(x)) \equiv \int_{-\infty}^{\infty} f(u)p_x(u)du \tag{80}$$

or more generally

$$E(f(x_1,x_2,\ldots,x_n)) \equiv \int_{-\infty}^{\infty}\int_{-\infty}^{\infty}\cdots\int_{-\infty}^{\infty} f(u_1,u_2,\ldots,u_n)p_{x_1,x_2,\ldots,x_n}(u_1,u_2,\ldots,u_n)du_1du_2\ldots du_n. \tag{81}$$

By convention, we write $E(f(x))^2$ to mean $E([f(x)]^2)$ and $E^2(f(x))$ to mean $[E(f(x))]^2$. Thus

$$E(x) = \mu_x, \tag{82}$$

and

$$E(x - \mu_x)^2 = E(x - E(x))^2 = \sigma_x^2. \tag{83}$$

Note that E is a linear operator, that is, for all a, b, f, and g,

$$E(af(x) + bg(x)) = aE(f(x)) + bE(g(x)), \tag{84}$$

where a and b are constants with respect to the probability distribution in question. We show this using the definition of E as follows:

$$\begin{aligned} E(af(x) + bg(x)) &= \int_{-\infty}^{\infty} (af(u) + bg(u))p_x(u)du \\ &= a\int_{-\infty}^{\infty} f(u)p_x(u)du + b\int_{-\infty}^{\infty} g(u)p_x(u)du \\ &= aE(f(x)) + bE(g(x)). \end{aligned} \tag{85}$$

The linearity of E is a very important property. It gives to measures which can be expressed as expected values (for example, mean and variance) some mathematical properties that are very useful and also conceptually reasonable for such measures.

Expected value can also be defined for a function of many random variables. In this case,

$$E(f(x_1,x_2,\ldots,x_n)) \equiv \int_{-\infty}^{\infty}\int_{-\infty}^{\infty}\cdots\int_{-\infty}^{\infty} f(u_1,u_2,\ldots,u_n)p_{x_1,x_2,\ldots,x_n}(u_1,u_2,\ldots,u_n)du_1du_2\ldots du_n. \tag{86}$$

Similarly, the conditional expected value can be defined:

$$E(f(x_1,x_2,\ldots,x_n)|x_m,x_{m+1},\ldots,x_n) \equiv \int_{-\infty}^{\infty}\int_{-\infty}^{\infty}\cdots\int_{-\infty}^{\infty} f(u_1,u_2,\ldots,u_n)$$

$$p_{x_1,x_2,\ldots,x_{m-1}|x_m,x_{m+1},\ldots,x_n}(u_1,u_2,\ldots,u_{m-1}|u_m,u_{m+1},\ldots,u_n)du_1du_2\ldots du_{m-1}. \tag{87}$$

When we consider measures for two random variables, we ask not only how each variable behaves individually but also how the two vary together. For example, we know that if a person's height tends to be greater than average, then his weight also tends to be greater than average. Full information about the way the variables vary together is given by their joint probability distribution, but here again it is useful to develop some measures which generally characterize the behavior of the joint distribution without fully describing it. The behavior we wish to characterize is that mentioned above: How do the variables vary together, or covary? We wish to know to what extent the fact that x increases by a given amount tells us how much y is going to increase (or perhaps decrease). More precisely, we can ask to what extent the value of y can be predicted from the value of x by a linear relationship, $y = ax + b$. What is their *linear covariation*?

Let us construct a graph with the random variable x on one axis and the random variable y on the other axis. Then the joint occurrence of $x = a$ and $y = b$ corresponds to the point (a,b) on this graph. A joint probability distribution over these two random variables is specified by giving a probability density for each point in the x,y plane. If we take a set of samples according to this probability distribution and graph the points corresponding to the samples, we obtain a graph like that in figure 1-4. As the number of sampled points becomes infinite, the point density at any position approaches a number proportional to the probability density at that position. The measure of linear covariance that we are interested in is a measure of how well a straight line fits this point distribution as the number of points becomes infinite.

Consider the line $y_L(x) = ax + b$ which fits the multiple-variable point distribution best in the same sense that the mean fits a distribution of one variable best, namely, that the average squared deviation from that line is minimum. We must define the *squared deviation* of a point from the line. Possibilities which come to mind (see figure 1-5) are (1) the squared y deviation at the given x: $(y - (ax + b))^2$, (2) the squared x deviation at the given y: $(x - (1/a)(y - b))^2$, which can be written $(1/a^2)(y - (ax + b))^2$, (3) the squared perpendicular distance from (x,y) to the line: $(1/(1 + a^2))(y - (ax + b))^2$, or (4) some weighted sum of

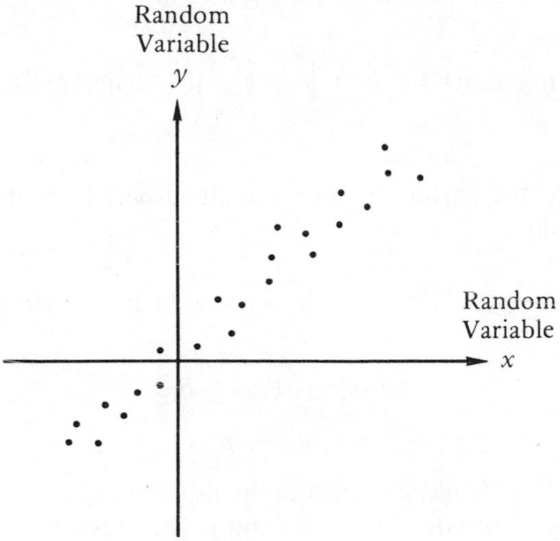

FIG. 1-4 Samples from highly correlated random variables, showing significant linear covariation

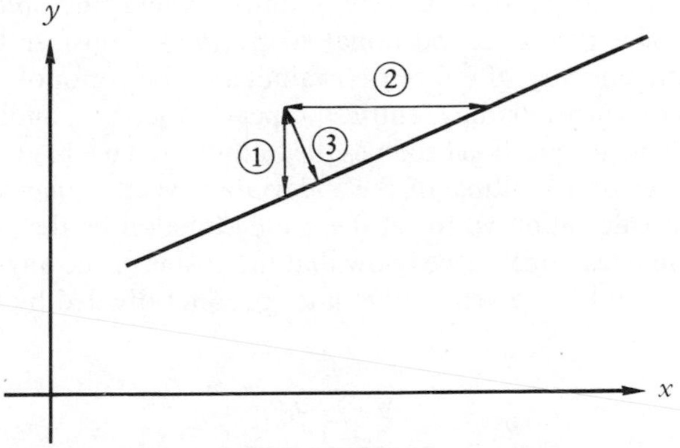

FIG. 1-5 Possible measures of deviation of point from line

these values. Thus we are interested in the line which minimizes $E[f(a)(y-(ax+b))^2]$ for some function $f(a)$.

$$E[f(a)(y-(ax+b))^2] = f(a) \int_{-\infty}^{\infty} \int_{-\infty}^{\infty} (v-(au+b))^2 p_{x,y}(u,v)du\, dv. \tag{88}$$

Setting to zero the partial derivative with respect to b of this expected value, we obtain

$$0 = -2f(a) \int_{-\infty}^{\infty} \int_{-\infty}^{\infty} (v-(au+b))p_{x,y}(u,v)du\, dv, \tag{89}$$

which implies

$$E(y) = aE(x) + b, \tag{90}$$

or

$$b = \mu_y - a\mu_x. \tag{91}$$

Thus the best line should go through the point (μ_x, μ_y).

Substituting equation 91 for b into the equation for the line, $y_L = ax + b$, we obtain the line

$$y_L = ax + E(y) - aE(x), \tag{92}$$

or

$$y_L - \mu_y = a(x - \mu_x). \tag{93}$$

Our question of the covariation of x and y has come down to the question: To what extent are the deviations of y from the mean of y multiples of the corresponding deviations of x from the mean of x? Let us consider the joint probability distribution of x and y as represented by an infinite set of $(u-\mu_x, v-\mu_y)$ pairs, where the frequency of occurrence of a pair is proportional to $p_{x,y}(u,v)$. Consider the vector (one-dimensional array) of x deviation values and the vector of y deviation values, where corresponding entries appear in corresponding vector elements. Then our question becomes the following: To what extent is the y-deviation vector a multiple of the x-deviation vector, that is, to what extent is the x-deviation vector in the same direction as the y-deviation vector? From linear algebra we know that the cosine of the angle between two vectors a and b is given by the inner product divided by the length of both vectors:

$$\frac{a^T b}{|a|\,|b|} = \frac{\sum_{i=1}^{N} a_i b_i}{\left(\sum_{i=1}^{N} a_i^2\right)^{1/2} \left(\sum_{i=1}^{N} b_i^2\right)^{1/2}}.\dagger \tag{94}$$

† See section 2.2.2 for discussion of this relationship.

Generalizing this definition to our vectors for which the x elements and y elements vary continuously from element to element and where each element must be weighted by its probability density (the number of times it occurs in the vector), we obtain

$$\rho_{xy} = \frac{\displaystyle\int_{-\infty}^{\infty}\int_{-\infty}^{\infty} (u-\mu_x)(v-\mu_y)p_{x,y}(u,v)\,du\,dv}{\left(\displaystyle\int_{-\infty}^{\infty} (u-\mu_x)^2 p_x(u)\,du\right)^{1/2}\left(\displaystyle\int_{-\infty}^{\infty} (v-\mu_y)^2 p_y(v)\,dv\right)^{1/2}}, \tag{95}$$

where ρ_{xy} is the measure of the cosine of the angle between our x-deviation and y-deviation vectors.

$$\rho_{xy} = \frac{E[(x-\mu_x)(y-\mu_y)]}{[E(x-\mu_x)^2]^{1/2}[E(y-\mu_y)^2]^{1/2}} \tag{96}$$

is called the *correlation coefficient* of x and y. Note that ρ_{xy} varies from -1 to $+1$ and measures the *linearity* of the relationship between x and y. $\rho_{xy} = 1$ implies $y = ax + b$ where $a > 0$, and $\rho_{xy} = -1$ implies $y = ax + b$ where $a < 0$. $\rho_{xy} > 0$ implies y tends to increase with x, and $\rho_{xy} < 0$ implies y tends to decrease as x increases. $\rho_{xy} = 0$ implies that on the average the tendency of y and that of x are not related. In this situation, x and y are said to be *uncorrelated*. This can happen only when the numerator of ρ_{xy}, that is, $E[(x-\mu_x)(y-\mu_y)]$, is equal to zero. This numerator, the *covariance* of x and y, is written σ_{xy} or $cov(x,y)$. It has the following interesting properties:

1. $\sigma_{xy} = \rho_{xy}\sigma_x\sigma_y$. (97)

2. $\sigma_{xx} = \sigma_x^2$. (98)

3. $\sigma_{xy} = E[(x-\mu_x)(y-\mu_y)] = E(xy - \mu_x y - \mu_y x + \mu_x\mu_y)$

$$= E(xy) - \mu_x\mu_y - \mu_y\mu_x + \mu_x\mu_y = E(xy) - E(x)E(y). \tag{99}$$

Equation 97 states that the covariance between x and y involves both the variability of each as measured by the standard deviation, and the way the two vary together, as measured by the correlation coefficient. From equations 97 and 99 we can conclude that two variables x and y are uncorrelated if and only if $E(xy) = E(x)E(y)$. This relation is often taken as the definition of uncorrelated.

Note that the independence of two random variables requires that their joint probability distribution factors into the product of their individual probability distributions *at every value* of the random variables, whereas the lack of correlation of two random variables requires only that *the expected value* of their product factors into the product of their expected values. We can show that independence implies lack of correlation but not vice versa (see exercise 1.4.7).

1.4.3 Functions of Random Variables; Estimators of Random Variables

Often we deal with a function of one or more random variables, $z = f(x_1, x_2, \ldots, x_n)$, and we need to know the probability distribution or measures of the resulting random variable z, given the probability distributions or measures of the input variables x_i. For example, we may be interested in the probability distribution of the sum of three times the height and the square of the weight of the next person entering some office, knowing the probability distributions of height and weight.

Let us consider the single-variable case first, $z = f(x)$. In the discrete case, the probability that $z = v$ is simply the sum of the probabilities of all x values such that $f(x) = v$. The continuous case is more complicated. Then the probability that $z \in [v, v + dv)$ is the sum of the probabilities of all x values such that $f(x) \in [v, v + dv)$, where dv is positive (see figure 1-6). Let u be any value of x such that $f(u) = v$. Then we need to know the value of the largest δu in magnitude such that for all $x \in [u, u + \delta u)$, $f(x) \in [v, v + dv)$. For infinitesimal δu,

$$v + dv = f(u + \delta u) = f(u) + f'(u)\,\delta u = v + f'(u)\delta u. \tag{100}$$

Thus

$$dv = |f'(u)|\,|\delta u|. \tag{101}$$

Therefore, the probability that $x \in [u, u + \delta u)$ is

$$p_x(u)|\delta u| = \frac{p_x(u)}{|f'(u)|}\,dv. \tag{102}$$

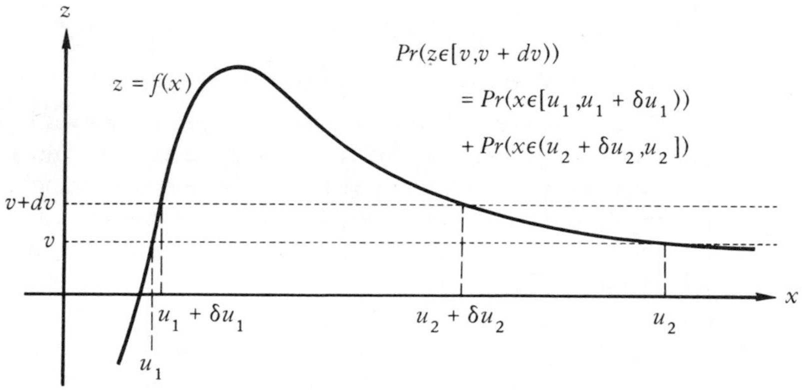

FIG. 1-6 Probability for a function of a continuous random variable

Summing all such probabilities over all u such that $f(u) = v$, we obtain

$$p_z(v)dv = \sum_{\substack{u\ni:\\ f(u)=v}} \frac{p_x(u)}{|f'(u)|}dv, \tag{103}$$

so

$$p_z(v) = \sum_{\substack{u\ni:\\ f(u)=v}} \frac{p_x(u)}{|f'(u)|}. \tag{104}$$

A common instance of the above is when $z = f(x) = ax + b$, where a and b are constants. In this case, for each value of v there is only one value of u for which $f(u) = v$, so

$$p_z(v) = \frac{p_x(u)}{|a|}. \tag{105}$$

From this result follow the relationships

$$\mu_z = a\mu_x + b \tag{106}$$

and

$$\sigma_z^2 = a^2\sigma_x^2 \tag{107}$$

(see exercise 1.4.8).

Functions of two random variables are of particular interest, because they include the simple arithmetic functions. For example, when we add two random variables, each with a given probability distribution, we need to know how to determine the probability distribution of the sum. Or if we have only measures of the original probability distributions, we would like to be able to derive the corresponding measures of the distribution of the sum.

The linearity of the expected value operator leads to some interesting results about the sum of random variables. Let $z = x + y$. Then

$$\mu_z = E(z) = E(x + y) = E(x) + E(y) = \mu_x + \mu_y. \tag{108}$$

The mean of a sum is the sum of the means. Also,

$$\begin{aligned}\sigma_z^2 &= E(z - \mu_z)^2 = E(x - \mu_x + y - \mu_y)^2\\ &= E(x - \mu_x)^2 + 2E[(x - \mu_x)(y - \mu_y)] + E(y - \mu_y)^2\\ &= \sigma_x^2 + 2\sigma_{xy} + \sigma_y^2.\end{aligned} \tag{109}$$

If and only if two variables are uncorrelated, the variance of the sum is the sum of the variances. Note that results 108 and 109 are true for any probability distribution of x and y.

Let us consider how to compute the probability distribution of the sum, or more generally, of any function of two random variables, x and

y. Given $z = f(x,y)$, $p_x(u)$, and $p_y(v)$, we wish to compute $p_z(w)$. Using the fact that

$$P_t(a) = \sum_b P_{t,s}(a,b) = \sum_b P_s(b)P_{t|s}(a|b), \tag{110}$$

we obtain

$$\begin{aligned} p_z(w)dw &= Pr(w \leq z < w + dw) \\ &= \sum_{\text{all } u} Pr(u \leq x < u + du) Pr(y \in \text{interval} \ni : f(x,y) \in [w, w + dw) | x = u) \\ &= \sum_{\text{all } u} (p_x(u)du) \sum_{\substack{\text{all } v \ni: \\ f(u,v) = w}} (p_y(v)|\delta v|), \end{aligned} \tag{111}$$

where dw is positive and $f(u,v + \delta v) = w + dw$. For an infinitesimal δv,

$$w + dw = f(u,v + \delta v) = f(u,v) + \delta v \frac{\partial f(u,v)}{\partial v}. \tag{112}$$

Thus, since $w = f(u,v)$,

$$|\delta v| = \frac{dw}{\left| \dfrac{\partial f(u,v)}{\partial v} \right|}, \tag{113}$$

so that

$$p_z(w)dw = \sum_{\text{all } u} p_x(u)du \sum_{\substack{\text{all } v \ni: \\ f(u,v) = w}} \frac{p_y(v)dw}{\left| \dfrac{\partial f(u,v)}{\partial v} \right|}, \tag{114}$$

or since

$$\sum_{\text{all } u} g(u)du \equiv \int_{-\infty}^{\infty} g(u)du, \tag{115}$$

$$p_z(w) = \int_{-\infty}^{\infty} \left[p_x(u) \sum_{\substack{\text{all } v \ni: \\ f(u,v) = w}} \frac{p_y(v)}{\left| \dfrac{\partial f(u,v)}{\partial v} \right|} \right] du. \tag{116}$$

Thus, for example, if $z = f(x,y) = x + y$, then $\frac{\partial f}{\partial v}(u,v) = 1$, so

$$p_z(w) = \int_{-\infty}^{\infty} p_x(u)\, p_y(w - u)du. \tag{117}$$

If $z = f(x,y) = xy$, then $\frac{\partial f}{\partial v}(u,v) = u$, so

$$p_z(w) = \int_{-\infty}^{\infty} p_x(u) p_y(w/u)(1/|u|)du. \tag{118}$$

Another important concept of probability is that of an *estimator* of an unknown value. For example, a measurement produces an estimation of the true value of the object being measured—an estimator because some error is made in the measurement. Similarly, we often estimate the mean of a distribution occurring in nature by taking some samples from that distribution and calculating the average value of the samples. One property we would like an estimator to have is that the average estimate over many experiments of estimating the same value is the true value. Such an estimator is said to be *unbiased*. Thus, if x^* is an estimator for x, x^* is unbiased if and only if $E(x^*) = x$. It can be shown that

$$\bar{x} \equiv \frac{1}{N} \sum_{i=1}^{N} x_i \tag{119}$$

is an unbiased estimator for μ_x if the x_i are independent samples from the distribution with probability density $p_x(u)$ (see exercise 1.4.10). Similarly, it can be shown that

$$s_x^2 \equiv \frac{1}{N-1} \sum_{i=1}^{N} (x_i - \bar{x})^2 \tag{120}$$

is an unbiased estimator for σ_x^2, as is $\frac{1}{N} \sum_{i=1}^{N} (x_i - \mu_x)^2$ (again, see exercise 1.4.10). The theory of producing these estimators is discussed in section 4.3.1.

1.4.4 Common Probability Distributions

Before we leave our introduction to probability, it is useful to discuss a few common probability distributions, namely, the uniform, normal, binomial, and Poisson distributions.

The *uniform distribution* is that for which the probability density is constant in the interval in which it is nonzero (every value in the interval is equally likely):

$$p_x(u) = \begin{cases} 1/(b-a) & \text{if} \quad u \in [a,b] \\ 0 \quad \text{if} \quad u \notin [a,b]. \end{cases} \tag{121}$$

The uniform distribution appears in real life as a result of cyclic processes. For example, assume we have arbitrarily marked a 0° point on every automobile tire. If the random variable is the point on a tire which is touching the ground when the North Carolina state line is crossed, and that point is measured in degrees, the variable can be expected to have a uniform distribution on [0°,360°).

Most noncyclic processes in nature have a higher probability density near the mean than at the edges of the distribution. However, another use of the uniform distribution is as an assumed *a priori* distribution when nothing is known about a random variable except its range. Information theory has shown that for a given range for the random variable, the probability distribution for which there is the greatest uncertainty about the value of the variable is the one for which all possible values have equal probability—the uniform distribution.

The probability distribution most commonly encountered in nature is the *normal distribution,* also called the *Gaussian distribution:*

$$p_x(u) = \frac{1}{\sigma_x\sqrt{2\pi}}\, e^{-\frac{1}{2}\left(\frac{u-\mu_x}{\sigma_x}\right)^2}, \quad -\infty \leq x \leq \infty. \tag{122}$$

The bell shape of this distribution, about the mean μ_x with variance σ_x^2, is shown in figure 1-7. Because this distribution is so common, people too often assume a normal distribution without proper reason. Nevertheless, it is true that the normal distribution is the most common distribution for measurement errors and many other random variables. This stems from the fact stated by the very important *central limit theorem*: The sum of any n independent random variables, each with any probability distribution, has a normal distribution as $n \to \infty$. This theorem can be generalized to show that for many other common symmetric functions (for example, $1/(x_1 + x_2 + \cdots + x_n)$), the result is normal as the number of independent variables $\to \infty$. Since most natural processes involve a number of independent causes, many natural random variables either have or at least approximately have a normal distribution.

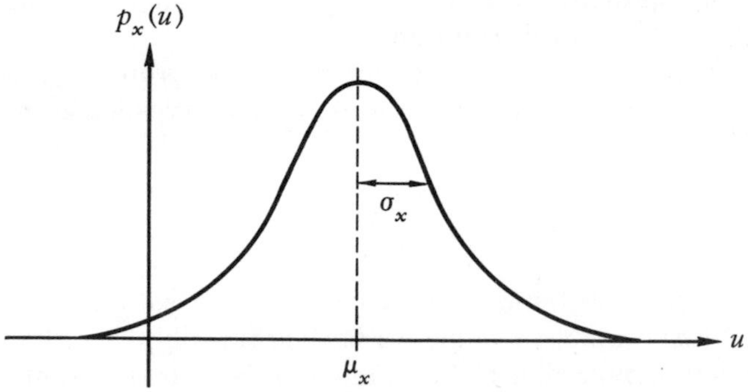

FIG. 1-7 Normal probability distribution

One also commonly encounters the normal distribution of many variables:

$$p_{\underline{z}}(\underline{u}) = \frac{1}{(2\pi)^{n/2} \prod_{i=1}^{n} \sigma_{z_i}} e^{-\frac{1}{2}(\underline{u}-\underline{\mu}_{\underline{z}})^T R^{-1}(\underline{u}-\underline{\mu}_{\underline{z}})}, \tag{123}$$

where $\underline{z} = (z_1, z_2, \ldots, z_n)^T$, $\underline{\mu}_{\underline{z}} = (\mu_{z_1}, \mu_{z_2}, \ldots, \mu_{z_n})^T$, and R is the $n \times n$ covariance matrix such that $R_{ij} = \sigma_{z_i z_j}$ (where $\sigma_{z_i z_i} = \sigma_{z_i}^2$).

A commonly encountered discrete distribution is the *binomial distribution*:

$$P_x(i) = \binom{n}{i} p^i (1-p)^{n-i}, \tag{124}$$

for i an integer between 0 and n, and $0 \leq p \leq 1$. This is the distribution of the number of occurrences, i, of a particular event which occurs independently with probability p each time among n experiments. For example, let i be the number of times in n throws of a fair die that 5 comes up ($p = \frac{1}{6}$). $\binom{n}{i} \equiv n!/(i!(n-i)!)$ is the number of different ways that i indistinguishable objects can be placed in n slots. Thus $P_x(i)$ is the number of distinct groups of i experiments which can produce the event in question, $\binom{n}{i}$, times the probability a particular group of i did produce the event, p^i, times the probability the remaining $n-i$ experiments did not produce the event, $(1-p)^{n-i}$. The mean and variance of the binomial distribution are np and $np(1-p)$ respectively (see exercise 1.4.11).

The *Poisson distribution* is a discrete distribution defined at an infinite number of points, the integers ≥ 0. It arises from a "random arrival process": If the probability of an event occurring in any small time interval is independent of the probability of an event occurring in any other small time interval, then the probability of receiving i events in a given time interval $[t_1, t_2]$ follows the Poisson distribution:

$$P_x(i) = \frac{e^{-\mu_x} \mu_x^i}{i!}, \quad i = 0, 1, 2, \ldots, \tag{125}$$

where μ_x is the average number of events expected in $[t_1, t_2]$.

The Poisson distribution can be shown to be the limit of the binomial distribution as the number of possible events, $n, \to \infty$ but the mean number of occurrences of the event in question in a set of experiments (time interval) is constant (see exercise 1.4.12); that is, $\mu_x = np$ is constant. An interesting feature of the Poisson distribution is that it is completely defined by its mean. In fact, its variance is equal to its mean (see exercise 1.4.11, part c). Examples of processes following the Poisson distribution are the emission of radioactivity, the arrival of customers at a market, and the detection of photons on a film (where the interval is in space as well as time).

1.4.5 Probabilistic Analysis of Error

We will now apply ideas from our discussion of probability to the analysis of computational error. For a given set of accurate inputs and a given algorithm and computer, the computational error is both deterministic and discrete. However, because the accurate inputs are not known and because we desire to do the analysis for a general set of inputs, it is useful to model the process probabilistically. Also, because the possible error values are numerous and closely packed, it is convenient to use a continuous random variable to model the error. Thus, we assume that each number represented in the computer has an associated continuous probability distribution of error (error distribution). Any such number is the result of measurements and computations, and the error distribution associated with the number depends on the error distributions of the contributing measurements and on the computations carried out.

A measurement error distribution can be approximated by carrying out the associated experiment many times, but this approach is not common. Rather some (sometimes unjustified) assumptions are made about the input error distributions (for example, that they are normal, on the basis of arguments from the central limit theorem). The variances of input errors may be estimated by sampling from the distribution, but often these measures are also assumed. It should be emphasized that input error distributions or at least their variances must be estimated accurately if the final results of a computation are to be trusted.

The mean of the measurement error distribution has not been mentioned, because we normally do not measure errors. Rather, we measure input parameter values, and we assume (and hopefully so design our measurement procedures) that our estimation procedures are unbiased. Thus if $x^* = x + \varepsilon_x$ is a measured value for x, we assume $E(x^*) = x$ (that $E(\varepsilon_x) = 0$). However, if x^* is the result of a calculation or computer representation of a number, the assumption of unbiased estimation is not valid if we practice truncation of low-order digits rather than symmetric rounding. In this case the error mean must be estimated as well as the variance.

Each computation done with inputs with associated error distributions produces an output with an error distribution that can be determined using the error propagation and generation properties of the computation. The propagated error distribution can be determined by applying the techniques of section 1.4.3 for computing the probability distribution of a function of random variables. The generated error distribution can be determined by experiment with the computation in question or by analysis of the computation as it is carried out in the computer. For example, assume we are adding two positive floating-point numbers in a base-b, n-digit floating-point computer, and we apply truncation rather than symmetric rounding. In this case, we might expect the generated error in the fraction part to be uniformly distributed over $[0,b^{-n})$.

Since the overall error in a computed number is the sum of the generated and propagated errors, the distribution of this overall error can be computed by the techniques of section 1.4.3 applied to $f(\varepsilon^{gen},\varepsilon^{prop}) = \varepsilon^{gen} + \varepsilon^{prop}$.

Usually, instead of computing an output error probability distribution, we compute a measure of the errors defined by that distribution. Our previous discussion leads us to the measure

$$\sigma_{x^*} = \sigma_{\varepsilon_x} \tag{126}$$

(see exercise 1.4.14). Then given a probability distribution for ε_x, we can compute σ_{x^*}. Or, if we are not given the probability distribution, we can estimate σ_{x^*} by taking many independent samples from the x^* distribution and using the estimator $s_{x^*}^2$ for $\sigma_{x^*}^2$ (recall equation 120). Thus we have a measure of the generated error.

For any operation, we can compute the measure of the propagated error given the measure of the input errors. For example, if x^* and y^* are two uncorrelated unbiased estimators for x and y respectively and $z^* = x^* + y^*$, then

$$\sigma_{z^*}^2 = \sigma_{x^*}^2 + \sigma_{y^*}^2, \tag{127}$$

or

$$\sigma_{z^*} = (\sigma_{x^*}^2 + \sigma_{y^*}^2)^{1/2}. \tag{128}$$

Note also that if x^* and y^* are uncorrelated,

$$\sigma_{x^*-y^*}^2 = \sigma_{x^*}^2 + \sigma_{y^*}^2 \tag{129}$$

(see exercise 1.4.15).

From exercise 1.4.16, we see that if x^* and y^* are independent,

$$\sigma_{x^*\cdot y^*}^2 = \sigma_{x^*}^2\sigma_{y^*}^2 + x^2\sigma_{y^*}^2 + y^2\sigma_{x^*}^2. \tag{130}$$

In parallel with error bounds, we can talk about the variance of the relative error in a given value of x. This variance is defined to be $E(\varepsilon_x/x)^2 - E^2(\varepsilon_x/x)$. For a given (nonrandom) number x, assuming unbiased estimators,

$$E\left(\frac{\varepsilon_x}{x}\right) = \frac{1}{x}E(\varepsilon_x) = 0, \tag{131}$$

and thus

$$var\left(\frac{\varepsilon_x}{x}\right) = E\left(\frac{\varepsilon_x}{x}\right)^2 = \frac{1}{x^2}E(\varepsilon_x^2) = \frac{\sigma_{x^*}^2}{x^2}. \tag{132}$$

The standard deviation of the relative error in x is therefore $\sigma_{x^*}/|x|$.

Dividing equation 130 by $(xy)^2$ (which is equal to $\mu^2_{x^*y^*}$ if x^* and y^* are uncorrelated, as shown by exercise 1.4.16),

$$\left(\frac{\sigma_{x^*\cdot y^*}}{xy}\right)^2 = \left(\frac{\sigma_{x^*}}{x}\right)^2\left(\frac{\sigma_{y^*}}{y}\right)^2 + \left(\frac{\sigma_{x^*}}{x}\right)^2 + \left(\frac{\sigma_{y^*}}{y}\right)^2. \tag{133}$$

If $\sigma_{x^*} \ll |x|$ and $\sigma_{y^*} \ll |y|$, the first term is negligible. We obtain

$$\left(\frac{\sigma_{x^*\cdot y^*}}{xy}\right)^2 \approx \left(\frac{\sigma_{x^*}}{x}\right)^2 + \left(\frac{\sigma_{y^*}}{y}\right)^2; \tag{134}$$

relative variances add under multiplication.

To develop a similar rule for division, we need only show that $\sigma^2_{1/y^*}/\mu^2_{1/y^*} \approx \sigma^2_{y^*}/y^2$ and $\mu_{1/y^*} \approx 1/y$ and then apply equation 134 to $(\sigma^2_{x^*/y^*})/(\mu^2_{x^*/y^*}) = (\sigma^2_{x^*\cdot(1/y^*)})/(\mu^2_{x^*\cdot(1/y^*)})$. If $Pr(|\varepsilon_y| \ll |y|) = 1$, that is, if the errors in y are small compared to y (an assumption we are already making), then $y^* = y + \varepsilon_y$ does not come much closer to zero than y does and

$$1/y^* = 1/(y + \varepsilon_y) = (1/y)(1 - \varepsilon_y/y + O(\varepsilon_y/y)^2). \tag{135}$$

Neglecting the term $O(\varepsilon_y/y)^2$, we obtain

$$\mu_{1/y^*} \approx (1/y)E(1 - \varepsilon_y/y) = 1/y, \tag{136}$$

since we are assuming ε_y is zero mean. Then

$$\sigma^2_{1/y^*} = E(1/y^* - \mu_{1/y^*})^2 \approx E(-\varepsilon_y/y^2)^2 = \sigma^2_{y^*}/y^4. \tag{137}$$

We now have, for the variances of uncorrelated variables with zero mean error which is small relative to the variable value, the same four relations that we have for error bounds: in addition and subtraction, variances add; in multiplication and division, relative variances add.

Note from equation 127 that, assuming no error generation, the sum of n independent values with the same standard deviation produces an output with a standard deviation $\sqrt{n}$ times that of the inputs. Compare this situation to that when the measure of error was the absolute error bound b_{ε_x}. In the latter case, the error measure b_{ε_z} of the sum of n independent values with the same error bound b_{ε_x} was $b_{\varepsilon_z} = n\, b_{\varepsilon_x}$. Thus, the standard deviation is a less conservative error measure than the absolute error bound in the sense that propagation does not cause the former to grow as fast as the latter.

Error propagation for operations on correlated random variables is also of interest. Consider addition and multiplication when the operands are equal.

We saw in equation 109 that if $z^* = x^* + y^*$,

$$\sigma_{z^*}^2 = \sigma_{x^*}^2 + \sigma_{y^*}^2 + 2\sigma_{x^*y^*}. \tag{138}$$

This rule for error propagation under addition can be applied to $y = x$, using

$$\sigma_{x^*x^*} \equiv \sigma_{x^*}^2, \tag{139}$$

to produce

$$\sigma_{z^*}^2 = \sigma_{x^*}^2 + \sigma_{x^*}^2 + 2\sigma_{x^*}^2 = 4\sigma_{x^*}^2. \tag{140}$$

This result agrees with that produced by applying equation 107 to $z^* = 2x^*$. It is off by a factor of 2 from the result obtained if the summands were assumed to be uncorrelated.

A similar result is obtained for multiplication. If $z^* = x^*y^*$, then

$$\mu_{z^*} = xy + \sigma_{x^*y^*} \tag{141}$$

(the computed value is not an unbiased estimator if $\sigma_{x^*y^*} \neq 0$), and

$$\sigma_{z^*}^2 = y^2\sigma_{x^*}^2 + x^2\sigma_{y^*}^2 - \sigma_{x^*y^*}^2 + 2xy\sigma_{x^*y^*} + E(\varepsilon_x^2\varepsilon_y^2) + 2yE(\varepsilon_x^2\varepsilon_y) + 2xE(\varepsilon_x\varepsilon_y^2) \tag{142}$$

(see exercise 1.4.16, part c). Applying equation 142 to $y^* = x^*$ and ignoring terms in third and higher powers of ε_x/x produces

$$\sigma_{z^*}^2/z^2 \approx 4\sigma_{x^*}^2/x^2. \tag{143}$$

Again, our result is off by a factor of 2 from the result obtained if the factors were assumed to be independent (see exercise 1.4.16, part d). Similarly, results of both subtraction and division of appropriately correlated variables can have greater relative variance than that predicted by the rules for independent variables.

In the above, we have discussed techniques for computing the variance of the propagated error in the result of an arithmetic operation and for estimating the variance of the generated error by experiment or analysis of the operation. From these variances we must produce the variance of the overall error. We will assume that the propagated and generated errors are uncorrelated. Then, since

$$\varepsilon^{overall} = \varepsilon^{prop} + \varepsilon^{gen}, \tag{144}$$

$$(\sigma^{overall})^2 = (\sigma^{prop})^2 + (\sigma^{gen})^2. \tag{145}$$

There are error measures other than bounds and standard deviation. These measures vary in the rate of growth caused by propagation as well as in their mathematical tractability, the reasonability of their conceptual basis, and the reasonability of the propagation formulas which follow from them (see exercise 1.4.17). An example is $E(|\varepsilon_x|)$.

Thus far, we have considered the error in x for a given x. We have discussed relative error, but only ε_x was a random variable; x itself remained fixed. We now let x be a random variable also, and we ask: Over all x and all ε_x, what is the value of a measure of the relative error, for example, $\sigma^2_{\varepsilon x}/x^2$? This question may be of interest when analyzing a method to be used with many different input values. It is of less interest when analyzing a particular result, because then we have an approximate value x^* for the result, so we can treat x as fixed.

If ε_x and x are independent random variables, and for each x, $E(\varepsilon_x|x) = 0$, then $E(\varepsilon_x/x) = 0$, so

$$\sigma^2_{\varepsilon x/x} = E(\varepsilon_x/x)^2 = \sigma^2_{\varepsilon x}E(1/x^2), \tag{146}$$

where the expected value is over the two random variables (equation 86).

If ε_x and x are not independent, as is often the case, the expected value which follows from equation 86 is more complicated than that derived in equation 146.

Whether or not ε_x and x are independent, it is clear that the measure of relative error over all ε_x and x is affected by the distribution of x. With error due to truncation of floating-point numbers in the computer, ε_x/x, represented as a floating-point number, tends to have an exponent which is independent of the exponent of x. However, the fraction part of ε_x is largely uncorrelated with x and is approximately uniformly distributed, so the measure of relative error for a given x is larger if the fraction part of x is small than if it is large. A surprising result is that if x is a base-b number resulting from an arithmetic operation, the probability that the leading digit, z, of x is i decreases strongly as i increases, with a distribution given by

$$P_z(i) = \frac{1}{ln(b)}\int_i^{i+1}\frac{1}{u}\,du.† \tag{147}$$

Thus when we average over x, a given truncation error ε_x causes a larger relative error on the average than might otherwise have been expected.

Probabilistic error analysis is not a well-developed subject. Even the results of this introductory treatment are not commonly available. However, such analysis can be fruitful, because it produces much more realistic results than error bounds, the common measure of error. Much useful research remains to be done in probabilistic error analysis.

---

† See R. W. Hamming, *Numerical Methods for Scientists and Engineers,* chapter 2.

PROBLEMS

1.4.1. Let $P_x(1) = \frac{1}{2}$, $P_x(2) = \frac{1}{4}$, $P_x(3) = \frac{1}{8}$.

a) Compute $P_x(1 \text{ or } 2)$, $P_x(1 \text{ or } 3)$, and $P_x(1 \text{ or } 2 \text{ or } 3)$.

b) Assume x must be an integer between 1 and 4. Compute $P_x(4)$.

c) Let x_1 and x_2 be independent values drawn from the above distribution. Compute $P_{x_1,x_2}(1,3)$.

d) Let x_1 and x_3 be drawn from the above distribution, but let $P_{x_1|x_3}(i|j) = 0$ if $i < j$. Compute $P_{x_1|x_3}(4|4)$.

c) Assume: $P_{x_1|x_3}(1|1) = 1$
$P_{x_1|x_3}(2|2) = \frac{1}{2}$
$P_{x_1|x_3}(3|2) = \frac{1}{4}$
$P_{x_1|x_3}(4|2) = \frac{1}{4}$
$P_{x_1|x_3}(3|3) = \frac{1}{2}$
$P_{x_1|x_3}(4|3) = \frac{1}{2}$

Compute $P_{x_1,x_3}(i,j)$ and $P_{x_3|x_1}(j|i)$ for all integers i and j between 1 and 4.

1.4.2. Using Bayes' rule, show that if x is independent of $y (\forall i \forall j P_{x|y}(i|j) = P_x(i))$, then y is independent of x. This result verifies that we may talk about the independence of x and y without ambiguity.

1.4.3. a) Is the following statement always true, sometimes true but sometimes not true, or never true?

If $p_{x,y,z}(u,v,w)$ is a probability density function and $p_{x,y,z}(u,v,w) = p_{x,y}(u,v)p_z(w)$ for all u,v,w, then x and z are independent random variables.

b) If it is always true or never true, give a proof. If it is sometimes true but sometimes not true, give a counterexample.

1.4.4. Show that $\int_{-\infty}^{\infty} (u - \mu_x)p_x(u)du = 0$; the average deviation about the mean is zero.

1.4.5. Show that the median of a probability distribution is that value $m = m_0$ such that $\int_{-\infty}^{\infty} |u - m| p_x(u)du$ is minimum over all $m \in [-\infty, \infty]$.

1.4.6. Let $P_{x,y}(i,j)$ be given by the following table:

$j \backslash i$	1	2	3
1	$\frac{1}{2}$	$\frac{1}{32}$	$\frac{1}{64}$
2	$\frac{1}{16}$	$\frac{1}{8}$	$\frac{1}{8}$
3	$\frac{1}{128}$	$\frac{1}{128}$	$\frac{1}{8}$

Compute $cov(x,y)$, ρ_{xy}.

1.4.7. a) Show that if two random variables x and y are independent, they are uncorrelated.

b) Give a counterexample for the converse.

1.4.8. Show that if $z = ax + b$, where x is a random variable with mean μ_x and variance σ_x^2 and a and b are constants, then $\mu_z = a\mu_x + b$, and $\sigma_z^2 = a^2\sigma_x^2$.

1.4.9. Let x, y, and z be decimal numbers. Let $z = xy$. Let x', y', and z' be the fraction parts of decimal floating-point representations of x, y, and z, respectively. Assume x' and y' are uniformly distributed in $[0.1,1.0]$. Show that

$$p_{z'}(w) = \begin{cases} (ln(10) - 9\,ln(w))/8.1, & w \in [0.1,1.0] \\ 0 \text{ otherwise.} \end{cases}$$

This result implies that multiplication tends to increase the relative abundance of numbers with small fraction parts.

1.4.10. a) Show that $x = \frac{1}{N}\sum_{i=1}^{N} x_i$ is an unbiased estimator for μ_x, where the x_i are independent samples from a probability distribution with mean μ_x.

b) Show that, under the same conditions, $s_x^2 = \frac{1}{N-1}\sum_{i=1}^{N}(x_i - x)^2$ is an unbiased estimator for σ_x^2, the variance of the probability distribution of the x_i.

1.4.11. Show:

a) If x is the number of occurrences of an event in n experiments where the experiments are independent and the probability of the event occurring in any given experiment is p (x is binomially distributed with parameters n and p), then $E(x) = np$ and $var(x) = np(1 - p)$.

b) If x is Poisson-distributed with parameter μ ($P_x(i) = e^{-\mu}\mu^x/i!$), then $E(x) = \mu$ and $var(x) = \mu$.

c) If x is normally distributed with parameters μ and σ, that is,

$p_x(u) = \frac{1}{\sigma\sqrt{2\pi}} e^{-\frac{1}{2}\left(\frac{u-\mu}{\sigma}\right)^2}$, then $E(x) = \mu$ and $var(x) = \sigma^2$.

d) If x is uniformly distributed over $[a,b]$, then $E(x) = (b + a)/2$ and $var(x) = (b - a)^2/12$.

1.4.12. Consider an interval $[t_1,t_2]$ divided into n equal subintervals of length $(t_2 - t_1)/n$. Consider the occurrence of some event such that the event occurs in each subinterval at most once and with probability p_n. Assume that the occurrences of the event in each subinterval are independent. If x is the number of occurrences of the event in $[t_1,t_2]$, then x has a binomial distribution with parameters n and p_n, and therefore with mean np_n. Let the mean number of counts in $[t_1,t_2]$ remain constant but let $n \to \infty$. That is, let n grow large in such a way that np_n remains constant, equal to μ. *Stirling's approximation for large factorials* states that as k gets large, $k! \approx \sqrt{2k\pi}(k/e)^k$. Use Stirling's approximation in this example to show that as $n \to \infty$, x becomes Poisson-distributed with mean μ.

1.4.13. Show that the sum of independent Poisson-distributed random variables is Poisson-distributed.

1.4.14. If $x^* = x + \varepsilon_x$ and x^* is an unbiased estimator for x, show that $\sigma_{x^*} = \sigma_{\varepsilon_x}$.

1.4.15. Show that if x^* and y^* are uncorrelated random variables, then $\sigma^2_{x^*-y^*} = \sigma^2_{x^*} + \sigma^2_{y^*}$.

1.4.16. Let $z^* = x^*y^*$ and assume x^* and y^* are unbiased estimators for x and y, which are not random variables.

a) Show that $\mu_{z^*} = xy + \sigma_{x^*y^*}$. (What is required of the probability distributions of the error in x and the error in y so that $E(z^*) = E(x^*)E(y^*)$?)

b) Let x^* and y^* be independent. Compute $var(z^*)$ in terms of the means and variances of x^* and y^*.

c) Show that in general

$$\sigma^2_{z^*} = y^2\sigma^2_{x^*} + x^2\sigma^2_{y^*} - \sigma^2_{x^*y^*} + 2xy\sigma_{x^*y^*} + E(\varepsilon_x^2\varepsilon_y^2) + 2yE(\varepsilon_x^2\varepsilon_y) + 2xE(\varepsilon_x\varepsilon_y^2).$$

d) Show that if $y^* = x^*$ and third- and higher-power terms in ε_x/x are ignored, $\sigma^2_{z^*}/z^2 = 4\sigma^2_{x^*}/x^2$.

1.4.17. Let $z^* = x^* + y^*$. Let ε_x and ε_y be independent random variables, distributed according to a uniform distribution on $[-a_x,a_x]$ and $[-a_y,a_y]$ respectively. Without loss of generality we can assume $a_x \geq a_y$. We know that

$$\sigma_{x^*} = \sigma_{\varepsilon_x} = a_x/\sqrt{3}$$
$$\sigma_{y^*} = \sigma_{\varepsilon_y} = a_y/\sqrt{3}$$
$$\sigma_{z^*} = (1/\sqrt{3})(a_x^2 + a_y^2)^{1/2} = (\sigma^2_{x^*} + \sigma^2_{y^*})^{1/2}$$

Let us take a different measure of width, say, $w_{x^*} = E(|\varepsilon_x|)$, and compare its behavior to the measure of width, σ_{x^*}.

a) Show $w_{x^*} = a_x/2$ (and similarly $w_{y^*} = a_y/2$).

b) Show $w_{z^*} = ((a_y^2/3) + a_x^2)/2a_x = ((w^2_{y^*}/3) + w^2_{x^*})/w_{x^*}$.

c) What are the differences between the two measures?

d) Why would you choose either of these measures over the other?

1.4.18. In the absence of generated error, iterative methods produce a sequence of values $x_0,x_1,x_2,\ldots$ converging to a desired result, z. Let $\varepsilon_i \equiv x_i - z$, the error in x_i. Some methods produce so-called simple linear convergence for propagated error; if no error is generated, as $x_i \to z$, $\varepsilon_{i+1} = k\varepsilon_i$, for some constant k. Assume the propagated error in an iterative step is given by this relation. Assume an error ε_i^{gen} with variance $(\sigma^{gen})^2$ is generated at the ith step for each i.

a) If ε_0 and ε_i^{gen}, $i = 0,1,2,\ldots$, have zero mean, show ε_i, $i = 1,2,\ldots$, have zero mean.

b) Assume the ε_i^{gen}, $i = 0,1,2,\ldots$, are independent random variables. Let σ_n^2 be the variance of ε_n. Show $\lim\limits_{n\to\infty} \sigma_n^2 = (\sigma^{gen})^2/(1-k^2)$.

c) As x_i becomes close to z, it is reasonable to assume that the generated error is approximately the same at every step. Assume it is exactly the same. Show in this case $\lim\limits_{n\to\infty} \sigma_n^2 = (\sigma^{gen})^2/(1-|k|)^2$.

1.5 ERROR DUE TO SUBTRACTION OF APPROXIMATELY EQUAL NUMBERS

One of the largest sources of error in computing has not yet been treated: A large relative error can result from the propagation of error in the subtraction of two numbers of approximately the same value. If $\sigma_x/|x| \approx a$, $\sigma_y/|y| \approx a$, and $y \approx x$, so

$$z = y - x = bx, \tag{148}$$

where $|b| \ll 1$, then

$$\sigma_z \approx \sqrt{2}a|x|, \tag{149}$$

so
$$\sigma_z/|z| \approx \sqrt{2}a|x|/|bx| = \sqrt{2}a/|b| \gg a. \tag{150}$$

A large relative error has resulted, not from the absolute error increasing rapidly, but by the computed value decreasing rapidly. For floating-point numbers, with an n-digit fraction, the most significant digits, the presence of which has caused truncation errors n digits beyond, are made zero by subtraction; the value of apparently non-significant digits which were truncated becomes of much greater relative significance. For example, consider the result of subtracting 1.00011 truncated to 1.0001 from 1.00029 truncated to 1.0002. Relative errors of the order of 10^{-4} become a relative error of $\frac{8}{10}$ or 80%. Note that the problem is error propagation, not error generation. In our discussion of error generation in floating-point addition, no important difference resulted from the summands being of opposite signs and approximately equal magnitudes.

Because of the susceptibility to large error propagation noted above, we must design computing methods so as not to involve subtraction of approximately equal values. Often this can be accomplished by analytical manipulation of the problem. Three examples follow.

1. $e^x - 1$ for $|x| \ll 1$. Instead of computing e^x and then subtracting 1, note

$$e^x = 1 + x + x^2/2! + \cdots, \tag{151}$$

so
$$e^x - 1 = x + x^2/2! + x^3/3! + \cdots. \tag{152}$$

Our term with largest magnitude is x, but we know that x is much smaller than 1. Not keeping the 1 allows us to keep many more digits of the other terms in a given-size floating-point fraction.

2. $\sqrt{x+1} - \sqrt{x}$ for $x \gg 0$. Note

$$\sqrt{x+1} - \sqrt{x} = 1/(\sqrt{x+1} + \sqrt{x}), \tag{153}$$

a form which does not require the subtraction of two approximately equal values.

3. $x = (-b + \sqrt{b^2 - 4ac})/2a$ when $b^2 \gg |4ac|$ and $b > 0.$ (154)

This is the formula for the root of smaller magnitude of a quadratic equation $ax^2 + bx + c = 0$. Using our knowledge that $ax^2 + bx + c = a(x\text{-}root_1)(x\text{-}root_2)$, which implies that $c = a\,root_1\,root_2$, and thus the product of the two roots is c/a, we find

$$x = -2c/(b + \sqrt{b^2 - 4ac}), \tag{155}$$

a form which produces less error propagation.

PROBLEMS

1.5.1. a) Find the smaller absolute root of the equation $x^2 + 80x + 1$ using equation 154 and symmetrically rounding all results (including intermediate results) to three significant decimal digits.

b) Repeat part a but use equation 155.

c) Compare your results in parts a and b with the correct answer to three significant digits: $-.0125$.

1.6 SUMMARY

We have seen that numerical computations and numerical methods are evaluated on the basis of accuracy and efficiency. One other criterion is sometimes important: the amount of computer storage required by the method.

We have seen that accuracy and efficiency are not unrelated. Extra steps not only require extra time, but also generate extra error and tend to propagate previous error more. An exception to this statement is when the method operates by making a guess, improving it by some computational procedure, and using the improved result as the next guess to be improved by the procedure. In this case, the accuracy tends to decrease with the number of steps in the improvement procedure but not with the number of applications of the procedure.

Sources of error have been identified. They are: (1) generated errors due to measurement, rounding, or truncation in the computer, and approximations of numerical methods; and (2) propagation of generated errors by the computations within numerical methods. In some very strong sense, this text is a book about error generation and propagation. The problem of finding good numerical methods differs from the problem of finding analytic methods precisely because numerical methods are designed and chosen on the basis of their error generation and propagation properties! Since analytic methods are not chosen on this basis, it should not be surprising that many analytic methods are not translatable directly to successful numerical methods.

In this chapter we have emphasized error generation and propagation due to computations. This is not a subject to be disposed of here and ignored throughout the rest of the book. Computational error is an important concern in the design of every numerical method discussed in this book. Because the approximational error of a numerical method cannot be discussed out of context with the numerical method, it has not been discussed here and thus must be discussed at length with the numerical method. Though this matter may take up the majority of the remaining part of the book, the student should not assume that the matter of computational error is negligible compared to that of approximational error. Both matters are very important in determining the final accuracy of a numerical method, and both will be discussed with respect to the numerical methods we develop.

Another important subject that we have discussed in this chapter is measures of error. Absolute and relative error have been defined and their usefulness discussed. Two measures of these errors were discussed at length: error magnitude bounds and error standard deviation (or variance). In both cases, we determined rules for computing measures of the error in a result given the measures of the error in the inputs and of the generated error. For any measure, such a set of rules exists. We choose our measures on the basis of the fact that they measure what we want to measure and have properties that are reasonable for such measures, and on the basis of their mathematical tractability.

REFERENCES

Clark, A. B., and Disney, R. L. *Probability and Random Processes for Engineers and Scientists.* New York: Wiley, 1970.

Cramér, H. *The Elements of Probability Theory.* New York: Wiley, 1955.

Feller, W. *An Introduction to Probability Theory and Its Applications.* 3rd ed. Vol. 1. New York: Wiley, 1967.

Hamming, R. W. *Numerical Methods for Scientists and Engineers.* New York: McGraw-Hill, 1962.

Parzen, E. *Modern Probability Theory and Its Applications.* New York: Wiley, 1960.

Wilkinson, J. H. *Rounding Errors in Algebraic Processes.* Englewood Cliffs, N.J.: Prentice-Hall, 1963.

2

SYSTEMS OF LINEAR EQUATIONS

2.1 INTRODUCTION

Systems of linear equations have the form:

$$\begin{aligned}
A_{11}x_1 + A_{12}x_2 + \cdots + A_{1n}x_n &= b_1 \\
A_{21}x_1 + A_{22}x_2 + \cdots + A_{2n}x_n &= b_2 \\
\vdots \qquad \vdots \qquad\quad\; \vdots \quad &\;\;\; \vdots \\
A_{m1}x_1 + A_{m2}x_2 + \cdots + A_{mn}x_n &= b_m.
\end{aligned} \tag{1}$$

Such linear equations appear again and again when we try to solve problems mathematically. Why is this true? First, many real-life situations are linear. For example, the total cost of purchases at a supermarket is a linear function of the amount of each item we buy (assuming no quantity discount), as is the total weight of our purchases. Second, since nonlinear problems are difficult to solve, we often approximate nonlinear systems by linear systems. Third, many algorithms of the type developed later in the book (for example, for solution of nonlinear equations, for approximation, and for solution of differential equations) require us to solve linear equations as subroutines. Fourth, many of the methods developed for solving linear equations are extendible with some modification to the solution of nonlinear equations. Finally, you will find that many of the concepts developed in our study of linear systems are applicable later in the book.

People often get the impression that the problem of solving linear systems is much more complex than that of solving nonlinear systems of equations. This is certainly not the case, but the reason for the impression is that linear equations are so restricted and easy to solve that we find ourselves able to consider solving a set of 100 linear equations in 100 unknowns, whereas we would not even conceive of trying to solve a set of 100 nonlinear equations. That is, the apparent complexity in our study of linear systems is possible because of the very simplicity of the basic problem.

2.2 LINEAR ALGEBRA

2.2.1 Determinants

The reader should be familiar with the notion of a determinant of a square matrix and how it is computed.† The following properties of determinants will be useful to us. Their proofs are quite algebraic and unenlightening.‡ In all cases, the matrices referred to are assumed to be square $(n \times n)$.

1. $det(A) = det(A^T)$, where A^T (*A transpose*) is the matrix such that
$$A^T_{ij} = A_{ji}. \tag{2}$$
2. Interchanging two rows in a matrix, or interchanging two columns, reverses the sign of the determinant.
3. If two rows or two columns of a matrix are identical, its determinant is zero.
4. If a row or column of a matrix is multiplied by a constant, then the determinant is multiplied by that constant.
5. If a multiple of one row (column) is subtracted from another row (column) of a matrix, the determinant is unchanged.
6. The determinant of upper-triangular and lower-triangular and diagonal matrices is simply the product of the diagonal elements.
7. Determinants can be computed by expanding on a row or column as follows. Let $A^{(ij)}$ be the matrix obtained by eliminating row i and column j from the matrix A. Then expanding on row i, we get
$$det(A) = \sum_{j=1}^{n} (-1)^{i+j} A_{ij}\, det(A^{(ij)}), \tag{3}$$
and expanding on column j, we get
$$det(A) = \sum_{i=1}^{n} (-1)^{i+j} A_{ij}\, det(A^{(ij)}). \tag{4}$$
8. The determinant of the product of two matrices is the product of the determinants.
9. Let I be the $n \times n$ *identity matrix* for which $I_{ij} = 0$ if $i \neq j$ and $I_{ii} = 1$ for $i = 1, 2, \ldots, n$. For any $n \times n$ matrix A,
$$IA = AI = A. \tag{5}$$

---

† See J. N. Franklin, *Matrix Theory*, chapter 1.
‡ Ibid.

Also, for any vector x, $$Ix = x. \tag{6}$$

By property 6, $$det(I) = 1. \tag{7}$$

10. A matrix A has a unique inverse A^{-1} (*A inverse*) such that

$$AA^{-1} = A^{-1}A = I \tag{8}$$

if and only if $det(A) \neq 0$. If $det(A) = 0$ there exists no matrix B such that either $AB = I$ or $BA = I$. We call a matrix which has no inverse a *singular matrix.*

From properties 8, 9, and 10, the following important theorem is proved:

Theorem: Let A be a square matrix. Then $Ax = b$ is solvable for every b if and only if $det(A) \neq 0$.

Proof: Assume the equation has a solution for every b. Then in particular there is a solution x for every equation $Ax^i = e^i$, $1 \leq i \leq n$, where e^i is the unit vector with 0s in every position but the ith position, which contains a 1:

$$e^i \equiv \begin{bmatrix} 0 \\ 0 \\ \cdot \\ \cdot \\ \cdot \\ 0 \\ 1 \\ 0 \\ \cdot \\ \cdot \\ \cdot \\ 0 \end{bmatrix} \begin{matrix} \\ \\ \\ \\ \\ \\ \leftarrow i \\ \\ \\ \\ \\ \\ \end{matrix} \tag{9}$$

Consider the matrix X whose ith column is x^i. Since the operation $B = AX$ involves setting each column of B equal to A times the corresponding column of X, AX equals the matrix whose columns are $e^1, e^2, \ldots, e^n$ (the identity matrix I). Thus $det(AX) = det(I) = 1$. By property 8,

$$det(AX) = (det(A))(det(X)), \tag{10}$$

from which it follows that $det(A) \neq 0$.

Conversely, assume $det(A) \neq 0$. Then by property 10, A^{-1} exists. Consider the vector $x = A^{-1}b$ for any given b. Then

$$Ax = A(A^{-1}b) = (AA^{-1})b = b, \tag{11}$$

so x is a solution for $Ax = b$.

2.2.2 Vector Spaces and Linear Transformations

We are interested in solving linear equations of the form $Ax = b$. We will see that multiplying a vector x by a matrix A can be thought of as a transformation of that vector x into the vector b. Solving the equation involves answering the question: Which vector x is transformed into the vector b by the matrix A? To understand how error is propagated, we also ask: How is x affected if small errors are made in either A or b? Thus we must understand how A transforms sets of vectors, or, more generally, vector spaces. To do so, we need certain definitions and basic concepts.

A *vector space* is defined with respect to a field F whose elements are called *scalars*. A vector space is a set S, the elements of which are called *vectors*, together with definitions of an addition operation on the vectors and an operation of multiplication of a vector by a scalar. Further, this set must satisfy the following requirements:

1. The set is closed under vector addition and multiplication by scalars.
2. The addition operation is commutative $[x + y = y + x]$.
3. The addition operation is associative $[x+(y+z) = (x+y)+z]$.
4. There is a unique 0 vector for addition $[x+0 = x]$.
5. For every vector x there is a unique inverse vector under addition, called $-x$ $[x + (-x) = 0]$.
6. Multiplication by two scalars is associative $[\alpha(\beta x) = (\alpha\beta)x]$.
7. Multiplication by the unit scalar of F is an identity operation $[1x = x]$.
8. Multiplication is distributive over addition both of vectors and of scalars $[\alpha(x+y) = \alpha x + \alpha y; (\alpha + \beta)x = \alpha x + \beta x]$.

The structures most commonly thought of as vectors are one-dimensional arrays of members of a field (for example, the field of real numbers) where addition is element-by-element application of the field addition, and multiplication by a scalar involves field multiplication of each element by that scalar. For the most part, these are the structures we will deal with in our study of linear equations (if the number of elements in the array is n, we will call the structure an *n-vector*). Note, however, that many other commonly encountered structures are also vector spaces. Most of our results will be applicable to these as well. Examples are:

1. The set of polynomials
2. The set of Fourier series of a fundamental frequency ω, that is, linear combinations of functions of the form $sin(k\omega t)$ or $cos(k\omega t)$ where k is an integer
3. The set of all continuous functions of one variable
4. The set of solutions to the homogeneous equation $Ax = 0$
5. The set of all solutions to a homogeneous differential equation

Our notation will be as follows:

1. Capital letters $(A, B, \ldots)$ refer to matrices.
2. Small Roman letters $(x, y, \ldots)$ refer to vectors.
3. Small Greek letters $(\alpha, \beta, \ldots)$ refer to scalars.
4. Doubly subscripted capital letters $(A_{ij}, B_{22}, \ldots)$ refer to matrix elements.
5. Singly subscripted small Roman letters $(x_i, y_j, \ldots)$ refer to vector elements.
6. Superscripted small Roman letters $(x^i, y^j, \ldots)$ specify particular vectors among a set of vectors.

If we let

$$a^1 = \begin{bmatrix} A_{11} \\ A_{21} \\ \cdot \\ \cdot \\ \cdot \\ A_{m1} \end{bmatrix}, \quad a^2 = \begin{bmatrix} A_{12} \\ A_{22} \\ \cdot \\ \cdot \\ \cdot \\ A_{m2} \end{bmatrix}, \quad \ldots, \quad a^n = \begin{bmatrix} A_{1n} \\ A_{2n} \\ \cdot \\ \cdot \\ \cdot \\ A_{mn} \end{bmatrix}, \tag{12}$$

and

$$b = \begin{bmatrix} b_1 \\ b_2 \\ \cdot \\ \cdot \\ \cdot \\ b_m \end{bmatrix}, \tag{13}$$

then equation 1 in this chapter can be written

$$\sum_{i=1}^{n} x_i a^i = b. \tag{14}$$

Here, b is said to be a *linear combination* of the vectors a^i, $i = 1,2,\ldots,n$. Precisely, we say a vector x is a linear combination of the vectors $y^1,y^2,\ldots,y^m$ if $x = \sum_{i=1}^{m} \alpha_i y^i$, for some set of α_i, $1 \leq i \leq m$. We ask whether we can find a set of y^i such that any vector in a given vector space L can be written as a linear combination of the y^i. If so, we say the set of y^i *spans* the space L. If one of the y^i, say without loss of generality y^m, can be written as a linear combination of the remaining y^i,

$$y^m = \sum_{i=1}^{m-1} \alpha_i y^i, \tag{15}$$

then we say y^m is *linearly dependent* on the remaining y^i. We note that any vector which can be written as a linear combination of $y^1,\ldots,y^m$, say

$$x = \sum_{i=1}^{m} \beta_i y^i, \tag{16}$$

can also be written as a linear combination of $y^1,\ldots,y^{m-1}$:

$$x = \sum_{i=1}^{m-1} (\beta_i + \beta_m \alpha_i) y^i. \tag{17}$$

We would like to know how many y^i are necessary to span the vector space in question. To obtain this result, we need a more careful definition of linear independence.

We say a set of vectors is *linearly independent* if a linear combination of the vectors, $\sum_{i=1}^{m} \alpha_i y^i$, is zero if and only if all multipliers α_i are zero. In the following theorem, we note that this definition is equivalent to the informal definition we gave above.

Theorem: A set of vectors y^i, $1 \leq i \leq m$, is linearly dependent if and only if there exists one member of the set which can be written as a linear combination of the remaining members.

Proof: Assume the y^i are linearly dependent. Then there exists a linear combination of the y^i, $\sum_{i=1}^{m} \alpha_i y^i$, which equals zero where all α_i are not equal to zero. Consider an $\alpha_k \neq 0$. Then

$$y^k = \sum_{\substack{i=1 \\ i \neq k}}^{m} (-\alpha_i/\alpha_k) y^i. \tag{18}$$

Conversely, assume there exists a k such that

$$y^k = \sum_{\substack{i=1 \\ i \neq k}}^{m} \beta_i y^i. \tag{19}$$

Then

$$\sum_{\substack{i=1 \\ i \neq k}}^{m} \beta_i y^i - y^k = 0. \tag{20}$$

The left member of equation 20 is a linear combination of the y^i, $1 \leq i \leq m$ such that all the multipliers, in particular $\beta_k = -1$, are not equal to 0.

Returning to our question of how many vectors are required to span L, we define a *basis* for L as a linearly independent set of vectors which span L. That is, a basis is a set which has enough vectors to span the space but not too many. If we can show that, for a given vector space, the number of vectors in any basis is the same, we will have a notion of the *dimension* of a vector space: the number of vectors in a basis.

Theorem: If $\{x^i \mid 1 \leq i \leq m\}$ is a basis for L and $\{y^i \mid 1 \leq i \leq n\}$ is a basis for L, then $m = n$.

Sketch of Proof:† The proof depends on a lemma which states that one cannot, by taking various linear combinations of a given set of vectors, produce more linearly independent vectors than one started with. Given that lemma, we note that because the y^i are a basis, each of the x^i can be written as a linear combination of the y^i and thus $m \leq n$. Similarly each of the y^i can be written as a linear combination of the x^i and thus $n \leq m$. Therefore $m = n$.

The proof of the lemma proceeds by showing that if each of p linearly independent vectors is a linear combination of q linearly independent vectors, we can find a set of $p - 1$ linearly independent vectors, each of which is a linear combination of $q - 1$ vectors of the second set. With repeated application of this process, we find that if $p > q$, we end up with $p - q + 1$ (>1) linearly independent vectors, each of which is a linear combination of one vector, a clear impossibility.

---

† See Franklin, *Matrix Theory*, chapter 2, for full proof.

Two important corollaries of the above theorem are: (1) If n is the dimension of a vector space L, then any set of more than n vectors is linearly dependent; and (2) Every set of n linearly independent vectors in an n-dimensional space is a basis for that space. The proof of the first corollary should be obvious and that of the second proceeds by showing that if we add any vector x to the linearly independent set, we obtain a linearly dependent set. From this it follows that x must be a linear combination of the original set.

Let us obtain a better understanding of the concept of the linear transformation imposed by a matrix A. We say a transformation is linear if, for all α, β, x, and y,

$$A(\alpha x + \beta y) = \alpha(Ax) + \beta(Ay), \tag{21}$$

a property which holds for the transformation defined by matrix multiplication. Thus if we know a basis $\{x^i\}$ of a vector space L and we know the transformation Ax^i of each element of the basis, we can find the transformation of any vector in the space: For any vector y, write y as $\sum_{i=1}^{n} \alpha_i x^i$ and then

$$Ay = \sum_{i=1}^{n} \alpha_i (Ax^i). \tag{22}$$

Consider the vector space L_n of all n-vectors of real numbers (arrays with n elements). Let x be a column vector in L_n with elements $x_1, x_2, \ldots, x_n$. The set of n-vectors e^i $(1 \le i \le n)$ as described in equation 9 is a basis for L_n because, for any x,

$$x = \sum_{i=1}^{n} x_i e^i \tag{23}$$

and the e^i are linearly independent (see exercise 2.2.4).

Let A be an $m \times n$ matrix with columns a^i $(1 \le i \le n)$. Then A transforms n-vectors (members of the vector space L_n) into m-vectors (members of L_m). The set of all vectors produced by applying A to every vector in L_n forms a vector space L_A (see exercise 2.2.6). Note that the dimension of L_A is no more than m, and that the dimension of L_A is the number of linearly independent images of vectors in L_n under the mapping A. If the dimension of this image space L_A is less than m, we say L_A is a *subspace* of L_m.

We have seen that the image Ax of the transformation A applied to the vector x can be written $\sum_{i=1}^{m} x_i a^i$. That is, the transformation involves taking linear combinations of the column vectors making up A.

In particular, notice that $Ae^i = a^i$. Thus a matrix can be thought of as a listing of the images of the e^i in L_n under the linear transformation imposed by A.

Since every image under A is a linear combination of the columns of A, the a^i span L_A, and the number of linearly independent columns of A is the dimension of L_A. This dimension of L_A is called the *rank* of A. Thus $rank(A) =$ the number of linearly independent columns of A. Since the number of linearly independent columns of A is no more than the total number of columns of A, $rank(A) \leq n$. Since the number of linearly independent columns of length m can be no more than m, $rank(A) \leq m$.

If $m = n$ (that is, A is a square matrix), A maps L_n onto L_n or a subspace of L_n: A can be thought of as mapping the space L_n into itself. Considering the geometric effects of such linear transformations may help our intuition. These effects are listed below and illustrated in figure 2-1 (the examples are for $n = 2$, but the effects are the same for any n):

a) Stretching of the vector space about an axis through the origin. Consider the matrix $\begin{bmatrix} 3 & 0 \\ 0 & 1 \end{bmatrix}$. This matrix transforms the vector $\begin{bmatrix} 1 \\ 0 \end{bmatrix}$ into the vector $\begin{bmatrix} 3 \\ 0 \end{bmatrix}$ and transforms the vector $\begin{bmatrix} 0 \\ 1 \end{bmatrix}$ into itself. Thus the matrix stretches the vector space L into three times itself in the x direction but does not change the y direction. Every point in the half-plane to the right of the y axis is stretched right by a factor of three and every point to the left of the y axis is stretched left by a factor of three.

b) Rotation of the space about the origin by a given angle. Consider the matrix $\begin{bmatrix} \sqrt{3}/2 & -1/2 \\ 1/2 & \sqrt{3}/2 \end{bmatrix}$, which maps the vector $\begin{bmatrix} 1 \\ 0 \end{bmatrix}$ into the vector $\begin{bmatrix} \sqrt{3}/2 \\ 1/2 \end{bmatrix}$ and the vector $\begin{bmatrix} 0 \\ 1 \end{bmatrix}$ into the vector $\begin{bmatrix} -1/2 \\ \sqrt{3}/2 \end{bmatrix}$, thus rotating the space 30° in the positive direction.

c) Reflection of the space about an axis through the origin. Consider the matrix $\begin{bmatrix} 0 & -1 \\ -1 & 0 \end{bmatrix}$, which reflects the space about the axis $y = -x$. Every point to the right of the axis is reflected perpendicular to the axis, ending up on the left of the axis, and vice versa. Reflection can be thought of as stretching by -1.

d) Shearing of the space along a given axis through the origin. Consider the matrix $\begin{bmatrix} 1 & 1 \\ 0 & 1 \end{bmatrix}$. This matrix shears along the x axis as if the space were pulled right from $[\infty, \infty]$ and left from $[-\infty, -\infty]$.

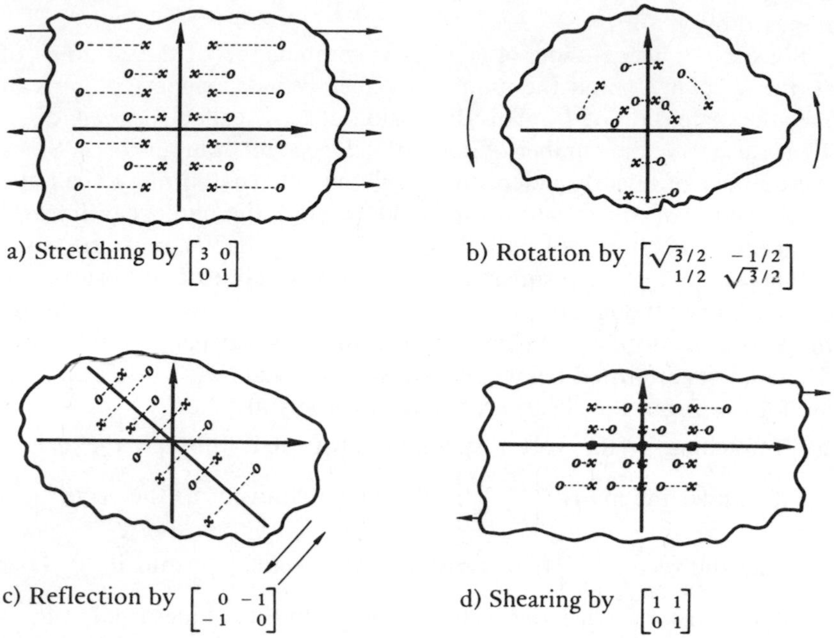

a) Stretching by $\begin{bmatrix} 3 & 0 \\ 0 & 1 \end{bmatrix}$

b) Rotation by $\begin{bmatrix} \sqrt{3}/2 & -1/2 \\ 1/2 & \sqrt{3}/2 \end{bmatrix}$

c) Reflection by $\begin{bmatrix} 0 & -1 \\ -1 & 0 \end{bmatrix}$

d) Shearing by $\begin{bmatrix} 1 & 1 \\ 0 & 1 \end{bmatrix}$

FIG. 2-1 Geometric effects of linear transformations (points defined by x are transformed into points defined by o)

It can be shown that any linear transformation is a composition of the four types of transformations listed above. In fact, reflection is just negative stretching, and shearing can be composed of stretchings and rotations, so every linear transformation can be composed of positive and negative stretchings and rotations, where the stretchings are perpendicular to an axis through the origin and rotation is about the origin, because all linear transformations hold the origin fixed.

Since geometric interpretation can help us to understand vectors and operations on them, let us continue by discussing the angle between two vectors. Let us define the *inner product* between the vector x and vector y as $x^T y \equiv \sum_{i=1}^{n} x_i y_i$.† We will show that

† We will see later that this is simply the most common inner product; many functions of two vectors satisfy the requirements formally specified for an inner product.

$$x^T y = |x|\,|y|\,cos(\theta), \tag{24}$$

where $|x|$ is the Euclidean length of the vector x defined as

$$|x| = \left(\sum_{i=1}^{n} x_i^2\right)^{1/2} \tag{25}$$

and θ is the angle between x and y (in the plane defined by the points 0, x, and y). Let $z = x - y$ be the vector connecting the point y to the point x (see figure 2-2). Then

$$z = (x_1 - y_1, x_2 - y_2, \ldots, x_n - y_n)^T. \tag{26}$$

By the law of cosines:

$$|z|^2 = |x|^2 + |y|^2 - 2|x|\,|y|\,cos(\theta). \tag{27}$$

Thus
$$\sum_{i=1}^{n} (x_i - y_i)^2 = \sum_{i=1}^{n} x_i^2 + \sum_{i=1}^{n} y_i^2 - 2|x|\,|y|\,cos(\theta). \tag{28}$$

Upon expanding the left side of equation 28 and canceling terms, we find

$$\sum_{i=1}^{n} x_i y_i = |x|\,|y|\,cos(\theta). \tag{29}$$

Thus the angle between the vectors x and y equals $cos^{-1}(x^T y/|x|\,|y|)$.

Two vectors x and y are said to be *orthogonal* if the angle between them is 90° (that is, if the cosine of the angle between them is 0). Thus, two nonzero vectors x and y are orthogonal if $x^T y$ equals 0.

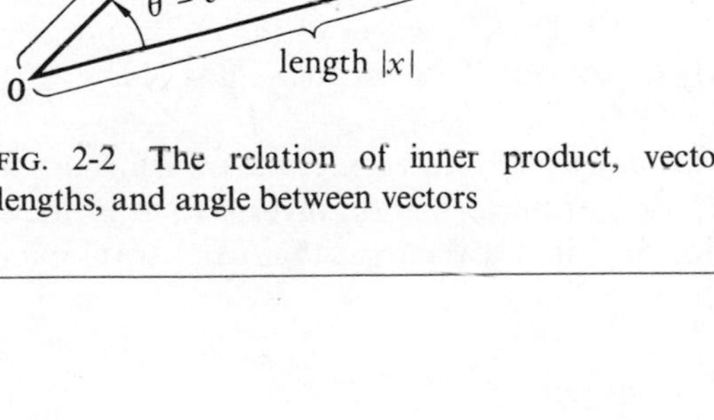

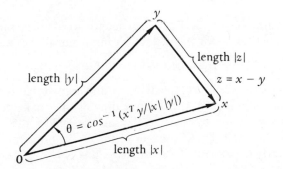

FIG. 2-2 The relation of inner product, vector lengths, and angle between vectors

Let us now interpret the determinant of a matrix geometrically. A matrix maps the unit hypercube (the cube with the e^i as sides) into a hyper-parallelepiped with sides equal to a^i (the images of the e^i). We can show that $|det(A)|$ is the volume of this hyperparallelepiped. A proof in two-space is given below, and a similar proof can be made for L_n for any n.

The parallelepiped in question for the mapping of L_2 by the matrix A is that with sides given by the vectors $a^1 = \begin{bmatrix} A_{11} \\ A_{21} \end{bmatrix}$ and $a^2 = \begin{bmatrix} A_{12} \\ A_{22} \end{bmatrix}$ (see figure 2-3). The area of the parallelogram is given by $|a^1|\,|a^2|\,sin(\theta)$ where θ is the angle between a^1 and a^2. Thus

$$Area = |a^1|\,|a^2|\,(1 - cos^2(\theta))^{1/2}$$

$$= |a^1|\,|a^2|\left(1 - \left(\frac{a^{1^T}a^2}{|a^1|\,|a^2|}\right)^2\right)^{1/2} = (|a^1|^2|a^2|^2 - (a^{1^T}a^2)^2)^{1/2}. \quad (30)$$

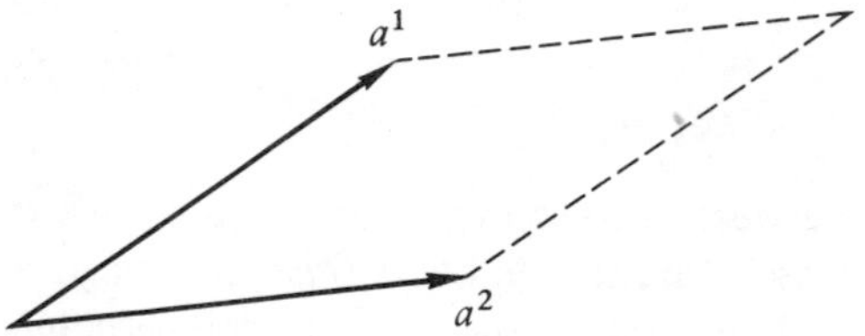

FIG. 2-3 Hyperparallelepiped formed by vectors a^1 and a^2

Using the definition of the Euclidean length and inner product given above and substituting for the vectors a^1 and a^2 as above, we find

$$Area = |A_{11}A_{22} - A_{12}A_{21}| = |det(A)|. \quad (31)$$

If $det(A) = 0$, the volume of the hyperparallelepiped that is the image of the unit hypercube is zero. It has fewer dimensions than the hypercube. That is, transformation by A has collapsed the space; there has been a stretching by factor 0. The a^i (images of the e^i) do not span L_n and thus they are linearly dependent. To summarize, if A is $n \times n$, then $det(A) = 0$ if and only if $rank(A) < n$.

Let us now geometrically interpret the transformation imposed by A^{-1}: A^{-1} imposes a transformation inverse to A in the sense that if A transforms a vector x into a vector y, then A^{-1} transforms y back to x.

Thus if A imposes a stretching by a factor c, then A^{-1} imposes a stretching in the same direction by factor $1/c$. Similarly, if A rotates the space by angle θ about a given axis, then A^{-1} rotates the space by angle $-\theta$ about the same axis.

If $det(A) = 0$ (A is singular), there are some vectors in L_n which are not in the image space. Furthermore we will soon show that every vector in the image space is the image of more than one vector in L_n. Thus A^{-1} is not defined.

For any $n \times n$ matrix A, if $rank(A) = r$, we can show that A maps a subspace of L_n with dimension $n-r$ into the 0 vector of L_n. This space of vectors in the domain of the transformation which are mapped into zero can be shown to be a vector space. This space is called the *null space* of A. The proof that the null space is a vector space involves checking that all vectors in this space satisfy the rules for a vector space (see exercise 2.2.3). The proof that the dimension of the null space is $n-r$ follows.

Theorem: If A is $m \times n$ and $rank(A) = r$, then the vector space of all x such that $Ax = 0$ has dimension n–r.

Proof: By definition of $rank(A)$, A has r independent columns. Order the columns so that $a^1, a^2, \ldots, a^r$ are independent and all remaining a^i are dependent on the first r columns. Then for $k > r$,

$$a^k = \sum_{i=1}^{r} \alpha_{ki} a^i, \tag{32}$$

which is equivalent to

$$a^k - \sum_{i=1}^{r} \alpha_{ki} a^i = 0. \tag{33}$$

Since for all i, A maps e^i into a^i and A is linear, A will map the vectors $d^k \equiv e^k - \sum_{i=1}^{r} \alpha_{ki} e^i$, $r < k \leq n$, into zero. Note that the d^k are linearly independent because for a given k_0, d^{k_0} is the only member of the set which has a nonzero component in e^{k_0}.

What remains to be proved is that the d^k span the null space of A. Let x be in the null space of A (that is, $Ax = 0$). Then

$$\sum_{i=1}^{n} x_i a^i = 0, \tag{34}$$

which implies $\sum_{i=1}^{r} x_i a^i + \sum_{i=r+1}^{n} x_i \sum_{j=1}^{r} \alpha_{ij} a^j = 0.$ (35)

Equation 35 implies

$$\sum_{i=1}^{r}\left(x_i + \sum_{j=r+1}^{n} x_j\alpha_{ji}\right)a^i = 0, \tag{36}$$

and by the linear independence of the a^i for $i \leq r$ we have

$$x_i = -\sum_{j=r+1}^{n} x_j\alpha_{ji}, \qquad 1 \leq i \leq r. \tag{37}$$

We wish to show that, for some set of β_k

$$x = \sum_{k=r+1}^{n} \beta_k d^k, \tag{38}$$

or equivalently, using the definition for d^k above, that

$$x = \sum_{k=r+1}^{n} \beta_k e^k - \sum_{i=1}^{r}\left(\sum_{k=r+1}^{n} \beta_k\alpha_{ki}\right)e^i. \tag{39}$$

But
$$x = \sum_{k=1}^{n} x_k e^k = \sum_{k=r+1}^{n} x_k e^k + \sum_{k=1}^{r} x_k e^k, \tag{40}$$

so using equation 37,

$$x = \sum_{k=r+1}^{n} x_k e^k + \sum_{k=1}^{r}\left(-\sum_{j=r+1}^{n} x_j\alpha_{jk}\right)e^k. \tag{41}$$

Comparing equations 39 and 41, we see that equation 39 is satisfied by $\beta_k = x_k$, $r < k \leq n$, that is,

$$x = \sum_{k=r+1}^{n} x_k d^k. \tag{42}$$

Finally, let us interpret geometrically the linear transformation imposed by A^T for any $m \times n$ matrix A (see figure 2-4). We can show that (1) if A^T maps L_m into L_{A^T} and A maps L_n into L_A, and (2) if A maps L_{A^T} into L_{AA^T} and A^T maps L_A into L_{A^TA}, then (3) A^T maps L_{AA^T} into L_A and (4) A maps L_{A^TA} into L_{A^T}. That is to say, there is a subspace N of L_n and a subspace M of L_m, each of dimension $rank(A)$, such that (5) A^T maps both M and L_m into N and (6) A maps both N and L_n into M (see exercise 2.2.8).

Since the dimension of $M = L_A$ is equal to the dimension of $N = L_{A^T}$, by definition

$$rank(A) = rank(A^T). \tag{43}$$

That is, since the rows of A are the columns of A^T, A has the same number of linearly independent rows and columns.

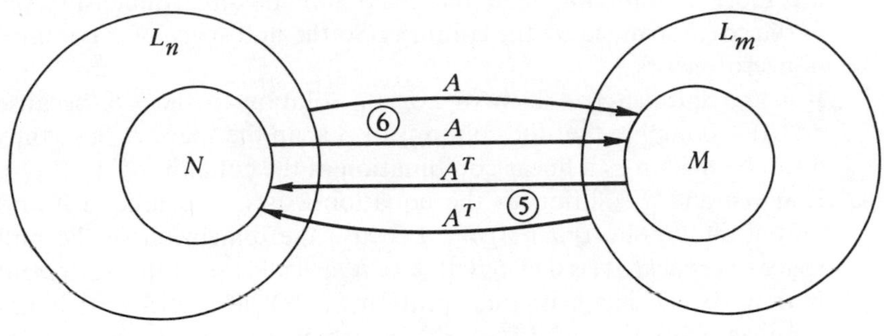

FIG. 2-4 Relations between transformation due to A and transformation due to A^T

2.2.3 Existence and Uniqueness of Solutions to Systems of Linear Equations

Since $Ax = b$ is equivalent to $\sum_{i=1}^{n} x_i a^i = b$, saying that a solution to $Ax = b$ exists is equivalent to saying that b is a linear combination of the columns of A. Conversely, if b is a linear combination of the columns of A, then the multiplying factors in that linear combination are the elements of the solution vector x. Thus there exists a solution to $Ax = b$ if and only if b is a linear combination of the columns of A. Equivalently, we can say there exists a solution if the concatenation of the column vector b to the columns of A produces a matrix with rank equal to that of A.

If a solution to $Ax = b$ exists, it is unique if and only if the dimension of the null space of A is 0 (contains only the 0 vector). Let x_0 be a solution to the equation and let z be a nonzero vector in the null space of A. Then $x_0 + z$ is a solution to the equation because

$$A(x_0 + z) = Ax_0 + Az = b + 0 = b. \tag{44}$$

If the dimension of the null space is greater than 0, we know there are an infinite number of different vectors in the null space, and thus an infinite number of solutions to the equation. We conclude that the system of linear equations has either no solutions, one solution, or an infinite number of solutions.

We can deduce the following corollaries of the above general result.

1. If A is $m \times n$ and $m > n$, there exists a vector b such that $Ax = b$ has no solution, because the columns of A do not span the space L_m.

2. If $m < n$, if there exists a solution, it is not unique, because there are more columns (n) than the maximum possible dimension (m) of the space spanned by the columns. So the null space of A contains nonzero vectors.
3. If $m = n$ and $det(A) \neq 0$, there exists a solution to $Ax = b$, because $det(A) \neq 0$ implies that the columns of A span the space L_m and thus b can be written as a linear combination of the columns of A.
4. If $m = n$ and a solution to the equation exists, it is unique if and only if (iff, for short) $det(A) \neq 0$ because the dimension of the null space ($n - rank(A)$) is 0 iff $det(A) \neq 0$. A special case of this statement is that $Ax = 0$ has a unique solution ($x = 0$) iff $det(A) \neq 0$; it has a nonzero solution iff $det(A) = 0$, in which case it has an infinite number of solutions.

PROBLEMS†

2.2.1. Show $det\begin{bmatrix} 1 & 7 & 13 \\ 3 & 9 & 15 \\ 5 & 11 & 17 \end{bmatrix} = 0$ without computing the determinant by expansion.

2.2.2. a) Let A be an $n \times n$ matrix of the form $A = \begin{bmatrix} A^{11} & A^{12} \\ 0 & A^{22} \end{bmatrix}$, where A^{ij} are $m_i \times m_j$ matrices and 0 is the $m_2 \times m_1$ matrix of zeros. That is, A is the "block upper triangular" matrix formed by concatenating the matrices A^{ij} and 0 as indicated. Show that $det(A) = det(A^{11}) \times det(A^{22})$.

b) Let A be a square matrix. Let A be a general block upper triangular matrix:

$$\begin{bmatrix} A^{11} & A^{12} & A^{13} \ldots A^{1q} \\ 0 & A^{22} & A^{23} \ldots A^{2q} \\ \cdot & \cdot & \cdot \\ \cdot & \cdot & \cdot \\ \cdot & \cdot & \cdot \\ 0 & 0 \ \ldots & A^{qq} \end{bmatrix},$$

where A^{ij} is an $m_i \times m_j$ matrix, and the 0 matrix in the ijth position is $m_i \times m_j$. Show by induction that $det(A) = \pi_{i=1}^{q} det(A^{ii})$.

2.2.3. a) Show that the set of polynomials with natural polynomial addition forms a vector space.

b) Show that the set of all vectors which solve $Ax = 0$ forms a vector space.

---

† Other suitable problems can be found in chapters 1 and 2 of Franklin.

c) Why is the set of all vectors which solve $Ax = b$ for $b \neq 0$ not a vector space?

2.2.4. Show that the set of unit n-vectors $e^1, e^2, \ldots, e^n$ is linearly independent.

2.2.5. Let V be the set of vectors in L_3 for which $x_1 - x_2 + 2x_3 = 0$. Show that V is a vector space. Find a basis for V.

2.2.6. Let A be an $m \times n$ matrix. Show that the set of images under transformation by A of vectors in L_n forms a vector space.

2.2.7. Show $rank(AB) \leq rank(A)$ and $rank(AB) \leq rank(B)$.

2.2.8. Let A be an $m \times n$ real matrix. Let $rank(A) = r$. Show the following:

a) $A^T Ax = 0$ iff $Ax = 0$.

b) Let $\{x^i | i = 1,2,\ldots,r\}$ be a set of linearly independent vectors in L_n such that $\{Ax^i | i = 1,2,\ldots,r\}$ is a basis for L_A. Using part a above, show that $\{A^T Ax^i | i = 1,2,\ldots,r\}$ is a linearly independent set of vectors. It follows that L_{A^TA}, the image under A^T of L_A, has the same dimension (r) as L_A, the image under A of L_n.

c) Show that $\{AA^T Ax^i | i = 1,2,\ldots,r\}$ is a set of linearly independent vectors in L_A. Conclude that these vectors are a basis for L_A. It follows that A maps L_{A^TA} into the same space that it maps L_n.

d) Let z be a vector in L_m such that z is not in L_A (that is, z is linearly independent of $\{Ax^i | i = 1,2,\ldots,r\}$). Show that $A^T z$ is in L_{A^TA} (that is, $A^T z$ is linearly dependent on $\{A^T Ax^i | i = 1,2,\ldots,r\}$). It follows that $L_{A^T} = L_{A^TA}$, so dimension of $L_A =$ dimension of $L_{A^TA} =$ dimension of L_{A^T}.

2.2.9. For which vectors b does the equation $\begin{bmatrix} 1 & 7 & 13 \\ 3 & 9 & 15 \\ 5 & 11 & 17 \end{bmatrix} x = b$ have a solution?

Give all the solutions of $\begin{bmatrix} 1 & 7 & 13 \\ 3 & 9 & 15 \\ 5 & 11 & 17 \end{bmatrix} x = 0$. Give all the solutions of $\begin{bmatrix} 1 & 7 & 13 \\ 3 & 9 & 15 \\ 5 & 11 & 17 \end{bmatrix} x = \begin{bmatrix} 22 \\ 30 \\ 38 \end{bmatrix}$.

2.2.10. Let A be $m \times n$. Show that for any m-vector y, there exists a solution to $A^T Ax = A^T y$. Give conditions on A such that for any such y, the solution x is unique.

2.2.11. Let A be an $n \times n$ matrix and let λ be a scalar. Give a condition on λ, in terms of a determinant, for the existence of a nonzero vector x such that $Ax = \lambda x$.

2.2.12. Let A be an $m \times n$ matrix, X be an $n \times p$ matrix, and B be an $m \times p$ matrix. Let $rank(A) = r$. Let the columns of X and B be x^i and b^i, respectively, $i = 1,2,\ldots,p$. Show $AX = B$ iff $Ax^i = b^i$, $i = 1,2,\ldots,p$. Then give a necessary and sufficient condition on B such that $AX = B$ has a solution X. If $AX^0 = B$, describe all solutions to $AX = B$.

2.3 AN ALGORITHM FOR SOLVING SYSTEMS OF LINEAR EQUATIONS

2.3.1 Gaussian Elimination

In high school we learned Cramer's rule for solving nonsingular systems of linear equations: $Ax = b \Rightarrow x_i = det(A^i)/det(A)$, where A^i is the matrix obtained by replacing the ith column of A with b. It involves evaluating $n + 1$ $n \times n$ determinants. To evaluate a determinant by expansion on a row or column involves more than $n!$ multiplications, a number which grows very rapidly with n. Let us examine a method which requires $n^3/3 + O(n^2)$ multiplications, where $O(n^2)$ means terms in n^2 and lower powers of n which are negligible compared to terms in n^3 for $n \gg 0$. This method, called *Gaussian elimination*, results from trying to transform the original system of equations into another system with the same solution but computationally easier to solve than the original one.

Assume we are given the following system of simultaneous equations:

$$x_1 - x_2 + x_3 = 2 \tag{45}$$

$$2x_1 + x_2 + x_3 = 7 \tag{46}$$

$$4x_1 + 2x_2 + x_3 = 11, \tag{47}$$

which has the solution $x_1 = 1$, $x_2 = 2$, $x_3 = 3$. We can accomplish the desired transformation by a multiple-stage process. In stage 1, we eliminate the first variable from equations 2 through n. In stage 2, we eliminate the second variable from equations 3 through n. In stage 3, we eliminate the third variable from equations 4 through n, and so on, until we arrive at a set of equations where the nth equation involves only x_n, the $(n-1)$th equation involves x_{n-1} and x_n, and so on. Once the equations are in this form, we can solve the nth equation for x_n. Substituting its value in the $(n-1)$th equation, we can solve for x_{n-1}, and so on, until we solve for x_1 in the first equation after substituting the already computed values for x_2 through x_n.

Consider how we can apply this idea to equations 45, 46, and 47. If we subtract twice equation 45 from equation 46, we have not changed the solution, and we obtain

$$3x_2 - x_3 = 3, \tag{48}$$

an equation in which x_1 has been eliminated. Similarly, if we subtract four times equation 45 from equation 47, we obtain

$$6x_2 - 3x_3 = 3. \tag{49}$$

Thus equations 45, 48, and 49 have the same solution as equations 45, 46, and 47, but x_1 has been eliminated from all but the first equation.

If we now subtract twice equation 48 from equation 49, we obtain

$$-1x_3 = -3. \tag{50}$$

We have eliminated x_2 from all equations below the second equation. It is simple to solve for x_3, x_2, and x_1 in our transformed equations, 45, 48, and 50, as follows. Equation 50 implies $x_3 = 3$. Substituting $x_3 = 3$ into equation 48, we compute $3x_2 - 3 = 3$, so $x_2 = 2$. Substituting $x_3 = 3$ and $x_2 = 2$ into equation 45, we obtain $x_1 - 2 + 3 = 2$, so $x_1 = 1$.

Let us describe the elimination process in general. The elimination proceeds at the ith stage with the last $n - i + 1$ equations in $n - i + 1$ unknowns (the first $i-1$ unknowns have been eliminated from these equations). The ith unknown is eliminated by subtracting an appropriate constant times the first equation of the group from the second, where the constant is chosen to make the resulting coefficient of the ith unknown equal to zero; then subtracting a different constant times the first equation from the third equation, choosing the constant to eliminate the ith unknown, and so on. This process results is an upper triangular matrix (only 0s below the diagonal) and is called the *triangularization* step of the solution. It is followed by the *substitution* step, in which for each equation we substitute the values of the variables already evaluated and solve for the remaining variable in the equation. (See program 2-1 for full specification of the algorithm.)

```
/* SOLVE AX = B: SET OF N LINEAR EQUATIONS IN N UNKNOWNS */

   DECLARE (A(N,N),          /* A IS N × N MATRIX */
            B(N),            /* B IS N-DIM VECTOR */
            X(N),            /* X IS N-DIM SOLUTION VECTOR */
            AMULT) FLOAT;    /* ROW MULTIPLIER */

/* TRIANGULARIZATION */
   L1:  DO I = 1 TO N - 1;          /* ELIMINATE ITH COL */
        L2:  DO J = I+1 TO N;       /* BELOW ITH ROW */
             AMULT = A(J,I) / A(I,I); /* MULT FOR JTH ROW */
        /* COMPUTE NONZERO ELEMENTS OF JTH ROW */
             L3:  DO K = I+1 TO N;
                  A(J,K) = A(J,K) - AMULT * A(I,K);
             END L3;
             B(J) = B(J) - AMULT * B(I); /* ALSO NEW B(J) */
        END L2;
   END L1;
```

PROGRAM 2-1 Elementary algorithm for Gaussian elimination
(*continued on next page*)

```
/* BACK SUBSTITUTION */
   L4:  DO I = N TO 1 BY -1;  /* LAST ROW FIRST */
        X(I) = B(I);
        /* SUBTRACT TERMS IN ALREADY COMPUTED X(J) */
        L5:  DO J = I+1 TO N;  /* NOT EXECUTED IF I = N */
                X(I) = X(I) - A(I,J) * X(J);
        END L5;
        X(I) = X(I) / A(I,I); /* SOLVE FOR X(I) */
   END L4;
```

PROGRAM 2-1 (*continued*)

Let us first analyze this Gaussian elimination algorithm with respect to efficiency—the number of multiplications and divisions required.

The ith stage of triangularization requires, for each of $n-i$ rows, one division to determine the multiplying constant and $n-i$ multiplications to determine the resulting elements in the row that are not necessarily equal to zero. Thus the total number of multiplications and divisions required in the triangularization is $\sum_{i=1}^{n-1} (n-i)(n-i+1)$. Using the formulas

$$\sum_{i=1}^{m} i = m(m+1)/2 \tag{51}$$

and

$$\sum_{i=1}^{m} i^2 = m(m+1)(2m+1)/6, \tag{52}$$

we find that $n^3/3 + O(n^2)$ multiplications and divisions are required.

The back substitution involves n stages, where the ith stage requires $i-1$ multiplications and one division. The total number of multiplications and divisions required by this back substitution is $\sum_{i=1}^{n} i = n(n+1)/2 = n^2/2 + O(n)$. Thus the complete Gaussian elimination process requires $n^3/3 + O(n^2)$ multiplications and divisions. This is much less than $(n+1)!$, which is in turn less than that required by Cramer's method.

2.3.2 Pivoting and Scaling

Before we can be satisfied with our algorithm, we must analyze it for error generation and propagation. Consider the basic computational step of the triangularization:

$$A_{jk} \leftarrow A_{jk} - A_{ik}(A_{ji}/A_{ii}). \tag{53}$$

We know that the terms in the subtraction should not be approximately equal (recall section 1.5). However, that situation is not unlikely. Many of the coefficients in a given row, or many of the coefficients in a given column, may be approximately the same size; and either of these circumstances may lead to subtracting approximately equal terms. How can we avoid the problem of large error propagation? That is, what simple transformations of the system of equations can we make so that this problem will arise as little as possible? Three simple transformations come to mind:

1. Multiply each equation (row) by an appropriate constant. This certainly does not change the solution.
2. Change the order of the equations (rows). As with 1 above, this does not change the solution.
3. Change the order of the variables (columns). This procedure changes only the order of the elements of the solution. If we remember how we change the columns, we can reorder the elements when we are finished with the computation.

Multiplying each row by a different constant (the first approach above) does not improve the error situation. Let us multiply the ith row by k_i for $i = 1,2,\ldots,n$. Let A'_{ij} be the matrix values after multiplication. Then the basic operation is

$$A'_{jk} \leftarrow A'_{jk} - A'_{ik}(A'_{ji}/A'_{ii}), \tag{54}$$

that is,
$$k_jA_{jk} \leftarrow k_jA_{jk} - k_iA_{ik}(k_jA_{ji}/k_iA_{ii}), \tag{55}$$

which is equivalent to
$$k_j(A_{jk}) \leftarrow k_j(A_{jk} - A_{ik}(A_{ji}/A_{ii})). \tag{56}$$

Thus multiplying the jth row by k_j simply results in the jth row at any step in the triangularization procedure being multiplied by k_j; any subtraction that caused error propagation difficulties in the unmultiplied matrix is unchanged in terms of the relative values of the terms being subtracted. And in floating-point subtraction, it is precisely this relative value that concerns us.

Switching either rows or columns can change the error propagation properties of the computation. Since multiplication of each row by a constant changes neither the roots of the equation nor the error properties of the computation, we can without loss of generality consider the set of equations where each equation has been scaled (multiplied by the reciprocal of the average magnitude of the coefficients of the equation) so that the average magnitude of each row of the resulting matrix, A'', is the same. Consider the general triangularization operation (relation 53) on the double-primed values. Assume that we are at the ith stage of the triangularization and that the matrix elements referred to are the transformed matrix values at that stage. Since the equations have been scaled, it is not unlikely that the kth element of the jth row (A''_{jk}) is approximately equal to the kth element of the ith row (A''_{ik}). To prevent the terms we subtract from being approximately equal, we would like the term $|A''_{ji}/A''_{ii}|$ to be as relatively far from 1 as possible. We can arrange this by properly choosing the new ith row, from among the present ith through nth rows, all of which have zeros in the first $i - 1$ columns. We would choose as the new row i that row m, such that A''_{mi} is as large in magnitude as possible. Thus, from among the ith through nth rows we find the row for which A''_{mi} is largest in magnitude, and we exchange the ith and mth rows. This process is called *pivoting by rows*, and the element A''_{ii} after the row exchange is called the *pivot element*. Experience shows that pivoting by rows can make a significant improvement in the error properties of the computation in Gaussian elimination.

Note that we also could have considered as the new row i the row such that A''_{mi} is as small as possible in magnitude, but that choice would have allowed us to choose a zero divisor and would have produced large multiplying terms rather than the small ones we have chosen. It would have produced no change in the amount of relative error but a larger amount of absolute error. Furthermore, small matrix elements should be avoided as divisors because they may have a large relative error—due to the fact that they may be the result of the subtraction of approximately equal numbers.

Note also that the argument for pivoting depends strongly on the assumption that the equations have been scaled. Pivoting without scaling makes relatively little sense. In fact, the argument for choosing the largest pivot after scaling is even stronger than that presented above. If A''_{ii} is larger in magnitude than A''_{ji}, then because the equations have been scaled, A''_{ik} tends to be smaller in magnitude than A''_{jk}. This fact contributes further to the objective that the rightmost term in relation 53 should be smaller in magnitude than A''_{jk}.

Suppose we apply Gaussian elimination with scaling and pivoting to equations 45, 46, and 47. First we scale each equation by multiplying it by the reciprocal of the sum of the magnitudes of its coefficients (3, 4, and 7, in that order). This produces three equations with the sum of the coefficient magnitudes of each equal to 1:

$$\tfrac{1}{3}x_1 - \tfrac{1}{3}x_2 + \tfrac{1}{3}x_3 = \tfrac{2}{3} \tag{57}$$

$$\tfrac{1}{2}x_1 + \tfrac{1}{4}x_2 + \tfrac{1}{4}x_3 = \tfrac{7}{4} \tag{58}$$

$$\tfrac{4}{7}x_1 + \tfrac{2}{7}x_2 + \tfrac{1}{7}x_3 = \tfrac{11}{7}. \tag{59}$$

Then we find the equation with the largest coefficient in magnitude in the first column (59), and we switch that equation and the first (57). After switching equations 59 and 57, we eliminate the first column of the second and third equations (58 and 57), producing

$$\tfrac{4}{7}x_1 + \tfrac{2}{7}x_2 + \tfrac{1}{7}x_3 = \tfrac{11}{7} \tag{60}$$

$$\tfrac{1}{8}x_3 = \tfrac{3}{8} \tag{61}$$

$$-\tfrac{1}{2}x_2 + \tfrac{1}{4}x_3 = -\tfrac{1}{4}. \tag{62}$$

We now choose the equation from among the second through the last for which the coefficient in the second column has maximum magnitude. We switch this equation with the second. In our example, this results in switching equations 62 and 61. Eliminating the coefficient of the second variable from all equations beyond the second, we find that in our example the third equation (61) is unchanged, because its x_2 coefficient is already zero. Had we not switched equations, that zero would have been a divisor and thus would have caused difficulty.

Our triangularized set of equations comprises 60, 62, and 61, in that order. Solving the last equation, we find $x_3 = 3$. Substituting this into equation 62, we obtain $-\frac{1}{2}x_2 + \frac{3}{4} = \frac{1}{4}$, so $x_2 = 2$. Substituting $x_3 = 3$ and $x_2 = 2$ into equation 60, we obtain $\frac{4}{7}x_1 + \frac{4}{7} + \frac{3}{7} = \frac{11}{7}$, so $x_1 = 1$.

We see that scaling the equations initially does not produce scaled equations after a stage of triangularization. We could rescale the last $n - i + 1$ equations after the ith stage of the triangularization for each i, but the unrescaled coefficients tend to be of the same order of magnitude because of the original scaling. Since this is all we require for the pivoting to work as desired to minimize error propagation, we do not normally rescale at every stage.

We need not explicitly carry out the scaling of the rows to accomplish the pivoting operation. Given a pivot row, the scaling does not affect the result. It only affects which is the pivot row. We can simply compute the scale factors by which each row should be multiplied to make its average coefficient magnitude equal to one. Then when we are choosing a pivot, instead of comparing the magnitudes of the candidate values directly, we compare the magnitudes of the product of each candidate value and the scale factor associated with its row. This way we can choose the correct pivot but not incur the computational error of multiplying each element of the original matrix by some number.

If instead of scaling each row, we had scaled each column (requiring rescaling of the solution elements at the end of the computation), and if we consider the basic operation on the scaled values (relation 53) in the form $A''_{jk} \leftarrow A''_{jk} - A''_{ji}(A''_{ik}/A''_{ii})$, a situation parallel to that discussed above with respect to rows pertains. Here we wish to choose as A''_{ii} that element of the ith row which is greatest and switch that column with the ith column. We see that such column pivoting and scaling causes some difficulty in rescaling and reordering the solution elements at the end of a computation. So, generally, row pivoting is preferred. There are cases in which we want to do row scaling followed by full pivoting (that is, switch rows and columns so that the largest element in the $n - i + 1$ by $n - i + 1$ right lower matrix of the partially triangularized A is placed at the ith position). But in most situations the amount of gain achieved by full pivoting over row pivoting is not great.

It should be noted that the switching of rows does not require physically switching the elements of the matrix in the computer. Rather one can create a permutation vector: a vector of length n where the ith element contains the original number of the row which is now considered to be the ith. The vector is initialized to $(1,2,3,\ldots,n)^T$, and at any point in the elimination the ith row is referenced via the ith element of the permutation vector. To switch two rows, j and k, one simply exchanges elements j and k of the permutation vector. The elimination algorithm with implicit scaling and pivoting by rows and no physical switching of elements is presented in program 2-2.

```
/* GAUSSIAN ELIMINATION TO SOLVE THE N LINEAR EQUATIONS */
/* AX = B WITH PIVOTING AND SCALING, DOING ROW PERMUTATION */
/* AND SCALING IMPLICITLY. THE DIMENSION N, MATRIX A, */
/* AND VECTOR B ARE GIVEN */

   DECLARE (A(N,N),B(N),  /* EQUATIONS MATRIX AND VECTOR */
           X(N),          /* SOLUTION VECTOR */
           P(N),          /* PERMUTATION VECTOR */
           SCALE(N),      /* SCALING VECTOR */
           MAX,JMAX,      /* LARGEST SCALED PIVOT MAG, ITS ROW */
           SCALEVAL,      /* MAG OF SCALED PIVOT CANDIDATE */
           AMULT,         /* ROW MULTIPLIER */
           TEMP)          /* PERM'N VECTOR EL'T BEING SWITCHED */
           FLOAT;
```

PROGRAM 2-2 Gaussian elimination with scaling and pivoting

(*continued on next page*)

```
/* INITIALIZATION */
   INIT: DO I = 1 TO N;
          P(I) = I;  /* INITIALIZE PERMUTATION VECTOR */
          SCALE(I) = 0;
   /* COMPUTE SUM OF ROW MAGNITUDES */
          SCALESUM: DO J = 1 TO N;
                 SCALE(I) = SCALE(I) + ABS(A(I,J) );
          END SCALESUM;
   END INIT;

/* TRIANGULARIZATION WITH PIVOTING */
   L1:   DO I = 1 TO N - 1;              /* AT THE ITH STEP */
         MAX = 0;
         FINDPIVOT: DO J = I TO N;  /* FIND PIVOT */
              SCALEVAL = ABS(A(P(J),I) / SCALE(P(J) ) );
              IF SCALEVAL < = MAX THEN GO TO NEXTROW;
              MAX = SCALEVAL;
              JMAX = J;
         NEXTROW: END FINDPIVOT;
   /* SWITCH ROWS */
         TEMP = P(JMAX);
         P(JMAX) = P(I);
         P(I) = TEMP;
         L2:  DO J = I +1 TO N;          /* BELOW ITH ROW */
         /* MULTIPLIER FOR JTH ROW */
              AMULT = A(P(J),I) / A(P(I),I);
         /* COMPUTE NONZERO ELEMENTS OF JTH ROW */
              L3:  DO K = I+1 TO N;
                   A(P(J),K) = A(P(J),K) - AMULT * A(P(I),K);
              END L3;
              B(P(J) ) = B(P(J) ) - AMULT * B(P(I) );
         END L2;
   END L1;

/* BACK SUBSTITUTION */
   L4:   DO I = N TO 1 BY -1;             /* LAST ROW FIRST */
         X(I) = B(P(I) );
         L5:  DO J = I +1 TO N;           /* NOT EXECUTED IF I = N */
         /* SUBTRACT TERMS IN ALREADY COMPUTED X(J) */
              X(I) = X(I) - A(P(I),J) * X(J);
         END L5;
         X(I) = X(I) / A(P(I),I);         /* SOLVE FOR X(I) */
   END L4;
```

PROGRAM 2-2 (*continued*)

2.3.3 Singular Matrices

Program 2-2 will work as long as A is nonsingular. If the matrix is singular, all elements that are candidates for the pivot in a particular column may be equal to zero, or at least indistinguishable from zero to within the error in those elements. If so, we must move to the next column and choose a pivot element from that column. Our procedure will terminate before we arrive at the nth row, because we will have arrived at the nth column first.

An algorithm which allows for singular matrices is described below. It performs triangularization with row pivoting (assuming scaled rows).

	1. Form the matrix $A:b$, $n \times (n+1)$.
	2. Let $c_1 = 1$.
	3. For $i = 1,2,\ldots,n-1$:
Find pivot row	a) Find $\max_{i \le j \le n} \lvert A_{jc_i} \rvert$. Let $r =$ the value of j giving this maximum.
Move to next column if all pivots = 0	b) If $A_{rc_i} = 0$, then set $c_i = c_i + 1$. If $c_i > n$, then stop; else go to step a.
Exchange ith row with pivot row	c) Exchange rows i and r in $A:b$.
Zero elements below pivot	d) For $j = i+1, i+2, \ldots, n$: (1) For $k = c_i, c_i + 1, \ldots, n+1$: (a) $A_{jk} = A_{jk} - (A_{jc_i}/A_{ic_i})A_{ik}$.
Go to next column	e) If $c_i \ge n$, stop; else $c_{i+1} = c_i + 1$.

To increase our understanding of the elimination procedure, we prove a few theorems showing that this algorithm produces $n - rank(A)$ rows of zeros at the bottom of the triangularized matrix.

Theorem: If $Ax = b$ produces $Cx = d$ after triangularization with row pivoting, and A is $n \times n$,

1. $Ax = b \Leftrightarrow Cx = d$.
2. The last $n - rank(A)$ rows of C are all zeros.

Proof: Define $A^{(1)} \equiv A$, $b^{(1)} \equiv b$. Then one step of triangularization is equivalent to

$$A^{(i+1)} = B^{(i)}P^{(i)}A^{(i)}, \tag{63}$$

$$b^{(i+1)} = B^{(i)}P^{(i)}b^{(i)}, \tag{64}$$

where:

$P^{(i)} = I$, or

$P^{(i)} =$ a permutation matrix of the form:

$$
\begin{array}{r}
\\ \\ \\ \\ \\ \\ i \rightarrow \\ \\ \\ \\ \\ \\ j \rightarrow \\ \\ \\ \\ \\ \\ \\
\end{array}
\begin{bmatrix}
1 & & & & & & & & & & & & & & \\
& 1 & & & & & & & & & & & & & \\
& & \cdot & & & & & & & & & & & & \\
& & & \cdot & & & & & & & & & & & \\
& & & & \cdot & & & & & & & & & & \\
& & & & & 1 & & & & & & & & & \\
& & & & & & 0 & & & & 1 & & & & \\
& & & & & & & 1 & & & & & & & \\
& & & & & & & & \cdot & & & & & & \\
& & & & & & & & & \cdot & & & & & \\
& & & & & & & & & & \cdot & & & & \\
& & & & & & & & & 1 & & & & & \\
& & & & & & 1 & & & & 0 & & & & \\
& & & & & & & & & & & 1 & & & \\
& & & & & & & & & & & & 1 & & \\
& & & & & & & & & & & & & \cdot & \\
& & & & & & & & & & & & & & \cdot \\
& & & & & & & & & & & & & & \cdot \\
& & & & & & & & & & & & & & 1
\end{bmatrix},
$$

$$
\qquad\qquad\qquad\qquad\quad \uparrow \qquad\qquad\quad \uparrow
$$
$$
\qquad\qquad\qquad\qquad\quad i \qquad\qquad\quad j
$$

$$
\qquad\qquad i
$$
$$
\qquad\qquad \downarrow
$$

$$
\text{and } B^{(i)} =
\begin{array}{r}
\\ \\ \\ \\ \\ i \rightarrow \\ \\ \\ \\ \\ \\ \\
\end{array}
\begin{bmatrix}
1 & & & & & & & & \\
& 1 & & & & & & & \\
& & \cdot & & & & & & \\
& & & \cdot & & & & & \\
& & & & \cdot & & & & \\
& & & & 1 & & & & \\
& & & & \dfrac{-A^{(i)}_{i+1,i}}{A^{(i)}_{ic_i}} & 1 & & & \\
& & & & \dfrac{-A^{(i)}_{i+2,i}}{A^{(i)}_{ic_i}} & & 1 & & \\
& & & & \cdot & & & \cdot & \\
& & & & \cdot & & & & \cdot \\
& & & & \cdot & & & & \cdot \\
& & & & \dfrac{-A^{(i)}_{ni}}{A^{(i)}_{ic_i}} & & & & 1
\end{bmatrix}, \qquad (65)
$$

where zeros appear in all of the blank locations in both $P^{(i)}$ and $B^{(i)}$.

Multiplying by $P^{(i)}$ effects the row exchange for pivoting, and multiplying the result by $B^{(i)}$ effects the zeroing of column i for rows $\geq i + 1$.

Lemma 1: $A^{(i+1)}x = b^{(i+1)} \Leftrightarrow A^{(i)}x = b^{(i)}$ for $i = 1,2,\ldots, n - 1$.

Proof: $\Leftarrow$: Assume $A^{(i)}x = b^{(i)}$.
Then $B^{(i)}P^{(i)}A^{(i)}x = B^{(i)}P^{(i)}b^{(i)}$.
$\Rightarrow$: Assume $B^{(i)}P^{(i)}A^{(i)}x = B^{(i)}P^{(i)}b^{(i)}$.

1. $|det(P^{(i)})| = 1,$ (66)
   since switching rows i and j produces no change in the magnitude of the determinant.
2. $det(B^{(i)}) = 1,$ (67)
   since $B^{(i)}$ is triangular and thus its determinant is the product of its diagonal elements.
3. Thus
   $$|det(B^{(i)}P^{(i)})| = |det(B^{(i)})det(P^{(i)})| = 1 \neq 0, \tag{68}$$
   so $B^{(i)}P^{(i)}$ has an inverse.
4. Thus $(B^{(i)}P^{(i)})^{-1}(B^{(i)}P^{(i)})A^{(i)}x$
   $$= (B^{(i)}P^{(i)})^{-1}(B^{(i)}P^{(i)})b^{(i)}, \tag{69}$$
   or $A^{(i)}x = b^{(i)}$.

Lemma 2: $Rank(A^{(i+1)}) = Rank(A^{(i)})$ for $i = 1,2,\ldots,n - 1$.

Proof: $A^{(i+1)}x = 0 \Leftrightarrow A^{(i)}x = 0$ by the same argument as in lemma 1. Thus $A^{(i+1)}$ and $A^{(i)}$ have the same null space. Thus they have the same rank.

We know $A^{(n)}$ has the same number of dependent rows as A. To show $A^{(n)}$ has $n - rank(A)$ rows of zeros, what remains to be proved is that any dependent row of $A^{(n)}$ is zero. Assume the mth row of $A^{(n)}$ is dependent on the 1st through $(m - 1)$th rows:

$$A_{mi}^{(n)} = \sum_{j=1}^{m-1} \alpha_j A_{ji}^{(n)}. \tag{70}$$

Notation: Let $A_{jK(j)}^{(n)}$ be the first nonzero element of the jth row of $A^{(n)}$. Then $K(j) > K(j-1)$ and $A^{(n)}_{qK(j)} = 0$ for $q > j$. Also, $A^{(n)}_{mi} = 0$, for $i = 1,2,\ldots,K(m) - 1$.

Consider $i = K(1) \leq K(m) - 1$. Then

$$0 = A^{(n)}_{mK(1)} = \alpha_1 A^{(n)}_{1K(1)} + \alpha_2 \cdot 0 + \alpha_3 \cdot 0 + \cdots + \alpha_{m-1} \cdot 0 = \alpha_1 A^{(n)}_{1K(1)}. \tag{71}$$

$A^{(n)}_{1K(1)} \neq 0$ by definition, so $\alpha_1 = 0$.

Next consider $i = K(2)$. Then

$$0 = A^{(n)}_{mK(2)} = 0 \cdot A^{(n)}_{1K(2)} + \alpha_2 A^{(n)}_{2K(2)} + \alpha_3 \cdot 0 + \alpha_4 \cdot 0 + \cdots + \alpha_{m-1} \cdot 0. \tag{72}$$

$A^{(n)}_{2K(2)} \neq 0$, so $\alpha_2 = 0$.

Similarly $\alpha_3 = \alpha_4 = \cdots = \alpha_{m-1} = 0$. So $A^{(n)}_{mi} = \sum_{j=1}^{m-1} \alpha_j A^{(n)}_j = 0$ for all i.

So we know $A^{(n)}$ must have $n - rank(A)$ zero rows. Since the pivoting procedure exchanges nonzero rows with zero rows, all the zero rows must be in the bottom of the matrix.

Corollary: $det(A^{(n)}) = (-1)^{n_0} det(A)$, where n_0 is the number of i such that $P^{(i)} \neq I$.

Proof: For all i, $det(B^{(i)}) = 1$. If $P^{(i)} \neq I$, $det(P^{(i)}) = -1$.

$$det(A^{(n)}) = det(A) \prod_{i=1}^{n-1} [det(B^{(i)}) \cdot det(P^{(i)})]. \tag{73}$$

Since $det(A^{(n)}) =$ the product of the pivots (the diagonal elements of $A^{(n)}$), we can compute the determinant of an arbitrary matrix in $n^3/3 + O(n^2)$ multiplications.

We have shown that we can detect singularity by detecting whether certain values in the triangularized matrix are zero. Because of error propagation and generation, true zeros are extremely rare in the results of computer solution. We must test whether the value in question is likely to be zero plus error or is not likely to be that. That is, at any stage in the computation, we must compute (based either on analytical results or on the computation itself) an estimate of the likely error in the result and compare the magnitude of the actual value computed to this error estimate. If it is significantly less than the error estimate, we are fairly certain that the result was zero or close enough to zero to be effectively zero. If the result is significantly greater in magnitude than the error estimate, we can be fairly certain that it does not represent a zero plus error. Because there is a large gray region in between, it is quite difficult to tell whether a matrix is actually singular. However, matrices that are close to singular are precisely those for which error is propagated particularly badly (see section 2.4). Unless we are specifically asking whether or not a matrix is singular, knowing that it is either singular or near singular is enough to tell us that a solution does not exist, is not unique, or may be very badly in error. In all of these cases, we want to report that fact with the solution.

We have seen that, with singular matrices, in the triangularization, all candidates for pivot in some column may be zero. However, error may cause these zero values to show up as small values in the range of likely error. We cannot distinguish valid small nonzero values from values which ought to be zero, so if all the candidates for pivot at any stage are in this gray area, we will do well to switch the present pivot column with the last column. This strategy avoids the difficulties associated with this column until the number of remaining operations is less and thus the error propagation is less. Alternatively, if we are confronted with this situation of very small pivot candidates, we can do full pivoting for the remainder of the triangularization rather than just pivoting by rows.

If we know a matrix is singular and we also know the rank of the matrix, we can do Gaussian elimination with $rank(A) - 1$ steps of triangularization using full pivoting and be sure that the last $n - rank(A)$ rows are zero. The existence of a solution is then determined by whether or not the transformed b vector also has zeros in its last $n - rank(A)$ elements. More precisely, to check a particular b_k, we should check whether $|b_k^{(\text{transformed})}/b_k^{(\text{original})}| \approx |A_{kk}^{(\text{transformed})}/A_{kk}^{(\text{original})}|$. If so, then we can say a solution exists. If not, then no solution exists.

2.3.4 Iterative Improvement

Improvements can be made on the elimination process given above. One of these allows us to iteratively improve our answer beyond the value found by the initial elimination procedure. Assume we have computed a solution to $Ax = b$, say x^0. Say the correct solution is $x^0 + \Delta x$. Then

$$A(x^0 + \Delta x) = b. \tag{74}$$

Thus

$$A\Delta x = b - Ax^0, \tag{75}$$

and we can solve this equation for Δx. Advantages of this second solution are that $b - Ax^0$ is very much smaller than b, and Δx is presumably very much smaller than x^0. We will not be involved with all of the significant digits of x^0, so our floating-point numbers will have digits in their (accurate) high-order parts that in the first solution were low-order digits and thus either truncated or contaminated by truncation error.

If we iterate this procedure, that is, at the ith step, solve equation 76 for Δx and substitute its value in equation 77 to obtain x^{i+1}, hopefully, we will converge to a very accurate solution to our original equation.

$$A\Delta x = b - Ax^i. \tag{76}$$

$$x^{i+1} = x^i + \Delta x. \tag{77}$$

What is the limit on the accuracy we can gain by the iteration technique, assuming A and b have no error? It is (1) the accuracy of our computations in the solution process and the properties of the matrix A in propagating our errors, and (2) the accuracy of the right-hand side of equation 76, $b - Ax^i$. Notice that the latter expression is the difference of two approximately equal vectors. This operation can propagate small relative errors in the terms into large relative errors. Therefore, we must compute the term Ax^i in double precision, and truncate to single precision only after the full sum of products is computed. Then, whether the iteration in fact produces more accurate results depends on the characteristics of the matrix A as regards the propagation of errors generated in the computation needed to solve the equation $Ax = b$. This matter is discussed in section 2.4.5. We can say now that in most cases this iterative technique improves the accuracy of our solution. Normally, most of the improvement is gained in the first few iterations.

2.3.5 Triangular Factorization

The second improvement we can make in the elimination process results from noting that in the iterative improvement scheme mentioned above we must again and again solve an equation of the form $Ax = b$, for different values of b but for the same A. This situation also arises in applied mathematical problems, in which A often corresponds to physical constraints of the problem and b to data values, and we wish to solve the same physical problem for many different sets of data.

Up until now, the value of b has been involved in our triangularization procedure. That is, at each triangularization step, we have subtracted a multiple of a given element of b from all elements of b below it. Since this triangularization is the most time-consuming part of the Gaussian elimination, requiring approximately $n^3/3$ multiplications, we would prefer not to repeat this step every time we change our b value. We can avert this repetition if, for each triangularization stage i, we save the multipliers by which b_i must be multiplied before subtracting it from

b_j for all $j > i$. We will show that a triangularization in which multipliers are saved is equivalent to a triangular factorization of the matrix A into a lower triangular matrix of the multipliers in question and an upper triangular matrix which was the result of the triangularization step of the elimination.

Assume the matrix A is in an order such that no permutation of rows need be done to accomplish pivoting by rows. In that case, by equation 63 the upper triangular matrix U (equal to $A^{(n)}$) which is the result of the triangularization equals $B^{(n-1)}B^{(n-2)}\dots B^{(1)}A$. Each $B^{(i)}$ is a matrix with 1s on the diagonal and the negative of the multipliers in question in the ith column below the diagonal (see equation 65). Thus

$$A = B^{(1)-1}B^{(2)-1}B^{(3)-1}\dots B^{(n-1)-1}U. \tag{78}$$

But $B^{(i)-1}$ is of the same form as $B^{(i)}$ except that all of the multipliers (elements below the diagonal) have opposite signs. Let L be the matrix $B^{(1)-1}B^{(2)-1}B^{(3)-1}\dots B^{(n-1)-1}$. We can show then that the matrix L is a lower triangular matrix with 1s on the diagonal such that for all i, the ith column of L is the ith column of $B^{(i)-1}$ (see exercise 2.3.4):

$$L = \begin{bmatrix} 1 & & & & & & \\ \frac{A_{21}^{(1)}}{A_{11}^{(1)}} & 1 & & & & & \\ \frac{A_{31}^{(1)}}{A_{11}^{(1)}} & \frac{A_{32}^{(2)}}{A_{22}^{(2)}} & 1 & & & & \\ \cdot & \cdot & \frac{A_{43}^{(3)}}{A_{33}^{(3)}} & \cdot & & & \\ \cdot & \cdot & \cdot & \cdot\;\cdot & & & \\ \cdot & \cdot & \cdot & \cdot\;\cdot\;\cdot & & & \\ \cdot & \cdot & \cdot & \cdot\;\cdot\;\cdot\;\cdot & & & \\ \frac{A_{n-1,1}^{(1)}}{A_{11}^{(1)}} & \frac{A_{n-1,2}^{(2)}}{A_{22}^{(2)}} & \frac{A_{n-1,3}^{(3)}}{A_{33}^{(3)}} & \cdot\;\cdot\;\cdot\;\cdot\;\cdot & 1 & \\ \frac{A_{n1}^{(1)}}{A_{11}^{(1)}} & \frac{A_{n2}^{(2)}}{A_{22}^{(2)}} & \frac{A_{n3}^{(3)}}{A_{33}^{(3)}} & \cdot\;\cdot\;\cdot\;\cdot\;\cdot & \frac{A_{n,n-1}^{(n-1)}}{A_{n-1,n-1}^{(n-1)}} & 1 \end{bmatrix}, \tag{79}$$

with zeros in the upper right.

Thus we have shown that if no row permutations need be done, A can be expressed as the product of a lower triangular matrix and an upper triangular matrix, where the lower triangular matrix has 1s on the diagonal and has the multipliers used in the triangularization below the diagonal, and the upper triangular matrix is the result of the triangularization step of elimination.

If we do not need to save the original A matrix, we can execute the triangular factorization without requiring extra space for the output. As we set a column element to zero, we can replace it by the corresponding multiplier. The 0s of the upper triangular matrix and the 1s on the diagonal of the lower triangular matrix are implicit.

We can do pivoting by the permutation vector approach given in the description of the triangularization of Gaussian elimination. That is, at the ith step, we can decide which row should have been the ith row and put that row number in the ith element of the permutation vector.

Having factored the A matrix, we can solve the equation for any given b vector as follows. If there had been no permutation of rows, we would have

$$LUx = b. \tag{80}$$

If we let $Ux = y$, equation 80 becomes

$$Ly = b. \tag{81}$$

This set of equations can be solved by substitution, starting with the first equation. Having solved for y, we now must solve the equation

$$Ux = y, \tag{82}$$

which can also be solved by substitution, starting with the last equation. Thus, given any b vector, two substitutions, each requiring approximately $n^2/2$ multiplications, will produce a solution. If the rows have been permuted and a permutation vector maintained, the only difference is that all *rows* of the L and U matrices as well as of the b and y vectors must be referenced via the permutation vector.

If we have factored our matrix as suggested above, then iterative improvement requires only $2n^2$ multiplications per iteration: n^2 to compute Ax^i, and n^2 for the two substitutions required.

Another way to solve $Ax = b^j$ for various vectors b^j without requiring retriangularization for each j is to compute A^{-1} once and then for each j to compute $x = A^{-1}b^j$. Multiplying A^{-1} by b^j requires n^2 multiplications, the same number as the two substitutions of the triangular factorization method, if it were used. However, computing A^{-1} involves simultaneously solving the n equations $Ax = e^i$ for $1 \leq i \leq n$, a process that itself requires the triangularization of A. Since finding this solution for the inverse requires n^3 multiplications, and triangular factorization requires only $n^3/3$, the algorithm involving triangular factorization and two substitutions is always more efficient and accurate than computing A^{-1} and multiplying b by it. Therefore, the only reason to compute A^{-1} is if the values of that matrix itself are of interest to us!

PROBLEMS

2.3.1. a) If A is $n \times n$ and B is $n \times p$, show $AX = B$ can be solved by elimination in $pn^2 + n^3/3 + O(n^2)$ multiplications and divisions. Thus if B is $n \times n$, the solution can be found by elimination in $\frac{4}{3}n^3 + O(n^2)$ multiplications and divisions.

b) If $B = I$, show that the solution can be found by elimination in "only" n^3 multiplications and divisions.

2.3.2. Let b_z be the least upper bound on the magnitude of the absolute error in z and let $r_z = b_z/|z|$. Let $y = az$.

a) Show $r_{z-y} = \dfrac{1}{|1-a|} r_z + \dfrac{|a|}{|1-a|} r_y + r_-$.

b) Assume $|a|$ is triangularly distributed over $[0,1]$ with mean $\frac{1}{2}$ ($p_{|a|}(u)$ rises linearly from 0 at $u = 0$ to a maximum at $u = \frac{1}{2}$ and then falls linearly to 0 at $u = 1$). Also assume $Pr(a > 0) = Pr(a < 0)$. This is meant as an approximation of the distribution in a for the subtractions in Gaussian elimination, assuming scaling and pivoting are done. Show $E(r_{z-y}) \approx 1.8r_z + 1.0r_y + r_-$.

c) Assume $r_{z-y} = 1.8r_z + 1.0r_y + r_-$ for every Gaussian elimination subtraction. Assume all elements of the original matrix are error free, and assume all operations are done in floating-point on a System/370 computer which has base-16 floating-point with six digits in the fraction part of the representation. Compute a relative error bound on $A_{nn}^{(n)}$, where $A^{(n)}$ is the triangularized matrix produced by Gaussian elimination.

2.3.3. a) Solve the following system of linear equations using Gaussian elimination, first with pivoting only and then with pivoting with implicit scaling so that the average magnitude of the elements in each row of the matrix is $\frac{1}{3}$. Do symmetric rounding to two significant decimal digits on the results of all arithmetic operations.

$$\begin{aligned} 11x + 12y + 16z &= 39 \\ 10x + 11y + 12z &= 33 \\ 1.2x + .1y + .1z &= 1.4 \end{aligned}$$

b) Compare your answers, using the fact that the correct solution is $x = y = z = 1$.

2.3.4. Let $C^{(k)}$ be an $n \times n$ matrix such that

$$C_{ij}^{(k)} = \begin{cases} 1 \text{ if } i = j \\ D_{ij} \text{ if } i = k \text{ and } i > j \\ 0 \text{ otherwise.} \end{cases}$$

Show that $(C^{(1)}C^{(2)} \ldots C^{(n)})_{ij} = \begin{cases} 1 \text{ if } i = j \\ D_{ij} \text{ if } i > j \\ 0 \text{ otherwise.} \end{cases}$

2.3.5. a) Write a program to solve $Ax = b$ for an arbitrary $n \times n$ matrix A by applying triangular factorization with pivoting by rows preceded by scaling of the rows.

Scale the rows of A so that the average magnitude of the elements in each row is $1/n$. You should not actually multiply the rows of A and b by scale factors; rather, you should store these factors in a "scaling vector" and use them in the choice of pivots. Similarly, the row interchanges in the pivoting procedure should not be done by physically moving matrix elements; they should be done implicitly using a "permutation vector." Do all arithmetic in single-precision floating point.

b) Apply the program that you wrote for part a to the equations for which

$$A = \begin{bmatrix} 3 & 2.88 & 2.52 & 1.92 \\ .72 & .691 & .3304 & .1666 \\ 1.26 & .6608 & 1.5 & .516 \\ .24 & .0833 & .129 & .375 \end{bmatrix} \quad \text{and} \quad b = \begin{bmatrix} 10.32 \\ 1.908 \\ 3.9368 \\ .8273 \end{bmatrix}.$$

c) Repeat part b, but use program 2-1 of this text (Gaussian elimination with no scaling or pivoting).

d) Compare the results of parts b and c, using the fact that the correct answer is $x = (1,1,1,1)^T$.

2.3.6. Consider the equation $Ax = b$ where A is an $n \times n$ matrix and b is an n-vector. Suppose you know the dimension of the null space of A is 2. *Specifically,* how does this knowledge influence your method for solving for x? Why? Be aware of the question of efficiency as well as the fact that the method must be realistic for implementation on a digital computer.

2.4 ERROR PROPAGATION IN SOLVING LINEAR EQUATIONS

We have treated the problem of organizing computations to generate as little error as possible and to computationally propagate generated errors or input errors as little as possible. However, we have not treated the problem of the propagation of input errors in the A matrix and b vector due to the fact that x is a function of A and b. Now we ask: Assuming no computational error in the solution process but initial errors in A and b, what will be the error in x?

2.4.1 Norms

When considering error propagation, we are largely interested in relative error (that is, we are not concerned with the fact that multiplying b and its error by a factor results in the multiplication of x and its error by the same factor). But we have no definition of the relative error of a vector because division of a vector by a vector is undefined. What we really want is the relative "size" of the error in x divided by the size of x. Thus we must define precisely what is meant by the notion of the size of the vector x, which we write "$\|x\|$".

Any reasonable definition of the vector size should at least satisfy the following properties:

1. The size of the 0 vector is 0: $\|0\| = 0$. (83)
2. The size of every nonzero vector is >0: $x \neq 0 \Rightarrow \|x\| > 0$. (84)
3. The multiplication of a vector by a constant multiplies the size of that vector by the magnitude of the constant: $\|\lambda x\| = |\lambda| \, \|x\|$. (85)
4. The triangle inequality holds: The size of the sum of two vectors is no larger than the sum of the sizes of the vectors:

$$\|x + y\| \leq \|x\| + \|y\|. \tag{86}$$

Any function of a vector which satisfies the above rules is called a *vector norm*. Any vector norm can be taken to measure vector size. The notation $\|x\|$ denotes the vector norm of x, with the particular vector norm denoted either understood or unspecified. A subscript can be appended to the norm to indicate which norm we are using.

These four rules are certainly necessary for any reasonable measure of vector size, but are they sufficient? It is quite convincing of their sufficiency that the following two relationships can be proved for any vector norm.†

1. $\|x\|$ depends continuously on x (that is, on the elements of x).
2. There exist positive constants α and β such that for all vectors x

$$\alpha\|x\|_\infty \leq \|x\| \leq \beta\|x\|_\infty, \tag{87}$$

where $\|x\|_\infty$ is a particular norm of x, namely, the $\mathscr{L}_\infty$ norm of x, defined below.

Relation 2 is especially convincing in that it says that all norms are reasonably closely related; if one norm of a vector is large, so also another norm of that vector, and so forth.

The most common vector norms, though by no means the only vector norms, are the $\mathscr{L}_p$ *norms* defined as

$$\|x\|_p \equiv \left(\sum_{i=1}^{n} |x_i|^p\right)^{1/p}, \quad \text{for } p \geq 1. \tag{88}$$

We saw these norms before, in the discussion of probability in chapter 1. The three most common $\mathscr{L}_p$ norms are: (1) the $\mathscr{L}_1$ norm (the sum of the absolute values of the elements), (2) the $\mathscr{L}_2$ norm (also called the *Euclidean norm*, which is the square root of the sum of the squares of the elements and is sometimes written "$|x|$"), and (3) the $\mathscr{L}_\infty$ norm (also called

† See chapter 6 of Franklin for proofs.

the *Tchebycheff norm*) which is equal to the magnitude of the element of greatest magnitude of the vector (see exercise 2.4.1). To convince yourself that all of the $\mathscr{L}_p$ norms indeed satisfy the rules for norms, see exercise 2.4.2.

We used an $\mathscr{L}_p$ norm earlier in this chapter to define the size of a vector when we dealt with scaling. We wanted to make constant the size of each row vector of the coefficient matrix. At that time we arbitrarily used the $\mathscr{L}_1$ norm but we could have used any norm which ensured that if one element in a row was relatively large, the others were likely to be small (that is, $\mathscr{L}_\infty$ would not be too good). The $\mathscr{L}_2$ norm in particular is also a good candidate. It is harder to compute than the $\mathscr{L}_1$ norm, but we will see some situations where we should scale using the $\mathscr{L}_2$ norm.

Now we can define the *relative error in a vector,* $x^* = x + \delta x$, as

$$\|\delta x\|/\|x^*\| \approx \|\delta x\|/\|x\| \quad \text{if} \quad \|\delta x\| \ll \|x\|.$$

Since we may also have errors in a matrix, we would also like to have a notion of the relative error in a matrix and thus of the size of a matrix. Because a matrix represents a linear transformation, its size needs to be related to the effect of the transformation. We ask: How big a vector does the matrix A transform the vector x into? To eliminate the linear effect of the size of x on its image, we are interested in the value of $\|Ax\|/\|x\|$, the factor by which the size of x is stretched under transformation by A. The maximum stretching factor over all possible x vectors can be taken as the definition of the size of the matrix:

$$\|A\| \equiv \max_{x \neq 0}(\|Ax\|/\|x\|). \tag{89}$$

Note first that the definition of a matrix norm depends on a vector norm; for each vector norm, there is a corresponding matrix norm. Second, note that for every vector $x \neq 0$, there is a corresponding *unit vector* (a vector whose norm is equal to 1), $x/\|x\|$, which under transformation by A stretches in size by the same factor as x. Thus an equivalent definition for $\|A\|$ is

$$\|A\| \equiv \max_{\|x\|=1} \|Ax\|. \tag{90}$$

Third, the maxima in equations 89 and 90 are achieved; that is, there exists an x such that

$$\|Ax\| = \|A\| \cdot \|x\|. \tag{91}$$

As an example of a matrix norm, let us prove that the matrix norm corresponding to the $\mathscr{L}_\infty$ vector norm is the maximum row sum of magnitudes in the matrix.

Theorem:

$$\|A\|_\infty = \max_k \sum_{j=1}^{n} |A_{kj}|. \tag{92}$$

Proof:

$$\|A\|_\infty \equiv \max_{\|x\|_\infty = 1} \|Ax\|_\infty. \tag{93}$$

Since $\|x\|_\infty = \max_k |x_k| = 1$, $|x_j| \leq 1$ for all j. So for any x,

$$\|Ax\|_\infty = \max_k |(Ax)_k| = \max_k |\sum_j A_{kj}x_j| \leq \max_k \sum_j |A_{kj}|\,|x_j| \leq \max_k \sum_j |A_{kj}|. \tag{94}$$

Thus

$$\|A\|_\infty \leq \max_k \sum_j |A_{kj}|. \tag{95}$$

Let k_0 be the row such that $\sum_j |A_{kj}|$ is maximum. Let x^0 be a vector such that

$$x_j^0 = \begin{cases} +1 \text{ if } A_{k_0 j} \geq 0 \\ -1 \text{ if } A_{k_0 j} < 0. \end{cases} \quad \text{Note } \|x^0\|_\infty = 1.$$

Then $$\|Ax^0\|_\infty = (Ax^0)_{k_0} = \sum_j |A_{k_0 j}| = \max_k \sum_j |A_{kj}|, \tag{96}$$

because for $k \neq k_0$, $$|(Ax^0)_k| \leq \sum_j |A_{kj}| \leq \sum_j |A_{k_0 j}|. \tag{97}$$

Since for a particular vector x^0 such that $\|x^0\|_\infty = 1$, $\|Ax^0\|_\infty = \max_k \sum_j |A_{kj}|$,

$$\|A\|_\infty = \max_{\|x\|_\infty = 1} \|Ax\|_\infty \geq \max_k \sum_j |A_{kj}|. \tag{98}$$

Combining equations 95 and 98, we obtain

$$\|A\|_\infty = \max_k \sum_j |A_{kj}|.$$

The matrix norms have properties corresponding to those of the vector norm:†

$$\|I\| = 1; \tag{99}$$

$$\|A\| \geq 0 \quad (\text{with } \|A\| = 0 \text{ if and only if } A = 0); \tag{100}$$

$$\|\lambda A\| = |\lambda|\,\|A\| \quad (\text{for all scalars } \lambda); \tag{101}$$

$$\|A + B\| \leq \|A\| + \|B\|; \tag{102}$$

$$\|AB\| \leq \|A\|\,\|B\|; \tag{103}$$

† The proof of these relations is given in chapter 6 of Franklin.

$$\|Ax\| \leq \|A\| \; \|x\| \quad \text{(for all vectors } x\text{);} \tag{104}$$

$$\|A\| \quad \text{depends continuously on } A; \tag{105}$$

$$\rho \max_{i,j}|A_{ij}| \leq \|A\| \leq \sigma \max_{i,j}|A_{ij}| \text{ for some}$$
$$\text{positive constants } \rho \text{ and } \sigma \text{ which are independent of A.} \tag{106}$$

These properties will be useful to us as we apply norms in the analysis of error propagation in the solution of linear equations.

2.4.2 A Determinant Measure of Conditioning

If a mapping propagates error badly, it is said to be *ill-conditioned.* If it does not increase relative errors much, it is said to be *well-conditioned.* That is, if we consider the mapping of b vectors into solution (x) vectors, a well-conditioned matrix corresponds to an inverse linear transformation which maps b vectors with small relative errors into solution vectors with small relative errors. An ill-conditioned matrix maps at least some vectors with small relative errors into solution vectors with relative errors that are not comparably small.

Let us develop a geometric feeling for conditioning by considering the equations

$$3x_1 + 2x_2 = 5 \tag{107}$$

$$4x_1 - 3x_2 = 1, \tag{108}$$

which have the solution $x = \begin{bmatrix} 1 \\ 1 \end{bmatrix}$. Let us use the $\mathscr{L}_\infty$ norm as a measure of the size of the vectors b and x. Then $\|b\|_\infty = 5$; $\|x\|_\infty = 1$. Consider making an error in b, $\delta b = \begin{bmatrix} 1 \\ 0 \end{bmatrix}$. If we solve the resulting equations

$$3x_1 + 2x_2 = 6 \tag{109}$$

$$4x_1 - 3x_2 = 1, \tag{110}$$

we arrive at the solution $x^* = \begin{bmatrix} 20/17 \\ 21/17 \end{bmatrix}$. That is, $\delta x = \begin{bmatrix} 3/17 \\ 4/17 \end{bmatrix}$. The relative error in x equals $\frac{4}{21}$ and the relative error in b equals $\frac{1}{6}$, so the relative error in x is 14% greater than that in b, which is not a bad propagation.

On the other hand, consider the equations with the same solution as equations 107 and 108 but with a sign reversed in the second equation:

$$3x_1 + 2x_2 = 5 \tag{111}$$

$$4x_1 + 3x_2 = 7. \tag{112}$$

With the same δb as before, we obtain

$$3x_1 + 2x_2 = 6 \tag{113}$$

$$4x_1 + 3x_2 = 7, \tag{114}$$

which has the solution $x^* = \begin{bmatrix} 4 \\ -3 \end{bmatrix}$. Here $\delta x = \begin{bmatrix} 3 \\ -4 \end{bmatrix}$ and the relative error in x is 1.0, six times the relative error in b.

What is the difference between these systems of equations, which appear so similar on the surface? Consider figure 2-5 in which equations 107–110 are graphed in the top graph and equations 111–114 in the bottom graph. The difference between the two graphs is clear. In the first case, the lines corresponding to the two equations are nearly perpendicular, so that a small change in b, which results in a small translation of one of the lines parallel to itself, shifts the solution approximately the same amount as the line is shifted. In the second case, the equations are very nearly parallel, and a small translation of one line parallel to itself results in a large change in the solution. Generalizing our discovery to n equations in n unknowns where each equation constrains x to a hyperspace of dimension $n - 1$, we find that the more close to orthogonal are the $(n - 1)$-dimensional hyperspaces corresponding to each equation, the more well-conditioned is our matrix. Similarly, the more nearly dependent are the hyperspaces (the more nearly singular is the matrix), the more ill-conditioned is our matrix.

We would like to have a measure of the orthogonality of the n hyperspaces specified by our n equations. Each equation is of the form

$$x^T a^{(i)} = b_i, \tag{115}$$

where $a^{(i)}$ is a row vector of the coefficient matrix A. Changing the b_i values changes the relation between the intercepts of the space in which the equation constrains x to lie, but it does not change the slopes. Thus setting $b_i = 0$ does not change the direction relations of the space, and each equation can be thought of as constraining x to lie in a space perpendicular to the row vector corresponding to that equation. Therefore, the orthogonality of these hyperspaces is equivalent to the orthogonality of the vectors normal (orthogonal) to the spaces, namely, the row vectors.

We now have to find a measure of the orthogonality of the n row vectors of our matrix A. Clearly, this measure should be independent of the size of the vectors, so we should divide each row vector by its Euclidean norm (because our concept of orthogonality is related to the Euclidean norm, as we will see in section 2.5) to make each vector of unit size. A reasonable measure of the orthogonality of n unit vectors is the volume of the hyperparallelepiped of which they are the edges. From section 2.2.2,

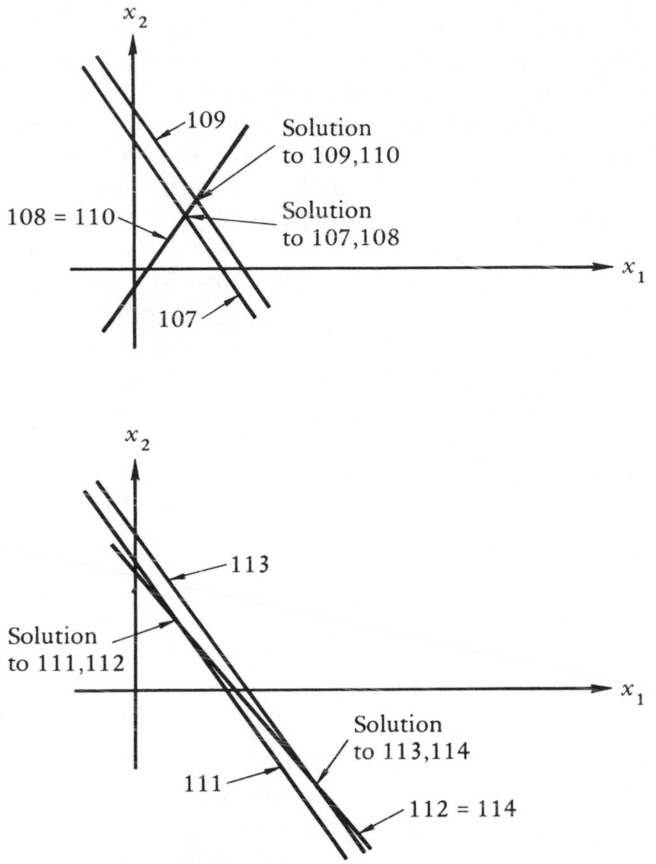

FIG. 2-5 Graphs of linear equations 107–114

we know that this is simply the determinant of the matrix with these unit vectors as columns or, since $det(B) = det(B^T)$, the determinant of the matrix with these unit vectors as rows. But the latter matrix is simply A except that each row, $a^{(i)}$, has been multiplied by the reciprocal of its Euclidean norm. Since multiplying a row of a matrix by a constant simply multiplies its determinant by the constant, our measure of orthogonality of the rows of our matrix A is given by the *normalized determinant* of A, which is defined as $|det(A)| \Big/ \prod_{i=1}^{n} |a^{(i)}|$. This normalized determinant is a reasonable measure of the conditioning of the matrix A.

The normalized determinant has a value between 0 and 1. It is 1 when all of the row vectors of A are orthogonal; it is 0 when some of them are linearly dependent. A value of .1 or greater represents a reasonably well-conditioned matrix, and a value of .01 or less represents a very ill-conditioned matrix. Conte states that a given relative error in an element of A can be propagated into a relative error in x by a factor on the order of the square of the reciprocal of the value of the normalized determinant.† In our example, equations 107, 108 and 109, 110 have normalized determinants .942 and equations 111, 112 and 113, 114 have normalized determinants .056.

Let us use the relationship between the rows of a matrix A and the solutions of $Ax = b$ to get another geometric interpretation of why the number of linearly independent columns of a matrix is equal to the number of linearly independent rows. We said that each equation requires the x vector to be in a space perpendicular to the row vector. If a given row vector is a linear combination of previous row vectors, then it imposes no additional constraint on the direction of the solution hyperspace. If there are m such linearly dependent rows, then if there exists a solution, there will be only $n - m$ constraints on that solution. That is, there will be m dimensions remaining for the solution, or the dimension of the null space of A will be m.

2.4.3 The Condition Number

The difficulty with the normalized determinant as a measure of conditioning is that it does not give us a direct measure of the propagation of error in systems of linear equations. We would like a number C, with which we could say that if the input vector b had relative error r_b, the solution vector x would have relative error no greater than Cr_b. That is, we would like a number C which gives the largest factor by which a relative error in the input can be magnified in the transformation into the output. Such a number is called the *condition number* of the transformation from b to x. Note that the idea of the condition number is applicable to any transformation of input into output. We will develop formulas for the condition number of the particular transformation defined by the solution of a system of linear equations. Since the relative error involves norms, this condition number will depend upon norms.

Assume we have an equation

$$Ax = b \tag{116}$$

† S. D. Conte, *Elementary Numerical Analysis*, chapter 5.

where A has no error and b has error δb. That is,

$$A(x + \delta x) = b + \delta b. \tag{117}$$

Assume $\|\delta b\|/\|b\|$ is given. We wish to find a least upper bound on $\|\delta x\|/\|x\|$ over all δb and b which have the relative error relation above. Equation 117 together with equation 116 implies

$$\delta x = A^{-1}\,\delta b. \tag{118}$$

Using relation 104 on equations 116 and 118, we obtain the relations

$$\|\delta x\| \le \|A^{-1}\|\,\|\delta b\| \tag{119}$$

and

$$\|b\| \le \|A\|\,\|x\|. \tag{120}$$

Inequalities 119 and 120 imply

$$\frac{\|\delta x\|}{\|A\|\,\|x\|} \le \frac{\|A^{-1}\|\,\|\delta b\|}{\|b\|}, \tag{121}$$

or

$$\frac{\|\delta x\|}{\|x\|} \le \|A\|\,\|A^{-1}\|\frac{\|\delta b\|}{\|b\|}. \tag{122}$$

Thus the product of the norm of A and the norm of A^{-1} is an upper bound on the factor by which relative errors in b can be multiplied when propagating into relative errors in x.

If we can find, for any given matrix A, a vector b and an error δb such that inequality 122 is an equality, we will have shown that $\|A\|\,\|A^{-1}\|$ is the least upper bound for the factor in question. We have seen that there exists a vector x^0 such that $\|Ax^0\| = \|A\|\,\|x^0\|$. Similarly, there is a vector x^1 such that $\|A^{-1}x^1\| = \|A^{-1}\|\,\|x^1\|$. Let b be the image of x^0 under A, and let δb be equal to a constant times x^1. Then inequalities 119 and 120 become equalities, and thus inequality 122 becomes an equality. Thus $\|A\|\,\|A^{-1}\|$ is the least upper bound of the relative error multiplying factor. Called the condition number of A, it is written $cond(A)$. Its value depends on which norm is used, and if an $\mathscr{L}_p$ norm is used, the value of p is used as a subscript to *cond*. Thus $cond_2(A)$ is the condition number of A using the Euclidean norm.

$Cond(A)$ is more generally useful than we have so far indicated, because if b is error free but A has error δA, we can show that

$$\frac{\|\delta x\|}{\|x^*\|} \le cond(A)\frac{\|\delta A\|}{\|A\|}; \tag{123}$$

the condition number is the largest factor by which a relative error in A can be multiplied when propagating into a relative error in x (see exercise 2.4.8). (The bound for $\|\delta x\|/\|x\|$, which is approximately equal to $\|\delta x\|/\|x^*\|$ if $\|\delta x\| \ll \|x^*\|$, is somewhat more complicated. It can be deduced from equation 124 with $\delta b = 0$.)

If both A and b have errors, the relationship between their relative errors and that in δx is somewhat more complicated:

$$\frac{\|\delta x\|}{\|x\|} \leq \frac{cond(A)}{1 - cond(A)\|\delta A\|/\|A\|}\left(\frac{\|\delta b\|}{\|b\|} + \frac{\|\delta A\|}{\|A\|}\right). \tag{124}$$

Let us geometrically interpret equation 122, which gives the largest propagation factor, given A has no error and b has error δb. The largest relative error in x will occur when the absolute error in x is as large as possible and x is as small as possible. By equation 118, δx will be as large as possible when we have as δb a vector in that direction such that the image under A^{-1} is increased in size by the largest factor. This is precisely the direction such that $\|\delta x\| = \|A^{-1}\|\,\|\delta b\|$. To get the smallest $\|x\|$ for a given size $\|b\|$, where $x = A^{-1}b$, we wish that b such that the size of the image under A^{-1} relative to its original size is smallest. This smallest stretch factor of A^{-1} is the reciprocal of the largest stretch factor of A (if x is stretched most into b by A, then b is stretched least into x by A^{-1}). Thus the condition number is the maximum stretch due to A^{-1} divided by the minimum stretch due to A^{-1}: $\|A^{-1}\|/(1/\|A\|) = \|A\|\,\|A^{-1}\|$.

By the geometric interpretation above, it is clear that $cond(A) \geq 1$, and there are matrices (for example, I) that have condition number 1. Well-conditioned matrices have condition numbers between 1 and 10; that is, they propagate relative errors by a factor no larger than 10. Ill-conditioned matrices have condition numbers greater than 100. The most ill-conditioned matrix is the singular matrix, for which $\|A^{-1}\| = \infty$, so $cond(A) = \infty$. Note however that if $cond(A)$ is large, it does not mean relative errors propagate badly for all b and δb, only for some b and δb.

2.4.4 Evaluation of Conditioning

We would like to use a measure of conditioning of a matrix to decide before we try to solve a set of linear equations how difficult obtaining accuracy in that solution is likely to be. Use of the normalized determinant measure requires computing a determinant. Since this determinant is most easily computed by multiplying the diagonal elements of the triangularized matrix in the elimination process, we do not usually obtain its values until after we have done the major part of the solution. In the case of the condition number, it appears that we must compute A^{-1}, a process which is not only time-consuming but also very inaccurate if conditioning is bad (precisely what we are trying to decide). We will learn a more accurate way to compute $cond_2(A)$, but it will still be time-consuming. Thus, unless we are going to compute many solutions using the matrix A and therefore can afford the time to investigate the conditioning of A carefully, the conditioning of a matrix is normally evaluated via the normalized determinant only after a major part of the solution (triangularization) has been done, and the condition number is used more to design methods to solve systems of linear equations than to evaluate the conditioning of particular matrices. For example, one can show that pivoting and scaling tends to improve the condition of a matrix and thus produces less error propagation. Note here that scaling a matrix produces a matrix with a changed condition number but it does not change the normalized determinant; $cond(A)$ is a more direct and more sensitive measure of conditioning than the normalized determinant. On the other hand, if we assume scaling has been done, the two measures of conditioning agree closely in their prediction of error propagation.

How do we recognize bad conditioning during a solution? As we have said, we watch to see whether the normalized determinant is small. Since the normalized determinant is equal to the product over all rows of the pivot in that row divided by the $\mathscr{L}_2$ scale factor for that row, we can watch to see whether that quotient becomes small for any row (the large values will come first because we choose the largest scaled pivots first). If the scaled pivots get small quickly, the matrix is ill-conditioned.

We are interested in the value of the pivots divided by the $\mathscr{L}_2$ scale factor, for each row. If we do $\mathscr{L}_2$ scaling in the first place (actually we would compute a scaling vector made up of the $\mathscr{L}_2$ norm of each row vector of the coefficient matrix), we can use the scale factors for both the scaling and the analysis of the conditioning. Alternatively, we can approximate the above by computing the scaling vector using the $\mathscr{L}_1$ norm and see whether the pivot in a given row divided by the corresponding $\mathscr{L}_1$ scale factor is small. If a pivot divided by an $\mathscr{L}_2$ scale factor is small so will be the pivot divided by the corresponding $\mathscr{L}_1$ scale factor.

Another method of detecting ill-conditioning is to make small changes in the elements of A and b, re-solve the equation, and note the change in x. If the relative change in x is large compared to the relative changes in A and b, the matrix is ill-conditioned.

2.4.5 Effectiveness of Iterative Improvement

We can evaluate the effectiveness of iterative improvement using the condition number. Even if we assume no computational error in solving

$$A\Delta x^{i+1} = (b - Ax^i), \tag{125}$$

we have error in computing $b - Ax^i$. Therefore, we are actually solving

$$A(\Delta x^{i+1} + \delta(\Delta x^{i+1})) = (b - Ax^i) + \delta(b - Ax^i). \tag{126}$$

Thus

$$\frac{\|\delta(\Delta x^{i+1})\|}{\|\Delta x^{i+1}\|} \leq cond(A)\frac{\|\delta(b - Ax^i)\|}{\|b - Ax^i\|}. \tag{127}$$

Since

$$\Delta x^{i+1} = A^{-1}b - x^i = x - x^i \equiv -\delta x^i \tag{128}$$

and

$$\delta(\Delta x^{i+1}) = \delta x^{i+1} \tag{129}$$

(there would be no error in x^{i+1} except for propagated error), we obtain

$$\|\delta x^{i+1}\| \leq cond(A)\frac{\|\delta(b - Ax^i)\|}{\|b - Ax^i\|}\|\delta x^i\|. \tag{130}$$

Thus iterative improvement is helpful as long as

$$cond(A)\frac{\|\delta(b - Ax^i)\|}{\|b - Ax^i\|} < 1, \tag{131}$$

producing $\|\delta x^{i+1}\| < \|\delta x^i\|$. Of course, we do make computational error in solving equation 125, but relation 131 is still a good guideline.

PROBLEMS

2.4.1. Show that $\lim_{p\to\infty}\left(\sum_{i=1}^{n}|x_i|^p\right)^{1/p} = \max_{1\leq i\leq n}|x_i|$.

Thus the $\mathscr{L}_\infty$ or Tchebycheff norm of x is simply the magnitude of the element of x with largest magnitude.

2.4.2. a) Show that $\|x\|_p$ satisfies the rules for a norm for any real number $p \geq 1$.
Hint: Show that the triangle inequality is satisfied for 2-vectors by letting $x = \begin{bmatrix} x_1 \\ x_2 \end{bmatrix} = x_1\begin{bmatrix} 1 \\ x_2/x_1 \end{bmatrix}$, $y = y_1\begin{bmatrix} 1 \\ y_2/y_1 \end{bmatrix}$, $z = x + y = z_1\begin{bmatrix} 1 \\ z_2/z_1 \end{bmatrix}$; defining $\lambda = x_1/z_1$, $t_u = u_2/u_1$ for any u, $D(\lambda) \equiv \|x\|_p + \|y\|_p - \|z\|_p$; and showing $0 \leq \lambda \leq 1$ and $D(\lambda) \geq 0$ for $0 \leq \lambda \leq 1$. Then let x, y be M-vectors and carry out proof by induction on M.

b) Which rule is not satisfied if $0 < p < 1$?

2.4.3. a) Show that the weighted average with positive weights of two norms is a norm.

A weighted average of a set of values x_i, $i = 1,2,\ldots,n$, is defined as $\sum_{i=1}^{n} w_i x_i$ where the sum of the weights $\sum_{i=1}^{n} w_i = 1$.

b) Show that for $x = \begin{bmatrix} x_1 \\ x_2 \end{bmatrix}$, if the norm $\|x\| = (\|x\|_1 + 2\|x\|_\infty)/3$ is used as an estimator for $\|x\|_2$, the mean magnitude of the relative error in the estimator is less than 3.5% if the vectors x are distributed according to a uniform probability distribution in $\theta = tan^{-1}(x_2/x_1)$. Use ε_z/z, not ε_z/z^*, as the definition of relative error.

2.4.4. If δx is an error in a vector x^*, $\|\delta x\|/\|x^*\|$ is a better definition for the relative error in x^* than $\|y\|$ where $y_i = \delta x_i/x_i$. Why? Give an example when the latter is not satisfactory.

2.4.5. Prove that the matrix norm related to $\|x\| = \sum_k |x_k|$ (the $\mathscr{L}_1$ norm) is $\|A\| = \max_j \sum_i |A_{ij}|$. Compare your method of proof to that used above to prove that the matrix norm related to $\|x\| = \max_k |x_k|$ (the $\mathscr{L}_\infty$ norm) is $\|A\| = \max_i \sum_j |A_{ij}|$.

2.4.6. a) Is the following statement always true, sometimes true but sometimes not true, or never true?

If $A^{(0)}, A^{(1)}, A^{(2)}, \ldots$ are a sequence of matrices, then $\lim_{i\to\infty} A^{(i)} = 0$ iff, for all possible norms, $\lim_{i\to\infty} \|A^{(i)}\| = 0$.

b) If it is always true or never true, give a proof. If it is sometimes true but sometimes not true, give a counterexample.

2.4.7.† a) If $\|x\|$ is a vector norm for vectors in L_n and K is any nonsingular $n \times n$ matrix, show that $\|x\|' = \|Kx\|$ is also a norm.
b) Show that $\|A\|' = \|KAK^{-1}\|$ is the related matrix norm.

2.4.8. Show that if b is exact and A is in error by δA in the equation $Ax = b$, then $\|\delta x\|/\|x + \delta x\| \leq cond(A)(\|\delta A\|/\|A\|)$ for any norm.

2.4.9. Consider the system of equations 107 and 108 and the system of equations 111 and 112.

a) For each set, compute both the normalized determinant and $cond_2(A)$. Note the relation between the conditioning of these systems of equations as shown in section 2.4.2 and the condition measures that you have computed.

b) For each set, execute the triangularization of Gaussian elimination with $\mathscr{L}_1$ scaling and pivoting. Note the size of the scaled diagonal element in the second equation after triangularization.

2.4.10. Show that $cond(AB) \leq cond(A)cond(B)$.

---

† J. N. Franklin, *Matrix Theory*, section 6.9, problem 12.

2.5 INNER PRODUCTS

2.5.1 Definitions and Basic Properties

To deepen our understanding of conditioning, we need to have a more careful and general understanding of inner products. Until now we have defined the inner product of two vectors x and y as $x^T y$. Let us generalize this definition so that an inner product more generally captures the notion of the product of the size of one vector and the size of the projection of the other vector onto the first vector, times -1 if the projection of the second vector is in the opposite direction from the first vector (see our geometric interpretation of $x^T y$ in section 2.2.2). We will find it useful for this definition to apply to complex vectors as well as real vectors.

Consider a vector space S. Given a function of two vectors x and y in S, we say that the function, written (x,y), is an *inner product* on S if the following rules hold for all x and y in S.

$$(x,y) = \overline{(y,x)}, \text{ where } \bar{\alpha} \text{ is the complex conjugate of } \alpha \tag{132}$$

[the function is symmetric for real vectors];

$$(\lambda x,y) = \lambda(x,y),\ (x,\lambda y) = \bar{\lambda}(x,y) \text{ for any scalar } \lambda \tag{133}$$

[changing the size of a vector by a real factor changes the function by the same factor];

$$(x + y,z) = (x,z) + (y,z) \tag{134}$$

[the function is distributive];

$$(x,x) > 0 \text{ if } x \neq 0 \tag{135}$$

[the projection of a nonzero vector on itself is the original vector; size$^2 > 0$].

From these rules we can prove that $(x,x) = 0$ if and only if $x = 0$ (see exercise 2.5.1) and that $(x,x)^{1/2}$ is a norm known as the "norm induced by the inner product" (see exercise 2.5.2). We see that for the particular inner product $(x,y) \equiv x^T y$, the norm induced is the Euclidean norm.

We also see from equations 133 and 134 that for real vectors and scalars, the inner product is linear in its arguments, the vectors.

We have said that we want the magnitude of the inner product to carry the notion of the product of the size of one vector and the size of the projection of the other vector onto the first vector. One requirement which the inner product must satisfy if it is to represent this notion is that the magnitude of the inner product of two vectors is less than the product of the sizes of the vectors (defined by the induced norm) unless the two vectors are in the same direction or exactly opposite directions (are linearly dependent). If the vectors are dependent, the inner product should be equal to the product of the vector sizes. This relation, called the *Cauchy-Schwarz inequality*, can be proved from our inner product rules.

Theorem: $|(x,y)|^2 \leq (x,x)(y,y)$, with equality if and only if x and y are linearly dependent.

Proof: Assume x and y are linearly independent. Then $y \neq 0$, which implies $(y,y) \neq 0$. By relation 135,

$$0 < \left(x - \frac{(x,y)}{(y,y)}y,\ x - \frac{(x,y)}{(y,y)}y\right) \tag{136}$$

because

$$x \neq \frac{(x,y)}{(y,y)}y \tag{137}$$

due to the independence of x and y. Using equation 134 to expand 136, we obtain

$$0 < (x,x) - \frac{|(x,y)|^2}{(y,y)}, \tag{138}$$

which implies

$$(x,x)(y,y) > |(x,y)|^2. \tag{139}$$

Conversely, assume x and y are linearly dependent. Then there exists a scalar α such that either $x = \alpha y$ or $y = \alpha x$. If $x = \alpha y$, by equation 133

$$|(x,y)|^2 = |(\alpha y,y)|^2 = |\alpha|^2\,|(y,y)|^2. \tag{140}$$

Also by equation 133

$$(x,x)(y,y) = (\alpha y,\alpha y)(y,y) = |\alpha|^2\,|(y,y)|^2. \tag{141}$$

Thus

$$|(x,y)|^2 = (x,x)(y,y). \tag{142}$$

Similarly if $y = \alpha x$.

2.5.2 Orthogonality

Two vectors x and y are said to be *orthogonal* according to an inner product if the projection of one on the other is zero, that is, if $(x,y) = 0$.† This notion of orthogonality is very important in numerical analysis.

Let us show that for any basis of L_n there exists an inner product by which the basis vectors are orthogonal. Let the basis vectors be b^i,

† If the inner product is not mentioned, $(x,y) = x^T y$ is assumed.

$1 \le i \le n$. Then the following definition satisfies both the rules for an inner product and the orthogonality condition:

$$(x,y) \equiv \sum_{i=1}^{n} \xi_i \eta_i, \tag{143}$$

where

$$x = \sum_{i=1}^{n} \xi_i b^i, \qquad y = \sum_{i=1}^{n} \eta_i b^i \tag{144}$$

(see exercise 2.5.3).

One useful property of a set of orthogonal vectors is that it is linearly independent. This property is proved in the following theorem.

Theorem: Given a vector space in which an inner product is defined, any set of nonzero orthogonal vectors, x^i, $1 \le i \le m$, is linearly independent.

Proof: Assume not. Then there exists a set of α_i, not all of which are equal to 0, such that

$$y = \sum_{i=1}^{m} \alpha_i x^i = 0. \tag{145}$$

Consider (x^k,y). By the linearity of the inner product,

$$(x^k,y) = \sum_{i=1}^{m} \alpha_i (x^k,x^i), \tag{146}$$

so by the orthogonality of the x^i,

$$(x^k,y) = \alpha_k (x^k,x^k). \tag{147}$$

But $y = 0$, so $(x^k,y) = 0$, and since $x^k \neq 0$, $(x^k,x^k) \neq 0$. Thus $\alpha_k = 0$. The above argument is true for any k, contradicting our assumption that not all $\alpha_i = 0$.

A set of vectors is said to be *orthonormal* according to a given inner product if the vectors are orthogonal and each vector in the set has unit length according to the norm induced by the inner product. An orthonormal basis is a basis made up of a set of orthonormal vectors.

We often wish to find an orthonormal basis for a vector space. We now show that, for any vector space with an inner product defined, such an orthonormal basis exists and we can find one. The proof is by a construction called *Gram-Schmidt orthogonalization.* Our approach is to pick a basis $\{b^i | 1 \le i \le n\}$ and to let the first vector in the orthonormal basis be a normalized vector in the direction of b^1; then to obtain the second vector in the orthonormal basis by removing from b^2 any component in the direction of the first vector in the orthonormal basis and normalizing the result;...; then to obtain the kth vector by subtracting from b^k all of the components in the direction of the first $k-1$ members of the orthonormal basis and normalizing the result. That is, we produce the orthonormal set of vectors u^k by the following scheme:

$$v^1 = b^1, \qquad u^1 = v^1/(v^1,v^1)^{1/2}; \tag{148, 149}$$

for $k = 2,3,\ldots,n$,

$$v^k = b^k - \sum_{i=1}^{k-1} (b^k,u^i)u^i, \qquad u^k = v^k/(v^k,v^k)^{1/2}. \tag{150, 151}$$

If we can show that the v^k are nonzero and the u^k are orthogonal, then clearly the u^k are n orthonormal vectors and (by the theorem above) are independent; therefore, they form an orthonormal basis.

To show $v^k \neq 0$, we note that if $v^k = 0$, then

$$b^k = \sum_{i=1}^{k-1} (b^k,u^i)u^i. \tag{152}$$

But the u^i, $1 \leq i \leq k-1$, are linear combinations of the b^i, $1 \leq i \leq k-1$, and thus we have b^k as a linear combination of the b^i, $1 \leq i \leq k-1$, a contradiction to the linear independence of the b^i.

We show that the u^k are orthogonal by induction. First we show that $(u^2,u^1) = 0$.

$$(u^2,u^1) = \left(\frac{b^2 - (b^2,u^1)u^1}{(v^2,v^2)^{1/2}}, u^1\right) = \frac{1}{(v^2,v^2)^{1/2}}[(b^2,u^1) - (b^2,u^1)(u^1,u^1)]$$

$$= \frac{1}{(v^2,v^2)^{1/2}}[(b^2,u^1) - (b^2,u^1)] = 0. \tag{153}$$

Now assume u^i is orthogonal to u^j for all i and $j \leq k-1$ and $i \neq j$. We make the inductive step by showing that for any $j \leq k$, $(u^k,u^j) = 0$.

$$(u^k,u^j) = \left(\frac{b^k - \sum_{i=1}^{k-1} (b^k,u^i)u^i}{(v^k,v^k)^{1/2}}, u^j\right)$$

$$= \frac{1}{(v^k,v^k)^{1/2}}\left[(b^k,u^j) - \sum_{i=1}^{k-1} (b^k,u^i)(u^i,u^j)\right]. \tag{154}$$

But by our inductive hypothesis $(u^i,u^j) = 0$ if $i \neq j$, since both i and j are less than or equal to $k-1$. Thus we have our desired result,

$$(u^k,u^j) = \frac{1}{(v^k,v^k)^{1/2}}[(b^k,u^j) - (b^k,u^j)(u^j,u^j)] = 0. \tag{155}$$

2.5.3 Choice of Norms

We have seen many norms so far, including the $\mathscr{L}_p$ norms and the inner product induced norms. These are the most common norms, though others are sometimes used. How can we decide which norm to use in a given situation?

First, if a particular norm gives a measure of size which is especially significant in the context of the source from which the problem came, we use that norm. It helps us to measure what we want to measure. For example, if we are measuring distance in a vector space where the Pythagorean theorem holds, the $\mathscr{L}_2$ norm measures distance properly but other norms do not. Or, if we are dealing with vectors made up of deviations in a number of variables and are concerned with avoiding large deviations from the average, we need a norm that grows quickly as the largest deviation grows and is not much affected by smaller deviations. Here an $\mathscr{L}_p$ norm for large p, or possibly the $\mathscr{L}_\infty$ norm, which ignores all deviations but the largest, is useful.

Second, we may choose the norm that is most theoretically useful. For example, if we are working in a theoretical framework in which a particular inner product is involved, it is often useful to use the norm induced by that inner product. Since this norm is compatible with the rest of the theoretical construct, it is likely to make the mathematics more straightforward than any other norm.

Third, we may be concerned with computing a norm. In this case, we want to choose a norm which is efficiently computed in the particular situation.

We can choose a norm which is most convenient theoretically or computationally because all norms satisfy our requirements for a reasonable definition of size and because, as we have stated, all norms are approximately the same in the sense that if one norm of a vector is known, any other norm of the vector can be bounded both above and below by constant multiples of the original norm, where the constants do not depend on the vector in question. Thus, it is most often assumed that if some criterion is satisfied according to one norm, it will be approximately satisfied for any norm.

PROBLEMS

2.5.1. Let x and y be vectors and let (x,y) be an inner product. Show $(x,x) = 0$ iff $x = 0$.

2.5.2. With the definitions of exercise 2.5.1, show $(x,x)^{1/2}$ is a norm.

2.5.3. Let $\{b^i\}$ be a basis for L_n. Let $x = \sum_{i=1}^{n} \xi_i b^i$ and $y = \sum_{i=1}^{n} \eta_i b^i$, both real vectors in L_n. Let $f(x,y) = \sum_{i=1}^{n} \xi_i \eta_i$. Show $f(x,y)$ is an inner product on L_n. Then show that b^i and b^j are orthogonal according to this inner product if $i \neq j$.

2.5.4. Let K be any $n \times n$ nonsingular real matrix. Show $(x,y) \equiv (Kx)^T Ky$ is an inner product on L_n.

2.5.5. Consider the linearly independent set of vectors on L_3:

$$x^1 = \begin{bmatrix} 1 \\ 2 \\ 3 \end{bmatrix}, \quad x^2 = \begin{bmatrix} 4 \\ 5 \\ 6 \end{bmatrix}, \quad x^3 = \begin{bmatrix} 1 \\ -2 \\ 0 \end{bmatrix}.$$

Using Gram-Schmidt orthogonalization, find an orthogonal set of vectors in L_3 where the first member of the set is x^1.

2.5.6. Consider the vector space of polynomials with real coefficients and degree $\leq n$. Define an inner product $(p,q) \equiv \int_{-1}^{1} p(x)q(x)dx$. Let $p_i(x) = x^i + \sum_{j=0}^{i-1} a_{ij}x^j$ be an ith-degree polynomial in an orthonormal basis for the vector space according to the above inner product. Using Gram-Schmidt orthogonalization on the polynomials x^i, $i = 0,1,2,\ldots,n$, which are a basis for the vector space, find the first four $(i = 0,1,2,3)$ of these orthonormal polynomials, called the *Legendre polynomials.*

2.6 EIGENTHEORY

2.6.1 Usefulness of Eigenvectors and Eigenvalues

We have seen that the conditioning of a matrix involves the factor by which a matrix stretches the size of a vector. To get a deeper feeling for this stretching, we can resolve the effect of a linear transformation into "basic stretchings." We define a direction of basic stretching for a linear transformation as one in which the vector space is truly stretched in the sense of section 2.2.2. Any vector in a given direction is stretched into a vector in the same direction but of proportionate size. The directions of these basic stretchings are called *eigenvectors,* and the stretch factors in these directions are called *eigenvalues.*† Thus x is an eigenvector of A and λ is an eigenvalue of A if

$$Ax = \lambda x; \tag{156}$$

the image of x is in the same direction as x but stretched by the factor λ.

† In German, *eigen* means "one's own" or "characteristic." That is, these vectors and values are specifically associated with the particular matrix in question.

Eigentheory is helpful in many areas including numerical analysis (and with vector spaces other than L_n) because it allows us to decompose complex transformations or situations into simple ones. In particular, it allows us to uncouple the variables in a problem, as shown in the following examples.

1. Consider the set of linear equations $\sum_{j=1}^{n} A_{ij}x_j = b_i$, $1 \leq i \leq n$. The x_j are coupled in each equation in the sense that every equation involves many of the x_j. If we can find a set of eigenvectors, v^i, which are a basis for L_n, then we can write x as

$$x = \sum_{i=1}^{n} \xi_i v^i \tag{157}$$

and b as

$$b = \sum_{i=1}^{n} \beta_i v^i; \tag{158}$$

the equation $Ax = b$ becomes

$$\sum_{i=1}^{n} \xi_i \lambda_i v^i = \sum_{i=1}^{n} \beta_i v^i. \tag{159}$$

Because the eigenvectors are assumed to be linearly independent, equation 159 implies that for each i,

$$\xi_i \lambda_i = \beta_i. \tag{160}$$

The n equations of this form make up an uncoupled set of equations for the ξ_i.

2. We often come up against linear simultaneous differential equations: $x_i'(t) = \sum_{j=1}^{n} A_{ij}x_j(t)$, $1 \leq i \leq n$. The functions $x_j(t)$ are coupled in each equation. If as in equation 157 we can write the vector x in terms of n linearly independent eigenvectors where the coefficients ξ_i are functions of t, the differential equations resolve into n differential equations of the form $\xi_i'(t) = \lambda_i \xi_i(t)$. Each of these equations is uncoupled and we know each has the solution $\xi_i(t) = c_i e^{\lambda_i t}$, for some c_i.
3. Assume the random variables x_i have the joint probability distribution $p(\underline{x})$. Then in general the variables are correlated, or coupled; the covariance matrix R ($R_{ij} = cov(x_i, x_j)$) has nonzero values for $i \neq j$. Since we can show that R has a set of unit-length orthogonal eigenvectors, v^i, we can write the vector $x - \mu_x$ as $\sum_{i=1}^{n} \zeta_i v^i$ wherein the random variables ζ_i are uncorrelated. To show this, we note that

$$\zeta_i = (x - \mu_x)^T v^i = v^{i^T}(x - \mu_x), \tag{161}$$

and thus

$$E(\zeta_i \zeta_j) = E(v^{i^T}(x - \mu_x)(x - \mu_x)^T v^j) = v^{i^T} E((x - \mu_x)(x - \mu_x)^T) v^j. \quad (162)$$

But

$$E((x - \mu_x)(x - \mu_x)^T) \equiv R, \quad (163)$$

so

$$E(\zeta_i \zeta_j) = \lambda_j v^{i^T} v^j. \quad (164)$$

By the orthogonality of v^i and v^j for $i \neq j$, we see that ζ_i and ζ_j are uncorrelated random variables. The ζ_i are called the *principal factors* in statistics.

Eigenvalues and eigenvectors also appear in the theory of vibrations where vectors (of uncountably infinite dimension, giving positions y at each value of x) represent the positions of a vibrating system at a given time. The eigenvectors (or eigenfunctions) represent the natural vibrations —those vibrations in which the positions, when transformed by forces in the system over some period of time, become as they were at the beginning of the period with the result that the system remains in that mode of vibration.

2.6.2 Theory for Eigenvectors and Eigenvalues

Now we consider the eigentheory. As noted above, we want to find the vectors x and values λ such that

$$Ax = \lambda x. \quad (165)$$

Equivalently, we can write

$$(A - \lambda I)x = 0; \quad (166)$$

x is in the null space of $A - \lambda I$. We know equation 166 has a nonzero solution if and only if

$$det(A - \lambda I) = 0. \quad (167)$$

Equation 167 is called the *characteristic equation* for A. Rewriting that equation,

$$det \begin{bmatrix} A_{11} - \lambda & A_{12} & A_{13} & \dots & A_{1n} \\ A_{21} & A_{22} - \lambda & A_{23} & \dots & A_{2n} \\ A_{31} & A_{32} & A_{33} - \lambda & \dots & A_{3n} \\ \cdot & \cdot & \cdot & & \cdot \\ \cdot & \cdot & \cdot & & \cdot \\ \cdot & \cdot & \cdot & & \cdot \\ A_{n1} & A_{n2} & A_{n3} & \dots & A_{nn} - \lambda \end{bmatrix} = 0; \quad (168)$$

we see the characteristic equation is an nth-degree polynomial equation in λ with $(-1)^n$ as the coefficient of λ^n (see exercise 2.6.1). It has n roots: $\lambda_1, \lambda_2, \ldots, \lambda_n$. We conventionally order these eigenvalues so that $|\lambda_1| \geq |\lambda_2| \geq \cdots \geq |\lambda_n|$. For each distinct λ_i we know there exists a nontrivial solution $x = v^i$ (the null space of $(A - \lambda_i I)$ has dimension > 0). In fact, corresponding to λ_i there are $n - rank(A - \lambda_i I)$ linearly independent eigenvectors, where we can show that $n - rank(A - \lambda_i I) \leq$ the multiplicity of the root $\lambda = \lambda_i$ in $det(A - \lambda I)$.

The following relationships hold if A is real.

1. If λ_i is complex, then $\bar{\lambda}_i$ is also an eigenvalue. This result follows from the fact that complex roots of polynomial equations with real coefficients come in conjugate pairs.
2. If λ_i is complex, then v^i is complex. Assume not. Then $\lambda_i v^i$ would be complex and Av^i would be real—a contradiction.
3. If v^i is a complex eigenvector associated with complex eigenvalue λ_i, then $\bar{v}^i$ is a complex eigenvector associated with the eigenvalue $\bar{\lambda}_i$. This results from taking the conjugate of both sides of the equation $Av^i = \lambda_i v^i$ and noting that the conjugate of a product is the product of the conjugates and the conjugate of the real matrix A is A.
4. If λ_k is real, there exists a real associated eigenvector. Assume not. Then $v^k = u^k + iw^k$, where u^k and w^k are real. Then

$$Av^k = \lambda_k v^k \tag{169}$$

$$\Rightarrow Au^k + iAw^k = \lambda_k u^k + i\lambda_k w^k. \tag{170}$$

   Equating the real and imaginary parts of equation 170, we see that both u^k and w^k are real eigenvectors associated with λ_k, a contradiction.

We see that real eigenvalues correspond to true stretchings of the real vector space. And it can be shown that complex eigenvalues correspond to rotations of the vector space, and, possibly, stretchings.

If v^i is an eigenvector associated with λ_i, then for any constant α, αv^i is also an eigenvector associated with λ_i by the linearity of the transformation specified by A. Thus there are an infinite number of eigenvectors associated with a given eigenvalue. In particular, there is associated with λ_i at least one unit eigenvector ($\|v^i\| = 1$), obtained by dividing any eigenvector by its norm.

If $\lambda_i \neq \lambda_j$, then $v^i \neq v^j$, because Av^i cannot be both $\lambda_i v^j$ and $\lambda_j v^j$. We can make a stronger statement:

Theorem: Let $\{\lambda_{i_k} | k = 1,2,\ldots,m\}$ be a set of distinct eigenvalues of A and let λ_{i_k} have associated eigenvector v^{i_k}. Then the v^{i_k}, $k = 1,2,\ldots,m$ are linearly independent.

Proof: Assume not. Then if we order the eigenvalues appropriately, there exists a smallest j such that $1 \leq j \leq m$ and

$$\sum_{k=1}^{j} \alpha_k v^{i_k} = 0 \tag{171}$$

and for all k, $\alpha_k \neq 0$. Multiply equation 171 by $A - \lambda_{i_q} I$ where $1 \leq q \leq j$. Then we have

$$\sum_{k=1}^{j} \alpha_k (\lambda_{i_k} - \lambda_{i_q}) v^{i_k} = 0. \tag{172}$$

We have assumed distinct eigenvalues, so for each $k \neq q$, $\alpha_k(\lambda_{i_k} - \lambda_{i_q}) \neq 0$, but $\alpha_q(\lambda_{i_q} - \lambda_{i_q}) = 0$. Thus we have a set of $j-1$ dependent eigenvectors with nonzero multipliers, contrary to our assumption that there are no fewer than j. Therefore, the eigenvectors must be linearly independent.

Our previous result is particularly important if all n eigenvalues of A $(n \times n)$ are distinct. In this case, we know that A has n linearly independent eigenvectors, that is, the eigenvectors of A form a basis for our vector space L_n. Thus we can expand any vector in terms of the eigenvectors of A. Many uses of eigenvectors depend on this property.

We will now show that every $n \times n$ *symmetric matrix* ($A = A^T$), whether or not it has distinct eigenvalues, has n linearly independent eigenvectors. The restriction to symmetric matrices is not as artificial as might first appear. In physical situations, the matrix element A_{ij} often stands for the relation between the ith and jth elements of a system, for example, the force between masses connected by springs or the correlation between random variables. This relationship is often symmetric: The force between the ith mass and the jth mass is the same as that between the jth mass and the ith mass; the correlation between the ith and jth random variables is the same as that between the jth and ith.

The first step is to characterize the eigenvalues of a symmetric matrix.

Theorem: The eigenvalues of a real symmetric matrix are real.

Proof: If x is an eigenvector with eigenvalue λ, then

$$Ax = \lambda x. \tag{173}$$

Taking the complex conjugate of both sides of this equation and using the fact that the conjugate of a product is the product of conjugates,

$$\bar{A}\bar{x} = \bar{\lambda}\bar{x}. \tag{174}$$

Since A is real, equation 174 implies

$$A\bar{x} = \bar{\lambda}\bar{x}. \tag{175}$$

Transposing both sides of equation 173,

$$(Ax)^T = \lambda x^T, \tag{176}$$

from which follows

$$(Ax)^T\bar{x} = \lambda x^T\bar{x}. \tag{177}$$

Using equation 175, we obtain

$$(Ax)^T\bar{x} = x^T A^T \bar{x} = x^T A\bar{x} = x^T \bar{\lambda}\bar{x} = \bar{\lambda} x^T \bar{x}. \tag{178}$$

Equations 177 and 178 imply

$$(\lambda - \bar{\lambda})x^T\bar{x} = 0. \tag{179}$$

Since $x \neq 0$,

$$x^T\bar{x} = \sum_{i=1}^{n} |x_i|^2 \neq 0. \tag{180}$$

Therefore, $\lambda = \bar{\lambda}$, that is, λ is real.

Note that the trick of this proof involves treating $(Ax)^T\bar{x}$ in two different ways: (1) taking the transpose first and then applying eigenrelations, and (2) taking the eigenrelations first and the transpose second. By applying the same trick to $(Av^i)^T v^j$ we can prove that the eigenvectors of a real symmetric matrix with distinct eigenvalues are orthogonal (see exercise 2.6.4).

Now let us discuss the orthogonality of the eigenvectors of the general symmetric matrix.

Theorem: Every $n \times n$ symmetric matrix has a set of n orthogonal eigenvectors.

Sketch of Proof: Let A be an $n \times n$ symmetric matrix. Let $A_\varepsilon \equiv A + \varepsilon K$, where K is a symmetric matrix and ε is a scalar. We can choose the matrix K so that there exists a circular neighborhood about $\varepsilon = 0$ such that A_ε has n distinct eigenvalues, $\lambda_i(\varepsilon)$, $1 \leq i \leq n$, if ε is in the neighborhood and $\varepsilon \neq 0$. This statement is reasonable in view of the fact that K has $n(n+1)/2$ independent elements and these elements are to be chosen to satisfy only $n(n-1)/2$ constraints: $\lambda_i(\varepsilon) \neq \lambda_j(\varepsilon)$ for $i \neq j$.

The eigenvectors and eigenvalues of a matrix are a continuous function of the matrix element values, so in the limit as ε goes to zero the eigenvectors and eigenvalues of A_ε are eigenvectors and eigenvalues of A. Since for all nonzero ε in the neighborhood, A_ε is a symmetric matrix with distinct eigenvalues, each such A_ε has n orthogonal eigenvectors. Thus the limiting eigenvectors, which are eigenvectors of A, are orthogonal.

Note that we have shown that A has a set of n orthogonal eigenvectors, not that any set of eigenvectors of A is orthogonal.

What we have said is that any symmetric matrix imposes a linear transformation which is made up of n orthogonal stretchings of L_n. If $\lambda_i < 0$, the stretching is negative; it includes a reflection of the space. If $\lambda_i = 0$, then $Av^i = 0v^i = 0$; v^i is a nonzero element of the null space of A. Thus if $\lambda_i = 0$, A is singular. Similarly, if A is singular, there exists a vector x such that $Ax = 0 \cdot x$ and x is an eigenvector of A, with an eigenvalue 0.

Let us further geometrically interpret the linear transformation imposed by a symmetric matrix. We can show that the unit sphere under the Euclidean norm ($\{x \mid |x| = 1\}$) is transformed by A into a hyperellipsoid with principal axes in the direction of the eigenvectors and with principal radii equal to the magnitude of the eigenvalues (see figure 2-6).

Theorem: Let A be a real symmetric matrix. Then the set of vectors $\{Ax \mid |x| = 1\}$ forms a hyperellipsoid.

Proof: A hyperellipsoid is a surface defined by the set of vectors y to the surface such that for some set of n orthogonal directions (the principal axes), if z_i is the component of any particular y in the ith orthogonal direction,

$$\sum_{i=1}^{n} z_i^2/a_i^2 = 1, \tag{181}$$

where the a_i are independent of the particular y chosen.

Since A is symmetric, its unit eigenvectors $\{v^i\}$ form an orthogonal basis. Consider any x such that $|x| = 1$. Then

$$|x|^2 = x^T x = 1. \tag{182}$$

Expand x along the basis of eigenvectors:

$$x = \sum_{i=1}^{n} \xi_i v^i. \tag{183}$$

Then

$$1 = x^T x = \left(\sum_{i=1}^{n} \xi_i v^i\right)^T \left(\sum_{j=1}^{n} \xi_j v^j\right) = \sum_{i=1}^{n} \sum_{j=1}^{n} \xi_i \xi_j v^{i^T} v^j = \sum_{i=1}^{n} \xi_i^2. \tag{184}$$

Note $$Ax = A \sum_{i=1}^{n} \xi_i v^i = \sum_{i=1}^{n} \xi_i A v^i = \sum_{i=1}^{n} \xi_i \lambda_i v^i. \tag{185}$$

Thus if we resolve Ax along the orthogonal directions defined by the eigenvectors v^i, $1 \leq i \leq n$, the component z_i of Ax in the ith direction is $\xi_i \lambda_i$. But by equation 184

$$\sum_{i=1}^{n} (z_i/\lambda_i)^2 = \sum_{i=1}^{n} (\xi_i \lambda_i/\lambda_i)^2 = \sum_{i=1}^{n} \xi_i^2 = 1. \tag{186}$$

Setting $a_i = \lambda_i$, we have

$$\sum_{i=1}^{n} (z_i/a_i)^2 = 1. \tag{187}$$

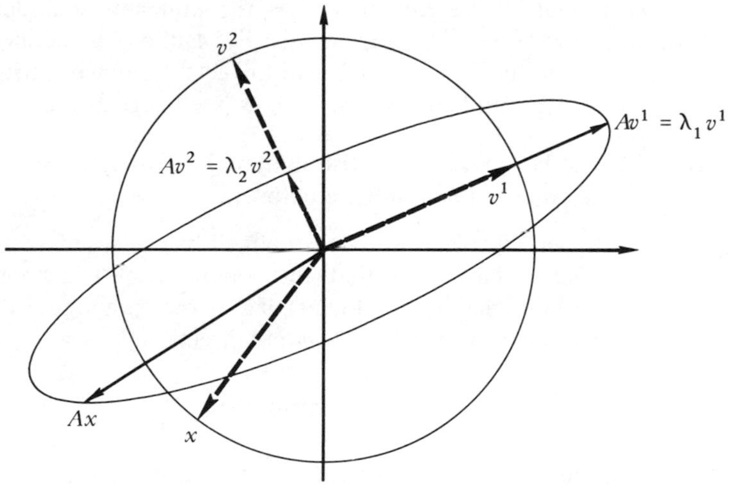

FIG. 2-6 Image of unit sphere under transformation corresponding to a symmetric matrix

We can now interpret the equal eigenvalue situation more clearly. If $\lambda_i = \lambda_j$ for $i \neq j$, the cross section of the hyperellipsoid in the plane of the orthogonal eigenvectors v^i and v^j is a circle. Since any linear combination of m eigenvectors with equal eigenvalues is also an eigenvector with that eigenvalue, any vector on the unit circle in the plane of v^i and v^j is a unit eigenvector with the same eigenvalue as that of v^i and v^j. Thus there exist nonorthogonal sets of eigenvectors associated with the multiple eigenvalue in question, but there also exists an orthogonal set.

If an eigenvalue is zero, the radius in the direction of the associated eigenvector is zero. The hyperellipsoid is flat in that dimension; we say that it has a *degeneracy* in that dimension.

As a result of the above geometric interpretation, we can produce a simple formula for the $\mathscr{L}_2$ condition number of a symmetric matrix.

Theorem: If A is symmetric with eigenvalues λ_i, then

$$cond_2(A) = \max_i |\lambda_i| / \min_i |\lambda_i|. \tag{188}$$

Proof:

$$cond_2(A) = |A|\,|A^{-1}|. \tag{189}$$

$|A|$ = the maximum $\mathscr{L}_2$ norm of an image under A of the unit sphere, and $|A^{-1}|$ = the reciprocal of the minimum $\mathscr{L}_2$ norm of an image under A of the unit sphere. But the image under A of the unit sphere is a hyperellipsoid, and we know that for a hyperellipsoid the most distant point and the least distant point from the center are along principal axes (eigenvectors), that is:

$$|A| = \max_i |\lambda_i v^i| = \max_i |\lambda_i|, \tag{190}$$

and

$$|A^{-1}| = 1/\min_i |\lambda_i|. \tag{191}$$

If A is not symmetric, A still corresponds to a transformation that maps the unit hypersphere into a hyperellipsoid. However, the eigenvectors of A are not all in the directions of the principal axes of A, and they are not all real and orthogonal, so the above results about $cond_2(A)$ do not hold. However, there is an equation for $cond_2(A)$ analogous to equation 188. It depends on the fact that there is a relation between the eigenvalues and eigenvectors of A and those of its transpose.

Let A be a not necessarily symmetric $n \times n$ matrix with n linearly independent eigenvectors. Let V be the $n \times n$ matrix with columns $v^1, v^2, \ldots, v^n$, where v^i is the eigenvector of A associated with eigenvalue λ_i. Then

$$AV = \begin{bmatrix} \uparrow & \uparrow & & \uparrow \\ \lambda_1 v^1 & \lambda_2 v^2 & \ldots & \lambda_n v^n \\ \downarrow & \downarrow & & \downarrow \end{bmatrix} = V\Lambda,\dagger \tag{192}$$

where

$$\Lambda = \begin{bmatrix} \lambda_1 & & & & & \\ & \lambda_2 & & & & \\ & & \cdot & & & \\ & & & \cdot & & \\ & & & & \cdot & \\ & & & & & \lambda_n \end{bmatrix}, \tag{193}$$

with $\Lambda_{ij} = 0$ if $i \neq j$.

Since V has n linearly independent columns, it is nonsingular, so V^{-1} exists. Premultiplying and postmultiplying equation 192 by V^{-1}, we obtain

$$V^{-1}A = \Lambda V^{-1}, \tag{194}$$

and transposing both sides of this equation, using the fact that Λ is diagonal, produces

$$A^T V^{-1^T} = V^{-1^T}\Lambda. \tag{195}$$

Equation 195, by analogy to equation 192, states that the columns of V^{-1^T} are eigenvectors of A^T and that A^T has the same eigenvalues as A. More completely, it states that for every $n \times n$ matrix A with n linearly independent eigenvectors v^i, $1 \leq i \leq n$, there exists a set of eigenvectors u^i, $1 \leq i \leq n$, of A^T such that the following hold:

1. u^i is defined as the ith column of V^{-1^T}, that is, if U is the matrix whose ith column is u^i, $1 \leq i \leq n$, then $U^T = V^{-1}$. (196)
2. The u^i, $1 \leq i \leq n$, are linearly independent, since U^T is nonsingular, and thus so is U.
3. u^i and v^i correspond to the same eigenvalue, $1 \leq i \leq n$.
4. u^i is orthogonal to v^j for $i \neq j$, since

$$u^{i^T} v^j = (U^T V)_{ij} = (V^{-1}V)_{ij} = 0. \tag{197}$$

We say that the u^i, $1 \leq i \leq n$, and the v^i, $1 \leq i \leq n$, are *biorthogonal*.

The preceding discussion has shown that a set of eigenvectors of A^T having the above properties can be computed from those of A. It has not

† The arrows in equation 192 indicate that the vector entries form columns of the matrix.

shown that every set of eigenvectors of A^T has these properties. However, if the eigenvalues of A are distinct, except for normalization only one set of linearly independent eigenvectors of A^T exists, so this set must have the specified properties.

Let us develop some further implications of equation 192 under the assumption that A has n linearly independent eigenvectors, a condition which holds when A is symmetric or has n distinct eigenvalues, and may hold even when neither of these situations pertains. Postmultiplying equation 192 by V^{-1} produces

$$A = V\Lambda V^{-1}, \tag{198}$$

which with equation 196 implies

$$A = V\Lambda U^T. \tag{199}$$

Equation 198 states an important fact: An $n \times n$ matrix which has n linearly independent eigenvectors is completely defined by its eigenvectors and eigenvalues. If two $n \times n$ matrices have the same eigenvalues and the same n linearly independent eigenvectors, they are the same matrix.

Another way to write equation 199 is

$$A = \sum_{i=1}^{n} \lambda_i v^i u^{iT}. \tag{200}$$

Note that $v^i u^{iT}$ is an $n \times n$ matrix, the jkth element of which is $v_j^i u_k^i$.

Equation 199 is an instance of a more general relation:

$$A^m = \sum_{i=1}^{n} \lambda_i^m v^i u^{iT} = V\Lambda^m U^T, \tag{201}$$

which can be proved for $m > 0$ using equation 200 and the biorthogonality of the v^i and the u^i (see exercise 2.6.8). Moreover, if A is nonsingular, inverting both sides of equation 198 produces

$$A^{-1} = V\Lambda^{-1}V^{-1} = V\Lambda^{-1}U^T = \sum_{i=1}^{n} (1/\lambda_i) v^i u^{iT}. \tag{202}$$

We see that A^{-1} has the same eigenvectors as A. Further, the eigenvalues of A^{-1} are the reciprocals of those of A, a relationship that can also be seen by noting that if $Ax = \lambda x$, then

$$A^{-1}Ax = A^{-1}\lambda x, \tag{203}$$

that is,

$$A^{-1}x = (1/\lambda)x. \tag{204}$$

Also,

$$A^0 \equiv I = VU^T = V\Lambda^0 U^T, \tag{205}$$

and thus equation 201 holds for any integer m, positive or negative. That is, if A has eigenvalues λ_i and eigenvectors v^i, $1 \leq i \leq n$, then A^m has eigenvalues λ_i^m and eigenvectors v^i, $1 \leq i \leq n$, for any positive or negative integer m.

If A is symmetric, we know it has n orthogonal eigenvectors, v^i. If these are normalized so that $v^{iT}v^i = 1$, $1 \leq i \leq n$, then

$$V^T V = I, \tag{206}$$

so

$$V^T = V^{-1}. \tag{207}$$

A matrix that satisfies equation 207, that is, a matrix with orthogonal columns of unit Euclidean length, is called an *orthogonal matrix*, although it would be clearer to call it an orthonormal matrix.

From equation 207 we see that if A is symmetric,

$$U = V, \tag{208}$$

and thus

$$A = V\Lambda V^T = \sum_{i=1}^{n} \lambda_i v^i v^{iT}. \tag{209}$$

Lest one assumes that nothing can be said about $n \times n$ matrices which do not have n linearly independent eigenvectors, it should be noted that the above results can be thought of as special cases of more general results, produced by Jordan.† It can be shown that any $n \times n$ matrix A can be written

$$A = DJD^{-1}, \tag{210}$$

where J and D are $n \times n$ matrices as described below. Let $det(A - \lambda I)$ have k distinct roots (eigenvalues of A) λ_i with multiplicities m_i, $1 \leq i \leq k$, respectively. Then

1. J is made up of k square matrices $\Lambda^{(i)}$ on the diagonal and zeros elsewhere (J is a block diagonal matrix):

$$J = \begin{bmatrix} \Lambda^{(1)} & & & & & \\ & \Lambda^{(2)} & & & & \\ & & \cdot & & & \\ & & & \cdot & & \\ & & & & \cdot & \\ & & & & & \Lambda^{(k)} \end{bmatrix}. \tag{211}$$

† J. N. Franklin, *Matrix Theory*, chapter 5.

Furthermore, each submatrix $\Lambda^{(i)}$ is $m_i \times m_i$ and has the form

$$\Lambda^{(i)} = \begin{bmatrix} \lambda_i & & & & & \\ \alpha_{i1} & \lambda_i & & & & \\ & \alpha_{i2} & \lambda_i & & & \\ & & \alpha_{i3} & \cdot & & \\ & & & \cdot & \lambda_i & \\ & & & & \alpha_{i,m_i-1} & \lambda_i \end{bmatrix}, \tag{212}$$

where each α_{ij} has a value of either 0 or 1, and zeros appear in all of the blank locations.

2. D is a nonsingular matrix made up of k matrices $D^{(i)}$, where $D^{(i)}$ is $n \times m_i$:

$$D = \left[D^{(1)} D^{(2)} \ldots D^{(k)} \right], \tag{213}$$

where the last column of $D^{(i)}$ is an eigenvector of A associated with λ_i and the columns of $D^{(i)}$ are a basis for the null space of $(A - \lambda_i I)^{m_i}$.

It can be shown that A maps vectors in the null space of $(A - \lambda_i I)^{m_i}$ into vectors in the same space. That is to say, each eigenvalue has an m_i-dimensional space associated with it such that A maps members of that space into the same space. Furthermore, the space with a basis which is the union of the bases of the k such spaces is L_n.

Since D is nonsingular, its columns, d^i, form a basis for L_n, and any vector $x \in L_n$ can be written $x = \sum_{j=1}^{n} \xi_i d^i$. The d^i are called *principal vectors* of A. In both earlier and later sections of this chapter, vectors are expanded in terms of the eigenvectors of a matrix. In most cases, if the matrix does not have n linearly independent eigenvectors, the expansion should be in terms of the principal vectors of the matrix and the behavior shown to hold in the case of n linearly independent eigenvectors can be shown to hold in the more general case.

When A has n distinct eigenvalues, we have the special case of the above in which $k = n$ and $m_i = 1, 1 \leq i \leq k$. Each d^i becomes an eigenvector v^i and each $\Lambda^{(i)}$ is the 1×1 matrix $[\lambda_i]$, so $J = \Lambda$. When $m_i \neq 1$ for some value(s) of i but A has n independent eigenvectors, all d^i are eigenvectors and all $\alpha_{ij} = 0$, so the $\Lambda^{(i)}$ are diagonal, so $J = \Lambda$.

2.6.3 Similarity Transformations

Equation 198 and the equivalent relationship

$$\Lambda = V^{-1}AV \tag{214}$$

are examples of a more general relationship called *similarity*: A is said to be similar to C if there exists a nonsingular matrix B such that

$$C = B^{-1}AB. \tag{215}$$

Thus we have shown above that if A has n independent eigenvectors, it is similar to a diagonal matrix whose elements are its eigenvalues, and any $n \times n$ matrix is similar to a matrix J of the form given by equations 211 and 212.

To interpret the concept of similarity geometrically, we will show that two matrices are similar if and only if they represent the same linear transformations in a different basis. This characteristic in turn will allow us to say that the eigenvalues of similar matrices are the same and that the eigenvectors are related in a relatively simple way.

Let x be an n-vector. Let x' be the image of x under transformation by A. If we express x in terms of the e^i basis, we have

$$x = \sum_{i=1}^{n} x_i e^i. \tag{216}$$

The same can be done for x'.

Because B is nonsingular, its columns are independent; they form a basis for L_n. We express x and x' in terms of the b^i, the columns of B:

$$x = \sum_{i=1}^{n} y_i b^i; \qquad x' = \sum_{i=1}^{n} y'_i b^i. \tag{217, 218}$$

The vector $y = (y_1, y_2, \ldots, y_n)^T$ is the same vector as the vector x except that the elements of y express x in terms of the b^i basis, whereas the elements of x express the vector x in terms of the e^i basis. Equations 217 and 218 can be written

$$x = By; \qquad x' = By'. \tag{219, 220}$$

Then

$$x' = Ax \Rightarrow By' = ABy, \tag{221}$$

which implies

$$y' = B^{-1}ABy = Cy. \tag{222}$$

That is, if A is a transformation which takes x expressed in terms of the e^i basis into x' expressed in the e^i basis, then $B^{-1}AB$ is the matrix which takes x expressed in the b^i basis into x' expressed in the b^i basis.

Since similar matrices impose the same linear transformation on a different basis, measures of the transformation not directly related to a particular basis are the same for these matrices. Thus for example, similar matrices have the same eigenvalues (vector stretch factors) and the same determinant (space compression factor). That the eigenvalues are the same follows from

$$\begin{aligned} det(B^{-1}AB - \lambda I) &= det(B^{-1}(A - \lambda I)B) \\ &= det(B^{-1})\, det(A - \lambda I)\, det(B) \\ &= det(A - \lambda I)\, det(BB^{-1}) \\ &= det(A - \lambda I), \end{aligned} \tag{223}$$

because the left side of the equation is equal to zero if and only if the right side of the equation is zero. That the determinant is the same is shown by

$$\begin{aligned} det(B^{-1}AB) &= det(B^{-1})\, det(A)\, det(B) \\ &= det(A)\, det(BB^{-1}) \\ &= det(A). \end{aligned} \tag{224}$$

What happens to the eigenvectors under a similarity transformation? They are the same except represented in different bases. In particular, if v_C^i is an eigenvector of C associated with λ_i and v_A^i is the corresponding eigenvector of A associated with λ_i, then

$$v_A^i = Bv_C^i \tag{225}$$

(see exercise 2.6.12).

Of what use are similarity transformations? If we wish to determine the eigenvalues of a matrix, the best technique is often to find a similarity transformation which transforms the matrix into some special form from which the eigenvalues can be computed easily. For example, we saw in equation 198 that a matrix A with n linearly independent eigenvectors is similar to a diagonal matrix Λ, and we know that the eigenvalues of a diagonal matrix are its diagonal elements. Thus the diagonal elements of Λ are the eigenvalues of A, a relation already known from the definition of Λ. This particular similarity transformation is not especially useful, because to find the matrix V used in the similarity transformation, we must find the eigenvectors of the matrix A. Once we have them, we do not need to go through a similarity transformation to find the eigenvalues. However, there are similarity transformations to forms other than diagonal matrices whose eigenvalues are easy to find. Examples of these are triangular matrices and tridiagonal matrices.

We often meet problems where we wish to modify a matrix while holding constant some of its transformational properties such as its eigenvalues. In these cases, similarity transformations are useful. Particularly noteworthy is the *unitary*, or *orthogonal, transformation*:

$$C = U^{-1}AU, \tag{226}$$

where U is a matrix which preserves Euclidean length;

for all x
$$|Ux| = |x|. \tag{227}$$

It can be shown that the only $n \times n$ real matrices which preserve Euclidean length are orthogonal matrices—those matrices with orthonormal columns.† Furthermore, products of such matrices are also orthogonal. If we use the inner product which induces the Euclidean norm $[(x,y) = x^T y]$, these matrices preserve inner products $[(Ux)^T(Uy) = x^T y]$. It can also be shown that every $n \times n$ matrix A can be made triangular by a similarity transformation with some orthogonal matrix.†

2.6.4 Normal and Positive Definite Matrices

We have found and will continue to find it useful for a matrix to have orthogonal eigenvectors. Besides symmetric matrices, what other matrices have this property? It can be shown that a real matrix A has this property (is *normal*) if and only if

$$AA^T = A^T A.\ddagger \tag{228}$$

Equation 228 is satisfied by symmetric matrices, unitary matrices, and antisymmetric matrices (those such that $A_{ij} = -A_{ji}$).

It is sometimes useful for a symmetric matrix to have all eigenvalues greater than zero. Such a matrix is called a *positive definite matrix*.

Theorem: A symmetric matrix A is positive definite if and only if, for all $x \neq 0$,

$$x^T Ax > 0. \tag{229}$$

Proof: A common strategy when proving theorems related to eigenvectors and eigenvalues is to express all vectors in the problems in terms of the eigenvectors of the matrix. If we do so for x, we have

† See Franklin, *Matrix Theory*, chapter 4.

‡ Ibid.

$$x = \sum_{i=1}^{n} \xi_i v^i. \tag{230}$$

Then
$$x^T A x = \sum_{i=1}^{n} \sum_{j=1}^{n} \xi_i \lambda_j \xi_j v^{iT} v^j. \tag{231}$$

Since A is symmetric, its eigenvectors are orthogonal. Thus

$$x^T A x = \sum_{i=1}^{n} \xi_i^2 \lambda_i, \tag{232}$$

where we have assumed that the eigenvectors are of unit Euclidean length. If $x \neq 0$, there exists a $\xi_i \neq 0$, and if A is positive definite, all $\lambda_i > 0$. Thus by equation 232, $x^T A x > 0$. Conversely, if there exists a $\lambda_i < 0$, if we take $x = v^i$, then $x^T A x = \lambda_i < 0$.

An example of a positive definite matrix is the matrix $A^T A$ for any $n \times n$ nonsingular matrix A. We show this as follows:

$$x^T(A^T A)x = (Ax)^T(Ax) \geq 0, \tag{233}$$

and $(Ax)^T(Ax) = 0$ iff $Ax = 0$ iff $x = 0$.

2.6.5 Conditioning in Terms of Eigenvalues

We can now express $cond_2(A)$ in terms of eigenvalues of $A^T A$ for non-singular square matrices A. We note first that by definition

$$cond_2(A) = |A|\,|A^{-1}| = (|A|^2\,|A^{-1}|^2)^{1/2}. \tag{234}$$

Also,

$$|A|^2 = \max_{|x|=1} |Ax|^2 = \max_{|x|=1} [(Ax)^T(Ax)] = \max_{|x|=1} [x^T(A^T A)x]. \tag{235}$$

If λ_i, $1 \leq i \leq n$, are the eigenvalues of $A^T A$ and $x = \sum_{i=1}^{n} \xi_i v^i$, then from equations 232 and 235 we have

$$|A|^2 = \max_{|x|=1} \sum_{i=1}^{n} \xi_i^2 \lambda_i, \tag{236}$$

where we know that since $A^T A$ is symmetric and positive definite, it has n orthogonal eigenvectors and positive eigenvalues. Since the constraint $|x| = 1$ is equivalent to $\sum_{i=1}^{n} \xi_i^2 = 1$, we conclude from equation 236 that

$$|A|^2 = \text{max eigenvalue of } A^T A. \tag{237}$$

$|A|$, the maximum factor by which A can increase the Euclidean norm of a vector, is the square root of the maximum eigenvalue of $A^T A$.

As argued in section 2.4.3,

$$|A^{-1}| = 1 / \min_{|x|=1} |Ax|, \tag{238}$$

so

$$|A^{-1}|^2 = 1 / \min_{|x|=1} |Ax|^2. \tag{239}$$

By an argument similar to that above,

$$\min_{|x|=1} |Ax|^2 = \text{min eigenvalue of } A^T A, \tag{240}$$

so

$$|A^{-1}|^2 = 1/\text{min eigenvalue of } A^T A. \tag{241}$$

Finally, we have

$$\begin{aligned} cond_2(A) &= \left[\frac{\text{max eigenvalue of } A^T A}{\text{min eigenvalue of } A^T A} \right]^{1/2} \\ &= \frac{(\text{max stretch of } A^T A)^{1/2}}{(\text{min stretch of } A^T A)^{1/2}}. \end{aligned} \tag{242}$$

The close relationship between the $\mathscr{L}_2$ norm and eigenvalues and eigenvectors is worth emphasizing. We have seen that the eigenvectors of a symmetric matrix are orthogonal according to the inner product which induces the $\mathscr{L}_2$ norm. Also, the $\mathscr{L}_2$ norm of a symmetric matrix is equal to the magnitude of its dominant eigenvalue; the $\mathscr{L}_2$ norm of an arbitrary matrix A (not even necessarily square) is the square root of the maximum eigenvalue of $A^T A$. We will continue to use this close relationship as we deal with eigenvectors and eigenvalues in linear systems.

PROBLEMS

2.6.1. Show that the coefficient of λ^n in $det(A - \lambda I)$ is $(-1)^n$, where A is an $n \times n$ matrix.

2.6.2. a) Find the eigenvalues and eigenvectors of the matrix $\begin{bmatrix} 1 & 2 \\ 3 & 2 \end{bmatrix}$.

b) Repeat part a for the singular matrix $\begin{bmatrix} 1 & 2 \\ 2 & 4 \end{bmatrix}$.

2.6.3.† Let

$$A = \begin{bmatrix} 0 & 1 & 0 & 0 \\ 0 & 0 & 1 & 0 \\ 0 & 0 & 0 & 1 \\ -\alpha_0 & -\alpha_1 & -\alpha_2 & -\alpha_3 \end{bmatrix}, \quad x = \begin{bmatrix} x_1 \\ x_2 \\ x_3 \\ x_4 \end{bmatrix}.$$

a) Can x be an eigenvector of A if $x_1 = 0$?

b) If λ is an eigenvalue of A and $x_1 = 1$, what is the form of the eigenvector x?

c) What is the characteristic equation for A?

d) Give a necessary and sufficient condition for A to have four linearly independent eigenvectors.

e) Generalize all of the above, replacing 4 by n. (The matrix A is known as the *companion matrix* of the polynomial $\lambda^n + \sum_{i=0}^{n-1} \alpha_i \lambda^i$.)

2.6.4. Show that the eigenvectors of a real symmetric matrix with distinct eigenvalues are orthogonal.

2.6.5. Let A be a nonsymmetric matrix with nondistinct eigenvalues. Apply the argument of section 2.6.2 to produce a sequence of matrices $A_\varepsilon = A + \varepsilon K$ whose eigenvalues are distinct for all suitably small $\varepsilon \neq 0$. The eigenvectors of A_ε are linearly independent for such ε. What happens to these eigenvectors as $\varepsilon \to 0$ if A does not have a full set of independent eigenvectors?

2.6.6. Let A be an $n \times n$ symmetric matrix with eigenvectors v^i.

a) Show there exists a solution to $Ax = b$ if and only if $b^T v^i = 0$ for all i such that $\lambda_i = 0$.

b) Using the result of part a, show that if $det(A - \lambda I) = \lambda^k p_{n-k}(\lambda)$ where $p_{n-k}(\lambda)$ is a polynomial in λ of degree $n-k$ such that $p_{n-k}(0) \neq 0$, then $rank(A) = n-k$.

2.6.7. By showing that A and A^T satisfy the same characteristic equation, give an alternative proof to that given in the text of the fact that A and A^T have the same eigenvalues.

2.6.8. Let A be an $n \times n$ matrix with distinct eigenvalues λ_i, $1 \leq i \leq n$, and associated eigenvectors v^i. Let A^T have eigenvectors u^i associated with eigenvalues λ_i. If $A^m = A \cdot A \cdot \ldots \cdot A$, m times, show that $A^m = \sum_{i=1}^{n} \lambda_i^m v^i u^{iT}$.

2.6.9. a) Show that similarity is a transitive relation: If A is similar to B and B is similar to C, then A is similar to C.

b) Show that similarity is a symmetric relation: If A is similar to B, then B is similar to A.

c) Show A is similar to A.

---

† J. N. Franklin, *Matrix Theory*, section 4.1, problems 3 and 4.

From parts a, b, and c, we can conclude that similarity is an equivalence relation on the set of all $n \times n$ matrices for a given n.

2.6.10. Show that if A is similar to B, then A^k is similar to B^k for all integers $k \geq 0$, and for all negative integers k if A is nonsingular.

2.6.11. Show that if A is a nonsingular matrix, A^TA is similar to AA^T.

2.6.12. Let $C = B^{-1}AB$ and let $Cv^i = \lambda_i v^i$. Show that $A(Bv^i) = \lambda_i(Bv^i)$, that is, Bv^i is an eigenvector of A associated with eigenvalue λ_i.

2.6.13. Show that if A is a unitary matrix, all of its eigenvalues have magnitude 1.

2.6.14. Let A be an $n \times n$ matrix. Say whether each of the following statements is always true, sometimes true but sometimes not true, or never true. If the statement is always true or never true, give a proof. If it is sometimes true but sometimes not true, give a counterexample.

a) $cond(A^{-1}) = cond(A)$.

b) $cond_1(A^T) = cond_1(A)$.

c) $cond_2(A^T) = cond_2(A)$.

d) A^{-1} has the same number of independent eigenvectors as A.

e) A^T has the same number of independent eigenvectors as A.

f) If A has orthogonal columns, $cond_2(A) = 1$ iff the columns have equal size according to the $\mathcal{L}_2$ norm.

2.6.15. Use equation 242 to find $cond_2(A)$, where A is the coefficient matrix of equations 107 and 108; of 111 and 112.

2.7 NUMERICAL METHODS TO FIND EIGENVECTORS AND EIGENVALUES

2.7.1 Computing the Eigenvector Corresponding to a Given Eigenvalue

Sometimes we know an eigenvalue of a matrix but we do not know the associated eigenvector, and we would like to. Assume we know that the multiplicity of the known eigenvalue, λ_i, is 1 ($\lambda_j \neq \lambda_i$ if $j \neq i$). Our task is to find a vector x such that $Ax = \lambda_i x$, or, equivalently,

$$(A - \lambda_i I)x = 0. \tag{243}$$

We know that $A - \lambda_i I$ has rank $n-1$ and that x is unique up to a multiplicative scalar. This means we can set any nonzero element of x equal to 1 and have the remaining elements of x fully determined. If we wish to solve equation 243 by Gaussian elimination, our problem is two-fold:

1. We want to triangularize the matrix $A - \lambda_i I$ to a form that is as well-conditioned as possible and has a zero row at the bottom.
2. We need to find a nonzero element of x so that we can set it equal to 1.

To solve the first part of this problem, we wish to use pivots that are as large as possible. We must avoid the possibility of searching for a pivot in a column that has only zeros from the diagonal down except for error, but we in fact cannot recognize this situation. We can avoid the possibility and produce the large pivots we seek by doing row scaling followed by full pivoting. Since we know that the matrix will end up with exactly one zero row, we are assured that triangularization with full pivoting will produce a matrix of the form

$$A' = \begin{bmatrix} A'_{11} & A'_{12} & A'_{13} & \cdots & A'_{1,n-1} & A'_{1n} \\ & A'_{22} & A'_{23} & \cdots & A'_{2,n-1} & A'_{2n} \\ & & A'_{33} & \cdots & A'_{3,n-1} & A'_{3n} \\ & & & \ddots & \vdots & \vdots \\ & & & & A'_{n-1,n-1} & A'_{n-1,n} \\ 0 & 0 & 0 & \cdots & 0 & 0 \end{bmatrix} \tag{244}$$

with no zero pivots (diagonal elements) in the first $n-1$ rows (there are zeros in all locations below the diagonal). The row multiplication and subtraction involved in the full pivoting procedure has no effect on the right-hand side of equation 243 because it is zero, so the multipliers need not be saved. The column switching involved in full pivoting permutes the elements of the solution vector x, so we must keep track of this switching to allow reordering of these elements. When using the computer, we should handle row and column switching with two permutation vectors. We should refer to elements of x via a column permutation vector and to elements of the scaling vector via a row permutation vector (see program 2-3).

We can now show that our triangularization with full pivoting finds an element of x which is nonzero: The element of x which ends up in the nth position after the permutations of the x elements due to full pivoting is precisely the element we seek. If not, then the modified equation is of the form

$$A'' \begin{bmatrix} x_{i_1} \\ x_{i_2} \\ \cdot \\ \cdot \\ \cdot \\ x_{i_{n-1}} \end{bmatrix} = 0, \tag{245}$$

where A'' is the upper left $n-1 \times n-1$ submatrix of A'. A'' is nonsingular, so $A''x' = 0$ iff $x' = 0$. Therefore, if $x_n = 0$, then $x = 0$. But this result cannot be true because x is given to be a (nonzero) eigenvector. Since the last element in the transformed x vector cannot be zero, we can set it equal to 1. The rest of the solution becomes a back substitution on

$$A''x'' = -a'^n, \tag{246}$$

where x'' consists of the first $n-1$ elements of the transformed x vector, and a'^n consists of the first $n-1$ elements of the nth column of A'.

Finally, note that if we do not know that the eigenvalue in question is of multiplicity 1, we are not sure that the $(n-1)$th row in A' is nonzero; if we find it to be zero, we are not sure that the $(n-2)$th row is nonzero; and so on. We have a basis for determining the number of rows which are zero except for error. Namely, we know that the last row must be zero, but if we compute A'_{nn}, it will turn out to be only approximately zero. Then we can use A'_{nn} as an estimate of the error being made and check whether the value of $A'_{n-1,n-1}$ is of the same magnitude. If so, we can assume that it is truly a zero and go on to the next-previous row, comparing the pivot in that row with the pivot in the present row. We continue this procedure until we find a pivot element that is not small enough to be considered a zero. We have achieved a large advantage by using the computation itself to give us an error estimate.

Now let us consider an example. Assume we have the following matrix:

$$A = \begin{bmatrix} -1 & 2 & -3 \\ 4 & 3 & 6 \\ 6 & 9 & -2 \end{bmatrix}.$$

Further assume that we know this matrix has an eigenvalue $\lambda = -2$. In solving this problem, we will carry out the calculations in parallel in two ways: (1) to full precision (following "$=$") and (2) rounding all results to one significant decimal digit (following "$\approx$"). We will explicitly do row and column permutations.

First we compute $B = A - \lambda I$ using the known eigenvalue -2:

$$B = A + 2I = \begin{bmatrix} 1 & 2 & -3 \\ 4 & 5 & 6 \\ 6 & 9 & 0 \end{bmatrix}.$$

Then we compute a scaling vector s comprising a scale factor s_i for each row of B, which we find by taking the $\mathscr{L}_1$ norm of the row. Since we will physically permute rows and columns (unlike in program 2-3), we initialize only a single permutation vector q, in which q_j indicates the original column number of the present jth column:

$$s = \begin{bmatrix} 6 \\ 15 \\ 15 \end{bmatrix}, \quad q = \begin{bmatrix} 1 \\ 2 \\ 3 \end{bmatrix}.$$

Now we search for the element of B such that the element divided by the scale factor for its row is largest in magnitude. In this case the 9 in position B_{32} is the element desired. We switch rows and columns to put this element in the pivot position, being careful to switch the permutation and scaling vectors accordingly. Specifically, we switch rows 1 and 3 of the matrix and scaling vector, and columns 1 and 2 of the matrix and elements 1 and 2 of the permutation vector, producing

$$B' = \begin{bmatrix} 9 & 6 & 0 \\ 5 & 4 & 6 \\ 2 & 1 & -3 \end{bmatrix}, \quad s = \begin{bmatrix} 15 \\ 15 \\ 6 \end{bmatrix}, \quad q = \begin{bmatrix} 2 \\ 1 \\ 3 \end{bmatrix}.$$

We then eliminate the first column to produce

$$B'' = \begin{bmatrix} 9 & 6 & 0 \\ 0 & \frac{2}{3} & 6 \\ 0 & -\frac{1}{3} & -3 \end{bmatrix} \approx \begin{bmatrix} 9 & 6 & 0 \\ 0 & .7 & 6 \\ 0 & -.3 & -3 \end{bmatrix}.$$

We now search for the scaled element that is largest in magnitude in the last two rows; it is -3. We switch rows 2 and 3 and thus elements 2 and 3 of s, and columns 2 and 3 and thus elements 2 and 3 of q, producing

$$B''' = \begin{bmatrix} 9 & 0 & 6 \\ 0 & -3 & -\frac{1}{3} \\ 0 & 6 & \frac{2}{3} \end{bmatrix} \approx \begin{bmatrix} 9 & 0 & 6 \\ 0 & -3 & -.3 \\ 0 & 6 & .7 \end{bmatrix}, \quad s = \begin{bmatrix} 15 \\ 6 \\ 15 \end{bmatrix}, \quad q = \begin{bmatrix} 2 \\ 3 \\ 1 \end{bmatrix}.$$

Eliminating the second column, we obtain

$$A' = \begin{bmatrix} 9 & 0 & 6 \\ 0 & -3 & -\frac{1}{3} \\ 0 & 0 & 0 \end{bmatrix} \approx \begin{bmatrix} 9 & 0 & 6 \\ 0 & -3 & -.3 \\ 0 & 0 & .1 \end{bmatrix}.$$

```
/* FIND EIGENVECTOR OF MATRIX A, GIVEN EIGENVALUE LAMDA OF MULTIPLICITY ONE */

   DECLARE (A(N,N), LAMDA) FLOAT,         /* MATRIX, EIGENVALUE */
           X(N) FLOAT,                    /* EIGENVECTOR */
           (SCALE(N), SCALEVAL) FLOAT,    /* SCALE VECTOR, SCALED PIVOT CANDIDATE */
           (PROW(N), PCOL(N)) FIXED,      /* ROW & COLUMN PERMUTATION VECTORS */
           (AMULT, PIVOT) FLOAT,          /* ROW MULTIPLIER, PIVOT */
           TEMP FIXED,                    /* TEMPORARY ROW AND COLUMN NUMBERS */
           (ROWPIV, COLPIV) FIXED;        /* ROW AND COLUMN OF PIVOT */

/* SUBTRACT LAMDA FROM DIAGONAL OF A */
/* AND SET UP PERMUTATION VECTORS AND SCALE VECTOR */
   INIT: DO I = 1 TO N;
         A(I,I) = A(I,I) - LAMDA; /* SUBTRACT LAMDA FROM DIAGONAL */
         PROW(I) = I;             /* INITIALIZE PERMUTATION VECTORS */
         PCOL(I) = I;
         SCALE(I) = Ø;            /* COMPUTE SCALE FACTORS */
         SCALECOMP: DO J = 1 TO N;
                    SCALE(I) = SCALE(I) + ABS(A(I,J));
         END SCALECOMP;
   END INIT;
/* TRIANGULARIZE A USING FULL PIVOTING */
   TRIANG: DO I = 1 TO N - 1;
           PIVOT = -1;
           FINDPIVOT: DO J = I TO N; /* FIND MAXIMUM CANDIDATE AFTER SCALING */
                KLOOP: DO K = I TO N;
                       SCALEVAL = ABS(A(PROW(J),PCOL(K))) / SCALE(PROW(J));
                       IF SCALEVAL > PIVOT THEN NEWMAX: DO;
                                                        PIVOT = SCALEVAL;
                                                        ROWPIV = J;
                                                        COLPIV = K;
                                                    END NEWMAX;
                END KLOOP;
           END FINDPIVOT;
```

```
            TEMP = PROW(I);          /* SWITCH ROWS I AND ROWPIV */
            PROW(I) = PROW(ROWPIV);
            PROW(ROWPIV) = TEMP;
            TEMP = PCOL(I);          /* SWITCH COLUMNS I AND COLPIV */
            PCOL(I) = PCOL(COLPIV);
            PCOL(COLPIV) = TEMP;

            PIVOT = A(PROW(I),PCOL(I));
            COLZERO: DO J = I+1 TO N; /* ZERO ITH COLUMN BELOW DIAGONAL */
                     AMULT = A(PROW(J),PCOL(I)) / PIVOT;
                     ROWCOMP: DO K = I+1 TO N; /* COMPUTE NONZERO ROW ELEMENTS */
                              A(PROW(J),PCOL(K)) = A(PROW(J),PCOL(K))
                                                 - AMULT * A(PROW(I),PCOL(K));
                     END ROWCOMP;
            END COLZERO;
    END TRIANG;

/* SET NORMALIZING ELEMENT OF EIGENVECTOR */
    X(PCOL(N)) = 1;

/* BACK SUBSTITUTION */
    SUBST: DO I = N-1 TO 1 BY -1;
           X(PCOL(I)) = -A(PROW(I),PCOL(N));
           SUBTRACT: DO J = I+1 TO N-1; /* SUBTRACT TERMS IN ALREADY COMPUTED X(J) */
                     X(PCOL(I)) = X(PCOL(I)) - A(PROW(I),PCOL(J)) * X(PCOL(J));
                                        /* NOT DONE IF J = N-1 */
           END SUBTRACT;
           X(PCOL(I)) = X(PCOL(I)) / A(PROW(I),PCOL(I));
    END SUBST;
```

PROGRAM 2-3 Compute eigenvector, given eigenvalue

Having finished the triangularization, we check whether A'_{33} (which we know must truly be zero) is of the same magnitude as A'_{22}. It is not, so we conclude that the eigenvector is unique up to a multiplicative factor. We fix that factor by setting to 1 the last element of the permuted eigenvector x'. This produces the equation

$$\begin{bmatrix} 9 & 0 \\ 0 & -3 \end{bmatrix} x'' = \begin{bmatrix} -6 \\ \frac{1}{3} \end{bmatrix} \approx \begin{bmatrix} -6 \\ .3 \end{bmatrix}.$$

Solving this equation by back substitution, we obtain

$$x'' = \begin{bmatrix} -\frac{2}{3} \\ -\frac{1}{9} \end{bmatrix} \approx \begin{bmatrix} -.7 \\ -.1 \end{bmatrix}, \text{ so } x' = \begin{bmatrix} -\frac{2}{3} \\ -\frac{1}{9} \\ 1 \end{bmatrix} \approx \begin{bmatrix} -.7 \\ -.1 \\ 1 \end{bmatrix}.$$

Permuting x' according to q, we find the eigenvector x as follows: $x_2 = -\frac{2}{3} \approx -.7$, $x_3 = -\frac{1}{9} \approx -.1$, $x_1 = 1$; that is,

$$x = \begin{bmatrix} 1 \\ -\frac{2}{3} \\ -\frac{1}{9} \end{bmatrix} \approx \begin{bmatrix} 1 \\ -.7 \\ -.1 \end{bmatrix}.$$

2.7.2 Power Method for Computing Eigenvectors and Eigenvalues

We now move on to a method for finding both eigenvalues and eigenvectors. It exemplifies concepts involved in many of the numerical methods we will see.

1. It is an iterative method, one in which we make a guess at the solution and execute a procedure which improves our guess.
2. The iteration procedure operates by producing the domination by the member of a set which is largest in magnitude.
3. Techniques of accelerating the convergence of the iteration are applicable.
4. Similarity transformations are used to produce matrices of a desired form while holding the eigenvalues constant.

Iteration is a useful technique because it is often possible to use considerably fewer arithmetic operations in an iterative step than would be required in a full direct solution. The advantage is not that the total number of operations per solution is necessarily less but rather that intuitively the error generated in the computation is propagated only during a single iterative step. At each step, we start afresh with a new "error-free" guess. As long as the error generated in an iterative step is less than the amount of improvement due to convergence of the iterative procedure (the decrease in the propagated error), the step produces an overall improvement (see exercise 2.7.2). And we can make this generated error small, because the number of computational steps in the iteration can be small.

We produce an iterative method to find an eigenvector and eigenvalue of a matrix by the common approach of arranging the computation so that the largest element in magnitude of a set dominates more and more. Let A be a matrix with n linearly independent eigenvectors v^i. Then any vector x can be written

$$x = \sum_{i=1}^{n} \xi_i v^i. \tag{247}$$

Then

$$A^m x = \sum_{i=1}^{n} \lambda_i^m \xi_i v^i. \tag{248}$$

Assume $|\lambda_1| > |\lambda_2|$. Then as m gets large, the direction of $A^m x$ converges to that of v^1 because

$$A^m x = \lambda_1^m \left(\xi_1 v^1 + \sum_{i=2}^{n} (\lambda_i/\lambda_1)^m \xi_i v^i \right) \tag{249}$$

and $|\lambda_i/\lambda_1| < 1$ for $i > 1$, so the sum becomes quite small compared to the first term in the parentheses.†

As a first step in the iteration procedure for finding an eigenvector and its eigenvalue, we take a guess x^0 at the eigenvector (the guess need not be a good one). We compute $x^1 = Ax^0$; $x^2 = Ax^1$;...; $x^{m+1} = Ax^m$. Since

$$x^m = A^m x^0, \tag{250}$$

x^m will converge in direction to that of v^1.

How can we decide when the convergence is close enough for our needs, that is, when Ax^m is approximately equal to a constant times x^m? We define "approximately equal" to mean that the size (norm) of the difference between $Ax^m \equiv x^{m+1}$ and a constant times x^m is appropriately small. The constant we should use is the one that makes this size as small as possible. This constant will be an estimate of the eigenvalue. That is, we wish to choose, as our estimate $\hat{\lambda}$ of the eigenvalue, that value which makes $\|Ax^m - \hat{\lambda}x^m\|$ as small as possible. If in particular we use the $\mathscr{L}_2$ norm,

† If A does not have n linearly independent eigenvectors, the iterative method to be developed still works. The proof of this fact is somewhat complicated. It depends on expanding x in terms of the principal vectors defined in section 2.6.2 instead of in terms of the eigenvectors of A (see exercise 2.7.3).

minimizing the above expression is equivalent to minimizing $(Ax^m - \hat{\lambda}x^m)^T(Ax^m - \hat{\lambda}x^m)$. Differentiating with respect to $\hat{\lambda}$ and setting the result to zero, we obtain

$$x^{m^T}(Ax^m - \hat{\lambda}x^m) = 0, \tag{251}$$

which implies

$$\hat{\lambda} = \frac{x^{m^T}Ax^m}{x^{m^T}x^m} = \frac{x^{m^T}x^{m+1}}{x^{m^T}x^m}. \tag{252}$$

The second derivative of the form we are minimizing is $2x^{m^T}x^m > 0$, and thus our value of $\hat{\lambda}$ gives a minimum for that form.

What we have said is that we will stop our iteration when the value of $\left|x^{m+1} - \frac{x^{m^T}x^{m+1}}{x^{m^T}x^m}x^m\right|$ is appropriately small, that is, when

$$\left(x^{m+1} - \frac{x^{m^T}x^{m+1}}{x^{m^T}x^m}x^m\right)^T\left(x^{m+1} - \frac{x^{m^T}x^{m+1}}{x^{m^T}x^m}x^m\right)$$

is appropriately small. Expanding, we wish

$$\frac{(x^{m^T}x^m)(x^{m+1^T}x^{m+1}) - (x^{m^T}x^{m+1})^2}{x^{m^T}x^m}$$

appropriately small. What is "appropriately small"? One definition is: "when the eigenvalue estimate is different from the value to which it is converging by less than some given relative error ε." In the following theorem, we show that if A is normal and the relative difference between Ax^m and $\hat{\lambda}x^m$ is less than ε (using the $\mathscr{L}_2$ norm), then $\hat{\lambda}$ is in error from an eigenvalue of A with relative error less than ε.

Theorem: If A is a normal matrix and $|Ax - \hat{\lambda}x|/|Ax| < \varepsilon$, then there exists an eigenvalue λ_i of A such that $|\hat{\lambda} - \lambda_i|/|\lambda_i| < \varepsilon$.

Proof: Write x in terms of the orthonormal eigenvectors, v^i, of A:

$$x = \sum_{i=1}^{n} \xi_i v^i. \tag{253}$$

Then

$$\frac{|Ax - \hat{\lambda}x|^2}{|Ax|^2} = \frac{\sum_{i=1}^{n}(\lambda_i - \hat{\lambda})\xi_i v^{iT} \sum_{j=1}^{n}(\lambda_j - \hat{\lambda})\xi_j v^{jT}}{\sum_{i=1}^{n}\lambda_i\xi_i v^{iT}\sum_{j=1}^{n}\lambda_j\xi_j v^{jT}}$$

$$= \left(\sum_{i=1}^{n}|\lambda_i - \hat{\lambda}|^2\xi_i^2\right)\Big/\left(\sum_{i=1}^{n}|\lambda_i|^2\xi_i^2\right) \tag{254}$$

by the orthonormality of the v^i.

Assume that for all $1 \le i \le n$,

$$|\hat{\lambda} - \lambda_i| \ge \varepsilon |\lambda_i|. \tag{255}$$

Then
$$\sum_{i=1}^{n} |\lambda_i - \hat{\lambda}|^2 \xi_i^2 \ge \varepsilon^2 \sum_{i=1}^{n} |\lambda_i|^2 \xi_i^2, \tag{256}$$

so
$$|Ax - \hat{\lambda} x|^2 / |Ax|^2 \ge \varepsilon^2, \tag{257}$$

a contradiction. Thus for some $1 \le i \le n$,

$$|\hat{\lambda} - \lambda_i| / |\lambda_i| < \varepsilon. \tag{258}$$

If A is not normal but has n linearly independent eigenvectors, we can use the norm induced by the inner product which makes these eigenvectors orthogonal (see exercise 2.5.3) instead of the $\mathscr{L}_2$ norm. With that change, the above proof goes through as before. When we are solving our problem, we do not know what this norm is, because we do not know what the eigenvectors are, but we can use the Euclidean norm together with the argument that if the criterion holds for the Euclidean norm, it approximately holds for our unknown norm.

In our discussion of the power method, we have handled only the case in which $|\lambda_1| > |\lambda_2|$. In this case λ_1 is real, because if it were not, its complex conjugate would also be an eigenvalue and thus we would have $|\lambda_1| = |\lambda_2|$. Let us analyze the iteration for the case $|\lambda_1| = |\lambda_2| > |\lambda_3|$. This situation can occur if $\lambda_1 = \lambda_2$, $\lambda_1 = -\lambda_2$, or $\lambda_1 = \bar{\lambda}_2$. In each case the analysis for the case $|\lambda_1| > |\lambda_2|$ goes through except that x^m approaches the vector $\lambda_1^m \xi_1 v^1 + \lambda_2^m \xi_2 v^2$.

If $\lambda_1 = \lambda_2$, then x^m approaches

$$w = \lambda_1^m (\xi_1 v^1 + \xi_2 v^2), \tag{259}$$

and $\xi_1 v^1 + \xi_2 v^2$ is an eigenvector corresponding to λ_1. That is, x^m approaches an eigenvector just as in the case where $\lambda_1 \ne \lambda_2$, and all techniques applicable in that case are applicable here.

If $\lambda_1 = -\lambda_2$, $x^m \to \lambda_1^m (\xi_1 v^1 + (-1)^m \xi_2 v^2)$. In this case x^m does not approach a fixed direction but rather oscillates between the direction of $\xi_1 v^1 + \xi_2 v^2$ and that of $\xi_1 v^1 - \xi_2 v^2$. Equivalently we can say x^{2m} approaches a fixed direction, and if we can detect this behavior, we can find the eigenvalues and eigenvectors in question. However, we will see that we can deal with this situation more easily using precisely the same analysis that we use when $\lambda_1 = \bar{\lambda}_2$.

If λ_1 is complex and thus $\lambda_1 = \bar{\lambda}_2$, then $\lambda_1 = \rho e^{i\theta}$ and $\lambda_2 = \rho e^{-i\theta}$. We know the eigenvector associated with λ_2 is $\bar{v}^1$ and thus $x^m \to \rho^m (e^{im\theta} \xi_1 v^1 + e^{-im\theta} \xi_2 \bar{v}^1)$. Let

$$v^1 = t^1 + i s^1, \tag{260}$$

where t^1 and s^1 are real vectors. As m gets large, x^m approaches a linear combination of t^1 and s^1. That is, after enough iterations such that the terms associated with eigenvalues $\lambda_3, \lambda_4, \ldots, \lambda_n$ are negligible, x^{m+1}, x^m, and x^{m-1} are each a linear combination of t^1 and s^1 and thus must be linearly dependent. Therefore, there exist numbers γ, α, and β, not all equal to zero, such that

$$\gamma x^{m+1} + \alpha x^m + \beta x^{m-1} \approx 0 \text{ compared to } x^{m+1}. \tag{261}$$

If $\gamma = 0$, then x^m is approximately a scalar times x^{m-1}, that is, x^{m-1} is approximately an eigenvector, which is only true if λ_1 is real. Since we are assuming λ_1 is complex, $\gamma \neq 0$, and we can divide equation 261 by γ, producing an equation of the form

$$x^{m+1} + \alpha x^m + \beta x^{m-1} \approx 0. \tag{262}$$

We note here that for the case $\lambda_1 = -\lambda_2$, as m gets large, x^{m+1} is approximately a multiple of x^{m-1}:

$$x^{m+1} - \lambda_1^2 x^{m-1} \approx 0. \tag{263}$$

In other words, we have equation 262 where $\alpha = 0$ and $\beta = -\lambda_1^2 = \lambda_1\lambda_2$.

Returning to the general case where $|\lambda_1| = |\lambda_2|$ but $\lambda_1 \neq \lambda_2$, let

$$x^{m-1} \approx \xi_1' v^1 + \xi_2' v^2. \tag{264}$$

We know $\xi_1' \neq 0$ and $\xi_2' \neq 0$.

Then
$$x^m \approx \lambda_1 \xi_1' v^1 + \lambda_2 \xi_2' v^2, \tag{265}$$

and
$$x^{m+1} \approx \lambda_1^2 \xi_1' v^1 + \lambda_2^2 \xi_2' v^2. \tag{266}$$

Using these equations in equation 262, we obtain

$$(\lambda_1^2 + \alpha\lambda_1 + \beta)\xi_1' v^1 + (\lambda_2^2 + \alpha\lambda_2 + \beta)\xi_2' v^2 \approx 0. \tag{267}$$

Since $\lambda_1 \neq \lambda_2$, v^1 and v^2 are linearly independent; so equation 267 implies

$$\lambda_1^2 + \alpha\lambda_1 + \beta \approx 0 \tag{268}$$

and
$$\lambda_2^2 + \alpha\lambda_2 + \beta \approx 0, \tag{269}$$

which determine α and β in the limit independent of the iterative step number. To restate the above, we have said λ_1 and λ_2 are the roots of

$$\lambda^2 + \alpha\lambda + \beta = 0. \tag{270}$$

Since
$$\lambda^2 + \alpha\lambda + \beta = (\lambda - \lambda_1)(\lambda - \lambda_2), \tag{271}$$

$$\alpha \rightarrow -(\lambda_1 + \lambda_2), \tag{272}$$

and
$$\beta \rightarrow \lambda_1\lambda_2. \tag{273}$$

Our problem is similar to that of finding the eigenvalue λ in the real case: We wish to find an $\hat{\alpha}$ and $\hat{\beta}$ which minimize $\|x^{m+1} + \hat{\alpha}x^m + \hat{\beta}x^{m-1}\|$. Using the $\mathscr{L}_2$ norm and setting the partial derivatives with respect to $\hat{\alpha}$ and $\hat{\beta}$ of the square of that norm equal to zero, we obtain

$$\begin{bmatrix} \hat{\alpha} \\ \hat{\beta} \end{bmatrix} = \frac{1}{(x^{m-1},x^m)^2 - |x^m|^2|x^{m-1}|^2} \begin{bmatrix} |x^{m-1}|^2 & -(x^{m-1},x^m) \\ -(x^{m-1},x^m) & |x^m|^2 \end{bmatrix} \times \begin{bmatrix} (x^{m+1},x^m) \\ (x^{m+1},x^{m-1}) \end{bmatrix}. \tag{274}$$

We know the denominator $\neq 0$, because by the Cauchy-Schwarz inequality, a zero value can occur only when x^m is a constant times x^{m-1}, and we have assumed that is not the case.

At the mth iteration, having found $\hat{\alpha}$ and $\hat{\beta}$, our best estimates of α and β, we must now decide whether $|x^{m+1} + \hat{\alpha}x^m + \hat{\beta}x^{m-1}|$ is small enough. Again we base our decision on whether we are relatively close enough to the eigenvalues λ_1 and λ_2. We can show that if A is normal, and if

$$\frac{|x^{m+1} + \hat{\alpha}x^m + \hat{\beta}x^{m-1}|^2}{|x^m|^2|\hat{\alpha}^2 - 4\hat{\beta}|} < \varepsilon^2, \tag{275}$$

then

$$|\hat{\lambda}_1 - \lambda_1|/|\lambda_1| < \varepsilon(1 + \delta), \tag{276}$$

and

$$|\hat{\lambda}_2 - \lambda_2|/|\lambda_2| < \varepsilon(1 + \delta), \tag{277}$$

where

$$\hat{\lambda}_1 = (-\hat{\alpha} + \sqrt{\hat{\alpha}^2 - 4\hat{\beta}})/2, \tag{278}$$

$$\hat{\lambda}_2 = (-\hat{\alpha} - \sqrt{\hat{\alpha}^2 - 4\hat{\beta}})/2, \tag{279}$$

and

$$\delta = |\hat{\lambda}_1 - \hat{\lambda}_2|/|\lambda_1 - \hat{\lambda}_2| - 1 \approx 0. \tag{280}$$

The support for the above claim is parallel to that for the real case, except that we also need to show that $|\hat{\alpha}^2 - 4\hat{\beta}| = |\hat{\lambda}_1 - \hat{\lambda}_2|$ (see exercise 2.7.4). If A is not normal but has n independent eigenvectors, the argument involves defining a new norm just as for the real case, but actually using the $\mathscr{L}_2$ norm as an approximation, because we cannot produce the desired norm.

Having found $\hat{\alpha}$ and $\hat{\beta}$, we solve the equation

$$\lambda^2 + \hat{\alpha}\lambda + \hat{\beta} = 0 \tag{281}$$

to find the roots $\hat{\lambda}_1$ and $\hat{\lambda}_2$, which are our best estimates of λ_1 and λ_2. Then we compute values $\hat{v}^1$ and $\hat{v}^2$ for the corresponding eigenvectors by the following equations, which follow from equations 265 and 266.

$$\hat{v}_1 = x^{m+1} - \hat{\lambda}_2 x^m. \tag{282}$$

$$\hat{v}_2 = x^{m+1} - \hat{\lambda}_1 x^m. \tag{283}$$

Let us summarize the algorithm we have developed to find the largest eigenvalue(s) and the associated eigenvector(s) of an $n \times n$ matrix with n independent eigenvectors. First we choose an ε specifying the relative error within which we wish to determine the eigenvalue(s) of the matrix. Then at each iterative step we first check to see whether the iterate is converging to a real eigenvector by the test

$$|x^{m+1} - \hat{\lambda} x^m|^2 / |x^{m+1}|^2 < \varepsilon^2, \tag{284}$$

where $\hat{\lambda}$ is given by equation 252. If it is not, we check to see whether it is approaching a linear combination of two eigenvectors (condition 275, where $\hat{\alpha}$ and $\hat{\beta}$ are given by equation 274). If either test is satisfied, we compute the eigenvalue(s) and eigenvector(s). If not, we iterate again. To take care of computational problems resulting from the fact that x^{m+1} is proportional to $|\lambda_1|^{m+1}$, which approaches 0 or ∞ depending upon whether $|\lambda_1|$ is less than or greater than 1, producing either floating-point underflow or overflow, we check at each step whether $|x^{m+1}| > Q$ or $|x^{m+1}| < q$ for parameters Q and q chosen before we start the computation. If one of these conditions holds, we normalize x^{m+1} without changing its direction by dividing it by $|x^{m+1}|$. Then we do two iterations before we begin checking for convergence again.

We must finally take care of a few details which have been neglected. First, it should not be assumed that if x^{m+1} is approximately a linear combination of x^m and x^{m-1}, then necessarily $|\lambda_1| = |\lambda_2|$. If, for example, $|\lambda_1| > |\lambda_2| \gg |\lambda_3|$, condition 275 may be satisfied at an earlier iteration than condition 284, because $(\lambda_3/\lambda_1)^m$ will become negligible sooner than $(\lambda_2/\lambda_1)^m$ will, so x^{m+1}, x^m, and x^{m-1} will each be approximately a linear combination of v^1 and v^2 and thus themselves be approximately linearly dependent. This situation need not bother us, for it will simply produce an $\hat{\alpha}$ and $\hat{\beta}$ such that $\hat{\alpha}^2 - 4\hat{\beta} > 0$, that is, such that the roots of equation 281 ($\hat{\lambda}_1$ and $\hat{\lambda}_2$) are real. Equations 282 and 283 still apply for finding the eigenvectors corresponding to λ_1 and λ_2.

Second, we have assumed that our original eigenvector guess, x^0, has a component in the direction of v^1, that is, $\xi_1 \neq 0$. If this is not the case, the theory indicates that x^m converges to the eigenvector with the second largest eigenvalue in magnitude. This situation would not be bad if we were looking for all eigenvalues and eigenvectors of A, but it would be bad if we were looking for λ_1 and v^1. Fortunately, this problem will not occur, because, for once, computational error works in our favor. Even if x^0 has no component in v^1, x^1 will have some component in v^1 due to computational error. If we iterate long enough, this small component due to computational error will eventually cause the v^1 component to dominate.

2.7.3 Acceleration of Convergence

If x^m is converging in direction to the vector v^1, by observing a few of the iterates we should be able to make a pretty good guess at the vector to which the sequence is converging. Taking advantage of this fact is called "accelerating the convergence." To formalize this notion, we will show that if x^m is converging in direction to v^1, then in the limit

$$\delta^{m+1} \equiv \frac{x^{m+1}}{\hat{\lambda}^{m+1}} - kv^1 = \frac{\lambda_2}{\lambda_1}\left(\frac{x^m}{\hat{\lambda}^m} - kv^1\right) \equiv \frac{\lambda_2}{\lambda_1}\delta^m \tag{285}$$

for some constant k and for some eigenvector v^1, where $\hat{\lambda}$ is the estimate of λ_1 produced using x^{m+1} and x^m (see equation 252). We prove this relation by noting that if $|\lambda_1| > |\lambda_2| > |\lambda_3|$, then in the limit

$$\frac{x^m}{\hat{\lambda}^m} \to \frac{\lambda_1^m(\xi_1 v^1 + (\lambda_2/\lambda_1)^m \xi_2 v^2)}{\lambda_1^m}, \tag{286}$$

so with $k = \xi_1$ we have $x^m/\hat{\lambda}^m - \xi_1 v^1 \to (\lambda_2/\lambda_1)^m \xi_2 v^2$, from which follows our result. Because the error at the $(m+1)$th step is in the limit a constant times the error at the mth step, the power method for finding the eigenvector is said to be "simply linearly converging." As we iterate, the convergence becomes more and more simply linear. If the error at one step is truly a linear multiple of that at the previous step, then we have equation 285, where we do not know k, v^1, λ_2, or λ_1. But we do know that equation 285 holds for $m = i$ and $m = i-1$, and subtracting these two forms of that equation, using in both forms the most recent estimate $\hat{\lambda}$, we obtain

$$\frac{x^{i+1}}{\hat{\lambda}^{i+1}} - \frac{x^i}{\hat{\lambda}^i} = \frac{\lambda_2}{\lambda_1}\left(\frac{x^i}{\hat{\lambda}^i} - \frac{x^{i-1}}{\hat{\lambda}^{i-1}}\right), \tag{287}$$

or equivalently $$x^{i+1} - \hat{\lambda}x^i = (\lambda_2/\lambda_1)\hat{\lambda}(x^i - \hat{\lambda}x^{i-1}). \tag{288}$$

We can solve this equation for the constant $(\lambda_2/\lambda_1)\hat{\lambda}$. Then, rewriting equation 285 as

$$x^{i+1} - \hat{\lambda}k_1 v^1 = (\lambda_2/\lambda_1)\hat{\lambda}(x^i - k_1 v^1), \tag{289}$$

where $k_1 = k\hat{\lambda}^i$, and using our computed value for $(\lambda_2/\lambda_1)\hat{\lambda}$, we can solve for $\hat{v}^1$, a constant times v^1 (that is, for the eigenvector desired):

$$\hat{v}^1 = x^{i+1} - (\lambda_2/\lambda_1)\hat{\lambda}x^i. \tag{290}$$

The difficulty with the procedure described above is that equations 288 and 289 are approximations for any finite i. That equation 289 is an approximation has the effect only that our computed value for $\hat{v}^1$ is an approximation. When trying to use equation 288, however, we will find that due to the approximation, the vector on the left side of the equation is not a true multiple of the vector in parentheses on the right side. Thus, we must estimate the constant $(\lambda_2/\lambda_1)\hat{\lambda}$ by choosing that value which minimizes the norm of the difference between the left side and the right side of equation 288. If we use the Euclidean norm, we obtain

$$(\lambda_2/\lambda_1)\hat{\lambda} \approx \hat{c} = (x^{i+1} - \hat{\lambda}x^i)^T(x^i - \hat{\lambda}x^{i-1})/|x^i - \hat{\lambda}x^{i-1}|^2. \tag{291}$$

This value of $\hat{c}$ can be used in equation 290 to produce

$$\hat{v}^1 = x^{i+1} - \hat{c}x^i. \tag{292}$$

It is convenient and acceptable to compute at the ith iterative step $x^i - \hat{\lambda}x^{i-1}$ using the estimate $\hat{\lambda}$ produced at that step, then to compute at the $(i+1)$th iterative step $x^{i+1} - \hat{\lambda}x^i$ using the estimate $\hat{\lambda}$ produced at that step (slightly different from the previous $\hat{\lambda}$). Having done so, we can use these vectors to compute the inner products required by equation 291.

In all of the above, we have assumed that x^m is approaching an eigenvector and δ^{i+1} is approaching a multiple of δ^i. These facts can be determined by satisfying empirical tests that x^{i+1} is approaching a multiple of x^i and that $x^{i+1} - \hat{\lambda}x^i$ is approaching a multiple of $x^i - \hat{\lambda}x^{i-1}$. For the first test we simply use condition 284, substituting for ε, a value $\varepsilon_1 \gg \varepsilon$ which is small enough to convince ourselves that convergence to a real eigenvector is occurring but not so large that we have an accurate enough estimate for λ_1:

$$|x^{i+1} - \hat{\lambda}x^i|^2/|x^{i+1}|^2 < \varepsilon_1^2. \tag{293}$$

The second test is of the same form as the first:

$$|(x^{i+1} - \hat{\lambda}x^i) - \hat{c}(x^i - \hat{\lambda}x^{i-1})|^2/|x^{i+1} - \hat{\lambda}\bar{x}^i|^2 < \varepsilon_2^2, \tag{294}$$

where we choose $\varepsilon_2 > \varepsilon_1$, because this test involves even more approximations than that given by condition 293.

If test 1 fails, we may not be converging to a real eigenvector. If test 2 fails, the assumption that the convergence is simply linear is not even approximately being satisfied. In either case, acceleration should not be applied. If we satisfy the above tests, we may apply the acceleration technique and use the result $\hat{v}^1$ as a new x^0.

When we add the possibility of acceleration, the power method algorithm becomes the following.

1. Choose ε, ε_1, and ε_2, where ε is the required relative accuracy in λ_1; $\varepsilon_1 \gg \varepsilon$ is the relative error in the difference between x^{i+1} and

$\hat{\lambda}x^i$ that is small enough to convince us that x^{i+1} is approaching a multiple of x^i; and $\varepsilon_2 > \varepsilon_1$ is the relative error in the difference between $x^{i+1} - \hat{\lambda}x^i$ and $\hat{c}(x^i - \hat{\lambda}x^{i-1})$ that is small enough to convince us that $x^{i+1} - \hat{\lambda}x^i$ is approaching a multiple of $x^i - \hat{\lambda}x^{i-1}$.

2. Choose Q and q as upper and lower bounds for $|x^m|$ to prevent overflow and underflow.
3. Choose a guess x^0 at an eigenvector (the guess will probably be arbitrary).
4. Execute an iteration as follows:
   a) Given x^i, compute $x^{i+1} = Ax^i$.
   b) If x^i is either x^0, the result of a normalization, or the result of an acceleration, increment i and return to step a. If not, execute step c.
   c) Compute $\hat{\lambda}$ by equation 252 and execute test 293. If it is satisfied, go to step d. If not, execute step g.
   d) Execute test 284. If it is satisfied, stop with eigenvalue $\hat{\lambda}$ and eigenvector x^{i+1}. If not, execute step e.
   e) Compute $\hat{c}$ by equation 291 and execute test 294. If it is satisfied, go to step f. If not, execute step g.
   f) Compute $\hat{v}^1$ by equation 292 and set $x^{i+1} = \hat{v}^1$. Then begin another iteration.
   g) Compute $\hat{\alpha}$ and $\hat{\beta}$ by equation 274 and execute test 275. If it is satisfied, compute $\hat{\lambda}_1$ and $\hat{\lambda}_2$ by solving equation 281, and compute $\hat{v}^1$ and $\hat{v}^2$ by equations 282 and 283. Stop with these eigenvalues and eigenvectors. If test 275 is not satisfied, go to step h.
   h) Check if $q < |x^{i+1}| < Q$. If not, set $x^{i+1} = x^{i+1}/|x^{i+1}|$. In either case, begin another iteration.

As a final detail about the power method, we note that if the eigenvalues of A are known to be real (for example, if A is symmetric), then step 4g of the power method may be omitted. In that case, if n is large, it becomes very time-consuming to compute the inner products required by step 4c just to decide that acceleration is not applicable and another iterative step is required. In this case a simpler but less trustworthy test may be applied to step 4c. What we desire to know is whether x^{m+1} is approximately the same multiple of x^m as x^m is of x^{m-1}. We can roughly check this by seeing if for some k such that $x_k^{m+1} \not\approx 0$, $x_k^{m+1}/x_k^m \approx x_k^m/x_k^{m-1}$. The success of this "consistency test" is not enough to convince us that acceleration should be applied, but its failure is enough to convince us that we need not make the more complicated test but should go on to the next iteration.

Let us apply the power method with acceleration to the matrix $A = \begin{bmatrix} -5 & -1 & 5 \\ 1 & -3 & 1 \\ -1 & -1 & 1 \end{bmatrix}$. We make the arbitrary guess $x^0 = \begin{bmatrix} 1 \\ 0 \\ 0 \end{bmatrix}$ and set $\varepsilon = .01$, $\varepsilon_1 = .2$, and $\varepsilon_2 = .4$. We begin the process by iterating twice: $x^1 = Ax^0 = \begin{bmatrix} -5 \\ 1 \\ -1 \end{bmatrix}$; $x^2 = \begin{bmatrix} 19 \\ -9 \\ 3 \end{bmatrix}$. We apply test 293 to x^2 and x^1: $\hat{\lambda} = x^{1T}x^2/x^{1T}x^1 = -107/27$, $|x^2 - \hat{\lambda}x^1|^2/|x^2|^2 \approx .06 \not< \varepsilon_1^2$, so we apply test 275 to x^0, x^1, and x^2: by equation 274

$$\begin{bmatrix} \hat{\alpha} \\ \hat{\beta} \end{bmatrix} = \frac{1}{-2}\begin{bmatrix} 1 & 5 \\ 5 & 27 \end{bmatrix}\begin{bmatrix} -107 \\ 19 \end{bmatrix} = \begin{bmatrix} 6 \\ 11 \end{bmatrix},$$

so $|x^2 + \hat{\alpha}x^1 + \hat{\beta}x^0|^2/(|x^1|^2|\hat{\alpha}^2 - 4\hat{\beta}|) \approx .08 \not< \varepsilon^2$. Thus we iterate again:

$$x^3 = \begin{bmatrix} -71 \\ 49 \\ -7 \end{bmatrix}, \quad \hat{\lambda} = \frac{x^{2T}x^3}{x^{2T}x^2} = \frac{-1811}{451}, \quad |x^3 - \hat{\lambda}x^2|^2/|x^3|^2 \approx .03 < \varepsilon_1^2.$$

We seem to be converging to a real eigenvector, so we test to determine whether δ^i is also approaching an eigenvector. To do so, we first use equation 291 to compute

$$\hat{c} = \frac{(x^3 - \hat{\lambda}x^2)^T(x^2 - \hat{\lambda}x^1)}{(x^2 - \hat{\lambda}x^1)^T(x^2 - \hat{\lambda}x^1)} \approx -2.8.$$

Then we use our result in equation 294:

$$\frac{|x^3 - \hat{\lambda}x^2 - \hat{c}(x^2 - \hat{\lambda}x^1)|^2}{|x^3 - \hat{\lambda}x^2|^2} \approx .05 < \varepsilon_2^2.$$

Our test is satisfied, so we can apply acceleration to x^1, x^2, and x^3.

$$\hat{v}^1 \approx x^3 - \hat{c}x^2 \approx \begin{bmatrix} -18.2 \\ 24 \\ 1.3 \end{bmatrix}.$$

We now use $\hat{v}^1$ as the new guess at the eigenvector and iterate twice, producing $x^4 = \begin{bmatrix} 73.5 \\ -88.9 \\ -4.5 \end{bmatrix}$, $x^5 \approx \begin{bmatrix} -301 \\ 335 \\ 11 \end{bmatrix}$, and so on, converging to the eigenvector $\begin{bmatrix} 1 \\ -1 \\ 0 \end{bmatrix}$ with eigenvalue -4.

2.7.4 Reduction to Allow Finding Other Eigenvalues and Eigenvectors

Given that we have found an eigenvalue and an eigenvector of a matrix, how do we find the remaining eigenvalues and eigenvectors? Two approaches are open to us: (1) modify the matrix A so that all of the eigenvalues are altered in such a way that in its altered form the eigenvalue found is no longer dominant, or (2) find an $n-1 \times n-1$ matrix that has as its eigenvalues all of the remaining eigenvalues of A and whose eigenvectors are related in some known way to those of A. In either case we would expect error made in determining the first eigenvalues and eigenvectors to propagate, producing even larger error in the eigenvalues and eigenvectors determined from the modified matrix.

We will give two examples of the first approach. First we know that the eigenvalues of $A - \alpha I$ are α less than the eigenvalues of A and that the associated eigenvectors are not changed: If $\hat{\lambda}_1$ is real, then $A - \hat{\lambda}_1 I$ is real and the eigenvalue associated with v^1 is now $\lambda_1 - \hat{\lambda}_1 \approx 0$, which is no longer dominant. Iteration with $A - \hat{\lambda}_1 I$ produces another eigenvalue (to which we must add $\hat{\lambda}_1$ when we converge) and its associated eigenvector. We may not be able to repeat the matrix modification scheme because v^1 may again become the dominant eigenvector. However, this method does have the advantage that the modification is based on a fixed matrix I, not on a matrix related to our computed $\hat{v}^1$, so inaccuracy in $\hat{v}^1$ or $\hat{\lambda}_1$ does not affect the accuracy of the next eigenvalue and eigenvector found.

Another modification method using the first approach above is applicable to symmetric matrices. If A is symmetric, we know

$$A = \sum_{i=1}^{n} \lambda_i v^i v^{iT}. \tag{295}$$

Knowing λ_1 and v^1, we can compute

$$A' = A - \lambda_1 v^1 v^{1T}, \tag{296}$$

a matrix that has the same eigenvectors as A and whose eigenvalues are the same as those of A except for λ_1, which has been changed to zero. We can now iterate with A' to find λ_2 and v^2. With this method, errors in $\hat{\lambda}_1$ and $\hat{v}^1$ produce errors in both the remaining eigenvectors and the remaining eigenvalues, but the method can be repeated to find λ_i and v^i for $i > 2$.

If A is not symmetric, this modification method is not applicable because the equation parallel to equation 295 involves an eigenvalue of A^T, which we have not found. In this case, we need to use the second approach above, that is, to find an $n-1 \times n-1$ matrix which has all of the eigenvalues of A other than λ_1. Assume we had a matrix Q which had the same eigenvalues as A and was of the form

$$Q = \begin{bmatrix} \lambda_1 & * & * & \cdot & \cdot & \cdot & * \\ 0 & & & & & & \\ 0 & & & & & & \\ \cdot & & & A' & & & \\ \cdot & & & & & & \\ \cdot & & & & & & \\ 0 & & & & & & \end{bmatrix}, \tag{297}$$

where "*" stands for elements whose values we don't care about and A' is some matrix, not the A' used in the paragraph above. Expanding on the first column, we see

$$det(\lambda I - Q) = (\lambda - \lambda_1)\, det(\lambda I - A'). \tag{298}$$

Since
$$det(\lambda I - Q) = det(\lambda I - A) = \prod_{i=1}^{n} (\lambda - \lambda_i), \tag{299}$$

we have
$$det(\lambda I - A') = \prod_{i=2}^{n} (\lambda - \lambda_i), \tag{300}$$

so A' is the desired $n-1 \times n-1$ matrix.

To find the matrix Q which has the same eigenvalues as A, we must use a similarity transformation: $Q = T^{-1}AT$. Once we know T, if we use the power method to find a dominant eigenvalue of A' (which is also an eigenvalue of A) and its associated eigenvector, we can transform that corresponding eigenvector into the eigenvector of A. We first find the corresponding eigenvector of Q from the eigenvector of A' (see exercise 2.7.9). We then find the eigenvector of A by multiplying the corresponding eigenvector of Q by T.

Our job now is to find the matrix T for which $T^{-1}AT = Q$. Our requirement on T is that the first column of Q be the first column of $\lambda_1 I$:

$$(T^{-1}AT)_{j1} = \lambda_1 I_{j1}, \quad 1 \le j \le n. \tag{301}$$

This implies
$$\sum_{k=1}^{n} T_{jk}^{-1}(AT)_{k1} = \lambda_1 I_{j1}, \quad 1 \le j \le n. \tag{302}$$

Certainly this is true if

$$(AT)_{k1} = \lambda_1 T_{k1}, \quad 1 \le k \le n, \tag{303}$$

that is, if
$$At^1 = \lambda_1 t^1, \tag{304}$$

where t^1 is the first column of T. Thus we wish as the first column of T the eigenvector of A associated with λ_1. But this is precisely v^1, which we have found. Our requirement on T is satisfied if its first column is v^1. We may choose the rest of T as we like. To make the computation of T^{-1} easy, we set all remaining elements of T to 0 except for the diagonal elements, which we set to 1.

To make T easily invertible, we also need $v_1^1 = 1$. This is simple if $v_1^1 \neq 0$, because the eigenvector is defined only up to a multiplicative factor. We can multiply v^1 by $1/v_1^1$ and put that form of the eigenvector in the first column. If $v_1^1 = 0$, we know we can find some other element v_i^1 which is nonzero. If we switch the first and ith elements, we know this modified vector is an eigenvector of $P^{-1}AP$, where P is the nonsingular permutation matrix which switches the first and ith rows of a vector. That is, the matrix $B = P^{-1}AP$ is a matrix similar to A whose eigenvectors are the same as those of A except that the first and ith elements are switched. We can apply the similarity transformation $T^{-1}BT$, where T is as above except that it has an eigenvector of B in the first column, and the first element of this eigenvector is equal to 1.

Given a matrix of the form

$$T = \begin{bmatrix} 1 & 0 & 0 & \cdot & \cdot & \cdot & 0 \\ v_2^1/v_1^1 & & & & & & \\ v_3^1/v_1^1 & & & I_{(n-1 \times n-1)} & & & \\ \cdot & & & & & & \\ \cdot & & & & & & \\ \cdot & & & & & & \\ v_n^1/v_1^1 & & & & & & \end{bmatrix}, \tag{305}$$

its inverse is

$$T^{-1} = \begin{bmatrix} 1 & 0 & 0 & \cdot & \cdot & \cdot & 0 \\ -v_2^1/v_1^1 & & & & & & \\ -v_3^1/v_1^1 & & & I_{(n-1 \times n-1)} & & & \\ \cdot & & & & & & \\ \cdot & & & & & & \\ \cdot & & & & & & \\ -v_n^1/v_1^1 & & & & & & \end{bmatrix}. \tag{306}$$

Since we are going to multiply by the elements of T and T^{-1}, we want them as small as possible to minimize absolute error. That is, we should choose as the first element of the eigenvector of B the largest element in magnitude of the eigenvector of A, whether or not the first element of the eigenvector of A is equal to zero.

To summarize, our reduction method assumes we have an eigenvector v^1. First, we find k such that $|v_k^1| = \max_i |v_i^1|$, and we switch v_1^1 and v_k^1 to produce u^1. We let P be the permutation matrix

$$
\begin{array}{r}
\\ \\ \\ \\ \\ \\ \\ k \rightarrow \\ \\ \\ \\ \\ \\ \\ \\
\end{array}
\begin{bmatrix}
0 & 0 & 0 & \cdot & \cdot & \cdot & 0 & 1 & 0 & 0 & \cdot & \cdot & \cdot & 0 \\
0 & 1 & & & & & & & & & & & & \\
0 & & 1 & & & & & & & & & & & \\
\cdot & & & \cdot & & & & & & & & & & \\
\cdot & & & & \cdot & & & & & & & & & \\
\cdot & & & & & \cdot & & & & & & & & \\
0 & & & & & & 1 & & & & & & & \\
1 & & & & & & & 0 & & & & & & \\
0 & & & & & & & & 1 & & & & & \\
0 & & & & & & & & & 1 & & & & \\
\cdot & & & & & & & & & & \cdot & & & \\
\cdot & & & & & & & & & & & \cdot & & \\
\cdot & & & & & & & & & & & & \cdot & \\
0 & & & & & & & & & & & & & 1
\end{bmatrix}
$$
$$
\begin{array}{c} \uparrow \\ k \end{array}
$$

with zeros in all of the blank locations. We let

$$
T = I + \frac{1}{u_1^1}\begin{bmatrix} \uparrow & \\ u^1 & 0 \\ \downarrow & \end{bmatrix} - \begin{bmatrix} \uparrow & \\ e^1 & 0 \\ \downarrow & \end{bmatrix}. \tag{307}
$$

Then
$$
T^{-1} = I - \frac{1}{u_1^1}\begin{bmatrix} \uparrow & \\ u^1 & 0 \\ \downarrow & \end{bmatrix} + \begin{bmatrix} \uparrow & \\ e^1 & 0 \\ \downarrow & \end{bmatrix}, \tag{308}
$$

and A' = the $n-1 \times n-1$ submatrix of $Q = T^{-1}P^{-1}APT$, that is,

$$
\begin{aligned}
A'_{ij} &= (P^{-1}AP)_{i+1,j+1} - (u_i^1/u_1^1)(P^{-1}AP)_{1,j+1} \\
&= A_{f(i+1),f(j+1)} - (u_i^1/u_1^1)A_{k,f(j+1)},
\end{aligned} \tag{309}
$$

where
$$
f(i) = \begin{cases} i \text{ if } i \neq k \text{ and } i \neq 1 \\ 1 \text{ if } i = k \\ k \text{ if } i = 1. \end{cases}
$$

Thus, the reduction algorithm is as follows:

1. Find k such that $|v_k^1| = \max_i |v_i^1|$.

2. Compute A' from

$$A'_{ij} = A_{f(i+1),f(j+1)} - (v^1_{f(i)}/v^1_k)A_{k,f(j+1)}, \tag{310}$$

for $1 \leq i \leq n-1$, $1 \leq j \leq n-1$, where f is defined above.†

The methods to find eigenvalues and eigenvectors presented in this section are not always the best methods for finding eigenvalues and eigenvectors. They are useful methods, sometimes the best, and illustrate important points. Many other useful methods are available.‡ In many of these methods, similarity transformations are applied to produce a special matrix form which allows simple determination of the eigenvalues. Another method, called the QR method, deserves special mention because it has become the method of choice for many problems which require computing only the eigenvalues of a matrix.

PROBLEMS

2.7.1. Calculate an eigenvector of the matrix $\begin{bmatrix} -2 & 1 & -4 \\ 5 & -7 & -5 \\ 0 & -9 & -6 \end{bmatrix}$ corresponding to the eigenvalue $\lambda = 3$. Truncate all calculations to two decimal places.

2.7.2. Let $x^{(i)}$ be the scalar result of the ith iterative step of an iterative method: $x^{(i+1)} = g(x^{(i)})$. Assume the iteration converges simply linearly to z, that is, in the absence of generated error $\lim_{i\to\infty} \dfrac{x^{(i+1)} - z}{x^{(i)} - z} = k$, where $0 < |k| < 1$. Assume i is large enough so that the relation $(x^{(i+1)} - z) = k(x^{(i)} - z)$, giving the propagated error in the $(i+1)$th step, can be taken to be true in the absence of generated error. Assume error bounded by b is generated in each step. By arguing that convergence will continue until the overall error ceases to get smaller at some step, show that the iteration cannot be assured to get closer to z than $b/(1 - |k|)$. Compare this result to the result of exercise 1.3.9.

2.7.3. Let A be a matrix which does not have n independent eigenvectors. Let A have k distinct eigenvalues λ_i, $1 \leq i \leq k$, where $|\lambda_1| > |\lambda_2| \geq |\lambda_3| \geq |\lambda_4| \geq \cdots \geq |\lambda_k|$. By writing $A = DJD^{-1}$, where D and J are given by equations 211–213, and writing x as a linear combination of the principal vectors d^i of A, show that if x has a component in a principal vector associated with λ_1, then $\lim_{n\to\infty} (A^n x/\lambda_1^n) =$ an eigen-

† If we have computed two eigenvalues and their associated eigenvectors from the power method iteration, the reduction procedure is similar to the above but more complex. It is described in J. N. Franklin, *Matrix Theory,* chapter 7.

‡ J. N. Franklin, *Matrix Theory,* chapter 7.

vector of A associated with λ_1. That is, $A^n x$ approaches in direction an eigenvector of A associated with λ_1, and the norm of $A^n x$ is proportional to λ_1^n as $n \to \infty$.

2.7.4. Assume A is a normal matrix. Let $x^m = Ax^{m-1}$, $x^{m+1} = Ax^m$, and assume $|x^{m+1} + \hat{\alpha}x^m + \hat{\beta}x^{m-1}|^2/(|x^m|^2|\hat{\alpha}^2 - 4\hat{\beta}|) < \varepsilon^2$. Further assume A has maximum eigenvalues (in magnitude) λ_1 and $\lambda_2 = \bar{\lambda}_1$. Let $\hat{\lambda}_1$ and $\hat{\lambda}_2$ be the roots of $\lambda^2 + \hat{\alpha}\lambda + \hat{\beta} = 0$. Show that $\hat{\lambda}_1$ and $\hat{\lambda}_2$, taken in the appropriate order, are relatively close to λ_1 and λ_2 according to the following relations: $|\hat{\lambda}_1 - \lambda_1|/|\lambda_1| < \varepsilon(1+\delta)$ and $|\hat{\lambda}_2 - \lambda_2|/|\lambda_2| < \varepsilon(1+\delta)$, where $\delta = (|\hat{\lambda}_1 - \hat{\lambda}_2|/|\lambda_1 - \hat{\lambda}_2|) - 1$.

2.7.5. Let P be an $n \times n$ matrix having a single 1 in each column and a single 1 in each row and zeros in all other positions. P, called a permutation matrix, simply permutes the elements of a vector x when it multiplies x. $P_{ij} = 1$ means that $(Px)_i = x_j$.

a) Show that there exists an m such that $P^m = I$.

b) What happens if the power method is applied to P?

c) What can you say about the eigenvalues of P from this behavior?

2.7.6. Assume we have a matrix A with an eigenvalue λ_i which is known to be approximately $\hat{\lambda}_i$. Let $C = (A - \hat{\lambda}_i I)^{-1}$.

a) To what vector will the power method converge when applied to C?

b) Using the result of part a, give a method for computing λ_i.

c) Why is this procedure unacceptable for computing λ_i?

2.7.7. Let A be a matrix with n independent eigenvectors $v^1, v^2, \ldots, v^n$ and eigenvalues λ_i such that $|\lambda_1| > |\lambda_2| > |\lambda_3| \geq |\lambda_4| \geq \cdots \geq |\lambda_n|$. Let $\hat{\lambda}^{(k)}$ be the estimate of λ_1 produced by the kth iteration of the power method using the matrix A.

a) Show that the sequence of $\hat{\lambda}^{(k)}$ converges simply linearly to λ_i; that is, $\lim_{k\to\infty} \dfrac{\hat{\lambda}^{(k+1)} - \lambda_1}{\hat{\lambda}^{(k)} - \lambda_1} = c$ for a constant c such that $|c| < 1$.

b) Show that if v^1 is orthogonal to all other eigenvectors, then $|c| = (\lambda_2/\lambda_1)^2$ and if not, then $|c| = max[(\lambda_2/\lambda_1)^2, |\lambda_j/\lambda_1|]$, where j is the smallest integer >1 such that $v^{jT}v^1 \neq 0$.

c) Since the sequence of $\hat{\lambda}^{(k)}$ converges simply linearly, we could apply acceleration to three successive members of the sequence. Why would such acceleration not be helpful?

2.7.8. Consider iterative step number i of the power method: $x^{i+1} = Ax^i$; $\hat{\lambda}^{(i+1)} = x^{i+1T}x^i/x^{iT}x^i$. We wish to compare the relative variance of the *computational* error in $\hat{\lambda}^{(i+1)}$, $\sigma^2(\hat{\lambda}^{(i+1)})/(\hat{\lambda}^{(i+1)})^2$, for the following two algorithms. That is, assume A and x^i are represented with no error as base-b floating-point numbers with an m-digit fraction, and be concerned with the error generated by computation but not with the effect of the fact that x^i is only a guess at an eigenvector. In the first algorithm, all arithmetic operations are done in floating-point with an m-digit fraction and symmetric rounding after each operation. In the second algorithm, all sums of products are done by maintaining $2m$ digits in all products and partial sums of products, doing symmetric rounding to m digits only when a full sum of products is computed.

a) Assume that the truncation error in a single-precision (m-digit) floating-point addition or multiplication is uniformly distributed in the range $\pm\frac{1}{2}b^{-m+1}b^{\lfloor \log_b |x| \rfloor}$, where the symbols "⌊ ⌋" indicate the floor function. Briefly justify this assumption.

b) Assume that all elements of A are approximately equal to the positive number p and all elements of x^i are approximately equal to the positive number r (in your calculations assume equality, but assume all errors are uncorrelated). Give approximate values for the relative variance produced by each algorithm and show that for the first algorithm it is approximately n times the value for the second algorithm, where A is $n \times n$.

2.7.9. Assume we have a matrix Q of the form given in equation 297. Assume we have an eigenvector u of A'. Give an equation for the corresponding eigenvector of Q.

2.7.10. In reducing a matrix after finding an eigenvalue λ and eigenvector v, one computes the $n-1 \times n-1$ matrix A' by

$$A'_{ij} = A_{f(i+1),f(j+1)} - (u_i/u_1)A_{k,f(j+1)}$$

for some $1 \le k \le n$ (see equation 309). Assume the relative variance of u_i is r^2 for all i. Assume the relative variance of A_{ij} is s^2 for all i and j. Assume that the A_{ij} are approximately constant $= a$. Assume that all A_{ij} and u_i elements are independent and unbiased. Assume $s^2 \ll 1$ and $r^2 \ll 1$.

a) What is an approximate value for $var(A'_{ij})$?

b) Argue on the basis of your response to part a that $var(A'_{ij})$ is minimum if $|v_k| \equiv |u_1| = \max_{1 \le i \le n} |v_i|$.

2.7.11. Using the power method and reduction, find all eigenvalues and eigenvectors of

$$\begin{bmatrix} -1 & 0 & 3 \\ 4 & 2 & 1 \\ 0 & -2 & -3 \end{bmatrix}.$$

2.7.12. Let A be any real matrix. Say we wish to compute $cond_2(A)$. So we compute $A^T A$ and apply the power method to find its eigenvalue of maximum magnitude. Having done this, how would you go about finding its eigenvalue of minimum magnitude without computing $(A^T A)^{-1}$? (This is not a hard problem. Look for a clever, easy solution.)

2.8 ITERATIVE METHODS FOR SOLUTION OF SYSTEMS OF LINEAR EQUATIONS

We have argued that iterative methods can decrease the error in a solution. Now let us apply this technique to the solution of linear equations. In the ill-conditioned case, a reduction in computational error generation through iterative methods may be of considerable help.

2.8.1 The Iterative Form and Its Convergence

The iterative technique involves setting up an equation of the form

$$x^{i+1} = f(x^i) \tag{311}$$

where this relation converges to the desired solution. What we may do is rewrite $Ax = b$ in the form

$$x = f(x) \tag{312}$$

and then iterate using equation 311. We can rewrite $Ax = b$ in the form given by equation 312 by writing

$$A = (A - E) + E, \tag{313}$$

so that

$$(A - E)x + Ex = b. \tag{314}$$

This leads to

$$Ex = b - (A - E)x. \tag{315}$$

Solving for the x on the left side of equation 315, we obtain

$$x = E^{-1}b - (E^{-1}A - I)x. \tag{316}$$

Putting this form into equation 311, the iteration becomes

$$x^{i+1} = c + Bx^i, \tag{317}$$

where

$$c = E^{-1}b \tag{318}$$

and

$$B = I - E^{-1}A. \tag{319}$$

Note that, as desired, this iterative method requires fewer operations in an iterative step (n^2) than does the full Gaussian elimination ($\approx n^3/3$) and thus can be expected to provide reduced error generation.

We must find out under what conditions the iteration given by equation 317 converges. First note that if z is the solution to $Ax = b$, then z satisfies equation 316. Subtracting z from both sides of equation 317, we obtain

$$x^{i+1} - z = c + Bx^i - (c + Bz) = B(x^i - z). \tag{320}$$

Let

$$\varepsilon^i \equiv x^i - z, \tag{321}$$

the error in the ith iterate. Thus we have

$$\varepsilon^{i+1} = B\varepsilon^i, \tag{322}$$

which leads to

$$\varepsilon^n = B^n\varepsilon^0. \tag{323}$$

We are familiar with this form, because we encountered it in the power method of finding eigenvectors. We know that ε^n will approach an eigenvector of B in direction and its magnitude will be proportional to $|\lambda_1|^n$, where λ_1 is the largest eigenvalue in magnitude of B. Therefore, for convergence for any ε^0, we require $|\lambda_1| < 1$. Sometimes $|\lambda_1|$ is called the *spectral radius* of B. We say that for convergence the spectral radius of B must be less than 1.

If $|\lambda_1| < 1$, clearly all eigenvalues of B are less than 1 in magnitude. Furthermore, we have seen that the eigenvalues of $A - kI$ are k less than the eigenvalues of A and that the eigenvalues of $-A$ are the negative of the eigenvalues of A. Thus, we can state our convergence requirement as follows: If μ_i are the eigenvalues of $E^{-1}A$, then for $1 \le i \le n$,

$$|\mu_i - 1| < 1. \tag{324}$$

That is, all eigenvalues of $E^{-1}A$ must be less than distance 1 in the complex plane from the point 1.

2.8.2 Jacobi Method

We must choose a matrix E^{-1} (or equivalently, E) that is likely to satisfy the above convergence requirement. A matrix E that is a good approximation to A may do the job. Using such an E, in each iteration we will be approximately solving our original equation, so convergence might be expected. For example, let us choose $E = D_A$, the diagonal part of A. Then E^{-1} is the diagonal matrix with the reciprocal of the diagonal elements of A on the diagonal (we must order our equations so that the ith coefficient of the ith equation is not equal to zero). With this choice of E, we can rewrite equation 317 as

$$x_j^{i+1} = \frac{b_j}{A_{jj}} - \sum_{\substack{k=1 \\ k \neq j}}^{n} \frac{A_{jk}}{A_{jj}} x_k^i \quad \text{for} \quad j = 1,2,\ldots,n. \tag{325}$$

That is, we solve the first equation for x_1, the second equation for x_2, and so on, up through the nth equation for x_n, and with the jth equation we get a value for x_j^{i+1} by using the values of x_k^i on the right side of the equation. This method is called the *Jacobi method* (see program 2-4).

We can show that a sufficient condition for convergence when using the Jacobi method is that the diagonal coefficient in each row of A is in magnitude greater than the sum of the magnitudes of the remaining elements in the row. With such a matrix, which is said to be *row diagonally dominant*, the matrix D_A is indeed a good approximation to the matrix A. To prove the sufficiency of this condition, we must show that all eigenvalues of the associated matrix B are less than 1 in magnitude. For the Jacobi method, the elements of the matrix B are given by

```
/* SOLVE N LINEAR EQUATIONS IN N UNKNOWNS BY JACOBI METHOD */
/* EQUATION IS AX = B; A AND B ARE GIVEN */
/* AT START, XOLD HOLDS GUESS AT SOLUTION VECTOR */
/* TOL IS TOLERANCE FOR STOPPING ITERATION */

   DECLARE A(N,N) FLOAT,       /* INPUT MATRIX */
           B(N) FLOAT,         /* INPUT VECTOR */
           (XOLD(N),           /* OLD AND NEW                   */
           XNEW(N)) FLOAT,     /* SOLUTION VECTOR ESTIMATES */
           C(N,N) FLOAT,       /* NORMALIZED A MATRIX */
           D(N) FLOAT,         /* NORMALIZED B VECTOR */
           (MAXNEW, MAXDIF, NEW, DIFF, TOL) FLOAT;

/* NORMALIZE MATRIX */
   L1:  DO I = 1 TO N;
        L2:  DO J = 1 TO I - 1, I+1 TO N;
             C(I,J) = A(I,J) / A(I,I);
        END L2;
        D(I) = B(I) / A(I,I);
   END L1;

/* COMPUTE NEW ESTIMATE */
   MAXNEW = 0; MAXDIF = 0;
   L3:  DO I = 1 TO N;
        XNEW(I) = D(I);
        L4:  DO J = 1 TO I - 1, I+1 TO N;
             XNEW(I) = XNEW(I) - C(I,J) * XOLD(J);
        END L4;
   /* FIND MAXIMUM ABSOLUTE DIFFERENCE BETWEEN OLD & NEW */
   /* ELEMENTS */
        DIFF = ABS(XNEW(I) - XOLD(I));
        IF DIFF > MAXDIF THEN MAXDIF = DIFF;
   /* FIND MAXIMUM NEW ELEMENT VALUE */
        NEW = ABS(XNEW(I));
        IF NEW > MAXNEW THEN MAXNEW = NEW;
   END L3;

/* NEED TO ITERATE AGAIN? */
/* ONE POSSIBLE TEST: IS RELATIVE DIFFERENCE BETWEEN OLD AND */
/* NEW VECTORS SMALL ENOUGH, I.E., LESS THAN TOLERANCE VALUE */
/* PROVIDED AS INPUT? */
   IF MAXDIF / MAXNEW < TOL THEN GO TO DONE;
/* LET COMPUTED VALUES BE NEW ESTIMATE */
   L5:  DO I = 1 TO N;
        XOLD(I) = XNEW(I);
   END L5;
   GO TO L3;
   DONE:  /* XNEW HOLDS BEST SOLUTION */
```

PROGRAM 2-4 Jacobi method for solution of linear equations

$$B_{jk} = \begin{cases} 0 \text{ if } j = k \\ A_{jk}/A_{jj} \text{ if } j \neq k. \end{cases} \tag{326}$$

We need only the following lemma to prove the sufficiency of the above condition for the convergence of the Jacobi method.

Lemma: The maximum eigenvalue in magnitude of a matrix is less than or equal to the maximum over all its rows of the sum of the magnitudes of its row elements.

Proof: If x is an eigenvector of B, then for $1 \leq i \leq n$,

$$\left| \sum_{j=1}^{n} B_{ij} x_j \right| = |\lambda| \, |x_i|, \tag{327}$$

which implies $$|\lambda| \, |x_i| \leq \sum_{j=1}^{n} |B_{ij}| \, |x_j|. \tag{328}$$

Taking the maximum over i of both sides of relation 328, we obtain

$$\begin{aligned} |\lambda| \max_i |x_i| &\leq \max_i \left(\sum_{j=1}^{n} |B_{ij}| \, |x_j| \right) \\ &\leq \max_i \left(\sum_{j=1}^{n} |B_{ij}| \max_j |x_j| \right) \\ &= \max_j |x_j| \max_i \left(\sum_{j=1}^{n} |B_{ij}| \right). \end{aligned} \tag{329}$$

Since $x \neq 0$, we conclude from equation 329 that

$$|\lambda| \leq \max_i \left(\sum_{j=1}^{n} |B_{ij}| \right). \tag{330}$$

We can also show that the Jacobi method will converge if A is a matrix in which the diagonal element in each column is larger in magnitude than the sum of the magnitudes of the remaining elements in that column (is *column diagonally dominant*). (See exercise 2.8.1.) Note that the diagonal dominance conditions are sufficient but not necessary. However, matrices that are strongly nondiagonally dominant are not likely to produce convergence. It is often effective to modify the original set of equations to produce a matrix which is, if not diagonally dominant, as close to diagonally dominant as possible. For example, consider the following set of equations, which has the solution $x_1 = x_3 = 1, x_2 = 2$:

$$\begin{aligned} x_1 + 6x_2 + 2x_3 &= 15 \\ x_1 + x_2 - 6x_3 &= -3 \\ 6x_1 + x_2 + x_3 &= 9. \end{aligned} \tag{331}$$

Setting up the Jacobi method by solving the ith equation for x_i, we obtain

$$\begin{aligned} x_1 &= 15 - 6x_2 - 2x_3 \\ x_2 &= -3 - x_1 + 6x_3 \\ x_3 &= 9 - 6x_1 - x_2. \end{aligned} \tag{332}$$

If we guess $x^0 = 0$, we obtain $x^1 = \begin{bmatrix} 15 \\ -3 \\ 9 \end{bmatrix}$, $x^2 = \begin{bmatrix} 15 \\ 36 \\ 78 \end{bmatrix}$. The method is clearly diverging. On the other hand, if we switch equations 1 and 3 of the original set and then switch equations 2 and 3 of the result, we obtain

$$\begin{aligned} 6x_1 + x_2 + x_3 &= 9 \\ x_1 + 6x_2 + 2x_3 &= 15 \\ x_1 + x_2 - 6x_3 &= -3. \end{aligned} \tag{333}$$

This result is both row and column diagonally dominant. Solving equation i for x_i, we obtain

$$\begin{aligned} x_1 &= \tfrac{3}{2} - \tfrac{1}{6}x_2 - \tfrac{1}{6}x_3 \\ x_2 &= \tfrac{5}{2} - \tfrac{1}{6}x_1 - \tfrac{1}{3}x_3 \\ x_3 &= \tfrac{1}{2} + \tfrac{1}{6}x_1 + \tfrac{1}{6}x_2. \end{aligned} \tag{334}$$

Guessing $x^0 = 0$, we obtain

$$x^1 = \begin{bmatrix} \frac{3}{2} \\ \frac{5}{2} \\ \frac{1}{2} \end{bmatrix}, \quad x^2 = \begin{bmatrix} 1 \\ \frac{25}{12} \\ \frac{7}{6} \end{bmatrix}, \quad x^3 = \begin{bmatrix} \frac{23}{24} \\ \frac{35}{18} \\ \frac{73}{72} \end{bmatrix}.$$

We are clearly converging.

Thus we see that by switching the rows or columns of a matrix, we can sometimes produce either row or column diagonal dominance or close to it. Furthermore, we can scale the rows or columns by appropriate factors to improve the chance that switching the rows or columns will produce diagonal dominance.

2.8.3 Gauss-Seidel Method

The Jacobi method is an example of a *method of simultaneous displacements.* That is, at the end of each iteration, we replace all of the elements of x^i simultaneously with the elements of x^{i+1}. But as soon as we have computed x_1^i, we presumably have a better estimate of z_1; we would do well to use it immediately and not wait until the end of the iterative step. If we use the computed values as soon as we have them, we have a *method of successive displacements,* known as the *Gauss-Seidel method* (see program 2-5). For example, applying the Gauss-Seidel method to equation 333 involves solving the ith equation for x_i, $1 \le i \le n$, just as in the Jacobi method, to produce equation 334, and then iterating as follows. With the starting guess $x^0 = 0$, we compute a new estimate for x_1 from equation 1 (of 334), producing $x_1^1 = \frac{3}{2}$. Now using as our estimate $x = (\frac{3}{2}, 0, 0)^T$, we compute a new estimate for x_2 from equation 2, producing $x_2^1 = \frac{5}{2} - \frac{1}{6} \cdot \frac{3}{2} - \frac{1}{3} \cdot 0 = \frac{9}{4}$. Using as our new estimate $x = (\frac{3}{2}, \frac{9}{4}, 0)^T$, we compute a new estimate for x_3 from equation 3, thereby producing $x^1 = \frac{1}{2} + \frac{1}{6} \cdot \frac{3}{2} + \frac{1}{6} \cdot \frac{9}{4} = \frac{9}{8}$. Our estimate of x after the first iteration, $x^1 = (\frac{3}{2}, \frac{9}{4}, \frac{9}{8})^T$, is considerably closer to the solution $(1, 2, 1)^T$ than the result of the first iteration of the Jacobi method was. Using this estimate, we return to equation 1 to compute x_1^2, and so forth.

The formal equation for a stage of the Gauss-Seidel method is

$$x_j^{i+1} = \frac{b_j}{A_{jj}} - \sum_{k=1}^{j-1} \frac{A_{jk}}{A_{jj}} x_k^{i+1} - \sum_{k=j+1}^{n} \frac{A_{jk}}{A_{jj}} x_k^i \qquad (335)$$

Compare this equation to equation 325 (and program 2-5 to program 2-4). Because we are using the new values immediately, we would expect that if the Jacobi method converges, the Gauss-Seidel method will also converge and faster. We would also expect that if the Jacobi method diverges, our new values will be worse sooner, so we would expect the Gauss-Seidel method to diverge more quickly. This is in fact most often the case. Generally, we should use the Gauss-Seidel method instead of the Jacobi method. In the case of convergence, fewer iterative steps will be required; in the case of divergence, we will find out sooner that the method is diverging. For completeness, we should note that there exist matrices for which the Jacobi method converges but the Gauss-Seidel method does not, and vice versa. Furthermore, even if A is diagonally dominant, there exist cases for which the Jacobi method converges faster (in the limit) than the Gauss-Seidel method. It can be shown, however,

```
/* SOLVE N LINEAR EQUATIONS IN N UNKNOWNS BY GAUSS-SEIDEL METHOD. EQUATION IS AX = B; */
/* A AND B ARE GIVEN. AT START, X HOLDS GUESS AT SOLUTION VECTOR */
/* TOL IS TOLERANCE FOR STOPPING ITERATION */
   DECLARE A(N,N) FLOAT,            /* INPUT MATRIX */
           B(N) FLOAT,              /* INPUT VECTOR */
           (X(N), XCOMP) FLOAT,     /* SOLUTION VECTOR ESTIMATE, NEW ELEMENT ESTIMATE */
           C(N,N) FLOAT,            /* NORMALIZED A MATRIX */
           D(N) FLOAT,              /* NORMALIZED B VECTOR */
           (MAXNEW, MAXDIF, NEW, DIFF, TOL) FLOAT;
/* NORMALIZE MATRIX */
   L1:  DO I = 1 TO N;
        L2:  DO J = 1 TO I - 1, I+1 TO N;
             C(I,J) = A(I,J) / A(I,I);
        END L2;
        D(I) = B(I) / A(I,I);
   END L1;
/* COMPUTE NEW ESTIMATE */
   MAXNEW = Ø; MAXDIF = Ø;
   L3:  DO I = 1 TO N;
        XCOMP = D(I);
        L4:  DO J = 1 TO I - 1, I+1 TO N;
             XCOMP = XCOMP - C(I,J) * X(J);
        END L4;
        DIFF = ABS(XCOMP - X(I)); /* FIND MAXIMUM ABSOLUTE DIFFERENCE BETWEEN OLD AND NEW ELEMENTS */
        IF DIFF > MAXDIF THEN MAXDIF = DIFF;
        X(I) = XCOMP;              /* USE NEW VALUE IMMEDIATELY */
        NEW = ABS(XCOMP);          /* FIND MAXIMUM NEW ELEMENT VALUE */
        IF NEW > MAXNEW THEN MAXNEW = NEW;
   END L3;
/* NEED TO ITERATE AGAIN? */
   IF MAXDIF / MAXNEW > = TOL THEN GO TO L3;
   DONE:  /* X HOLDS BEST SOLUTION */
```

PROGRAM 2-5 Gauss-Seidel method for solution of linear equations

that if after each row of A has been divided by its diagonal element, all of the off-diagonal elements have the same sign, the predicted behavior must hold.†

It is useful to analyze the convergence of the Gauss-Seidel method geometrically. The jth stage of an iterative step (see equation 335) involves moving from the present estimate of the solution to the hyperplane defined by the jth equation in a direction such that all vector elements (co-ordinates) but x_j are held constant. For example, consider the two equations in two unknowns, x_1 and x_2, graphed in figure 2-7, part a, or the two equations graphed in figure 2-7, part b. From our starting guess, x^0, we move at stage 1 in the x_1 direction to the line defined by equation 1; we move at stage 2 in the x_2 direction to the line defined by equation 2. We repeat stage 1, stage 2, and so on, and we see from our graph that we are converging to the solution, which is the intersection of the two lines. On the other hand, if we simply switch the order of the equations, parts c and d of figure 2-7 show that the Gauss-Seidel method diverges.

We can also analyze the convergence of the Gauss-Seidel method algebraically. We rewrite equation 335 in the form

$$\sum_{k=1}^{j} A_{jk} x_k^{i+1} = b_j - \sum_{k=j+1}^{n} A_{jk} x_k^i, \qquad 1 \leq j \leq n, \tag{336}$$

and compare this equation with equation 315 written in its iterative form:

$$E x^{i+1} = b - (A - E) x^i. \tag{337}$$

We can see that for the Gauss-Seidel method

$$E = L_A + D_A, \tag{338}$$

where L_A is the strictly lower triangular part of A and D_A is the diagonal part of A. Thus the matrix E in the Gauss-Seidel method is an even better approximation to A than the matrix $E\ (=D_A)$ is in the Jacobi method.

Using the form of E for the Gauss-Seidel method given in equation 338, we can show that if A is symmetric, the Gauss-Seidel method converges if and only if A is positive definite (see exercise 2.8.2). The condition that A cannot have zero or negative eigenvalues if the Gauss-Seidel method is to converge is a very strong one indeed. It should indicate to us that even if row and column scaling and permutation are allowed, getting the Gauss-Seidel method to converge is a difficult, often impossible chore. Experience shows this is the case. If that is so, what is the usefulness of iterative methods of the form of equation 317 for solving linear equations?

† R. S. Varga, *Matrix Iterative Analysis,* chapter 3.

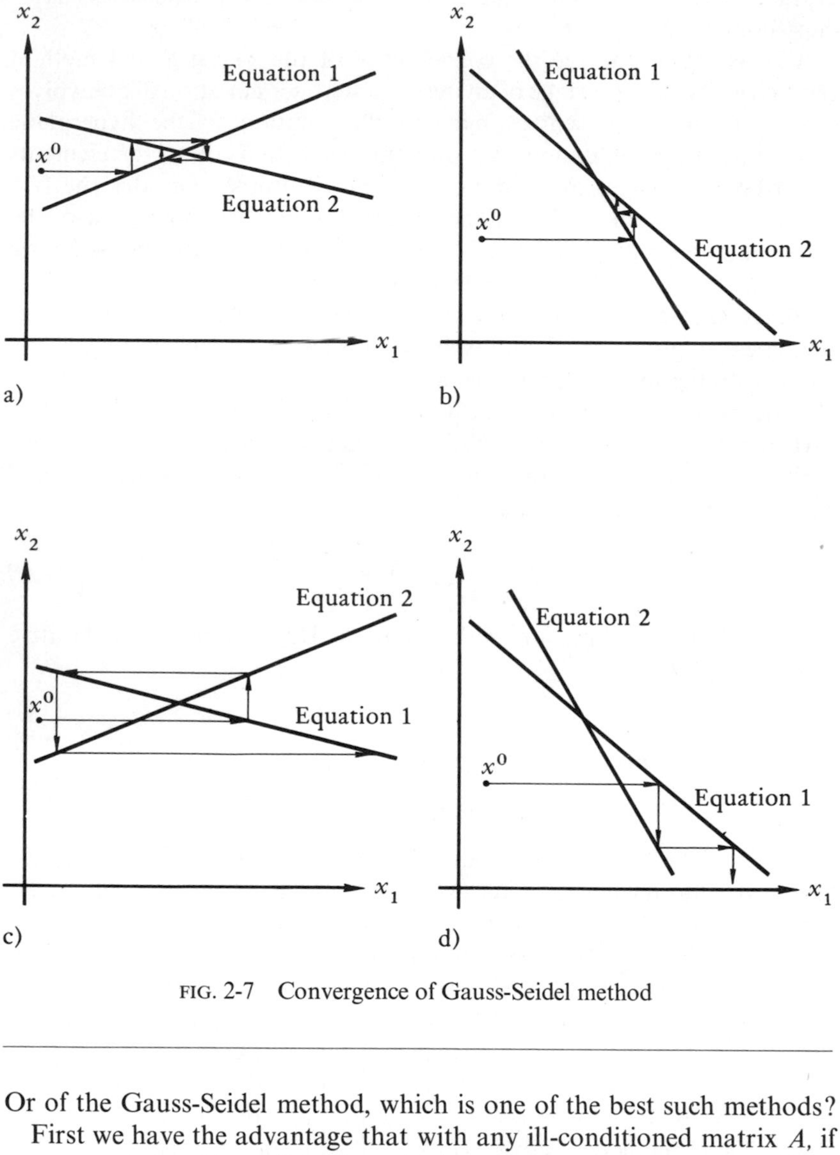

FIG. 2-7 Convergence of Gauss-Seidel method

Or of the Gauss-Seidel method, which is one of the best such methods?

First we have the advantage that with any ill-conditioned matrix A, if we can get the Gauss-Seidel method to converge, its answer will be considerably more accurate than that obtained using elimination. Even if the Gauss-Seidel method converges, however, it may do so very slowly, so that the total number of operations required is many more than that required by elimination. In that case it may be helpful to solve the system by elimination and use the result of that solution as the first guess in a Gauss-Seidel iteration.

The second advantage of the Gauss-Seidel method is that, unlike elimination, it does not change the coefficients of the equations by adding multiples of one equation to another. This advantage is important when there are many zeros in the original matrix because addition to zero and multiplication by zero are accurate and efficient (we need not actually do either operation). Sparse matrices, those with few nonzero elements, occur fairly often, particularly when the kth equation involves only unknowns with indices near k. Such a situation occurs, for example, in the linear equations produced by the numerical solution of partial differential equations, in the linear equations produced when approximating using cubic splines (discussed in section 4.2.1h), and in equations from mechanics where the kth element of the physical system being modeled is directly attached to only a few other elements of the system. With such sparse matrices, elimination undesirably changes the zero elements to nonzeros but the Gauss-Seidel method does not. Furthermore, as described above, the nonzero elements of sparse matrices are often near the diagonal, so it is reasonably likely that we can get the Gauss-Seidel method to converge.

Another reason for interest in iterative methods like Gauss-Seidel is that they extend straightforwardly to solving nonlinear equations, where direct methods like elimination do not exist. In section 2.8.6 we develop an iterative method for solving linear equations which not only has the advantages of the Gauss-Seidel method listed above, but also always converges. It should be selected over the Gauss-Seidel method except with strongly diagonally dominant coefficient matrices, where the Gauss-Seidel method is apt to converge more quickly. However, the always-convergent method does not extend to solving nonlinear equations.

When we apply an iterative method, we need a criterion by which we can decide to stop the iteration and take the latest iterate as our answer. We would like to stop when the iterate is close enough to the solution, but since we do not know the solution, a test to determine closeness to the solution is not practicable. Two other possibilities occur to us:

1. When we are close to the solution, $x^{i+1} \approx x^i$. So we should stop when $x^{i+1} - x^i$ is small, or better, small relative to x^{i+1}. That is, we should stop when

$$\|x^{i+1} - x^i\| / \|x^{i+1}\| < \varepsilon \tag{339}$$

for some tolerance ε (probably the most convenient norm to use is the $\mathscr{L}_\infty$ norm). Unfortunately, this test is not definitive—we may be far from the root when the difference between consecutive iterates is small.

2. If x^{i+1} is close to z, then Ax^{i+1} should be close to Az, that is, to b. So we should stop when the *residual* $Ax^{i+1} - b$ is small relative to b. That is, we should stop when

$$\|Ax^{i+1} - b\|/\|b\| < \varepsilon \tag{340}$$

for some tolerance ε. Again this test is not definitive—we may get smaller residuals relatively far from the solution than we do near it.

Still we must use some criterion for stopping the iteration, definitive or not, and the two above are the most common. In the case of the Jacobi method, in fact, the two criteria are equivalent, as we show in section 2.8.5. With the Gauss-Seidel method, they are close to equivalent, so criterion 1 is used because of its small computational requirements.

When using an iterative method that may diverge (such as Gauss-Seidel), a test to determine whether the method is diverging is useful. Since we do not wish to stop a process which may yet converge, we should detect only gross divergence, and stop iterating if that condition arises. At the very minimum, every iterative method should have as a parameter an upper bound on the number of iterations allowed. Alternatively, the iteration can be carried out in a conversational computing mode, where divergence can be detected by the user.

2.8.4 Acceleration of Convergence

If we decide to use the Gauss-Seidel method, we should consider the possibility of applying an acceleration technique to speed the convergence. From equation 324 we have seen that the error vector in Gauss-Seidel iteration behaves just as the eigenvector in the power method for finding eigenvectors. Thus, if the error vector is approaching a single eigenvector (that is, if the maximum eigenvalue in magnitude of $I - (L_A + D_A)^{-1}A$ is real and its negative is not also an eigenvalue), the acceleration procedure given in section 2.7.3 for the power method is applicable. There, it was applied to the iterates, which we knew, but here, it must be applied to the error vectors, which we do not know. Therefore, we must modify it somewhat.

If ε^n is approaching the direction of v^1, then

$$\varepsilon^n \approx \xi v^1, \tag{341}$$

so

$$\varepsilon^{n+1} \approx \lambda_1 \xi v^1 \tag{342}$$

and

$$\varepsilon^{n+2} \approx \lambda_1^2 \xi v^1. \tag{343}$$

Thus, $$\varepsilon^{n+1} - \lambda_1\varepsilon^n \approx 0, \tag{344}$$

and $$\varepsilon^{n+2} - \lambda_1\varepsilon^{n+1} \approx 0. \tag{345}$$

Since $$\varepsilon^n \equiv x^n - z, \tag{346}$$

equation 345 gives

$$(x^{n+2} - z) - \lambda_1(x^{n+1} - z) \approx 0. \tag{347}$$

Solving for z, we obtain

$$z \approx (x^{n+2} - \lambda_1 x^{n+1})/(1 - \lambda_1). \tag{348}$$

We need an estimate for λ_1 in terms of the x^i. This can be obtained by subtracting equation 344 from 345, giving

$$(x^{n+2} - x^{n+1}) - \lambda_1(x^{n+1} - x^n) \approx 0, \tag{349}$$

which indicates we should choose the value of λ_1 so that $|\Delta^{n+1} - \lambda_1\Delta^n|^2$ is minimized, where

$$\Delta^n \equiv x^{n+1} - x^n. \tag{350}$$

Then $$\hat{\lambda}_1 = (\Delta^{n+1}, \Delta^n)/|\Delta^n|^2, \tag{351}$$

which we substitute in equation 348 to produce $\hat{z}$, the result of the acceleration.

Note that acceleration requires three successive iterates, all of which are approximately in the direction of the dominant eigenvector of B. Since we have no reason to believe that $\hat{z}$ is in that direction, we must iterate a number of times more before we reapply the acceleration technique.

Even if the eigenvalue of B that is maximum in magnitude is complex, if the iterative method is converging, then x^m has a limit, and a form of acceleration is applicable. Since, in the limit, ε^m must approach a linear combination of two eigenvectors, we know

$$\varepsilon^{m+2} + \alpha\varepsilon^{m+1} + \beta\varepsilon^m \approx 0 \tag{352}$$

and $$\varepsilon^{m+3} + \alpha\varepsilon^{m+2} + \beta\varepsilon^{m+1} \approx 0, \tag{353}$$

where $$\alpha = -(\lambda_1 + \lambda_2) \tag{354}$$

and $$\beta = \lambda_1\lambda_2. \tag{355}$$

If we knew α and β, equation 353 would contain only one unknown, the solution z. We could solve for that to produce

$$\hat{z} = \frac{x^{m+3} + \hat{\alpha}x^{m+2} + \hat{\beta}x^{m+1}}{1 + \hat{\alpha} + \hat{\beta}}. \tag{356}$$

To find $\hat{\alpha}$ and $\hat{\beta}$, we subtract equation 352 from equation 353 to produce

$$\Delta^{m+2} + \alpha\Delta^{m+1} + \beta\Delta^{m} \approx 0. \tag{357}$$

Thus we should choose as $\hat{\alpha}$ and $\hat{\beta}$ those values of α and β which minimize $|\Delta^{m+2} + \alpha\Delta^{m+1} + \beta\Delta^{m}|^2$, resulting in a solution just like that in equation 274, where x^k is replaced by Δ^{k-1} for all k.

We should note that nowhere have we really used the assumption that $\varepsilon^m \rightarrow 0$ as $m \rightarrow \infty$. Rather we have used only the assumption that ε^m approaches either an eigenvector or a linear combination of two eigenvectors of B. Thus, the acceleration method is applicable even if the iterative method is diverging, and the result of the acceleration in such a situation may still be a good estimate of the solution.

Our algorithm for Gauss-Seidel iteration (without a divergence test) is summarized as follows:

1. Choose a tolerance ε for the stopping criterion, a tolerance δ for the criterion of when to apply acceleration, and a maximum number of iterations M.
2. Scale and arrange the variables and/or equations so that the matrix A is as close to diagonally dominant as possible.
3. Scale all equations so that $A_{ii} = 1$.
4. Make a guess at the solution vector $x^0 = b$ (after scaling).
5. For $i = 0,1,2,\ldots,M-1$,
   a) Execute one Gauss-Seidel iteration to compute x^{i+1}.
   b) Check whether the iterate x^{i+1} is close enough to the final solution by checking whether $\|x^{i+1} - x^i\|/\|x^{i+1}\| < \varepsilon$. It is probably simplest to use the $\mathscr{L}_\infty$ norm. If the condition holds, stop with x^{i+1} as the answer. If not, if $i < M-1$, go on; otherwise stop with no answer.
   c) If three Gauss-Seidel iterations have been carried out since the last acceleration, go on; otherwise go to the next iteration.
   d) Compute $\hat{\lambda} = (\Delta^i,\Delta^{i-1})/(\Delta^{i-1},\Delta^{i-1})$ and check whether the error vector is converging to a single eigenvector by checking whether $[1 - \hat{\lambda}(\Delta^i,\Delta^{i-1})/(\Delta^i,\Delta^i)] < \delta^2$. If so, compute the accelerated value $x^{i+1} = (x^{i+1} - \hat{\lambda}x^i)/(1 - \hat{\lambda})$. If not, compute

$$\begin{bmatrix}\hat{\alpha}\\ \hat{\beta}\end{bmatrix} = \frac{1}{(\Delta^{i-2},\Delta^{i-1})^2 - |\Delta^{i-1}|^2|\Delta^{i-2}|^2} \times \begin{bmatrix} |\Delta^{i-2}|^2 & -(\Delta^{i-1},\Delta^{i-2}) \\ -(\Delta^{i-1},\Delta^{i-2}) & |\Delta^{i-1}|^2 \end{bmatrix}\begin{bmatrix}(\Delta^i, & \Delta^{i-1}) \\ (\Delta^i, & \Delta^{i-2})\end{bmatrix}.$$

   Then check whether ε^i is approximately in the plane of eigenvectors by checking whether $\dfrac{|\Delta^i + \hat{\alpha}\Delta^{i-1} + \hat{\beta}\Delta^{i-2}|^2}{|\Delta^{i-1}|^2|\hat{\alpha}^2 - 4\hat{\beta}|} < \delta^2$. If so, compute the accelerated value $\hat{x}^{i+1} = \dfrac{x^{i+1} + \hat{\alpha}x^i + \hat{\beta}x^{i-1}}{1 + \hat{\alpha} + \hat{\beta}}$. If not, go to the next iteration.

e) Because the test of whether to apply acceleration is not definitive, we need a further test to see whether the acceleration was successful. Thus, if we are willing to put in the time to be careful, after each acceleration we might check to determine whether $\|Ax^{i+1}_{after\ accel} - b\| < \|Ax^{i+1}_{before\ accel} - b\|$, where the $\mathscr{L}_\infty$ norm is probably most efficient. If the test passes, we go on to the next iteration. If not, we replace $\hat{x}^{i+1}$ by its value before acceleration, and go on as though acceleration had taken place.

2.8.5 Overrelaxation

The Gauss-Seidel method is an example of a *relaxation method,* a method whereby at each stage of the iterative step a given element of the residual, $b_j - (Ax^i)_j$, is set to zero (relaxed). This relaxation corresponds to the movement to the hyperplane corresponding to the jth equation ($(Ax^i)_j = b_j$), illustrated in figure 2-7. As shown in figure 2-7, the amount of movement at the jth step is sometimes more than that desired (to produce $x_j = z_j$) and sometimes less. Sometimes the movement is in the wrong direction. If we could decide which one of these conditions exists, we could multiply the amount of movement by a number ω_j, greater than 1 if the movement produced by relaxation was too small, positive but less than 1 if the movement produced by relaxation was too large, and negative if the movement was in the wrong direction. Such a process is called *overrelaxation.* It is another technique to accelerate convergence or produce convergence in the case of divergence.

The amount of movement at the jth step is

$$x_j^{i+1} - x_j^i = b_j - \sum_{k=1}^{j-1} A_{kj}x_j^{i+1} - \sum_{k=j+1}^{n} A_{kj}x_j^i - x_j^i, \tag{358}$$

assuming each equation has been scaled so that its diagonal element $A_{jj} = 1$. From equation 358, we obtain

$$x_j^{i+1} - x_j^i = b_j - \sum_{k=1}^{n} A_{kj}x_j^{present} = (b - Ax^{present})_j, \tag{359}$$

where $x^{present}$ is the value of x before the jth stage of the $(i+1)$th iteration.

Note that equation 359 is correct for the Jacobi method as well as for the Gauss-Seidel method, because in the Jacobi method $x^{present} = x^i$ until the $(i+1)$th iteration is complete. Thus, for the Jacobi method, we have the relation

$$x^{i+1} - x^i = b - Ax^i, \tag{360}$$

to which we alluded near the end of section 2.8.3. For the Gauss-Seidel method, the relation is true at each stage if we replace x^i on both sides of the equation by $x^{present}$, but $x^{present}$ varies from stage to stage, that is, from element to element of x^{i+1}.

Equation 359 states that at the jth stage of the $(i+1)$th iteration, x_j^i is changed by the jth residual, $(b - Ax^{present})_j$. To overrelax, we should change x_j^i by $\omega_j(b - Ax^{present})_j$ using an appropriately chosen ω_j. That is,

$$x_j^{i+1} = x_j^i + \omega_j(b - Ax^{present})_j. \tag{361}$$

The theory of how to choose ω_j is complicated and will not be covered here.† However, we will discuss this technique further in chapter 3 in relation to the solution of nonlinear equations.

2.8.6 A Convergent Iterative Solution Method

Lest anyone get the impression that one cannot find an iterative method for the solution of linear equations that always converges, we shall now discuss such a method.‡ We will see, however, that the method is not extendible to nonlinear equations (unlike the relaxation methods) because it depends on the linearity of the equations.

Let us consider geometrically a set of n linearly independent equations in n unknowns. We have seen that each equation defines an $(n-1)$-dimensional hyperplane, orthogonal to the row vector corresponding to that equation. The solution is the point at which the hyperplanes intersect. Finding a vector which solves the ith equation is equivalent to moving to the ith hyperplane. We would like to find a method of moving from one hyperplane to the next such that at every step we get closer to the solution.

If we consider the problem in two dimensions (see figure 2-8), we note that if we start on hyperplane 1 and move to hyperplane 2 in a direction orthogonal to hyperplane 2, by the Pythagorean theorem we

---

† For details, see Varga, *Matrix Iterative Analysis,* chapter 4.

‡ To the best of the author's knowledge, this method was discovered by Kaczmarz and rediscovered by Dr. Lester Levy of Long Island Jewish Medical Center, who brought it to this author's attention.

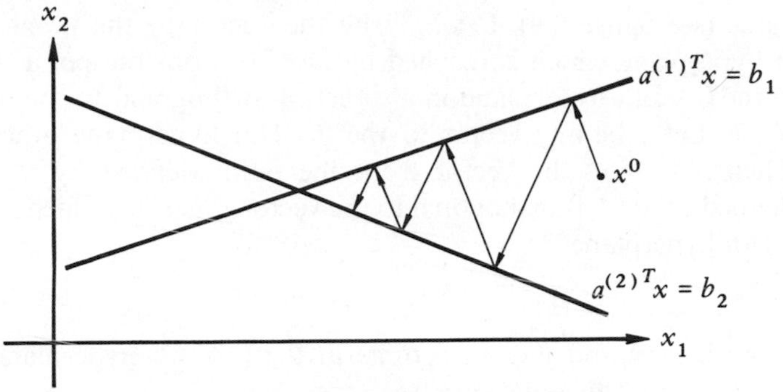

FIG. 2-8 Projection method in two dimensions

will be closer to the intersection of the two hyperplanes. Then if we move from the new point back to hyperplane 1 in a direction orthogonal to hyperplane 1, we will be still closer. By repeating the process, we will converge to the point of intersection, the solution. We can start the process by making any guess and then moving to hyperplane 1 in a direction orthogonal to hyperplane 1. We call this procedure a *projection method* because at each step we project our present estimate of the solution on some hyperplane.

If we generalize this process to n dimensions, we arrive at a process which starts at an arbitrary point, moves to hyperplane 1 in a direction orthogonal to it, then moves to hyperplane 2 in a direction orthogonal to it, then moves to hyperplane 3 in a direction orthogonal to it, and so on, up through movement to hyperplane n in a direction orthogonal to it. From that result, the process starts over by moving to hyperplane 1 in a direction orthogonal to it, thence to hyperplane 2 in a direction orthogonal to it, and so on. It is not as clear that the method converges in n dimensions as it was in two dimensions. When we move to hyperplane $i+1$ from hyperplane i (solve the $(i+1)$th equation), the Pythagorean theorem tells us that we move closer to the intersection of these two hyperplanes, but it is not clear that we may not move further from the intersection of all of the hyperplanes (the solution) by moving away from that point in a dimension along the intersection of the two hyperplanes in question. To convince ourselves that this projection method does converge, we must formulate it and analyze it algebraically.

Consider the equation $Ax = b$, where A is $n \times n$, nonsingular, and has row vectors $a^{(i)}$, $i = 1,2,\ldots,n$. Let x^i be a vector to a point in the ith hyperplane (see figure 2-9). Let x^{i+1} be the vector to the point in the $(i+1)$th hyperplane which is reached by moving from the point defined by x^i to the $(i+1)$th hyperplane in a direction orthogonal to the $(i+1)$th hyperplane. Let y be any vector to the $(i+1)$th hyperplane other than x^{i+1}. Then $x^{i+1} - x^i$, the vector from the point defined by x^i to the point defined by x^{i+1}, is orthogonal to the vector $y - x^{i+1}$, which is along the $(i+1)$th hyperplane:

$$(x^{i+1} - x^i)^T(x^{i+1} - y) = 0. \tag{362}$$

Since both x^{i+1} and y define points in the $(i+1)$th hyperplane, they both solve the $(i+1)$th equation:

$$a^{(i+1)T}x^{i+1} = b_{i+1}, \tag{363}$$

and

$$a^{(i+1)T}y = b_{i+1}, \tag{364}$$

so

$$a^{(i+1)T}(x^{i+1} - y) = 0. \tag{365}$$

Equations 362 and 365 are true for any y to a point in the $(i+1)$th hyperplane (y can range over $n-1$ dimensions). The direction orthogonal

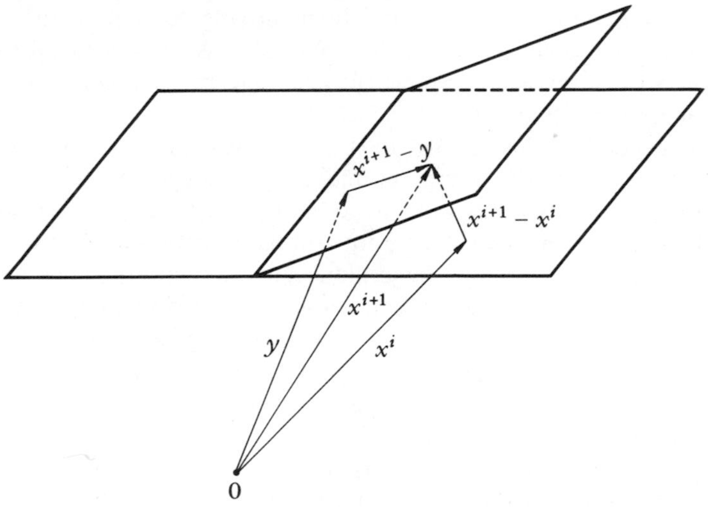

FIG. 2-9 Projection method step

to $x^{i+1} - y$ for all such y is uniquely defined. Thus the direction of $a^{(i+1)}$ is the same as the direction of $x^{i+1} - x^i$, so, for some scalar k,

$$x^{i+1} - x^i = ka^{(i+1)}, \tag{366}$$

or equivalently

$$x^{i+1} = x^i + ka^{(i+1)}. \tag{367}$$

We determine k using equation 363 by taking the inner product of $a^{(i+1)}$ with equation 367, producing

$$b_{i+1} = a^{(i+1)T}x^i + ka^{(i+1)T}a^{(i+1)}. \tag{368}$$

Solving for k, we have

$$k = (b_{i+1} - a^{(i+1)T}x^i)/(a^{(i+1)T}a^{(i+1)}). \tag{369}$$

If, to save time in the iteration, we scale each equation according to the $\mathscr{L}_2$ norm so that $a^{(i+1)T}a^{(i+1)} = 1$, the projection step becomes

$$x^{i+1} = x^i + (b_{i+1} - a^{(i+1)T}x^i)a^{(i+1)}. \tag{370}$$

We can show that if we apply this projection iteratively, the method converges. First, we should define carefully what we mean by applying the method iteratively (see program 2-6). Make a guess x^0. Apply equation 370 successively for $i = 0,1,2,\ldots,n-1$. These successive projections constitute a full iteration. Then we choose a new $x^0 = x^n$ and repeat another full iteration, and so on. So that we can talk of the successive full iterations, let us call x^{mn+i+1} the result of the projection to hyperplane $i+1$ in the $(m+1)$th iteration (where $m = 0$ initially). Then the projection step can be concisely written

$$x^{k+1} = x^k + (b_j - a^{(j)T}x^k)a^{(j)}, \tag{371}$$

where $j =$ the residue of $k+1$ modulo n.

Let z be the solution to $Ax = b$. Then

$$z = z + (b_j - a^{(j)T}z)a^{(j)}, \tag{372}$$

because the term in parentheses is zero. Subtracting equation 372 from equation 371 and defining $\varepsilon^k \equiv x^k - z$, we obtain

$$\varepsilon^{k+1} = \varepsilon^k - (a^{(j)T}\varepsilon^k)a^{(j)} = (I - a^{(j)}a^{(j)T})\varepsilon^k. \tag{373}$$

Thus $\varepsilon^{(m+1)n} = (I - a^{(n)}a^{(n)T})(I - a^{(n-1)}a^{(n-1)T})\ldots(I - a^{(1)}a^{(1)T})\varepsilon^{mn}$. (374)

Let $B = (I - a^{(n)}a^{(n)T})(I - a^{(n-1)}a^{(n-1)T})\ldots(I - a^{(1)}a^{(1)T})$. (375)

Then

$$\varepsilon^{(m+1)n} = B\varepsilon^{mn}. \tag{376}$$

As in section 2.8.1, we can show that $\lim_{m\to\infty} \varepsilon^{mn} = 0$ for all x^0 iff $|B| < 1$.

```
/* PROJECTION METHOD FOR N LINEAR EQUATIONS IN N UNKNOWNS */
/* EQUATION IS AX=B; A AND B ARE GIVEN */
/* AT START, X HOLDS GUESS AT SOLUTION VECTOR */
/* TOL IS RELATIVE ERROR TOLERANCE FOR STOPPING ITERATION */

   DECLARE (A(N,N),            /* INPUT MATRIX */
           B(N),               /* INPUT VECTOR */
           X(N),               /* SOLUTION VECTOR ESTIMATE */
           TOL, ROWMPR,        /* STOPPING TOL, ROW MULTIPLIER */
           PRESX, MAXX;        /* PRESENT AND MAX X(J) VALUES */
           DIFF(N),            /* CHANGE IN X(J) */
                               /* SINCE START OF FULL ITERATION */
           PRESDIF,            /* PRESENT AND */
           MAXDIF) FLOAT,      /* MAXIMUM X(J) CHANGES */
           (SUM,MULR,MULD)       FLOAT(16); /* DOUBLE PRECISION */

/* NORMALIZE MATRIX */
   NORML: DO I=1 TO N;
          SUM=Ø;                          /* COMPUTE IN DBLE PREC */
          ROWNORM: DO J=1 TO N; /* L2 NORM OF EACH ROW */
                   MULR=A(I,J);   /* MAKE A(I,J) DBLE PREC */
                   SUM=SUM+MULR * MULR;
          END ROWNORM;
          SUM=SQRT(SUM);                  /* L2 ROW NORM */
          ROWDIV: DO J=1 TO N;            /* NORMALIZE ROW */
                  MULR=A(I,J);            /* MAKE A(I,J) DBLE PREC */
                  A(I,J)=MULR / SUM;
          END ROWDIV;
   END NORML;

/* COMPUTE NEW ESTIMATE OF SOLUTION */
   NEWITER: DIFF=Ø; /* ZERO CHANGE VECTOR */
   ITER:    DO I=1 TO N;
            SUM=Ø;
   /* COMPUTE (ROW,X) INNER PRODUCT */
         INPROD: DO J=1 TO N;
                 MULR=A(I,J); /* MAKE A(I,J) DBLE PREC */
                 MULD=X(J);   /* MAKE X(J) DBLE PREC */
                 SUM=SUM+MULR * MULD; /* SUM DBLE PREC */
         END INPROD;
         ROWMPR=B(I)-SUM;
   /* COMPUTE NEW SOLUTION ESTIMATE */
         NEWEST: DO J=1 TO N;
                 PRESDIF=ROWMPR * A(I,J);
                 X(J)=X(J)+PRESDIF;
                 DIFF(J)=DIFF(J)+PRESDIF;
         END NEWEST;
   END ITER;
```

PROGRAM 2-6 Projection method for solution of linear equations

(*continued on next page*)

```
/* CHECK WHETHER STOPPING CRITERION IS MET */
   MAXDIF = Ø;
   MAXX = Ø;
   /* FIND MAXIMUM X(J) CHANGE AND MAXIMUM X(J) */
   MAXFIND: DO J = 1 TO N;
               PRESDIF = ABS(DIFF(J));
               IF PRESDIF > MAXDIF THEN MAXDIF = PRESDIF;
               PRESX = ABS(X(J));
               IF PRESX > MAXX THEN MAXX = PRESX;
   END MAXFIND;
   /* ITERATE AGAIN? */
   IF MAXDIF / MAXX > TOL THEN GO TO NEWITER;
   DONE:  /* X HOLDS BEST SOLUTION */
```

PROGRAM 2-6 (*continued*)

From equation 373 we see that

$$|\varepsilon^{k+1}|^2 = |\varepsilon^k|^2 - (a^{(j)T}\varepsilon^k)^2, \tag{377}$$

since $a^{(j)T}a^{(j)} = 1$ (j is as before the residue of $k+1$ modulo n). Since $|\varepsilon^k|^2 \geq 0$ for all k, we see that at every projection $|\varepsilon^k|^2$ gets no larger, and it gets smaller unless $a^{(j)T}\varepsilon^k = 0$. Furthermore, if $|\varepsilon^k|$ gets no smaller at a certain projection, then $\varepsilon^{k+1} = \varepsilon^k$ by equation 373.

Thus $|\varepsilon^{(m+1)n}| \leq |\varepsilon^{mn}|$ for any m, that is, $|B\varepsilon^{mn}| \leq |\varepsilon^{mn}|$ for any m. Assume there exists an m such that

$$|B\varepsilon^{mn}| = |\varepsilon^{mn}|, \text{ that is, } |\varepsilon^{(m+1)n}| = |\varepsilon^{mn}|. \tag{378}$$

This is true iff

$$|\varepsilon^{mn+i+1}| = |\varepsilon^{mn+i}| \quad \text{for } i = 0,1,\ldots,n-1. \tag{379}$$

We have shown this is true iff

$$\varepsilon^{mn+n} = \varepsilon^{mn+n-1} = \cdots = \varepsilon^{mn}, \tag{380}$$

which implies that

$$a^{(i)T}\varepsilon^{mn} = 0, \quad i = 1,2,\ldots,n, \tag{381}$$

that is,

$$A\varepsilon^{mn} = 0. \tag{382}$$

Since A is nonsingular, equation 382 implies that

$$\varepsilon^{mn} = 0. \tag{383}$$

We have shown that if $x \neq 0$, $|Bx| < |x|$. Therefore,

$$|B| = \max_{x \neq 0} (|Bx|/|x|) < 1, \tag{384}$$

from which it follows that

$$\lim_{m \to \infty} \varepsilon^{mn} = \lim_{m \to \infty} B^m \varepsilon^0 = 0. \tag{385}$$

Let us compare the efficiency of the always-convergent projection method to the seldom-convergent Gauss-Seidel method. Examining equation 371, we see that each projection requires $2n$ multiplications. This compares to the $n-1$ multiplications per stage of the Gauss-Seidel method (after all equations have been divided through by their diagonal coefficients). This decrease in efficiency is a small cost for the assurance of convergence. We see also that each new vector element in the projection method requires $n+1$ multiplications as compared to $n-1$ in the Gauss-Seidel method. It is this number of multiplications which determines the computationally generated and propagated error in the final result (if the error due to scaling the equation is negligible, as it is when we do the scaling in double precision). Thus the two methods are approximately equivalent with respect to computational accuracy. But the projection method always converges, normally even if the equation being solved is quite ill-conditioned, though then the convergence is very slow. Thus it is preferable to the Gauss-Seidel method for solving most linear equations.

We have shown that $\varepsilon^{(m+1)n} = B\varepsilon^{mn}$. By the argument used with the Gauss-Seidel method, in the limit as $m \to \infty$, ε^{mn} must approach an eigenvector associated with an eigenvalue of maximum magnitude of B, or a plane defined by two such eigenvectors. Therefore the acceleration technique applied to Gauss-Seidel iteration in section 2.8.4 is applicable here.

We commented before that the projection method cannot be extended to nonlinear equations. The method fails with nonlinear equations because it depends on the fact that each equation represents a hyperplane. If this is not the case, the determination of the direction orthogonal to the surface represented by an equation is very complex, and the method breaks down. It is unfortunate that this method, which is so useful for solving large sets of linear equations, cannot be applied to nonlinear equations as well.

PROBLEMS

2.8.1. Show that if A is an $n \times n$ column diagonally dominant matrix, then the Jacobi method applied to $Ax = b$ converges for any initial estimate x^0.

2.8.2.† Let A be a nonsingular real symmetric $n \times n$ matrix and assume all $A_{ii} > 0$.

a) If a vector u is given, show by calculus that the unique value of the real parameter α which minimizes $(A(u - \alpha e^k), u - \alpha e^k)$ is $\alpha = (1/A_{kk}) \sum_{j=1}^{n} A_{kj} u_j$.

b) Let x^i and x^{i+1} be successive iterates in the Gauss-Seidel process for solving $Az = b$. Let δ^i and δ^{i+1} be the successive errors, $\delta^i = x^i - z$ and $\delta^{i+1} = x^{i+1} - z$. From the result of part a, show that $(A\delta^{i+1}, \delta^{i+1}) \leq (A\delta^i, \delta^i)$ with equality if and only if $\delta^{i+1} = \delta^i = 0$. (Note that δ^{i+1} is formed from δ^i by a succession of n transformations of the form $u \to u - \alpha e^k$. Note also that $\delta^i = \delta^{i+1}$ implies $A\delta^i = 0$, which implies $\delta^i = 0$.)

c) From the result of part b, prove that the Gauss-Seidel method converges for any initial estimate x^0 if and only if A is positive definite.

2.8.3. Let us generalize the Jacobi method of iterative solution of systems of linear equations to apply to matrices of matrices. That is, let the system be $Ax = b$ where A is a square matrix:

$$A = \begin{bmatrix} C^{11} & C^{12} & \cdots & C^{1N} \\ C^{21} & C^{22} & \cdots & C^{2N} \\ \vdots & \vdots & & \vdots \\ C^{N1} & C^{N2} & \cdots & C^{NN} \end{bmatrix},$$

where the C^{ii} are nonsingular, square matrices with n_i rows and n_i columns and the C^{ij} are matrices with n_i rows and n_j columns.

$$\text{Let } b = \begin{bmatrix} b^1 \\ b^2 \\ \vdots \\ b^N \end{bmatrix} \text{ and } x = \begin{bmatrix} x^1 \\ x^2 \\ \vdots \\ x^N \end{bmatrix}$$

where b^i and x^i are vectors with n_i elements.

Method:

Let us iterate by relaxing n_i rows at a time. That is, choose $x^{(0)}$, an initial guess at the solution to $Ax = b$. The iterative step, to compute $x^{(k)}$ from $x^{(k-1)}$ is as follows:

1. Solve for $x^{1(k)}$ (using C^{11} as the coefficient matrix of the unknowns) by taking all of the x elements not in x^1 to be known to be equal to their guessed values in $x^{(k-1)}$.

---

† J. N. Franklin, *Matrix Theory*, section 7.6, problems 1–3.

2. Solve for $x^{2(k)}$ (using C^{22} as the coefficient matrix of the unknowns) by taking all of the x elements not in x^2 to be known to be equal to their guessed values in $x^{(k-1)}$.

.

.

.

N. Solve for $x^{N(k)}$ (using C^{NN} as the coefficient matrix of the unknowns) by taking all of the x elements not in x^N to be known to be equal to their guessed values in $x^{(k-1)}$.

Mathematically stated, the iterative step is:

Solve $C^{ii}x^{i(k)} = b^i - \sum_{j \neq i} C^{ij}x^{j(k-1)}$ by Gaussian elimination for $i = 1,2,\ldots,N$.

Problem:

a) Prove that a sufficient condition for convergence of this block Jacobi method for all starting guesses is

$$\sum_{\substack{j=1 \\ j \neq i}}^{N} \|C^{ij}\| < 1/\|(C^{ii})^{-1}\|.$$

b) Prove that a sufficient condition for convergence of this block Jacobi method for all starting guesses is

$$\sum_{\substack{i=1 \\ i \neq j}}^{N} \|C^{ij}\| < 1/\|(C^{jj})^{-1}\|.$$

c) Show that this block Jacobi iteration can be written $x^{i+1} = Bx^i + c$, where

$$-B = \begin{bmatrix} 0^1 & (C^{11})^{-1}C^{12} & (C^{11})^{-1}C^{13} & \cdots & (C^{11})^{-1}C^{1N} \\ (C^{22})^{-1}C^{21} & 0^2 & (C^{22})^{-1}C^{23} & \cdots & (C^{22})^{-1}C^{2N} \\ (C^{33})^{-1}C^{31} & (C^{33})^{-1}C^{32} & 0^3 & \cdots & (C^{33})^{-1}C^{3N} \\ \vdots & \vdots & \vdots & & \vdots \\ (C^{NN})^{-1}C^{N1} & (C^{NN})^{-1}C^{N2} & (C^{NN})^{-1}C^{N3} & \cdots & 0^N \end{bmatrix},$$

where 0^i is an $n_i \times n_i$ matrix of zeros. Thus, a necessary and sufficient condition for convergence of this block Jacobi method is that the eigenvalue of maximum magnitude of this matrix B is less than 1 in magnitude. Show directly that the condition of part a is sufficient to ensure this eigenvalue bound.

d) How would you modify this method (while maintaining the same block structure) to improve the rate of convergence for most matrices? Using the notation $A = D + E$, where the iteration is $x^{(k)} = E^{-1}(b - Dx^{(k-1)})$, what is the matrix E in this modified method?

2.8.4. Iterative methods for finding roots of systems of n linear equations in n unknowns depend on transforming the equation $Ax = b$ into $Ex = b - (A - E)x$, and then iterating by the formula

$$Ex^{(i+1)} = b - (A - E)x^{(i)}. \tag{1}$$

Let E be the tridiagonal part of A. That is, E is made up of the diagonal, super-diagonal, and subdiagonal of A:

$$E_{ij} = \begin{cases} A_{ij}, & |i-j| \le 1 \\ 0, & |i-j| > 1. \end{cases}$$

Assume E is nonsingular.

We desire to solve equation 1 for $x^{(i+1)}$, given $x^{(i)}$.

a) Show that with $2(n-1)$ multiplications and divisions we can triangularly factorize the tridiagonal matrix E into a lower triangular matrix L with only a diagonal and a subdiagonal and an upper triangular matrix U with 1s on the diagonal and only a superdiagonal besides.

b) Once we have done this factorization, how many multiplications and divisions are required to solve equation 1, that is, how many multiplications and divisions are required per iteration?

c) Instead we could invert E and multiply it by A and b to obtain the iteration above in the form

$$x^{(i+1)} = E^{-1}b - (E^{-1}A - I)x^{(i)}. \tag{2}$$

How many operations would the inversion require?

d) How many multiplications would computing $E^{-1}A$ and $E^{-1}b$ require?

e) How many multiplications per iteration would an iteration using equation 2 require?

f) How should the iteration be modified to obtain faster convergence for most matrices? Why might we choose to use the method using equation 2 over that using equation 1?

g) If we made the choice specified in part f, we would in effect be using an iteration with a different matrix E than the one specified above. Give the element values of the new matrix E.

2.8.5. Solve the equation $Ax = b$ using the Gauss-Seidel method with acceleration (see algorithm at end of section 2.8.4) for the following values of A and b.

a)
$$A = \begin{bmatrix} 3 & 2.88 & 2.52 & 1.92 \\ .72 & .691 & .3304 & .1666 \\ 1.26 & .6608 & 1.5 & .516 \\ .24 & .0833 & .129 & .375 \end{bmatrix} \text{ and } b = \begin{bmatrix} 10.32 \\ 1.908 \\ 3.9368 \\ .8273 \end{bmatrix}.$$

(Solution is $x = (1,1,1,1)^T$.) Use the value $\delta = .1$ for acceleration and $\varepsilon = 10^{-6}$ for final convergence.

b)

$$A = \begin{bmatrix} 2 & -4 & 1 \\ 4 & -5 & 2 \\ -2 & -5 & 1 \end{bmatrix} \quad \text{and } b = \begin{bmatrix} 11 \\ 7 \\ 4 \end{bmatrix}.$$

(Solution is $(2, -1, 3)^T$.) Use the value $\delta = .05$ for acceleration and $\varepsilon = 10^{-5}$ for final convergence.

2.8.6. Solve the equations of problem 2.8.5 using the projection method with acceleration. Use the same values for ε and δ as specified in problem 2.8.5.

REFERENCES

Forsythe, G. E., and Moler, C. B. *Computer Solution of Linear Algebraic Systems.* Englewood Cliffs, N.J.: Prentice-Hall, 1967.

Franklin, J. N. *Matrix Theory.* Third Printing, Englewood Cliffs, N.J.: Prentice-Hall, 1968.

Noble, B. *Applied Linear Algebra.* Englewood Cliffs, N.J.: Prentice-Hall, 1969.

Ralston, A. *A First Course in Numerical Analysis.* New York: McGraw-Hill, 1965.

Varga, R. S. *Matrix Iterative Analysis.* Englewood Cliffs, N.J.: Prentice-Hall, 1962.

3

NONLINEAR EQUATIONS

3.1 INTRODUCTION

We now move on to nonlinear equations, equations where at least one variable appears in a form more complicated than simply multiplied by a constant. For example, consider the following system of two nonlinear equations in two unknowns:

$$e^x \cos(y-1) + e^{xy} \cos(x^2 y) - 1.95 = 0$$

$$\tan^{-1}\left(\frac{y+e^x}{2}\right) - \tfrac{3}{4}(x + e^{y-1})^2 = 0. \tag{1}$$

In general we are interested in solving equations of the form $\underline{f}(\underline{x}) = 0$, where both $\underline{f}$ and $\underline{x}$ are vectors. What makes such equations difficult to solve is that we have no general algorithm for solving them directly. Since there is no closed form solution, we are forced to use an iterative technique.

We have seen that getting many linear equations in many unknowns to converge using an iterative technique may be difficult. Nonlinearity complicates matters, making it harder to find convergent methods, and even harder to find such methods which are reasonably efficient. So, we do not often attempt to solve nonlinear equations in many (more than ten) unknowns, and in fact we begin to worry if we see even more than one. Therefore, in this chapter, we will direct most of our attention to single nonlinear equations in one unknown,

$$f(x) = 0. \tag{2}$$

We will devise a number of methods for iteratively solving nonlinear equations, though our catalog will not be exhaustive. Whether or not a particular method should be used for a given equation (function f) depends on answers to the following questions.

1. Will the method work for this root of this function f?
   a) Is the method applicable?
   b) Does the method converge to the desired root?

2. How efficient is the method for this problem?
   a) How fast does it converge? That is, how many iterative steps are required to achieve the desired accuracy?
   b) How much computing time is required in each iterative step?
      (1) Due to the method itself?
      (2) Due to evaluations of the function f required by the method?
3. How accurate is the method?
   a) How many and which operations are required in each iterative step?
   b) Can we establish a criterion determining when the iterate is of desired accuracy?

Answers to these questions depend on the function f itself, in particular on its general shape. Therefore, before selecting a method, we must sketch a graph of the function.

3.2 GRAPHING FUNCTIONS OF ONE VARIABLE

An inefficient way to sketch a graph of a function is to calculate its value at many points in the interval in question. Instead of this approach, the following techniques should be used.

1. Evaluate the function at the endpoints of the interval in question (perhaps $+\infty$ and $-\infty$). Also evaluate the function at points in the interval where the computation is simple.
2. Find the singular points of the function, points at which it is infinite or undefined.
3. The first derivative and sometimes the second derivative give very useful shape information, more useful than computed value information. Thus, compute the first derivative at appropriate points: in particular, at points near the points at which the function has been evaluated, and at points where evaluation of the derivative is simple.
4. Find the approximate places where the derivative is zero, that is, where maxima and minima occur. At each of these points, determine whether the extremum in question is a maximum or a minimum, either as a result of knowing the sign of the slope on one side of the point or by evaluating the second derivative at the point.
5. Decompose the function into additive or multiplicative parts, sketch each part, and then build the graph of the overall function by combining the graphs of the parts.

Let us consider the function

$$f(x) = x - e^{1-x}(1 + ln(x)) \tag{3}$$

on the interval $[-\infty, +\infty]$. If we first analyze the function for singular points, we note that $ln(x)$ is defined only for $x > 0$, so $f(x)$ is defined only there. Thus we restrict the interval to $[0, \infty]$. As $x \to 0$, $ln(x) \to -\infty$, so $f(x) \to +\infty$. As $x \to \infty$, $ln(x) \to \infty$, but e^{1-x} approaches 0 faster, so the second term in f approaches 0. The first term (which approaches ∞) dominates and $f(x) \to \infty$. We conclude that $f(x) \to \infty$ as x approaches either 0 or ∞, and that f has no other singular points in $[0, \infty]$.

While trying to find other points at which to evaluate f, we notice that both the exponential and the logarithm are evaluated easily at $x = 1$, producing $f(1) = 0$.

Calculating the first derivative of f we obtain

$$f'(x) = 1 - e^{1-x}(1/x - 1 - ln(x)). \tag{4}$$

Evaluating f' at 1, we obtain $f'(1) = 1$. Our present knowledge of the function f is recorded in figure 3-1, part a. Since f is continuous, we can see that there must be another zero of the function f between zero and one. Are there any other zeros of f and what more can we say about the shape of f? We note that $f'(\infty) = 1$, and evaluating f'' at ∞ using

$$f''(x) = e^{1-x}(1/x^2 + 2/x - 1 - ln(x)), \tag{5}$$

we see that f'' is negative as x gets large, that is, the function f is concave downward as x gets large. Furthermore, we see $f''(1) = 2$, so f is concave upward at the root $x = 1$. Our present information is sketched in figure 3-1, part b.

We can get more information about f by examining its parts: x, e^{1-x}, and $1 + ln(x)$. Being familiar with the graphs of elementary functions is helpful when considering more complex functions. The parts of f are shown in figure 3-1, part c. Note that $1 + ln(x)$ is a graph of $ln(x)$ raised by one unit and e^{1-x} is e^{-x} set to the right one unit. Multiplying e^{1-x} and $1 + ln(x)$ (figure 3-1, part d), we see that the product is negative between 0 and some value less than 1, goes to 0 as x gets large, and stays below the graph of x for all values of $x > 1$. Since the derivative of the product is zero at $x = 1$, the product curve has a maximum there, and we can sketch the rest of the curve as in the figure. Note that the zeros of f are the two points where the curve $y = x$ and the curve representing the product of the other two functions intersect. This means there are exactly two zeros (roots) of our function. We can also conclude that since $e^{1-x}(1 + ln(x)) \to 0$ as x gets large, f approaches $y = x$

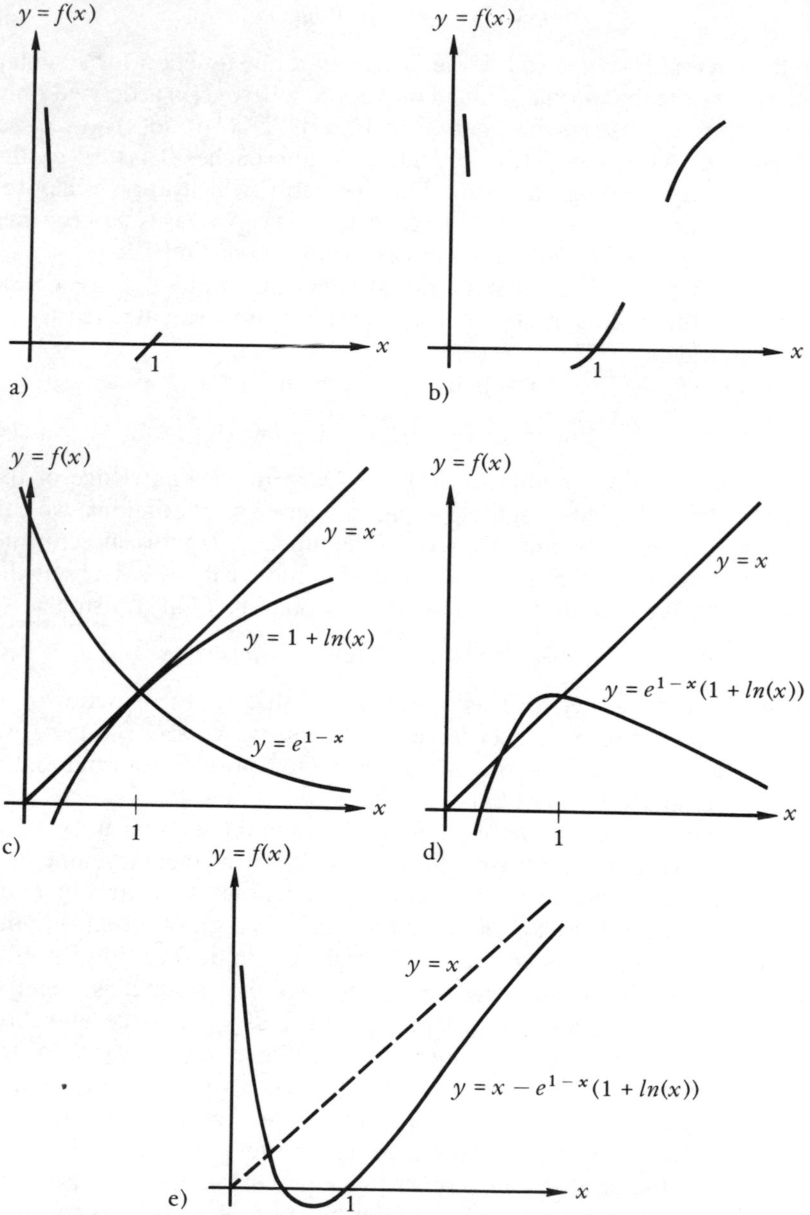

FIG. 3-1 Graphing $f(x) = x - e^{1-x}(1 + ln(x))$

asymptotically as x gets large, and that the slope of f is $-\infty$ at $x = 0$, because the slope of f is the slope of x $(=1)$ minus the slope of the product $(=\infty)$. Finally, we can sketch the curve as in figure 3-1, part e. There is a root at 1 and a root near $\frac{1}{2}$.

PROBLEMS

3.2.1. Graph the following functions in the intervals specified.

a)†	$log(3 - x)\, tan(\pi x) - (x - 2)^2$	(1.75, 2.25)
b)	$e^{.001 sec(x)}(x + \frac{1}{2}) + sin^{-1}(x)$ [where $sin^{-1}(-1) = -\pi/2$]	$(-1, 0)$
c)	$e^x - 1 + .02 sin(30\pi x + \pi/4)$	$(-2, 2)$
d)	$e^{x^2} + x - 1000$	(1, 10)
e)	$4(x - 2)^{1/3} + sin(3x)$	$(0, \pi)$
f)	$.1x^2 + tan(x+1)$	$(-2, -1)$
g)	$(x - 5)(e^{x/5} - e)$	(4, 6)
h)	$4(log(x^2+4) - e^{.3x}) - e^{x-1}$	(1, 2)
i)	$4(log(x^2+4) - e^{.3x})(1 + .02 sin(100x)) - e^{x-1}$	(1, 2)
j)	$4(log(x^2+4)\, cos(.001(x-1)) - e^{.3x} e^{sec((x-1)/100)-1}) - e^{(x-1)^6}$	(1, 2)
k)	$(e^x + 1)\, log(x+1) - 3x$	$(.1, \infty)$
l)	$log(sin(x)+2) - (x+1)^{1/2}/2 + \frac{1}{2}$	$(0, 2\pi)$
m)	$e^{sec(x)} + (x - 5)(sec(\pi x/4))^{.001}$	(0, 1)
n)	$tan(x-1) + e^{.001x}$	(0, 2)

3.3 ITERATIVE METHODS FOR SOLVING ONE NONLINEAR EQUATION IN ONE UNKNOWN

3.3.1 Properties of Roots and Sequences

Solving nonlinear equations in one unknown involves searching for roots of equations of the form $f(x) = 0$, where f is a continuous function on an interval of interest including the root. By definition, a *root* of

† Throughout this book, *log* with no subscript will be taken to be base e, that is, the same as *ln*.

$f(x) = 0$ (or equivalently, a root of $f(x)$) is a value $x = \xi$ such that $f(\xi) = 0$. We can find the root by designing iterations defined by

$$x_{i+1} = g(x_i), \tag{6}$$

or more generally by

$$x_{i+1} = g(x_i, x_{i-1}, \ldots, x_{i-n+1}), \tag{7}$$

that converge to the root ξ.

Intuitively, we have a notion that a root at ξ has *multiplicity m* if $(x - \xi)^m$ is a factor of $f(x)$ and $(x - \xi)^{m+\varepsilon}$ is not a factor of $f(x)$ for $\varepsilon > 0$. That is, the root is of multiplicity m if $f(x)/(x - \xi)^m$ is well defined as $x \to \xi$ and is not zero there. To handle cases like $f(x) = |x - \xi|^3$, which intuitively has multiplicity 3 but for which $\underset{x\to\xi}{lim} |x - \xi|^3/(x - \xi)^3$ is not well defined because there are two limit points ($+1$ and -1), we define multiplicity to involve all limit points of $f(x)/(x - \xi)^m$: The multiplicity m of the root at ξ is that value of $m > 0$ such that if $\{p\}$ is the set of limit points of $f(x)/(x - \xi)^m$ as $x \to \xi$, each element of the set is finite, and zero is not the only element.

Note that for any $n < m$,

$$f(x)/(x - \xi)^n = [f(x)/(x - \xi)^m][(x - \xi)^{m-n}]. \tag{8}$$

As $x \to \xi$, the first factor on the right is bounded while the second factor approaches zero, so $f(x)/(x - \xi)^n \to 0$, and thus n is not the multiplicity of $f(x)$ at ξ.

Similarly, for $n > m$,

$$f(x)/(x - \xi)^n = [f(x)/(x - \xi)^m][1/(x - \xi)^{n-m}]. \tag{9}$$

As $x \to \xi$, the first factor on the right has at least one nonzero limit point and the second factor on the right becomes unbounded, so $f(x)/(x - \xi)^n$ has an unbounded limit point, and thus n is not the multiplicity of $f(x)$ at ξ.

From equations 8 and 9, we conclude that m is unique.

Consider the function $g(x) = f(x)/(x - \xi)^m$. Assume g is continuous at ξ. Then $g(\xi) = k \neq 0$. Assume g is differentiable at ξ. Then wherever $f'(x)$ is defined,

$$f'(x) = \frac{d}{dx}[(x - \xi)^m g(x)] = m(x - \xi)^{m-1} g(x) + (x - \xi)^m g'(x), \tag{10}$$

so

$$f'(\xi) = \begin{cases} 0 \text{ if } m > 1, \\ k \text{ if } m = 1, \\ \text{is not defined if } m < 1. \end{cases}$$

Thus, if f has a root of multiplicity 1 (a single root) at $x = \xi$, the graph of $f(x)$ crosses the x axis at ξ with some slope not equal to zero. If $m > 1$, f has a zero derivative at ξ; f is tangent to the x axis and may or may not cross the axis. If $m < 1$, something peculiar happens to f at $x = \xi$; either the derivative is infinite or f is not defined on one side of the axis.

Another concept with a definition similar to that of the multiplicity of a root is the *order of convergence* of a sequence. In our case the sequence is produced by an iteration defined by equation 6 or 7 and initial guess(es) $x_0(x_1,\ldots,x_n)$, but the order of convergence is defined with respect to an arbitrary sequence. Given a sequence $\{x_i\}$ and assuming the sequence has a limit, we wish to characterize the speed of the convergence of the sequence to its limit, so that we may choose the method which is most efficient. The order of convergence is used in such a characterization.

Clearly, there is a qualitative difference between the sequence

a) $\{2,\frac{3}{2},\frac{5}{4},\frac{9}{8},\frac{17}{16},\ldots\}$

and the sequence

b) $\{2,\frac{3}{2},\frac{9}{8},\frac{129}{128},\ldots\}$.

Close inspection will show that in the first case $(x_{i+1} - 1) = .5(x_i - 1)$, whereas in the second case $(x_{i+1} - 1) = .5(x_i - 1)^2$. The latter behavior appears to be a preferred quality of convergence. As we shall see, these are cases of *first-order* (*linear*) *convergence* and *second-order* (*quadratic*) *convergence*, respectively. Because of inevitable special cases, the quantification of the distinction between these sequences has some complexity.

Let $\{x_i\}$ be a sequence converging to ξ. Intuitively, we would like to define α, the order of convergence of $\{x_i\}$, as the value of α such that as $x_i \to \xi$,

$$(x_{i+1} - \xi) \to k(x_i - \xi)^\alpha, \tag{11}$$

where k is a constant $\neq 0$. With such a definition, the notion is that the higher the order of convergence of a sequence, the quicker the convergence. Note, however, that even if equation 11 holds for all x_i, the order of convergence is related to the speed of convergence only when x_i is near ξ. Furthermore, just because a sequence has the property $(x_{i+1} - \xi) \to k(x_i - \xi)^\alpha$ as $x_i \to \xi$, it does not follow that it has that property for x_i arbitrarily far from ξ.

The above definition needs to be made more precise. We define the sequence $\{\rho_i(\alpha)\}$ by

$$\rho_i(\alpha) = (x_{i+1} - \xi)/(x_i - \xi)^\alpha. \tag{12}$$

The order of convergence is then the value of α such that $\{\rho_i(\alpha)\}$ has a finite, nonzero limit point. With such a definition, the distinction between the sequences above is drawn as $\alpha = 1$ and $\alpha = 2$, respectively.

Three problems arise with our second definition of the order of convergence:

1. The limit point is not necessarily unique for a particular α.
2. The value of α such that $\{\rho_i(\alpha)\}$ has a finite nonzero limit point may not be unique.
3. Such a limit point does not necessarily exist for *any* α.

As an example of problem 1, consider the convergent sequence

c) $\{2, \frac{3}{2}, \frac{9}{8}, \frac{17}{16}, \frac{65}{64}, \frac{129}{128}, \ldots\}$.

For this sequence $\{\rho_i(\alpha)\}$ has two limit points for $\alpha = 1$ (at .5 and at .25), zero limit for $\alpha < 1$, and no limit (or infinite limit, if you prefer) for $\alpha > 1$.

As an example of problem 2, consider the sequence

d) $\{2, \frac{3}{2}, \frac{9}{8}, \frac{17}{16}, \frac{513}{512}, \ldots\}$, for which as $i \to \infty$, $\rho_i(1) \to \begin{cases} \frac{1}{2}, & i \text{ even} \\ 0, & i \text{ odd} \end{cases}$ and

$$\rho_i(2) \to \begin{cases} \infty, & i \text{ even} \\ \frac{1}{2}, & i \text{ odd.} \end{cases}$$

As an example of problem 3, consider a sequence

e) $\{x_i\}$ converging to a value ξ in such a manner that $(x_{i+1} - \xi) = (\frac{1}{2})^i (x_i - \xi)$.

In this case $\rho_i(\alpha) = (x_{i+1} - \xi)/(x_i - \xi)^\alpha = (\frac{1}{2})^i (x_i - \xi)^{1-\alpha}$. When $\alpha \leq 1$ and x_i approaches ξ, the sequence $\{\rho_i(\alpha)\}$ converges to zero, and when $\alpha > 1$, it diverges.

All of these problems are accounted for in the definition that follows: Let $\{x_i\}$ be a convergent sequence, and let ξ be its limit point. Now let α_0 be the least upper bound of the numbers α such that $\{\rho_i(\alpha) = (x_{i+1} - \xi)/(x_i - \xi)^\alpha\}$ has only finite limit points. Then α_0 is called the *order of convergence* of $\{x_i\}$. The least upper bound of the limit points of $\rho_i(\alpha_0)$ is called the *convergence factor* of the sequence $\{x_i\}$.

The above definition has been made quite complicated to cover many peculiar cases. However, most commonly encountered sequences converging with order α and convergence factor k have the simple property that

$$\lim_{x_i \to \xi} [(x_{i+1} - \xi)/(x_i - \xi)^\alpha] = k \neq 0. \tag{13}$$

We will call such behavior *simple convergence* of order α with convergence factor k.

Table 3-1 gives the order of convergence and convergence factor for each of the sequences used as examples above.

TABLE 3-1 PROPERTIES OF EXAMPLE SEQUENCES

Sequence Identifier	Order of Convergence	Convergence Factor	Simple Convergence
a	1	$\frac{1}{2}$	Yes
b	2	$\frac{1}{2}$	Yes
c	1	$\frac{1}{2}$	No
d	1	$\frac{1}{2}$	No
e	1	0	No

By its definition as a least upper bound, the order of convergence of a sequence is unique if it exists, that is, if the order of convergence is not infinite. (An example of a sequence with infinite order of convergence is that defined by $x_i = (\frac{1}{2})^{i!}$.) A convergent sequence cannot have an order of convergence $\alpha < 1$, because for x_i sufficiently near ξ,

$$|x_{i+1} - \xi| \approx |k(x_i - \xi)^\alpha| = \left(\frac{|k|^{1/(1-\alpha)}}{|x_i - \xi|}\right)^{1-\alpha} |x_i - \xi|, \tag{14}$$

and thus for $|x_i - \xi| < |k|^{1-\alpha}$,

$$|x_{i+1} - \xi| > |x_i - \xi|, \tag{15}$$

a contradiction to the fact that the sequence converges.

When $\alpha < 1$, equation 14 defines a sequence diverging from ξ in a way that can be described as "with order of divergence α" for x_i sufficiently near ξ. This notion of *order of divergence* would be well defined if we had a concept of the "point from which the sequence is diverging." This concept is not well defined for general sequences, but it can be made so for the special sequences defined by an iteration function, g (equation 6 or 7).

The case given by equation 7, in which g is a function of n previous iterates, can be treated in much the same way as equation 6, in which g is a function of only the immediately preceding iterate. A means for doing so is illustrated in section 3.3.7. For now assume that a sequence is defined by the iteration function g in equation 6 and the initial value x_0. If the sequence converges to ξ, then

$$\xi = g(\xi). \tag{16}$$

A point ξ such that $g(\xi) = \xi$ is called a *fixed point* of g.

Theorem: If $x_{i+1} = g(x_i)$ defines a sequence $\{x_i\}$, where g is continuous at ξ, then if the sequence converges to ξ, ξ must be a fixed point of the iteration.

Proof: We assume $\lim_{i\to\infty} x_i = \xi$. Then by the continuity of g,

$$g(\xi) = g\left(\lim_{i\to\infty} x_i\right) = \lim_{i\to\infty} g(x_i) = \lim_{i\to\infty} x_{i+1} = \xi. \tag{17}$$

An iteration may have a fixed point to which it does not converge. That is, we may choose x_0 arbitrarily close to a fixed point ξ and still have the sequence diverge. However, we now have a notion of the point from which the sequence is diverging—the fixed point. Thus we can extend our notion of order of convergence of a *sequence* to cover cases of divergence if we define the order of convergence (divergence) of an *iteration function* g at fixed point ξ as the least upper bound of $\{\alpha|$ all limit points as $x\to\xi$ of $(g(x) - \xi)/(x - \xi)^\alpha$ are finite$\}$. Note that a given iteration function g may have a different order of convergence (divergence) at each fixed point. Furthermore, the iteration may (and often does) converge for some starting values x_0 and diverge for others.

We have already shown that any function with order of convergence $\alpha < 1$ at a fixed point ξ will produce a sequence diverging from ξ if x_0 is chosen appropriately close to ξ. By a similar argument we can show that if g has order of convergence $\alpha > 1$ at ξ, there exists a neighborhood of ξ such that if x_0 is in the neighborhood, the sequence will converge to ξ. Since

$$|x_{i+1} - \xi| \approx |k|\,|x_i - \xi|^\alpha = (|k|\,|x_i - \xi|^{\alpha-1})|x_i - \xi|, \tag{18}$$

we require only $|x_i - \xi| < |k|^{1/(1-\alpha)}$ for convergence.

If equation 18 holds for x_i far from ξ (and it need not, even if $\alpha > 1$), we see that for $|x_i - \xi|$ large enough, $|x_{i+1} - \xi| > |x_i - \xi|$, so methods producing sequences with order of convergence greater than 1 may diverge if x_0 is not sufficiently close to ξ.

We have so far seen two reasons for preferring methods which produce order of convergence > 1 to methods which produce first-order, or linear, convergence:

1. In contrast to linearly convergent methods, higher-order methods are certain to have an interval about the root for which convergence is assured if x_0 is in that interval.
2. Higher-order methods tend to be more efficient, because the sequence of iterates converges faster.

Let us analyze this efficiency in more detail for simply converging methods.

In general it is not possible to compute analytically the time requirements of an iterative method (iteration function) knowing only the order of convergence, convergence factor, and amount of computing time required per iteration. We cannot deduce from these three values the number of iterations required to achieve a desired error tolerance. However, given only these values for a method, it is possible to deduce the limiting behavior of the method when the value of the iterate is "very close" to the root, and the efficiency of a method is normally gauged on this basis.

Let $\{x_0,x_1,x_2,\ldots\}$ be a convergent sequence for which x_0 is sufficiently close to the limit such that

$$|x_{i+1} - \xi| = |k|\,|x_i - \xi|^\alpha \tag{19}$$

is approximately true for any i. Then

$$\begin{aligned} |x_{i+2} - \xi| &= |k|\,|x_{i+1} - \xi|^\alpha \\ &= |k|^{1+\alpha}|x_i - \xi|^{\alpha 2}, \end{aligned} \tag{20}$$

$$\begin{aligned} |x_{i+3} - \xi| &= |k|\,|x_{i+2} - \xi|^\alpha \\ &= |k|^{1+\alpha+\alpha 2}|x_i - \xi|^{\alpha 3}, \end{aligned} \tag{21}$$

and so

$$|x_{i+m} - \xi| = |k|^{\sum_{i=0}^{m-1} \alpha i}\,|x_i - \xi|^{\alpha m}. \tag{22}$$

That is, if $\alpha > 1$,

$$|x_{i+m} - \xi| = |k|^{(\alpha m - 1)/(\alpha - 1)}|x_i - \xi|^{\alpha m}. \tag{23}$$

Equation 22 states that the sequence $\{x_0,x_m,x_{2m},\ldots\}$ has order α^m; if the original sequence converges linearly, so does the new sequence obtained by applying the old iteration function m times per new iterative step. But if $\alpha > 1$, the new sequence has order of convergence $\alpha^m > \alpha$. In particular if $m = 1/log_2(\alpha)$, the new sequence has order of convergence 2. Thus, in the sense that any superlinear iteration function ($\alpha > 1$) is equivalent to some quadratic iteration function, there are only two types of convergent iteration functions: linear and superlinear.

With a linearly converging iteration function with convergence factor k_1, we obtain from equation 22

$$|x_m - \xi| = |k_1|^m |x_0 - \xi|, \tag{24}$$

so if δ is the required accuracy of the approximation to ξ, $log(|x_0 - \xi|/\delta) \,/\, log|k_1|$ iterations are required. If t_1 is the computing time per iteration required by the method, the total time T_1 required is given by

$$T_1 = t_1 \, log(|x_0 - \xi|/\delta) \,/\, log|k_1|. \tag{25}$$

With a superlinearly converging iteration function with order of convergence α_2 and convergence factor k_2, to find the number of iterations required, we must solve $|x_m - \xi| = \delta$ for m, which produces

$$m = \frac{1}{log(\alpha_2)} \, log\left(\frac{log(|k_2|)/(\alpha_2 - 1) + log(\delta)}{log(|k_2|)/(\alpha_2 - 1) + log(|x_0 - \xi|)} \right) \tag{26}$$

(see exercise 3.3.3, part a). If t_2 is the computing time per iteration for this iteration function, the total time T_2 required is given by

$$T_2 = \frac{1}{log(\alpha_2)} \, log\left(\frac{log(|k_2|)/(\alpha_2 - 1) + log(\delta)}{log(|k_2|)/(\alpha_2 - 1) + log(|x_0 - \xi|)} \right). \tag{27}$$

Note that the base of the logarithm is arbitrary. If it is taken to be 2, the first factor of T_2, namely, $t_2/log_2(\alpha_2)$, is the computing time required for a new iterative step which produces a quadratically converging sequence.

Since we are interested in limiting behavior, we can assume $|x_0 - \xi|$ and δ are as small as we like. Under this assumption, their logarithms dominate the numerator and denominator, respectively, of the argument of *log* in T_2. Thus in the limit, the efficiency of a superlinearly converging iteration does not depend on the convergence factor. It depends only on the order of convergence and the time per iterative step.

Finally, since neither δ nor $|x_0 - \xi|$ can be presumed to be a function of the particular iterative method chosen, the most efficient superlinear method is the one which minimizes $t_2/log(\alpha_2)$.

If we compare T_1 and T_2, the times required to achieve accuracy δ with linear and superlinear methods, we find that for any t_1, t_2, k_1, k_2, and α_2, in the limit as $x_0 \to \xi$ and $\delta \to 0$, $T_2 < T_1$, that is, the superlinear method is more efficient than the linear method (see exercise 3.3.3, part b).

The order of convergence and convergence factor of an iteration determine not only the efficiency of the iteration as the iterates approach the limit, but also the accuracy with which the limit (in this chapter, a root of an equation) can be determined, given that error is generated at each iterative step during the computation of g. We cannot be assured of getting any closer to the limit than the point where the improvement due to the fact that $g(x_i)$ is closer to the limit than x_i, is less than the generated error at that step (see exercises 2.7.2 and 3.3.2). This improvement, and thus the error bound for the computed limit, depends on the speed of convergence.

More specifically, we can show that the greatest achievable accuracy of a linearly converging method with convergence factor k is proportional to $1/(1 - |k|)$, which for $|k| < 1$ (a converging sequence) decreases as $|k|$ decreases (see exercise 2.7.2). Since for a simply converging sequence with order of convergence $\alpha_2 > 1$, in the limit

$$|x_{i+1} - \xi| = |k_2|\,|x_i - \xi|^{\alpha_2}$$
$$= (|k_2|\,|x_i - \xi|^{\alpha_2 - 1})|x_i - \xi| \tag{28}$$

and

$$\lim_{i\to\infty} |k_2|\,|x_i - \xi|^{\alpha_2 - 1} = 0, \tag{29}$$

superlinear convergence can be likened to linear convergence with convergence factor zero, thus minimizing the above-mentioned proportionality constant. Therefore, superlinear convergence is to be preferred not only on the grounds of efficiency and certainty of convergence for an appropriate starting value, but also on the grounds of accuracy.

3.3.2 Bisection Method

One technique for finding the root of a nonlinear equation is to trap the root in an interval and then make the interval smaller and smaller. If applicable, such a procedure always produces convergence to the root. The simplest form of this technique is to find two points x^+ and x^- such that $f(x^+) > 0$ and $f(x^-) < 0$. Since f is assumed to be continuous, there must be a root of f in the interval $[x^+,x^-]$.† We choose x at the midpoint of the interval and evaluate $f(x)$. If $f(x) > 0$, the root must be in the interval $[x,x^-]$. We can replace x^+ with x, resulting in an interval that is half the size of the previous interval. If $f(x) < 0$, we can replace x^- with x, again resulting in an interval that is half the size of the previous interval. Thus, at each iteration the root is trapped in an interval that is half the size of that produced by the previous iteration. We continue until the interval is smaller than the tolerance within which

† The notation $[a,b]$ will be used to denote the closed interval with endpoints a and b, with no requirement that $a \le b$.

we wish to know the root, or until $f(x)$ is so small that the computational error in $f(x)$ may be as large as $f(x)$, so we cannot trust the sign of the computed value of $f(x)$.

This technique for finding the root of a nonlinear equation is called the *bisection method.* It is carefully specified in program 3-1. We start with two argument values straddling the root. These two values, which we will call x_0^+ and x_0^-, must satisfy the relations $f(x_0^+) > 0$ and $f(x_0^-) < 0$. The existence of such points is necessary for applicability of the bisection method. The points exist if $f(x)$ crosses the x axis at $x = \xi$, which is certainly the case if ξ is a single root.

At the $(i+1)$th iteration, the algorithm begins with two interval endpoints, which we will call x_i^+ and x_i^-, and computes a new iterate, which we will call x_{i+1}, as the interval midpoint. For example, let $f(x) = x^2 - 2$, $x_0^+ = 2$, and $x_0^- = 1$. Then $x_1 = 1.5$, and since $f(1.5) > 0$, $x_1^+ = 1.5$ and $x_1^- = x_0^- = 1$. Then $x_2 = 1.25$, and since $f(1.25) < 0$, $x_2^- = 1.25$ and $x_2^+ = x_1^+ = 1.5$. Further iterations continue the convergence to $\xi = 1.414$.

```
/* BISECTION METHOD FOR FINDING ROOTS OF EQUATIONS */
/* ONE EQUATION IN ONE UNKNOWN: F(X) = Ø */
/* SUBROUTINE F IS GIVEN; LFTGUESS AND RTGUESS, */
/* WHERE LFTGUESS < ROOT AND RTGUESS > ROOT, ARE GIVEN */
/* TOL, A THRESHOLD FOR ERROR BOUND ON ESTIMATE, IS GIVEN */

    DECLARE (LFTGUESS, RTGUESS,      /* CLOSEST ESTIMATES */
                                     /* TO LEFT AND RIGHT OF ROOT */
             XNEW,                   /* NEW ROOT ESTIMATE */
             TOL) FLOAT, /* THRESHOLD FOR STOPPING ITERATION */
             LFTSIGN FIXED BINARY; /* SIGN OF F(LFTGUESS) */

/* INITIALIZE */
    LFTSIGN = SIGN(F(LFTGUESS));

/* COMPUTE IMPROVED ESTIMATE OF ROOT */
    IMPROVE: XNEW = (LFTGUESS + RTGUESS) / 2;

/* TEST WHETHER CLOSE ENOUGH TO ROOT */
             IF (RTGUESS - LFTGUESS) / 2 < TOL THEN GO TO DONE;

/* REPLACE WITH XNEW THE ONE OF LFTGUESS AND RTGUESS */
/* FOR WHICH F AGREES IN SIGN WITH F AT XNEW */
             IF SIGN(F(XNEW)) = LFTSIGN THEN LFTGUESS = XNEW;
                                        ELSE RTGUESS  = XNEW;
             GO TO IMPROVE;  /* ITERATE AGAIN */

    DONE:  /* XNEW HOLDS BEST ESTIMATE OF ROOT */
```

PROGRAM 3-1 Bisection method for solution of nonlinear equations

An advantage of the bisection method is that if appropriate starting points are found, the sequence of iterates always converges to the desired root. Let us analyze its efficiency. As far as computation per iterative step, the method is quite efficient. It requires no significant computation beyond a single evaluation of the function f at each step. However, that evaluation may be time-consuming. In fact, function evaluations are usually the most time-consuming component of the computation in iterative methods.

We must also analyze efficiency from the point of view of the number of iterative steps required to reach a desired accuracy. Considering the sequence of iterates, x_i, it can be shown that if $x_i \neq \xi$ for some i ($|x_0^+ - \xi| \neq |x_0^+ - x_0^-|(j/2^i)$ for all integers j and i), the convergence is linear. However, the proof is not instructive, and the convergence generally is not simple (does not have the property that as $i \to \infty$, $(x_{i+1} - \xi)/(x_i - \xi) \to k$). It is more instructive to consider the sequence of bounds on the error in the iterate x_i than the sequence of errors, $x_i - \xi$. Let b_i be the maximum possible error in x_{i+1}, the center of the interval $[x_i^+, x_i^-]$. Then

$$b_i = |x_i^+ - x_i^-|/2. \tag{30}$$

Because the interval is cut in half at each iteration,

$$b_{i+1} = b_i/2. \tag{31}$$

The error bound for the bisection method converges simply linearly to 0 with convergence factor $\frac{1}{2}$. Such convergence is relatively slow.

3.3.3 Method of False Position

How can we improve the rate of convergence of the bisection method? Can we take advantage of our knowledge of the function f to make a better estimate of the root in the interval in question than the midpoint of the interval? We will do so by applying a technique on which many of the methods for solving nonlinear equations are based: Using the present estimate(s) of the root, we will approximate the nonlinear equation by a linear equation and use the root of our approximation to produce a better estimate of the root of the nonlinear equation. That is, we can fit a straight line between the points $(x_i^+, f(x_i^+))$ and $(x_i^-, f(x_i^-))$ to obtain a linear approximation to the function f in the interval (see figure 3-2), and we can use the root of the line as an approximation of the root of f. Then we can proceed as with the bisection method, replacing the appropriate one of x_i^+ and x_i^- with the new estimate of the root. If f is reasonably well approximated by the straight line, we can expect this

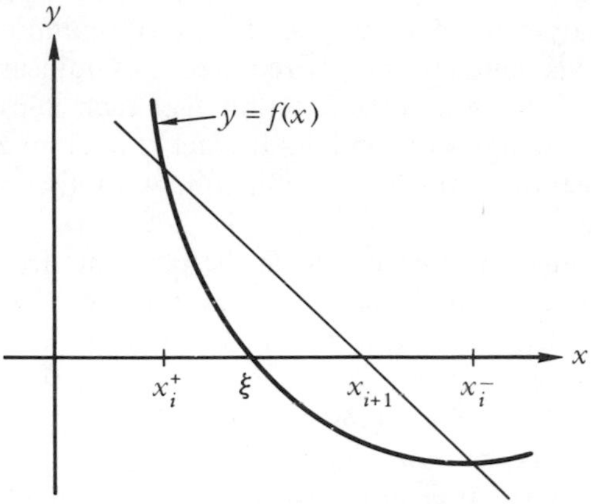

FIG. 3-2 A step of the method of false position

method to produce faster convergence than the bisection method because we are choosing the successive new interval endpoints in a more educated manner.

Let us analyze more carefully the convergence properties of this *method of false position,* which is specified in program 3-2. As with the bisection method, we start with two values x_0^+ and x_0^- straddling the root such that $f(x_0^+) > 0$ and $f(x_0^-) < 0$. At the $(i+1)$th iteration, the algorithm takes the interval endpoints, x_i^+ and x_i^-, and computes the new iterate x_{i+1} as the zero of the line going through f at x_i^+ and x_i^-:

$$x_{i+1} = \frac{x_i^+ f(x_i^-) - x_i^- f(x_i^+)}{f(x_i^-) - f(x_i^+)}. \tag{32}$$

Note that the subtractions in equation 32 do not cause large propagation of error, because the terms subtracted are of opposite signs, always in the denominator and normally in the numerator.

Before replacing one of the interval endpoints with x_{i+1}, we must check whether we have converged to within our desired tolerance. Because the interval width need not approach zero with this method, the test that we used with the bisection method is not applicable. The two most common convergence tests, which are used with all methods following in section 3.3, are of the form given in section 2.8 to decide whether a sequence of iterates which are approximations to the solution of an equation ends close enough to the solution.

```
/* METHOD OF FALSE POSITION FOR FINDING ROOTS OF EQUATIONS */
/* ONE EQUATION IN ONE UNKNOWN: F(X) = Ø; SUBROUTINE F IS GIVEN */
/* LFTGUESS AND RTGUESS, WHERE LFTGUESS < ROOT AND RTGUESS > ROOT, ARE GIVEN */
/* TOL, A THRESHOLD FOR RELATIVE DIFFERENCE BETWEEN SUCCESSIVE ROOT ESTIMATES, IS GIVEN */
   DECLARE (XNEW, XOLD,              /* NEW ROOT ESTIMATE, NEXT MOST RECENT ROOT ESTIMATE */
            LFTGUESS, RTGUESS,       /* CLOSEST ESTIMATES TO LEFT AND RIGHT OF ROOT */
            LFTVAL, RTVAL, NEWVAL,   /* F(LFTGUESS), F(RTGUESS), F(XNEW) */
            SLOPE,                   /* SLOPE OF LINE THROUGH F AT LFTGUESS AND RTGUESS */
            TOL) FLOAT,              /* THRESHOLD FOR STOPPING ITERATION */
            LFTSIGN FIXED BINARY;    /* SIGN OF F(LFTGUESS) */
/* INITIALIZE */
   XOLD    = LFTGUESS;
   LFTVAL  = F(LFTGUESS);
   RTVAL   = F(RTGUESS);
   LFTSIGN = SIGN(LFTVAL);
/* COMPUTE IMPROVED ESTIMATE OF ROOT */
   IMPROVE: SLOPE = (LFTVAL - RTVAL) / (LFTGUESS - RTGUESS);
            XNEW = LFTGUESS - LFTVAL / SLOPE;
/* TEST WHETHER CLOSE ENOUGH TO ROOT */
            IF ABS(XNEW - XOLD) / ABS(XNEW) < TOL THEN GO TO DONE;
/* REPLACE WITH XNEW THE ONE OF LFTGUESS AND RTGUESS FOR WHICH F AGREES IN SIGN WITH F AT XNEW */
            NEWVAL = F(XNEW);
            IF SIGN(NEWVAL) = LFTSIGN
               THEN LFTREPL: DO;
                             LFTGUESS = XNEW; LFTVAL = NEWVAL;
                    END LFTREPL;
               ELSE  RTREPL: DO;
                             RTGUESS = XNEW; RTVAL = NEWVAL;
                     END RTREPL;
            XOLD = XNEW;
            GO TO IMPROVE;          /* ITERATE AGAIN */
   DONE:  /* XNEW HOLDS BEST ESTIMATE OF ROOT */
```

PROGRAM 3-2 Method of false position for solving nonlinear equations

1. Test whether the successive iterates are relatively close enough together, that is, whether

$$|x_{i+1} - x_i|/|x_{i+1}| < \varepsilon. \tag{33}$$

2. Test whether
$$|f(x_{i+1})| < \delta. \tag{34}$$

Both of these tests can produce undesirable results with appropriately pathological functions. Test 1 is most commonly used, because it depends less on our accuracy in calculating f. Of course, we could require that the convergence satisfy both tests before we stopped.

As an example of the application of the method, again let $f(x) = x^2 - 2$, $x_0^+ = 2$, and $x_0^- = 1$. Then $x_1 = (2(-1) - 1(2))/(-1-2) = \frac{4}{3} \approx 1.33$. Since $f(\frac{4}{3}) < 0$, $x_1^- = \frac{4}{3}$ and $x_1^+ = x_0^+ = 2$. Then $x_2 = 1.4$, and since $f(1.4) < 0$, $x_2^- = 1.4$ and $x_2^+ = x_1^+ = 2$. And so on. As we hoped, x_2 from the method of false position (1.4) is closer to the root than x_2 from the bisection method (1.25) is.

We must now discuss the convergence of the sequence of iterates produced by the method of false position.

Theorem: Let $f(x)$ be a continuous function on $[x_0^+, x_0^-]$, and assume $f(x_0^+) > 0$ and $f(x_0^-) < 0$. Let $\{x_i\}$ be the sequence of iterates produced by this method of false position applied to f with starting values x_0^+ and x_0^-. Then the sequence $\{x_i\}$ converges to a root ξ of f.

Proof: We will proceed by showing that the sequence of x_i^+ values converges to a limit and that the sequence of x_i^- values converges to a limit, then showing that at least one of these limits is a root of f, and deducing from these results that the x_i values converge to ξ.

It should be intuitively clear that the zero-crossing point of a straight line fit between two points, one at which the ordinate is positive and one at which the ordinate is negative, must be between the two argument points in question. The formal proof of this statement is not difficult (see exercise 3.3.5). It follows that the sequences of x_i^+ and of x_i^- are monotonic in a closed interval $[x_0^+, x_0^-]$, and thus by a well-known theorem in mathematics, each must have a limit. Let these limits be ξ^+ and ξ^- respectively.

We must now show that at least one of these limits is a root of f. Assume not. Then

1. There exists an $\varepsilon > 0$ such that $f(x_k^+) > \varepsilon$ for all $k \geq$ some N_ε and $f(x_k^-) < -\varepsilon$ for all $k \geq N_\varepsilon$.
2. By the continuity of f on $[x_0^+, x_0^-]$ there exists a $\gamma > 0$ such that for all $k \geq 0$,

$$|f(x_k^+)| < \gamma \quad \text{and} \quad |f(x_k^-)| < \gamma, \tag{35}$$

so that
$$|f(x_k^+) - f(x_k^-)| < 2\gamma. \tag{36}$$

3. There exists a $\delta > 0$ such that

$$|\xi^+ - \xi^-| = \delta, \tag{37}$$

since if $\xi^+ = \xi^-$, the root, which is in $[\xi^+, \xi^-]$, must be equal to ξ^+.
4. For any $\beta > 0$ there exists an integer k_β such that for all $k \geq k_\beta$,

$$|x_k^- - \xi^-| < \beta \text{ and } |x_k^+ - \xi^+| < \beta. \tag{38}$$

From equation 32 we derive

$$x_{k+1} - \xi^+ = \frac{(x_k^+ - \xi^+)f(x_k^-) + (\xi^+ - x_k^-)f(x_k^+)}{f(x_k^-) - f(x_k^+)} \tag{39}$$

and

$$x_{k+1} - \xi^- = \frac{(\xi^- - x_k^-)f(x_k^+) + (x_k^+ - \xi^-)f(x_k^-)}{f(x_k^-) - f(x_k^+)}. \tag{40}$$

By condition 2 above and equations 39 and 40:

$$|x_{k+1} - \xi^+| > (1/2\gamma)|(x_k^+ - \xi^+)f(x_k^-) + (\xi^+ - x_k^-)f(x_k^+)| \tag{41}$$

and

$$|x_{k+1} - \xi^-| > (1/2\gamma)|(\xi^- - x_k^-)f(x_k^+) + (x_k^+ - \xi^-)f(x_k^-)|. \tag{42}$$

From conditions 2 and 4:

$$|(x_k^+ - \xi^+)f(x_k^-)| < \gamma\beta \tag{43}$$

and
$$|(\xi^- - x_k^-)f(x_k^+)| < \gamma\beta \tag{44}$$

if $k \geq k_\beta$.

Since
$$|\xi^+ - x_k^-| > |\xi^+ - \xi^-| = \delta \tag{45}$$

and
$$|x_k^+ - \xi^-| > |\xi^+ - \xi^-| = \delta, \tag{46}$$

from conditions 1 and 3:

$$|(\xi^+ - x_k^-)f(x_k^+)| > \delta\varepsilon \tag{47}$$

and

$$|(x_k^+ - \xi^-)f(x_k^-)| > \delta\varepsilon \tag{48}$$

if $k \geq N_\varepsilon$.

Let $\beta = \delta\varepsilon/2\gamma$ in relations 43 and 44. Then the absolute values on the right side of each of the relations 41 and 42 are greater than $\delta\varepsilon/2$ for all $k \geq M = max(N_\varepsilon, k_{\delta\varepsilon/2\gamma})$, so

$$|x_{k+1} - \xi^+| > \delta\varepsilon/(4\gamma) \tag{49}$$

and

$$|x_{k+1} - \xi^-| > \delta\varepsilon/(4\gamma) \tag{50}$$

for all $k \geq M$. Thus the new iterate cannot get closer than $\delta\varepsilon/(4\gamma)$ to ξ^+ and ξ^-. Since the new iterate becomes either x_{k+1}^+ or x_{k+1}^- and the other interval endpoint remains fixed, neither of these can get closer than $\delta\varepsilon/(4\gamma)$ to ξ^+ and ξ^-. But this is a contradiction to the fact that ξ^+ and ξ^- are their limits. We have a proof by contradiction that either ξ^+ or ξ^- is a root of f.

We must finally show that x_{k+1} approaches a root of f. If $\xi^+ = \xi^-$, the result has already been shown. If $\xi^+ \neq \xi^-$, assume ξ^+ is a root of f (the proof in the other case is symmetric). Then by equation 32:

$$\lim_{k\to\infty} x_{k+1} = \frac{\xi^+ f(\xi^-) - 0}{f(\xi^-) - 0} = \xi^+. \tag{51}$$

We now know that the method of false position always produces convergence. But we must analyze the speed of convergence. The amount of computation per step is only slightly more time-consuming than that of the bisection method. Like bisection, it requires the evaluation of f at only one new point per iterative step (the evaluation of f at the other interval endpoint has been done in a previous step). However, the evaluation must be somewhat more accurate for the method of false position because the value itself is used rather than simply its sign. The method of false position also requires the calculations given in equation 32: two multiplications and a division.

Considering the number of iterations required, we note that the method of false position is designed to produce faster convergence than the bisection method. For some functions it does, but for others it does not. Since the method of false position depends on approximating f by a straight line, we might expect that when the approximation is a good one, convergence will be fast. That expectation is true (see figure 3-3).

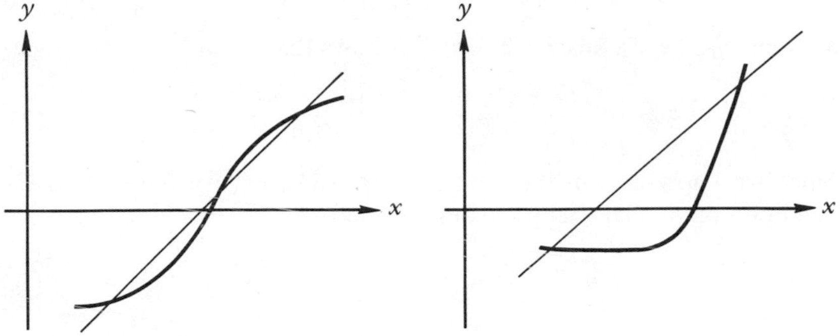

a) f fit well by line b) f fit poorly by line

FIG. 3-3 Speed of convergence of method of false position

We will see that the convergence is particularly slow when the interval endpoint on one side of the root becomes frozen far from the root, producing a straight line approximation that does not improve quickly.

Let us mathematically analyze the convergence of the method of false position. It can be shown that if f and its first and second derivatives are continuous at ξ, unless $f''(\xi) = 0$, after some iteration, one interval endpoint will be frozen and all subsequent x_i will be on one side of the root (see exercise 3.3.6). So we see that frozen-point convergence in the limit is the normal situation, and we evaluate the convergence under the assumption that such is the mode of convergence. If $f''(\xi) = 0$, the analysis given under the secant method in section 3.3.7 is applicable.

If $x = a$ is a frozen point in the false position iteration, then $a \neq \xi$ and thus $f(a) \neq 0$. Furthermore, each iteration is of the form.

$$x_{i+1} = \frac{af(x_i) - x_i f(a)}{f(x_i) - f(a)}. \tag{52}$$

Subtracting ξ from both sides of equation 52, we obtain

$$\begin{aligned} x_{i+1} - \xi &= [af(x_i) - x_i f(a) - (f(x_i) - f(a))\xi]/(f(x_i) - f(a)) \\ &= [(a - \xi)f(x_i) - f(a)(x_i - \xi)]/(f(x_i) - f(a)). \end{aligned} \tag{53}$$

Since we are trying to obtain an expression for the error at the $(i+1)$th step, $x_{i+1} - \xi$, in terms of the error at the ith step, we wish to expand $f(x_i)$ in the numerator as a Taylor series about ξ, producing

$$x_{i+1} - \xi = \frac{(a-\xi)(f(\xi) + (x_i - \xi)f'(\eta_i)) - f(a)(x_i - \xi)}{f(x_i) - f(a)} \tag{54}$$

for some $\eta_i \in [x_i, \xi]$. Since $f(\xi) = 0$, it follows that

$$\frac{x_{i+1} - \xi}{x_i - \xi} = \frac{(a-\xi)f'(\eta_i) - f(a)}{f(x_i) - f(a)}. \tag{55}$$

Since we know that in the limit, x_i approaches ξ, we have, from the condition on η_i, that η_i approaches ξ so that

$$\lim_{i\to\infty} \frac{x_{i+1} - \xi}{x_i - \xi} = \frac{(a-\xi)f'(\xi) - f(a)}{-f(a)}. \tag{56}$$

Expanding $f(a)$ in a Taylor series about the root, a technique we have used and will use again and again, we obtain

$$\lim_{i\to\infty} \frac{x_{i+1} - \xi}{x_i - \xi} = \frac{(a-\xi)(f'(\xi) - f'(\theta))}{-(a-\xi)f'(\theta)} \tag{57}$$

for some $\theta \in [a, \xi]$, which implies our final result:

$$\lim_{i\to\infty} \frac{x_{i+1} - \xi}{x_i - \xi} = \frac{f'(\theta) - f'(\xi)}{f'(\theta)} \text{ for some } \theta \in [a, \xi]. \tag{58}$$

Thus, in the limit, the error at the $(i+1)$th step is a constant times the error at the ith step, so the convergence is simply linear with convergence factor $(f'(\theta) - f'(\xi))/f'(\theta)$.

3.3.4 Acceleration of Convergence

Because the method of false position produces simple linear convergence if one interval endpoint remains frozen, the acceleration scheme presented in section 2.7.3 is applicable. In this case the (unknown) vector to which the sequence is converging has dimension 1. Whereas with a higher-dimensional case, certain equalities could only approximately be satisfied (and thus a norm minimized), in the one-dimensional case we can exactly satisfy these equations. There exists a constant k_i such that

$$x_{i+1} - \xi = k_i(x_i - \xi), \tag{59}$$

and in the limit $k_{i+1} = k_i$. Rewriting equation 59 for index $i+1$, we see that

$$x_{i+2} - \xi = k_{i+1}(x_{i+1} - \xi). \tag{60}$$

Assuming $k_i = k_{i+1} = k$, we can solve equations 59 and 60 for ξ, producing

$$\hat{\xi} = \frac{x_{i+2}x_i - x_{i+1}^2}{x_{i+2} - 2x_{i+1} + x_i}, \tag{61}$$

an accelerated estimate for the root ξ. This acceleration technique for a one-dimensional variable is called *Aitken's δ^2 acceleration,* because the denominator of equation 61 is a value called $\delta^2 x_{i+1}$ (see section 4.2.1g). A computationally more convenient and accurate form of equation 61 is

$$\hat{\xi} = x_{i+2} - \frac{(x_{i+2} - x_{i+1})^2}{(x_{i+2} - x_{i+1}) - (x_{i+1} - x_i)}. \tag{62}$$

Note that if the convergence factor k is a positive number very close to 1, the denominator of equation 61 or 62 is very, very small compared to the x_i values, so computational error can cause serious errors in $\hat{\xi}$. Thus, one must be wary of the acceleration method in that circumstance. Furthermore, we saw that the acceleration method depends on the assumption that $k_i \approx k_{i+1}$. If not, $\hat{\xi}$ can be a poorer estimate of the root than x_{i+2}. However, if we apply the technique to the iterates of false position, we can avoid this difficulty. We have the root trapped in an interval, and if $\hat{\xi}$ is outside of the interval, we simply reject the accelerated value.

To summarize, Aitken's δ^2 acceleration is applicable to three successive false-position iterates, x_i, x_{i+1}, and x_{i+2}, if one interval endpoint remains frozen. That is, if $f(x_i)$, $f(x_{i+1})$, and $f(x_{i+2})$ have the same sign, we can apply the acceleration to the three iterates. Then we check to determine whether the accelerated value falls between the latest iterate and the other endpoint value. If not, we iterate twice more and try again. If so, we treat the accelerated value just as any new value, replacing one or the other of the interval endpoints. Then we iterate twice more, attempt to accelerate, and so on.

Once acceleration produces an acceptable result, it is likely to do so for every succeeding pair of iterations, because the applicability of the acceleration technique depends on the convergence of the k_i and that in turn depends on the closeness to the root of the x_i. This situation is unlike that obtained with iterative solution of linear equations, where the applicability of acceleration depends on the direction of the error vector and the condition required for applicability is not more likely to be true near the root than far from the root. When using the latter method, we have to iterate a number of times after each acceleration to obtain the required conditions for reacceleration.

Assume we have converged far enough so that acceleration is applicable every two steps of the method of false position. Consider the sequence of accelerated values. Let x_i be the result of a previous acceleration and let x_{i+1} and x_{i+2} be the results of two successive iterations of our linearly convergent method:

$$x_{i+1} = g(x_i), \tag{63}$$

$$x_{i+2} = g(x_{i+1}), \tag{64}$$

where for the method of false position, g is given by equation 52.

Expanding equations 63 and 64 in a Taylor series about $x = \xi$, we obtain

$$\begin{aligned} x_{i+1} &= g(\xi) + g'(\theta)(x_i - \xi), \quad \theta \in [x_i, \xi], \\ &= g(\xi) + g'(\xi)(x_i - \xi) + (g''(\eta)/2)(x_i - \xi)^2, \quad \eta \in [x_i, \xi], \end{aligned} \tag{65}$$

and

$$\begin{aligned} x_{i+2} &= g(\xi) + g'(\phi)(x_{i+1} - \xi), \quad \phi \in [x_{i+1}, \xi], \\ &= g(\xi) + g'(\xi)(x_{i+1} - \xi) + (g''(\zeta)/2)(x_{i+1} - \xi)^2, \quad \zeta \in [x_{i+1}, \xi]. \end{aligned} \tag{66}$$

Since $x_{i+1} = g(x_i)$ is convergent to ξ, ξ is a fixed point of g, that is,

$$\xi = g(\xi). \tag{67}$$

Thus, equation 65 becomes

$$\begin{aligned} x_{i+1} &= \xi + g'(\theta)(x_i - \xi) \\ &= \xi + g'(\xi)(x_i - \xi) + (g''(\eta)/2)(x_i - \xi)^2. \end{aligned} \tag{68}$$

Using equations 67 and 68 in equation 66, we see

$$x_{i+2} = \xi + g'(\phi)(g'(\theta)(x_i - \xi)). \tag{69}$$

Let $\hat{x}_i$ be the result of δ^2 acceleration using x_i, x_{i+1}, and x_{i+2}. We can show that in the limit

$$\hat{x}_i - \xi = K(x_i - \xi)^2 \tag{70}$$

for some constant K. The proof follows:

$$\begin{aligned} \hat{x}_i &= \frac{x_i[\xi + g'(\phi)g'(\theta)(x_i - \xi)] - [\xi + g'(\theta)(x_i - \xi)]^2}{[\xi + g'(\phi)g'(\theta)(x_i - \xi)] - 2[\xi + g'(\theta)(x_i - \xi)] + x_i} \\ &= \frac{(x_i - \xi)[\xi + x_i g'(\phi)g'(\theta) - 2\xi g'(\theta) - (g'(\theta))^2(x_i - \xi)]}{(x_i - \xi)[1 - 2g'(\theta) + g'(\phi)g'(\theta)]}. \end{aligned} \tag{71}$$

Canceling $(x_i - \xi)$ in equation 71 and then subtracting ξ from both sides of this equation produces

$$\hat{x}_i - \xi = \frac{(x_i - \xi)(g'(\theta))(g'(\phi) - g'(\theta))}{1 - 2g'(\theta) + g'(\phi)g'(\theta)}. \tag{72}$$

From equations 65 and 66 we see that

$$g'(\theta) = g'(\xi) + (g''(\eta)/2)(x_i - \xi) \tag{73}$$

and

$$g'(\phi) = g'(\xi) + (g''(\zeta)/2)g'(\theta)(x_i - \xi). \tag{74}$$

Thus

$$g'(\phi) - g'(\theta) = \tfrac{1}{2}(x_i - \xi)(g''(\zeta)g'(\theta) - g''(\eta)), \tag{75}$$

so

$$\hat{x}_i - \xi = (x_i - \xi)^2 \frac{g'(\theta)}{1 - 2g'(\theta) + g'(\phi)g'(\theta)} \frac{g''(\zeta)g'(\theta) - g''(\eta)}{2}. \tag{76}$$

As $i \to \infty$, the values θ, ϕ, ζ, and $\eta \to \xi$,

so

$$\hat{x}_i - \xi \to (x_i - \xi)^2 \frac{g'(\xi)g''(\xi)}{2(g'(\xi) - 1)}. \tag{77}$$

As long as $g'(\xi) \neq 1$, our claim is true: As i gets large, successive results of acceleration (x_i and $\hat{x}_i$) are related such that the error in the second is proportional to the square of the error in the first. The successive results of acceleration are converging quadratically to ξ, a desirable situation.

When is a convergent sequence of results of δ^2 acceleration not quadratically convergent?

1. When $g'(\xi) = 0$ or $g''(\xi) = 0$, in which case the convergence has order >2.
2. When $g'(\xi) = 1$, in which case the convergence has order <2. For the acceleration of the method of false position after one point has become frozen, we see $g'(\xi) = 1$ iff $f'(\xi) = 0$. Thus, as we have seen before, we can expect δ^2 acceleration to misbehave with this method if f is nearly flat near the root.

3.3.5 Picard Iteration

In section 2.8 we developed the general notion of producing an iterative method by transforming the equation $f(x) = 0$ into $x = g(x)$ such that ξ satisfies the first equation if and only if it satisfies the second, and then using the iteration $x_{i+1} = g(x_i)$. For nonlinear equations, such a procedure is called *Picard iteration.* This iteration often produces simple linear convergence, in which case Aitken's δ^2 acceleration is applicable as is all of the mathematical analysis presented in section 3.3.4.

Let us set up a Picard iteration for the equation

$$f(x) = x^2 - 3x + sin(x). \tag{78}$$

Then $f(x) = 0$ iff $\quad x = (3x - sin(x))^{1/2} \equiv g_1(x).$ (79)

But note $\quad g_2(x) \equiv sin^{-1}(3x - x^2),$ (80)

$$g_3(x) \equiv (x^2 + sin(x))/3, \quad (81)$$

and $\quad g_4(x) \equiv 3 - sin(x)/x$ (82)

also do the job. In general, there are an infinite number of ways of transforming $f(x) = 0$ into $x = g(x)$. We need a rule by which we can judge whether the iteration corresponding to a given g will converge; and if so, whether the convergence will be to the root; and if so, how fast the convergence is.

Theorem: If $x = g(x)$ only if x is a root of f, and if $\lim_{i \to \infty} x_{i+1}[= g(x_i)] = \xi$, where g is continuous at ξ, then ξ is a root of $f(x)$.

Proof: $g(\xi) = \xi$ by the theorem that we proved in section 3.3.1, so by assumption ξ is a root of f.

Next we show that the order of convergence is at least linear if g is differentiable in $[x_i, \xi]$. This statement follows from noting that equations 65 and 67 are applicable, so

$$(x_{i+1} - \xi) = g'(\theta_i)(x_i - \xi) \text{ for some } \theta_i \in [x_i, \xi]. \quad (83)$$

If the iteration converges, then $\theta_i \to \xi$, so

$$(x_{i+1} - \xi) \to g'(\xi)(x_i - \xi), \quad (84)$$

and the convergence is simple and linear unless $g'(\xi) = 0$, in which case it is simple and of quadratic or higher order (see exercise 3.3.8).

When equation 83 holds, a sufficient condition for convergence is that for every i greater than some n,

$$|g'(\theta_i)| \leq M < 1, \quad (85)$$

since then $\quad |x_{n+i} - \xi| \leq M^i |x_n - \xi| \to 0 \text{ as } i \to \infty.$ (86)

We do not know the value of θ_i, except that it is restricted to an interval, so relation 85 is not testable. We must test whether for all x between x_i and ξ,

$$|g'(x)| \leq M < 1. \quad (87)$$

Because we do not know the value of ξ, we test by finding $x = z$ such that z is on the opposite side of ξ from x_i and then assuring that the condition holds in $[x_i, z]$. We have thus a testable sufficient condition for convergence. Note, however, that this condition is not necessary for convergence.

The reason for stating condition 85 as $|g'(\theta_i)| \leq M < 1$ rather than $|g'(\theta_i)| < 1$ is to remove the possibility that $\lim_{i\to\infty} |g'(\theta_i)| = 1$ from below in such a way that $\prod_{i=n}^{\infty} |g'(\theta_i)|$ does not approach 0.

To understand the convergence of Picard iteration more clearly, we will look at the method graphically. Each iterative step can be thought to be made up of two parts: (1) $y_i = g(x_i)$ and (2) $x_{i+1} = y_i$. Part 1 involves moving vertically to the curve $g(x)$, and part 2 involves moving horizontally from the result of the previous part to the curve $x = y$. If we sketch the curves for $y = x$ and $y = g(x)$ on the same graph, our iteration consists of a sequence of horizontal and vertical steps between the curves. As figure 3-4 indicates, we can converge or diverge either in a stairstep fashion or spirally.

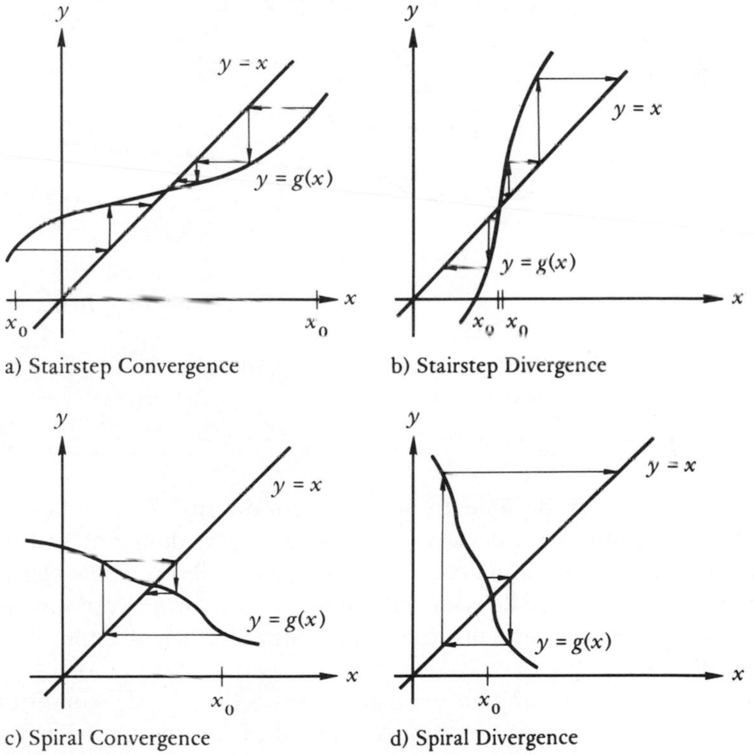

FIG. 3-4 Picard iteration

A Picard iteration can also fail to converge by making a sequence of moves which spiral in and out around the root, never going more than a fixed distance from the root but never converging. Such failure to converge is shown in figure 3-5. Generally, convergence takes place if the slope of $g(x)$ is less than 1 in magnitude near the root; the mode of the movement is stairstep if the slope is positive, and spiral if the slope is negative.

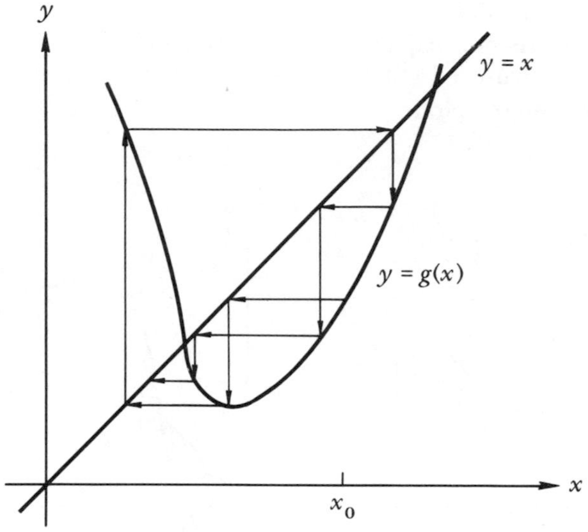

FIG. 3-5 Lack of convergence of Picard iteration without divergence to $\pm\infty$

We should now be able to see why condition 87 is sufficient for convergence, but not necessary. In figure 3-6, $g(x)$ does not satisfy the condition, but the iteration converges. As long as $g(x)$ remains below the line $y = x$ for $x > \xi$ and does not fall below $g(\xi) = \xi$ for $x > \xi$, convergence will take place. We can construct many other sufficient conditions for convergence.

We should also be able to see that if $g'(x) \geq 0$ in $[x_i, \xi]$, condition 87 need be satisfied on only a one-sided interval of ξ to obtain convergence for a starting point in that interval. On the other hand, if $g'(x) < 0$ in $[x_i, \xi]$, the condition must be satisfied in a symmetric interval about ξ.

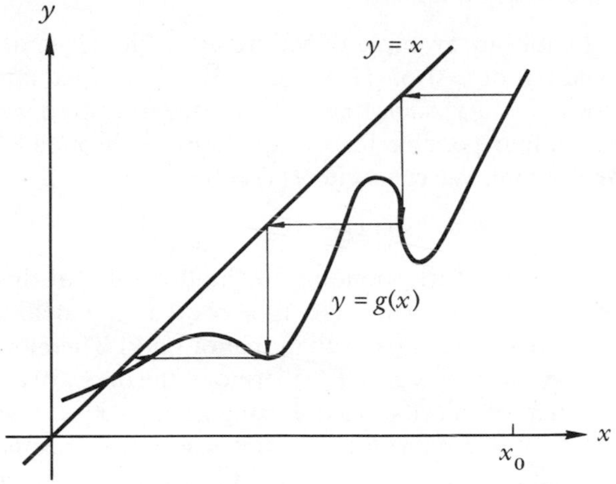

FIG. 3-6 Convergence of Picard iteration where $|g'(x)| > 1$ for some $x \in [x_i, \xi]$

Now that we understand the requirements on g for convergence, we must indicate how to choose a convergent formula $x = g(x)$, given an equation $f(x) = 0$. No strict rule can be given, but there are certain approaches which often work.

1. If $f'(x)$ is nonzero and of constant sign in an interval on one side of the root, choose m to be a constant with that sign and with a magnitude greater than the largest magnitude of $f'(x)$ in the interval. Then

$$g(x) = x - f(x)/m \tag{88}$$

has the property that $g(x) = x$ iff $f(x) = 0$ and $0 < g'(x) < 1$ in the interval.

2. Write $x = g(x)$ in some obvious way, and evaluate g' in an interval near ξ. If the derivative is always greater than 1 in magnitude in that interval, find g^{-1} if possible. That is, write $z = g(x)$ and solve for x in terms of z. Since

$$\frac{d}{dz} g^{-1}(z) = 1 \bigg/ \frac{d}{dx} g(x) \bigg|_{x=g^{-1}(z)}, \tag{89}$$

g^{-1} will have a derivative less than 1 near ξ, and thus g^{-1} should be used as the iteration function.

Continuing with our example of setting up a Picard iteration, let us find the smallest root >0 of $f(x) = x^2 - 3x + sin(x)$ (equation 78) by choosing a form $x = g(x)$ such that convergence to this root will occur. We have written four possible forms, equations 79 through 82. First we must localize the root. We can write $f(x) = 0$ as

$$sin(x) = 3x - x^2 \tag{90}$$

and sketch two curves corresponding to the two sides of this equation as in figure 3-7. We see that there is a root at $x = 0$ and that the slope of $3x - x^2$ at $x = 0$ is greater than that of $sin(x)$. Therefore, initially, $3x - x^2$ is above $sin(x)$ for $x > 0$. Furthermore, the derivative of $3x - x^2$ is greater than that of $sin(x)$ at least as long as $3x - x^2 < 1$, so $3x - x^2$, a parabola, stays above $sin(x)$ for x values where that inequality holds. It reaches a maximum at $x = \frac{3}{2}$ and falls to zero at $x = 3$, where $sin(x)$ is still positive, so the root must lie in $[\frac{3}{2}, 3]$. In fact, the root must be at a place where $3x - x^2 < 1$, which does not occur until $x > 2.6$, so the root must be in $[2.6, 3]$.

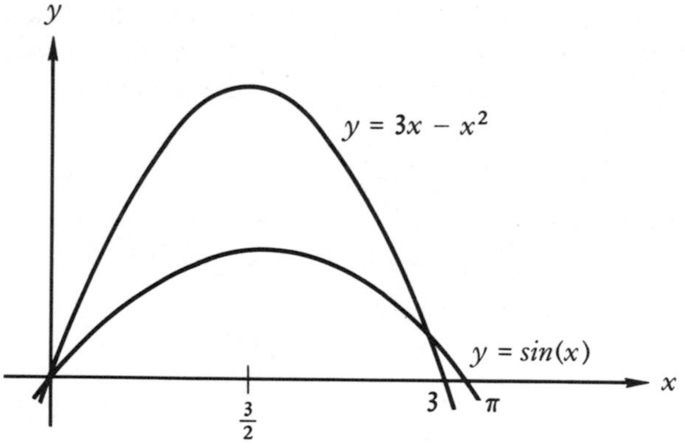

FIG. 3-7 Localizing the root of $x^2 - 3x + sin(x) = 0$

In $[2.6, 3]$, $g_1'(x) = \frac{1}{2}(3 - cos(x))/(3x - sin(x))^{1/2}$ is positive and less than $4/[2(3(2.6) - \frac{1}{2})^{1/2}] = .74 < 1$. Therefore $x_{i+1} = g_1(x_i)$ will converge in a stairstep fashion to the desired root. By a similar calculation, $|g_2'(x)|$ and $|g_3'(x)|$ are greater than 1 in $[2.6, 3]$, so they will not produce convergence to the desired root. $g_4'(x) = (x\,cos(x) - sin(x))/x^2$ is negative and less in magnitude than $(3(1) + \frac{1}{2})/(2.6)^2 = .52$ in [2.6,3], so $x_{i+1} = g_4(x_i)$ will converge spirally to the desired root, providing we assure that we have a symmetric interval where $|g_4'(x)| < 1$ by starting the iteration at the closer of 2.6 and 3 to the root. If we evaluate $sin(x)$ and $3x - x^2$ at $x = 2.8$, we find $3x - x^2 > sin(x)$, so 3 is closer to the root than 2.6.

Since our bound on $|g_4'(x)|$ is less than our bound on $|g_1'(x)|$ in [2.6,3], the iteration using g_4 will probably converge faster than that using g_1, so we should use $x_{i+1} = g_4(x_i)$.

Since Picard iteration produces simple linear convergence, Aitken's δ^2 acceleration is applicable as soon as the convergence begins to have close to its limiting simple linear behavior, that is, as soon as x_i becomes close to ξ. Since we have no way of detecting that the acceleration has produced a very wrong answer, as we do with the method of false position, we must be conservative in applying acceleration. We need a test which tells us whether the convergence is close to simply linear. Such a test can be obtained by noting that if the convergence is simply linear,

$$(x_{i+2} - x_{i+1})/(x_{i+1} - x_i) \to k, \tag{91}$$

where k is the convergence factor. We can tell whether the convergence is close to simply linear by seeing whether the value of k has settled down. That is, we test whether the value of k obtained at one step is close (within a few percent) to the value of k obtained in the previous step(s). If so, we can accelerate.

Note that in developing the acceleration method, we did not assume that the method is converging, only that the mode of movement toward or away from the root is close to linear. Thus, if we start close enough to the root with a divergent Picard iteration, we can still apply Aitken's δ^2 acceleration and produce quadratic convergence.

We note finally that if by chance $g'(\xi) = 0$ (the Picard iteration is itself quadratically convergent), the application of Aitken's δ^2 acceleration will result in a cubically convergent sequence. That is, the acceleration will help even if we apply it inadvertently to a quadratically convergent sequence. This result follows from our analysis of Aitken's δ^2 acceleration in section 3.3.4.

3.3.6 Newton's Method

Picard iteration is itself only in general linearly converging, that is, fairly slow, though with the application of Aitken's δ^2 acceleration the method becomes quadratically convergent. However, it is rather difficult to decide when acceleration should be applied, and it is often very difficult to find an initially convergent form for the iteration. We would like to design a method, the convergence of which is more predictable and is quadratic. Given a Picard iteration, we have noted that unless $g'(\xi) = 0$, the iterative step either takes us not far enough (not all the way to the root) if $g'(\theta_i) > 0$, or takes us too far if $g'(\theta_i) < 0$. We should apply the idea of overrelaxation which we saw in section 2.8.5. That is, when $g'(\theta_i) > 0$, the step size should be greater than the step size given by the Picard iteration, and if $g'(\theta_i) < 0$, the step size should be smaller than that given by the Picard iteration. How much larger or how much smaller? Precisely enough to provide quadratic convergence.

Consider the spirally convergent case. Since the root ξ is between x_i and $g(x_i)$, ξ is some positively weighted average of x_i and $g(x_i)$. Thus we have a new iteration function h given by

$$h(x) = \beta g(x) + (1 - \beta)x. \tag{92}$$

Note that we have introduced one parameter, β, precisely what we need to satisfy the one constraint that the linear convergence factor is 0, producing quadratic convergence. That is, we wish $h'(\xi) = 0$, which implies

$$\beta g'(\xi) + (1 - \beta) = 0, \tag{93}$$

leading to

$$\beta = 1/(1 - g'(\xi)). \tag{94}$$

Let us evaluate the iteration resulting from this choice of β. If $g'(\xi) < 0$, then $0 < \beta < 1$, so both weights in equation 93 are positive and we have underrelaxation, producing a value between x_i and $g(x_i)$ as we desired. Note that it is not necessary that the original spiral motion of the iteration $x_{i+1} = g(x_i)$ be convergent. Even if it is not, the underrelaxation produces a quadratically convergent sequence.

If $0 \le g'(\xi) < 1$, then $\beta > 1$, so

$$h(x_i) - x_i = \beta(g(x_i) - x_i). \tag{95}$$

Equation 95 implies that the new iteration moves in the same direction from x_i as the old iteration did but with a larger step: overrelaxation as we desired.

If $g'(\xi) = 1$, we have trouble, a matter we will deal with later. If $g'(\xi) > 1$, then $\beta < 0$, so the sign of $x_{i+1} - x_i$ in the new iteration is opposite from what it was in the old iteration; what was formerly stairstep divergence becomes quadratic convergence.

Since we do not know ξ, we cannot compute β perfectly, but we can approximate it by using x_i in place of ξ. When we do that, we obtain the iteration

$$x_{i+1} = \frac{g(x_i) - x_i g'(x_i)}{1 - g'(x_i)}. \tag{96}$$

We must now choose g. Any differentiable $g(x)$ which satisfies $x = g(x)$ iff $f(x) = 0$ will do. A simple g is

$$g(x) = x - f(x). \tag{97}$$

Equation 96 then becomes

$$x_{i+1} = x_i - f(x_i)/f'(x_i). \tag{98}$$

This iteration is called the *Newton-Raphson method,* or sometimes just *Newton's method* (see program 3-3).

```
/* NEWTON'S METHOD FOR FINDING ROOTS OF EQUATIONS */
/* ONE EQUATION IN ONE UNKNOWN: F(X) = Ø */
/* SUBROUTINES F AND FPRIME = DERIVATIVE OF F ARE GIVEN */
/* XGUESS, A GUESS AT THE ROOT, IS GIVEN */
/* TOL, A THRESHOLD FOR RELATIVE DIFFERENCE BETWEEN */
/* SUCCESSIVE ROOT ESTIMATES, IS GIVEN */

   DECLARE (XGUESS,          /* GUESS AT ROOT */
            XNEW,            /* IMPROVED ESTIMATE OF ROOT */
            TOL) FLOAT;      /* THRESHOLD FOR STOPPING ITERATION */

/* COMPUTE IMPROVED ESTIMATE OF ROOT */
   IMPROVE: XNEW = XGUESS - F(XGUESS) / FPRIME(XGUESS);

/* TEST WHETHER CLOSE ENOUGH TO ROOT */
            IF ABS(XNEW - XGUESS) / ABS(XNEW) < TOL
               THEN GO TO DONE;

/* MAKE GUESS = IMPROVED ESTIMATE */
            XGUESS = XNEW;
            GO TO IMPROVE;   /* ITERATE AGAIN */

   DONE:  /* XNEW HOLDS BEST ESTIMATE OF ROOT */
```

PROGRAM 3-3 Newton's method for solving nonlinear equations

To verify that Newton's method is quadratically convergent, we must take the derivative of

$$h(x) = x - f(x)/f'(x) \tag{99}$$

and show that it is 0 at $x = \xi$. The differentiation produces

$$h'(x) = f(x)f''(x)/[f'(x)]^2, \tag{100}$$

so if $f'(\xi) \neq 0$, then since $f(\xi) = 0$, $h'(\xi) = 0$. Note that if $f'(\xi) = 0$, $g'(\xi) = 1$, a condition which we have previously seen causes trouble.

Let us be careful about what is involved in quadratic convergence. Quadratic convergence means that *in the limit* we converge in such a way that the error at the end of one step is proportional to the square of the error at the beginning of the step. This does not mean that at all iterative steps such a relation holds. In fact, we have the iteration

$$x_{i+1} = h(x_i), \tag{101}$$

and by equation 83 we know that

$$x_{i+1} - \xi = h'(\theta_i)(x_i - \xi), \text{ for some } \theta_i \in [x_i, \xi], \tag{102}$$

where h' is given by equation 100. It follows that to have convergence at all and thus in the limit quadratic convergence, for all i greater than some n, $h'(\theta_i)$ must be less than 1 in magnitude.

Let us interpret Newton's method (equation 98) graphically. Rewriting equation 98, we obtain

$$0 = f(x_i) + f'(x_i)(x_{i+1} - x_i), \tag{103}$$

the equation of a line with slope $f'(x_i)$ at $y = f(x_i)$ such that x_{i+1} is the value of x where the line is 0 (see figure 3-8). That is, an iteration of Newton's method involves approximating f by a straight line, the line tan-

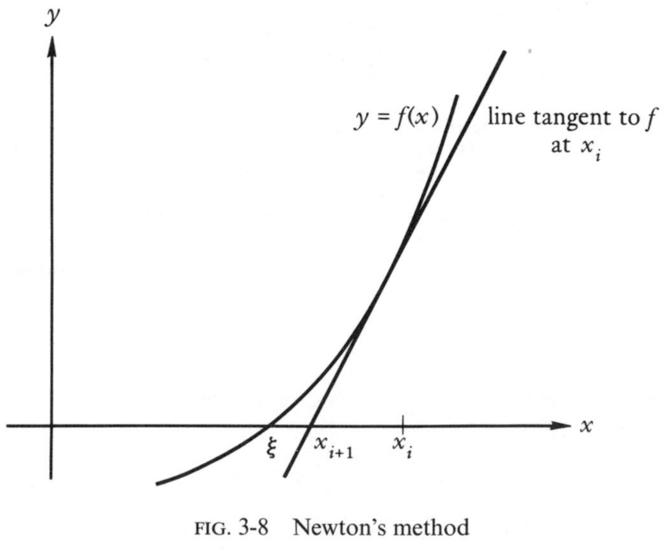

FIG. 3-8 Newton's method

gent to f at x_i, and approximating the root of f by the root of the straight line.

We can see why there is trouble when $f'(x) = 0$. In that case the approximating straight line is parallel to the x axis, so it has no root. Trouble also occurs when $f'(x)$ is approximately 0. In that case the value x_{i+1} may be in the wrong direction from x_i (see figure 3-9, part a), or an oscillation may take place about the zero derivative (see figure 3-9, part b). We must be very careful when considering application of Newton's method to a function which has zero or near-zero derivatives between the root and any point that may be reached by our iteration. In fact, we must establish beforehand that this behavior does not occur. Certainly, a function that has many oscillations between our estimate of the root and the root itself is a clear case where Newton's method should not be used.

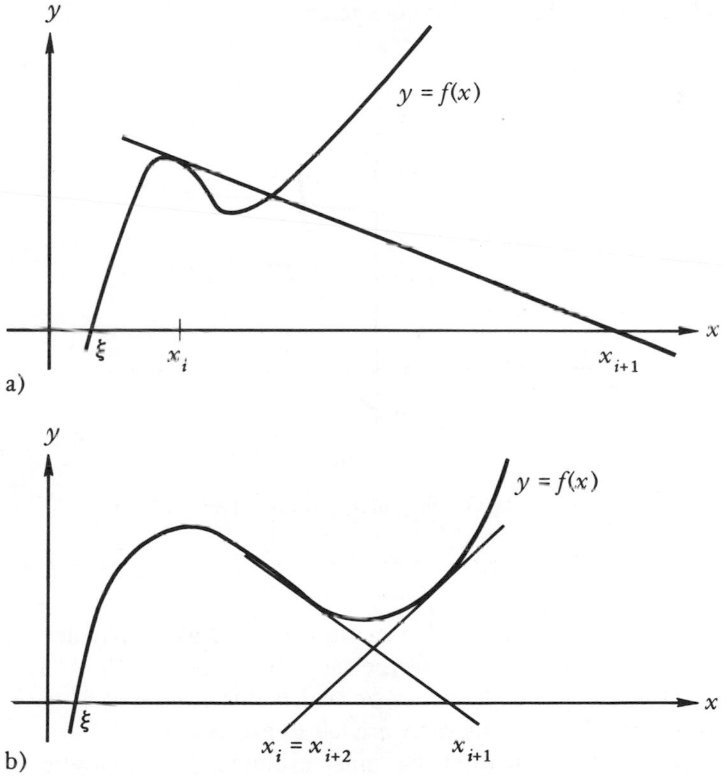

FIG. 3-9 Newton's method when $f'(x) \approx 0$

On the other hand, if the function $f(x)$ has no near-zero derivative near the root, Newton's method is a most effective method. For example, consider the operation of taking the square root of a. This involves solving the equation

$$f(x) = x^2 - a = 0, \tag{104}$$

which is interpreted graphically in figure 3-10. We see that for any initial guess greater than 0, there are no near-zero derivatives in the region of interest. Thus, Newton's method, given by

$$x_{i+1} = \tfrac{1}{2}(x_i + a/x_i), \tag{105}$$

converges quickly. Therefore, this method is used for finding square roots in most square root subroutines commonly provided. For example, if we apply this method to the equation used to illustrate previous methods, $x^2 - 2 = 0$, with $x_0 = 2$, we see $x_1 = \frac{1}{2}(2 + \frac{2}{2}) = 1.5$, $x_2 = \frac{1}{2}(1.5 + 2/1.5) = 1.42$, and so forth, a rapid convergence.

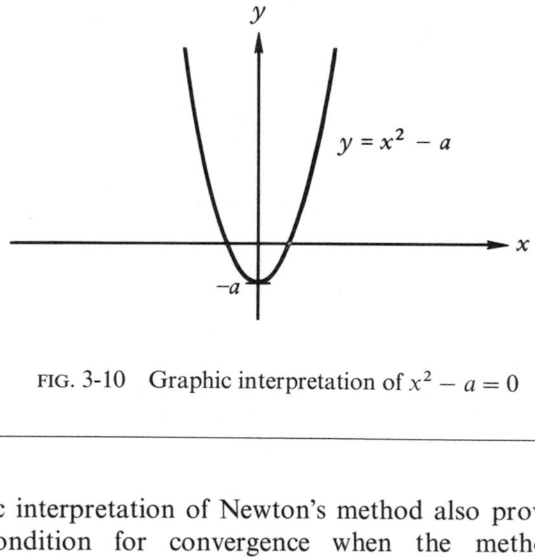

FIG. 3-10 Graphic interpretation of $x^2 - a = 0$

Geometric interpretation of Newton's method also provides another sufficient condition for convergence when the method is used: $f(x)f''(x) > 0$ in $[x_0, \xi)$. Sketches on which this condition is based and the formal proof of its sufficiency are left to exercise 3.3.14.

To use Newton's method, we must evaluate $f'(x)$ analytically and provide the formula for f' to the computer. If we have no way of formally differentiating f, for example if we have a subroutine for f but no analytic expression for f, Newton's method cannot be applied. Even if we can differentiate f, at each iteration we require two function evaluations, one of f and one of f' (as compared to one function evaluation for the method of false position). If the time required for the evaluation of $f'(x_i)$ is about the same as that for $f(x_i)$, one step of Newton's method requires approximately the same amount of computation as two steps of false position (including an application of δ^2 acceleration). Although Newton's method will normally require fewer steps for convergence than the method of false position if that method is unaccelerated, the total time needed to obtain a solution of the required accuracy may be less with false position than with Newton's method. And this time differential widens when we apply acceleration to the method of false position.

We should note that many common functions have derivatives which involve the same function or functions; the derivatives can be simply computed from the values of the original function. Examples are:

$$\frac{d}{dx} e^x = e^x, \tag{106}$$

$$\frac{d}{dx} tan(x) = sec^2(x) = 1 + tan^2(x). \tag{107}$$

In other cases, the derivative can be written in terms of the original function, but in a way that requires an amount of computation which although not great is not negligible. For example,

$$\frac{d}{dx} sin(x) = (1 - sin^2(x))^{1/2}. \tag{108}$$

Especially in the first cases, the extra computation involved in evaluating the derivative at each step does not significantly increase computation time once we have evaluated the function. In any case, the computation time per iterative step must be taken into account when deciding whether Newton's method should be used.

3.3.7 Variations on Newton's Method: Secant Method and Method of Constant Slope

Methods that attempt to approximate Newton's method without requiring the evaluation of a derivative at each step have been developed. One such method is called the *secant method*. It involves approximating the tangent by a secant through the two most recent iterates and using the zero of this line as the next iterate (see figure 3-11). This approach produces the iteration formula

$$x_{i+1} = (x_{i-1} f(x_i) - x_i f(x_{i-1}))/(f(x_i) - f(x_{i-1})). \tag{109}$$

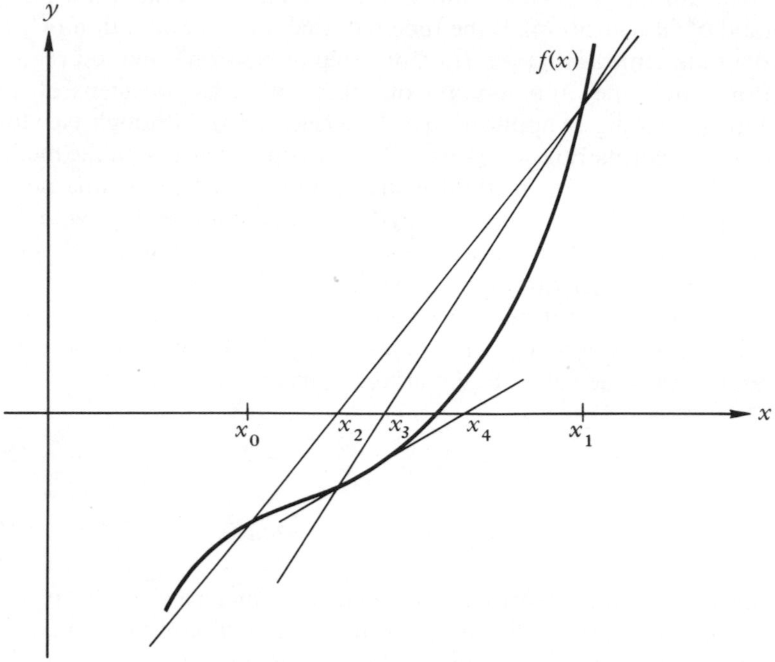

FIG. 3-11 Secant method

The same formula is used in the method of false position. The only difference is that in the secant method we always replace x_{i-1} with x_{i+1} (see program 3-4), whereas in the method of false position we replace either x_i or x_{i-1}, depending on the sign of f at these points. The secant method always uses the most recent information and thus cannot freeze one point. Because of this difference, the secant method does not always converge, but when it does converge, it converges faster than the method of false position.

When does the secant method get in trouble? Generally when Newton's method does: around zero derivatives.

What is the order of convergence of the secant method? We answer this question by subtracting ξ from both sides of equation 109, expanding all occurrences of f in a Taylor series about ξ, and taking the limit as $i \to \infty$, producing

```
/* SECANT METHOD FOR FINDING ROOTS OF EQUATIONS */
/* ONE EQUATION IN ONE UNKNOWN: F(X) = Ø */
/* SUBROUTINE F IS GIVEN */
/* XGUESS1 AND XGUESS2, TWO GUESSES AT THE ROOT, ARE GIVEN */
/* TOL, A THRESHOLD FOR RELATIVE DIFFERENCE BETWEEN SUCCESSIVE ROOT ESTIMATES, IS GIVEN */

   DECLARE (XGUESS1, XGUESS2,  /* TWO MOST RECENT ROOT ESTIMATES; XGUESS1 IS MOST RECENT */
            NEWVAL, OLDVAL,    /* VALUE OF FUNCTION F AT XGUESS1 AND XGUESS2 RESPECTIVELY */
            XNEW,              /* IMPROVED ESTIMATE OF ROOT */
            SLOPE,             /* SLOPE OF SECANT */
            TOL) FLOAT;        /* THRESHOLD FOR STOPPING ITERATION */

   OLDVAL = F(XGUESS2);

/* COMPUTE IMPROVED ESTIMATE OF ROOT */
   IMPROVE: NEWVAL = F(XGUESS1);
   /* COMPUTE SLOPE OF SECANT THROUGH XGUESS1 AND XGUESS2 */
            SLOPE = (NEWVAL - OLDVAL) / (XGUESS1 - XGUESS2);
   /* COMPUTE IMPROVED ROOT ESTIMATE */
            XNEW = XGUESS1 - NEWVAL / SLOPE;

/* TEST WHETHER CLOSE ENOUGH TO ROOT */
            IF ABS(XNEW - XGUESS1) / ABS(XNEW) < TOL THEN GO TO DONE;

/* MAKE XGUESS1 AND XGUESS2 THE TWO MOST RECENT ROOT ESTIMATES */
            XGUESS2 = XGUESS1;
            XGUESS1 = XNEW;
            OLDVAL = NEWVAL;
            GO TO IMPROVE; /* ITERATE AGAIN */

   DONE:  /* XNEW HOLDS BEST ESTIMATE OF ROOT */
```

PROGRAM 3-4 Secant method for solving nonlinear equations

$$\lim_{i \to \infty} (\varepsilon_{i+1}/(\varepsilon_i \varepsilon_{i-1})) = f''(\xi)/(2f'(\xi)), \tag{110}$$

where $\varepsilon_i \equiv x_i - \xi$. The details of this proof are left to exercise 3.3.18. We wish to know the value of α such that

$$\lim_{i \to \infty} (\varepsilon_{i+1}/\varepsilon_i^{\alpha}) = k \neq 0, \tag{111}$$

if such exists. Assume

$$f''(\xi)/(2f'(\xi)) = L \neq 0. \tag{112}$$

Then if we define

$$b_i \equiv ln(L\varepsilon_i), \tag{113}$$

in the limit equation 110 becomes

$$b_{i+1} - b_i - b_{i-1} = 0. \tag{114}$$

This is a second-order homogeneous difference equation for b_i, the method of solution of which is given in appendix A of this text. The solution is

$$b_i = c_1((1 + \sqrt{5})/2)^i + c_2((1 - \sqrt{5})/2)^i, \tag{115}$$

where c_1 and c_2 depend on the values of b_0 and b_1 (see exercise 3.3.19). As $i \to \infty$, the second term becomes negligible, producing in the limit

$$ln(L\varepsilon_i) = c_1((1 + \sqrt{5})/2)^i, \tag{116}$$

or equivalently

$$\varepsilon_i = (1/L)\, e^{c_1((1+\sqrt{5})/2)^i}. \tag{117}$$

From this we can show that

$$\varepsilon_{i+1}/\varepsilon_i^{(1+\sqrt{5})/2} = L^{(-1+\sqrt{5})/2} = a\ constant. \tag{118}$$

Thus, the order of convergence of the secant method is $(1 + \sqrt{5})/2 \approx 1.6$. The secant method has order of convergence faster than linear but not quite quadratic. On the other hand, two steps of the secant method, taken together, require the same number of function evaluations as one step of Newton's method and have order of convergence $(1.6^+)^2 \approx 2.6$.

To illustrate the secant method, we again use $x^2 - 2 = 0$ with $x_0 = 0$ and $x_1 = 2$. Then

$$x_2 = (0(2) - 2(-2))/(2 - (-2)) = 1,$$

$$x_3 = (2(-1) - 1(2))/(-1 - 2) = \tfrac{4}{3},$$

$$x_4 = (1(-\tfrac{2}{9}) - \tfrac{4}{3}(-1))/(-\tfrac{2}{9} - (-1)) = 1.43,$$

and so forth, a fairly fast convergence, but not as fast as with Newton's method.

Another approximation to Newton's method is to pick a constant slope m which is an approximation to the slope of f in the interval in question and iterate using the formula

$$x_{i+1} = x_i - f(x_i)/m \tag{119}$$

(see figure 3-12). This method of constant slope (see program 3-5) is one of the methods suggested for finding a convergent Picard iteration (see equation 88). The order of convergence is linear. Sufficient conditions for convergence are as follows:

1. Either
   a) $|f'(x)| < 2M|m|$ for some $0 \leq M < 1$ in a symmetric interval about ξ, or
   b) $|f'(x)| < |m|$ in a one-sided interval about ξ.
2. $|f'(x)| \geq \varepsilon$ in the interval in question for some $\varepsilon > 0$.
3. $f'(x)$ has the same sign as m in the interval in question.

Proof of the sufficiency of these conditions is left to exercise 3.3.21.

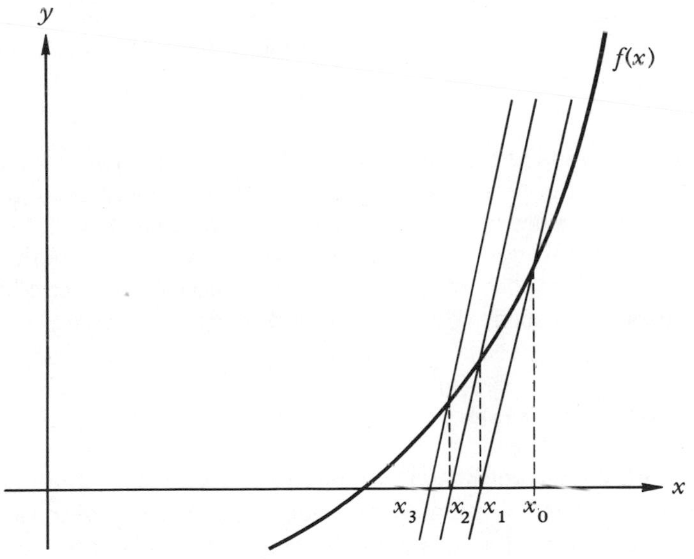

FIG. 3-12 Method of constant slope

```
/* METHOD OF CONSTANT SLOPE FOR FINDING ROOTS OF EQUATIONS */
/* ONE EQUATION IN ONE UNKNOWN: F(X) = Ø */
/* SUBROUTINE F IS GIVEN */
/* XGUESS, A GUESS AT ROOT, IS GIVEN */
/* SLOPE, AN ESTIMATE OF SLOPE OF F NEAR ROOT, IS GIVEN */
/* TOL, A THRESHOLD FOR RELATIVE DIFFERENCE BETWEEN */
/* SUCCESSIVE ROOT ESTIMATES, IS GIVEN */

   DECLARE (XGUESS,       /* GUESS AT ROOT */
            XNEW,         /* IMPROVED ESTIMATE OF ROOT */
            SLOPE,        /* ESTIMATE OF SLOPE OF F NEAR ROOT */
            TOL) FLOAT;   /* THRESHOLD FOR STOPPING ITERATION */

/* COMPUTE IMPROVED ESTIMATE OF ROOT */
   IMPROVE: XNEW = XGUESS - F(XGUESS) / SLOPE;

/* TEST WHETHER CLOSE ENOUGH TO ROOT */
             IF ABS(XNEW - XGUESS) / ABS(XNEW) < TOL
                THEN GO TO DONE;

/* MAKE GUESS = IMPROVED ESTIMATE */
             XGUESS = XNEW;
             GO TO IMPROVE;  /* ITERATE AGAIN */

   DONE:  /* XNEW HOLDS BEST ESTIMATE OF ROOT */
```

PROGRAM 3-5 Method of constant slope for solving nonlinear equations

Since the method of constant slope produces simple linear convergence, Aitken's δ^2 acceleration is applicable to make convergence quadratic after the convergence of the unaccelerated method has become close to simply linear. However, since acceleration can lead to a result on the opposite side of the root from the result before acceleration, condition 1a above must be satisfied; 1b is not sufficient.

3.3.8 Hybrid and Modified Methods

a) Modified False Position. Hamming has presented a modification of the method of false position which retains the property of always converging if the method is applicable but prevents one interval endpoint from becoming frozen.† The modification involves, at the end of each iterative step, replacing the value of f at the interval endpoint not changed

† R. W. Hamming, *Applied Numerical Analysis,* section 2.6.

at that step, by γ times its value, where γ is a positive number less than 1. Thus, if a particular interval endpoint is not changed for n steps, the value of f used there in fitting the line in the next step will be γ^n times the correct value of f there. This procedure maintains the convergence of the method because the line which is fit still crosses the x axis between the interval endpoints. A frozen interval endpoint is impossible because the multiplication by γ eventually causes the line to cross on the opposite side of the root from that on which it has been crossing, producing a new value for the interval endpoint on that side (see figure 3-13).

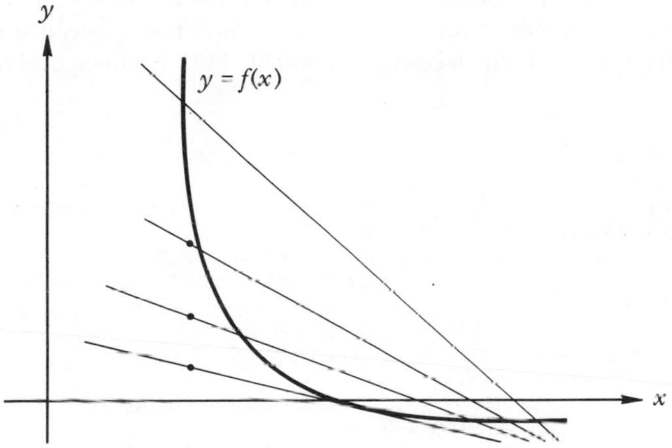

FIG. 3-13 Modified method of false position, $\gamma = \frac{1}{2}$

The modified false position method converges simply linearly (see exercise 3.3.23). If both interval endpoints are close to the root ξ, the value of γ should be near 1 for fastest convergence. On the other hand, γ should be small if one endpoint is far from ξ to prevent that endpoint from being frozen for many iterations. Hamming suggests using $\gamma = \frac{1}{2}$. Ideally γ should be even smaller for early iterations, increasing to 1 in the limit as ξ is approached. This behavior can be effected by letting γ be a monotonically decreasing function of $|x_i^+ - x_i^-|/tol$, where $x_i^+ - x_i^-$ is the width of the interval at the ith step and tol is the accuracy within which we desire to compute the root. Note, however, that if Aitken's δ^2 acceleration is to be applied, γ must be held constant for the steps providing values for a given acceleration.

b) Always-Convergent Superlinear Methods. We have seen that superlinear methods sometimes produce an iterate farther from the root than the preceding iterate (they may diverge). On the other hand, these methods always converge once the iterate is close enough to the root. In contrast, some linear methods always converge but only slowly. We can combine these methods, using the result of the superlinear method as long as that result is in the interval in which the root is trapped, and using the always-convergent linear method when it is not. The interval is made smaller at each step by taking the new iterate in place of the endpoint for which the function f evaluated at the iterate has the same sign. For example, we can produce a secant-false position hybrid (see program 3-6) which is superlinearly convergent and is also always-convergent if false position is applicable. Even better, we can produce a secant-modified false position hybrid which converges quickly both far from and near the root.

```
/* HYBRID SECANT-FALSE POSITION METHOD FOR FINDING ROOTS */
/* ONE EQUATION IN ONE UNKNOWN: F(X) = Ø */
/* SUBROUTINE F IS GIVEN; LFTGUESS AND RTGUESS, */
/* WHERE LFTGUESS < ROOT AND RTGUESS > ROOT, ARE GIVEN */
/* TOL, A THRESHOLD FOR RELATIVE DIFFERENCE BETWEEN */
/* SUCCESSIVE ROOT ESTIMATES, IS GIVEN */
   DECLARE (XNEW,                /* NEW ROOT ESTIMATE */
            XOLD, XPREV,
            /* NEXT TWO MOST RECENT ROOT ESTIMATES */
            LFTGUESS, RTGUESS,
            /* CLOSEST ESTIMATES TO LEFT & RIGHT OF ROOT */
            LFTVAL, RTVAL,           /* F(LFTGUESS), F(RTGUESS) */
            NEWVAL, OLDVAL,          /* F(XNEW), F(XOLD) */
            PREVAL,                  /* F(XPREV) */
            SLOPE,                   /* SLOPE OF LINE THROUGH F */
            TOL) FLOAT,              /* FOR STOPPING ITERATION */
            LFTSIGN FIXED BINARY; /* SIGN OF F(LFTGUESS) */

/* INITIALIZE */
   XOLD    = LFTGUESS;
   XPREV   = RTGUESS;
   LFTVAL  = F(LFTGUESS);
   RTVAL   = F(RTGUESS);
   LFTSIGN = SIGN(LFTVAL);
   OLDVAL  = LFTVAL;
   PREVAL  = RTVAL;
```

PROGRAM 3-6 Hybrid secant-false position method

(*continued on next page*)

```
/* COMPUTE IMPROVED ESTIMATE OF ROOT; SECANT METHOD */
  IMPROVE: SLOPE = (OLDVAL - PREVAL) / (XOLD - XPREV);
           XNEW = XOLD - OLDVAL / SLOPE;
           IF XNEW < RTVAL & XNEW > LFTVAL THEN GO TO TEST;

/* SECANT RESULT NO GOOD; USE FALSE POSITION */
  FALPOS:  SLOPE = (LFTVAL - RTVAL) / (LFTGUESS - RTGUESS);
           XNEW = LFTGUESS - LFTVAL / SLOPE;

/* TEST WHETHER CLOSE ENOUGH TO ROOT */
  TEST:    IF ABS(XNEW - XOLD) / ABS(XNEW) < TOL
              THEN GO TO DONE;

/* TAKE XNEW IN PLACE OF THE ONE OF LFTGUESS AND RTGUESS */
/* FOR WHICH F AGREES IN SIGN WITH F AT XNEW */
           PREVAL = OLDVAL;
           OLDVAL = NEWVAL;
           NEWVAL = F(XNEW);
           IF SIGN(NEWVAL) = LFTSIGN
              THEN LFTREPL: DO;
                               LFTGUESS = XNEW;
                               LFTVAL = NEWVAL;
                   END LFTREPL;
              ELSE RTREPL:  DO;
                               RTGUESS = XNEW;
                               RTVAL = NEWVAL;
                   END RTREPL;
           XPREV = XOLD;
           XOLD  = XNEW;
           GO TO IMPROVE;  /* ITERATE AGAIN */

  DONE:  /* XNEW HOLDS BEST ESTIMATE OF ROOT */
```

PROGRAM 3-6 (*continued*)

3.3.9 Choice of Method

As noted in section 3.3.1, the best method to use in finding a single root ξ of $f(x) = 0$ depends on f. If f behaves smoothly and monotonically near the root and $f'(x)$ is easily evaluated, Newton's method is a wise choice. If the first condition is satisfied but the second is not, the secant method may be the best choice. If there is some question about the smoothness of $f(x)$ near the root, a hybrid of one of these methods and false position (or modified false position) may be the best approach. If $f(x)$ is "wiggly" near the root, none of these methods should be applied.

The method of false position (probably in its modified form with the multiplier γ chosen dynamically) should be used with Aitken's δ^2

acceleration unless $f'(\xi) \approx 0$. In its accelerated form, it is quadratically convergent but slower than either Newton's method or the secant method per function evaluation. It is normally used when there is some question about the convergence of Newton's method or the secant method. The method of false position is rather slow, even in its modified form, when $f(x)$ is very convex near the root; f is not fit well by the approximating line of false position.

Picard iteration is chosen only in special cases where it is quadratically convergent. Otherwise, because transforming $f(x) = 0$ into a convergent form $g(x) = x$ is not straightforward and because the time at which acceleration is properly applicable is difficult to determine, this method is less desirable than others.

The bisection method is chosen primarily when it is important to know beforehand precisely how many iterations are required to achieve a desired accuracy in the value of the root. Otherwise, it is too slow and is not used unless $f(x)$ is very peculiarly behaved so that all of the other methods have a major defect.

PROBLEMS

3.3.1. In the bisection method, at the ith step the interval of width w_i is divided into two parts of widths γw_i and $(1 - \gamma)w_i$ respectively, where $\gamma = \frac{1}{2}$. But the choice of the value $\frac{1}{2}$ for γ is arbitrary. Discuss the advantages and disadvantages of using a fixed γ value which is between 0 and 1 but other than $\frac{1}{2}$.

3.3.2. Assume $x_{i+1} = g(x_i)$ produces a quadratically convergent sequence to ξ with convergence factor k. Assume an error bounded by b is generated at each iterative step.

a) By noting that no further improvement in the iterate value can be assured when the improvement in the propagated error in a step is not greater than the error generated in that step, show that the error in the result of such an iteration is bounded by $(1 - \sqrt{1 - 4b|k|})/2|k|$ if x_0 is chosen appropriately close to ξ and $4b|k| < 1$.

b) What happens when $4b|k| > 1$?

c) Give an approximate value for the bound if $4b|k| \ll 1$.

3.3.3. a) Using equation 23, solve $|x_m - \xi| = \delta$ for m to show that if an iterative method has order of convergence $\alpha_2 > 1$ and convergence factor k_2, and if $|x_0 - \xi|$ is small enough so that $|x_{i+1} - \xi| = |k_2|\,|x_i - \xi|^{\alpha_2}$ approximately holds for all i, then the number of iterations, m, needed to achieve accuracy δ, assuming no generated error, is given by equation 26.

b) Show that if m is the value given by equation 26 and $0 < \delta < |x_0 - \xi|$,

$$\lim_{\substack{\delta \to 0 \\ |x_0 - \xi| \to 0}} \left(\frac{t_2 m}{t_1(\log(|x_0 - \xi|) - \log(\delta))/\log(|k_1|)} \right) = 0.$$

From this result we can conclude that for small δ and $|x_0 - \xi|$, the time required to achieve accuracy δ using a superlinear method with order of convergence α_2, convergence factor k_2, and computing time per iteration t_2, is much less than that required when using a linear method with convergence factor k_1 and computing time per iteration t_1, for any values of these method parameters.

3.3.4. a) Refer to the convergence test defined by equation 33. Sketch a function for which when the method of false position is applied, the test will be satisfied when x_{i+1} is still far from the root.

b) Repeat part a, but for the convergence test defined by equation 34.

3.3.5. Show that if $f(x^+) > 0$ and $f(x^-) < 0$, the line passing through $(x^+, f(x^+))$ and $(x^-, f(x^-))$ has its zero in the interval (x^+, x^-).

3.3.6. Assume $f(x)$, $f'(x)$, and $f''(x)$ exist for $x \in [x_0^+, x_0^-]$. Further assume $f(x_0^+) > 0$ and $f(x_0^-) < 0$, so $f(x)$ must have a root $\xi \in [x_0^+, x_0^-]$. Prove that if $f''(\xi) \neq 0$, one interval endpoint of the method of false position must eventually remain frozen (there exists an integer $N > 0$ such that either for $i > N$, $x_{i+1}^+ = x_i^+$, or for $i > N$, $x_{i+1}^- = x_i^-$).

3.3.7. Apply the method of false position to find the root in $[-1,0]$ of the equation $x + \frac{1}{2} + \sin^{-1}(x) = 0$

a) Using the method of false position alone.

b) Applying Aitken's δ^2 acceleration every two iterations after one interval endpoint becomes frozen (and rejecting the accelerated value if it falls outside the present interval).

3.3.8. Show that if we have a Picard iteration, $x_{i+1} = g(x_i)$, where g, g', and g'' are defined in a neighborhood of the root ξ, then if $g'(\xi) = 0$, the order of convergence is quadratic or higher.

3.3.9. Put the following equations in a form such that Picard iteration will converge. In each case, support your conclusion with a graph.

a) $\tan(x+1) - 20x = 0$, for the root near $x = 0.1$.

b) $\tan(x+1) - 20x = 0$, for the root near $x = 0.5$.

c) $x^3 - 2x^2 + 2x - 5 = 0$, for its only real root.

3.3.10. Prove that if

$f(x) = 0$ has only one root and that root is at $x = \xi$,
$f(x) = 0$ is written as $g(x) = x$, where $g(x) = x \Leftrightarrow f(x) = 0$ and g is continuous,
$g'(x)$ is continuous at ξ,
$0 < g'(\xi) < 1$,
$g(x) > \xi$ for $x > \xi$, and
$x_0 > \xi$,

then Picard's method applied to $g(x)$ with initial value x_0 converges to ξ.

3.3.11. Say whether each of the following statements is always true, sometimes true but sometimes not true, or never true. If a statement is always true or never true, give a proof. If it is sometimes true but sometimes not true, sketch a counterexample.

a) In Picard's method, with the iteration in the form $x_{i+1} = g(x_i)$ where $x = g(x)$ has only one root, if $|g'(x_i)| > 1$ and $|g'(x_{i+1})| > 1$ for two successive iterates, then the iteration will not converge to the root.

b) Let ξ be a solution to $x = g(x)$. Let $g(x)$, $g'(x)$, and $g''(x)$ be continuous functions. Then since we know that Picard's method followed every two iterations by Aitken's δ^2 acceleration has quadratic convergence, there exists an interval about ξ such that if we choose x_0 in this interval and apply Picard's method with acceleration starting at x_0, the iteration will converge no matter what the value of $g'(\xi)$.

3.3.12. Apply Picard's method with Aitken's δ^2 acceleration, when it becomes applicable, to the equations in exercise 3.3.9.

3.3.13. a) Choose starting points for the following equations so that Newton's method will converge to the specified root.
(1) $ln(x) + x^2 - 8 = 0$, only root.
(2) $x - cos(2x) = 0$, smallest positive root.

b) Apply Newton's method to the equations in part a.

3.3.14. Assume ξ is a root of $f(x)$, and $f(x)$ is twice differentiable on $[x_0, \xi]$. Assume $f(x)f''(x) > 0$ if $x \in [x_0, \xi)$.

a) There are four possible general shapes which f can have on $[x_0, \xi]$. Sketch them. Argue geometrically that in each case Newton's method with starting value x_0 will converge to ξ.

b) Prove that Newton's method with starting value x_0 will converge to ξ.

3.3.15. a) Devise an algorithm based on Newton's method of finding the nth root of a positive number for n an integer > 1.

b) Under what conditions does the method converge?

3.3.16. Show that under the appropriate conditions (what are they?), the errors for Newton's method satisfy the equation $\lim_{i \to \infty} (\varepsilon_{i+1}/\varepsilon_i^2) = f''(\xi)/(2f'(\xi))$.

3.3.17. Consider the equation $x - e^{1-x}(1 + ln(x)) = 0$. It has roots at $x = \xi_1 = 1$ and $x = \xi_2 \approx .5$.

a) For the starting value, $x = 0.8$, set up a Picard iteration that will converge to the root at $x = 1$. Using a graph to illustrate your case, discuss the convergence properties of this iteration for each of the following intervals for the starting value x_0:
(1) $0 < x_0 < \xi_2$.
(2) $\xi_2 < x_0 < \xi_1$.
(3) $x_0 < \xi_1$.

b) For the starting value, $x = 0.8$, set up Newton's method and execute two steps to slide-rule accuracy. To which root is the iteration converging? How can you get Newton's method to converge to the other root?

c) Discuss the advantages and disadvantages of using Picard iteration for this equation.

d) Repeat part c but for Newton's method.

3.3.18. Show that equation 110 is true, that is, if the secant method converges to ξ and $\varepsilon_i \equiv x_i - \xi$, then $\lim_{i \to \infty} (\varepsilon_{i+1}/(\varepsilon_i \varepsilon_{i-1})) = f''(\xi)/(2f'(\xi))$.

3.3.19. Solve equation 114, the difference equation $b_{i+1} - b_i - b_{i-1} = 0$, which occurs in finding the order of convergence of the secant method. (See appendix A for the solution method.)

3.3.20. Apply the secant method to the equations in exercise 3.3.13.

3.3.21. Prove that the convergence conditions given in section 3.3.7 for the method of constant slope do indeed suffice for its convergence.

3.3.22. Apply the method of constant slope with Aitken's δ^2 acceleration, when it becomes applicable, to the equations in exercise 3.3.13.

3.3.23. Show that Hamming's modification of the method of false position converges simply linearly. What is the convergence factor? (*Hint:* First show that in the limit, $x_i, x_{i+2}, x_{i+4}, \ldots$ are on one side of the root and $x_{i+1}, x_{i+3}, \ldots$ are on the other side of the root.)

3.3.24. Consider the following iterative methods for the solution of nonlinear equations:
(1) Newton's method
(2) Secant method
(3) Method of false position (possibly with Aitken's δ^2 acceleration)
(4) Bisection method

a) For each of the functions $f(x)$ in exercise 3.2.1, tell which one of these methods you would use to find the root of $f(x)$ in the interval noted, and say why. Say why you would not use each of the other methods instead. Assume you must use one or both endpoints of the interval noted as initial estimate(s) of the root. If you use one, say which one and why. Do not equivocate about the method and endpoint you would choose!

b) Repeat part a for the equation $x^2 + G(x) = 0$, with the root known to be in (a,b), where you have only a calling routine for $G(x)$. That is, for any x_0, the computer can give you $G(x_0)$, but you have no analytic expression for $G(x)$. Assume $G(x)$ and its derivatives are continuous.

3.4 MULTIPLE ROOTS

So far, all of the proposed methods for solving nonlinear equations have been analyzed only with respect to single roots. What if f has a root of multiplicity greater than 1 at ξ?

If the root is of even multiplicity (the multiplicity is an even integer), $f(x)$ does not cross the x axis at ξ and neither the method of bisection nor false position is applicable. If the root is of odd multiplicity so that f crosses the axis at ξ, it does so with a zero derivative. Both methods are applicable but errors in evaluating the function may cause a change of sign while the error of the estimate is still large. The point to which the iterates converge will be ill-conditioned. Also, acceleration of the method of false position will often cause trouble, so acceleration should not be applied. Thus we can obtain only linear convergence and, even that, only if we are very careful about the precision of the evaluation of f.

Suppose we use Picard iteration. If $g(x) = x$ has the same multiplicity as $f(x)$ (as we expect it will), then if the root is of even multiplicity, g will have a derivative greater than 1 on one side of ξ and a derivative less than 1 on the other side of ξ (see figure 3-14). Picard iteration will converge from one side of ξ and diverge from the other side. Even if we pick a starting value on the convergent side, when the iterate gets close to ξ, a small computational error may cause the iterate to fall on the divergent side; further iteration will produce results farther away from ξ. If Picard iteration is applied to a multiple root of odd multiplicity such that the curve $g(x)$ is below $y = x$ for $x > \xi$, the iteration will converge but very slowly.

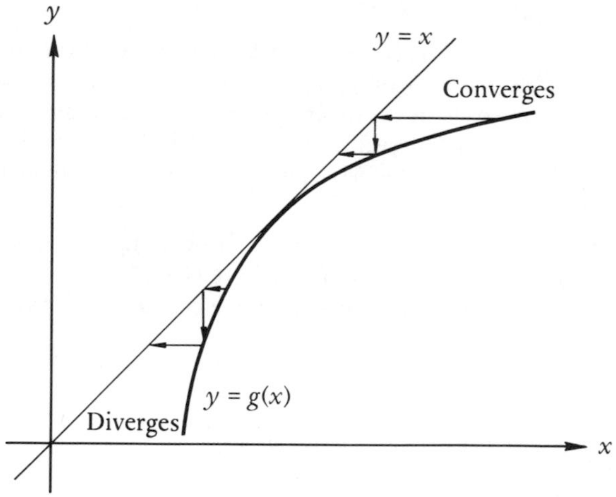

FIG. 3-14 Picard iteration with a root of even multiplicity

Newton's method converges linearly to ξ at a root of multiplicity $m > 1$, with convergence factor $1 - 1/m$. The proof, which follows from writing $f(x)$ as $(x - \xi)^m g(x)$, is left for exercise 3.4.1, part a.

The secant method may not converge at a root of even multiplicity, because the values of f at the two most recent iterates may be approximately the same. This method will converge to a root of odd multiplicity if the starting values are chosen close enough to the root. The proof of this statement is left to exercise 3.4.3.

The method of constant slope, which is a Picard iteration, has the behavior of Picard iterations at multiple roots as specified above.

Thus, at double roots, at least, the only method of those we have discussed that is clearly applicable is Newton's method.

PROBLEMS

3.4.1. a) Show that Newton's method converges linearly to a root of multiplicity $m > 1$ with convergence factor $1 - 1/m$.

b) Show that the iteration $x_{i+1} = x_i - mf(x_i)/f'(x_i)$ converges quadratically to the root.

3.4.2. A method that is quadratically convergent to roots of multiplicity greater than 1 can be devised by using the iteration $x_{i+1} = x_i - kf(x_i)/f'(x_i)$ and using the values of three successive iterates to choose the value of k to be used in the next two iterations. Devise such a method.

3.4.3. Prove that for any root of odd multiplicity of $f(x) = 0$, where f is continuous, there exists an interval about the root such that if starting values x_0 and x_1 are chosen in that interval, the secant method converges to the root.

3.5 SIMULTANEOUS EQUATIONS IN MANY UNKNOWNS

Let us consider a system of m nonlinear equations in n unknowns. As in the case of linear equations, if there are fewer equations than unknowns, if there exists a solution, there exist an infinite number, since we can let one or more of the variables be parameters and solve the equations for various values of the parameters. The only exception to this rule of either no solutions or an infinite number of them occurs when there exists a solution for only a finite number of values of the parameter variables. This is a very unusual situation.

If there are more equations than unknowns, it is likely that no solution exists. If each equation expresses an independent constraint, there is none.

If the number of equations is equal to the number of unknowns and none of the equations is either redundant to the remaining equations (like linear dependence when there are an infinite number of solutions) or contradictory to the remaining equations (like linear dependence when there does not exist a solution), then discrete solutions exist. We shall assume this case. We will develop iterative solution methods for these equations which are extensions of our methods for nonlinear equations in one unknown and involve ideas we applied in developing iterative methods to solve systems of linear equations.

Before we develop iterative methods for solution of systems of nonlinear equations in many unknowns, we should emphasize the difficulty of applying such methods. It is very difficult, if not impossible, to graph the multidimensional surfaces defined by these equations. Thus, finding an approximation to the root(s) is very difficult, and choosing a method that will converge is very difficult. Furthermore, methods like bisection and false position which always converge in the one-dimensional case are inapplicable in many dimensions. Finally, the nonlinearity of the equations makes prediction of the behavior of the solution method quite hard.

The difficulty in using the available iterative methods to find solutions of systems of many nonlinear equations increases as the number of equations and unknowns increases. It pays to try to reduce the number of equations and unknowns analytically before attempting a numerical solution. In particular, we should try to solve each equation for some unknown in terms of the others. Then we can substitute the resulting equation for that unknown in all remaining equations, thereby removing the solved equation and unknown from the system.

3.5.1 Picard Iteration for Many Unknowns

Consider a set of equations

$$
\begin{aligned}
f_1(x_1,x_2,\ldots,x_n) &= 0 \\
f_2(x_1,x_2,\ldots,x_n) &= 0 \\
\cdot \qquad\qquad & \quad \cdot \\
\cdot \qquad\qquad & \quad \cdot \\
\cdot \qquad\qquad & \quad \cdot \\
f_n(x_1,x_2,\ldots,x_n) &= 0,
\end{aligned}
\tag{120}
$$

where the x_j are variables, and the f_i are functions. Let the solution be $x = z$, where x and z are vectors.

As in Picard iteration for one unknown (section 3.3.5) and in the relaxation methods for solving linear equations (sections 2.8.2 and 2.8.3), we solve the first equation for x_1, the second for x_2, and so forth, producing

$$\begin{aligned} x_1 &= g_1(x_1,x_2,\ldots,x_n) \\ x_2 &= g_2(x_1,x_2,\ldots,x_n) \\ &\vdots \\ x_n &= g_n(x_1,x_2,\ldots,x_n). \end{aligned} \tag{121}$$

Note that unlike the relaxation methods for linear equations, we may not be able to eliminate x_j from the right side of the jth equation, so it appears there in equation 121.

An iteration is produced from equation 121 by using the variable values from the ith iteration on the right side of the equations to produce the variable values in the $(i+1)$th iteration on the left side of the equations (x_j^i is the value of the jth variable at the ith iteration):

$$\begin{aligned} x_1^{i+1} &= g_1(x_1^i,x_2^i,\ldots,x_n^i) \\ x_2^{i+1} &= g_2(x_1^i,x_2^i,\ldots,x_n^i) \\ &\vdots \\ x_n^{i+1} &= g_n(x_1^i,x_2^i,\ldots,x_n^i). \end{aligned} \tag{122}$$

By analogy to cases discussed previously, we can predict the behavior of this Picard iteration for many unknowns. Equations in one unknown are a special case of equations in many unknowns. Since we know that in the case of one unknown, the particular way we solve $f(x) = 0$ for $x = g(x)$ affects the convergence of the resulting iteration, we should expect that the way we solve the jth equation for $x_j = g_j(x)$ for $1 \leq j \leq n$ will affect the convergence of the iteration given by equation 122. In a similar manner, linear equations are a special case of general (nonlinear) equations. Since we know that the order of the equations affects the convergence in the linear case, we should expect that is also the case here. Also, because we had difficulty obtaining convergence with relaxation methods for linear equations, we should expect even more difficulty here, since nonlinear equations tend to be less well behaved and are less predictable than linear equations. Finally, because both relaxation methods for systems of linear equations and Picard iteration for one unknown (a relaxation method) had simple linear convergence when they converged, we should expect our extended Picard iteration to have this property.

The theorem for a sufficient condition for the convergence of Picard iteration in one unknown depends on a Taylor series expansion. So does that for Picard iteration for many unknowns. Furthermore, we note that the sufficient condition for convergence of Picard iteration for many unknowns becomes, for linear equations, precisely the row-diagonal dominance condition for the convergence of Jacobi iteration.

Theorem: If K is some constant <1 and if for $j = 1,2,\ldots,n$,

$$\sum_{k=1}^{n} \left| \frac{\partial g_j}{\partial x_k}(x) \right| \leq K \tag{123}$$

for all x in the neighborhood of the root z such that $|x_k - z_k| < R$, $1 \leq k \leq n$, for some R, then the Picard iteration given by equation 122 converges to z if an initial guess is made in that neighborhood.

Proof: By Taylor's theorem we know there exist vectors η^{ji} such that $\eta_k^{ji} \in [x_k^i, z_k]$ and

$$\begin{aligned} x_j^{i+1} - z_j &= g_j(x_1^i, x_2^i, \ldots, x_n^i) - z_j \\ &= g_j(z) + \sum_{k=1}^{n} \left[\frac{\partial g_j}{\partial x_k}(\eta^{ji})(x_k^i - z_k) \right] - z_j \\ &= \sum_{k=1}^{n} \left[\frac{\partial g_j}{\partial x_k}(\eta^{ji})(x_k^i - z_k) \right], \quad j = 1,2,\ldots,n. \end{aligned} \tag{124}$$

Defining the matrix G^i by

$$G_{jk}^i = \frac{\partial g_j}{\partial x_k}(\eta^{ji}) \tag{125}$$

and using

$$\varepsilon^i \equiv x^i - z, \tag{126}$$

we can rewrite equation 124 as

$$\varepsilon^{i+1} = G^i \varepsilon^i. \tag{127}$$

If x^i is in the required neighborhood, then for all j

$$\begin{aligned} |\varepsilon_j^{i+1}| &\leq \sum_{k=1}^{n} |G_{jk}^i|\, |\varepsilon_k^i| \\ &\leq \max_k |\varepsilon_k^i| \sum_{k=1}^{n} |G_{jk}^i| \\ &\leq K \max_k |\varepsilon_k^i|, \end{aligned} \tag{128}$$

since we have assumed $\sum_{k=1}^{n} |G^i_{jk}| \leq K$.

Therefore
$$\max_j |\varepsilon_j^{i+1}| \leq K \max_j |\varepsilon_j^i|, \tag{129}$$

so the maximum element error magnitude will decrease at the ith step. Also, since $\max_j |\varepsilon_j^{i+1}| < \max_j |\varepsilon_j^i|$, if $\max_j |\varepsilon_j^i| < R$, so will $\max_j |\varepsilon_j^{i+1}| < R$ and thus for each j, x_j^{i+1} will be in the required neighborhood. Therefore, the iteration will converge.

If Picard iteration converges,

$$\lim_{i\to\infty} \eta^{ji} = z \quad \text{for} \quad j = 1,2,\ldots,n. \tag{130}$$

Thus, if we define

$$G_{jk} = \frac{\partial g_j}{\partial x_k}(z), \tag{131}$$

in the limit

$$\varepsilon^{i+1} = G\varepsilon^i; \tag{132}$$

we have simple linear convergence.† Note that a necessary and sufficient condition for the existence of a neighborhood about z such that Picard iteration starting at a point in the neighborhood converges to z is that the maximum eigenvalue of G in magnitude be less than 1.

An alternative sufficient condition for convergence of Picard iteration is that if A is the neighborhood of z consisting of all x such that $\sum_{j=1}^{n} |x_j - z_j| \leq \sum_{j=1}^{n} |x_j^0 - z_j|$,

$$\sum_{j=1}^{n} \max_{x\in A} \left|\frac{\partial g_j}{\partial x_k}(x)\right| < 1, \quad \text{for} \quad k = 1,2,\ldots,n. \tag{133}$$

This condition is quite difficult to satisfy or test because the maximum is inside the summation. The formal proof of its sufficiency is left to exercise 3.5.1.

It is difficult to find a set of functions $\{g_j\}$ such that equation 121 is equivalent to equation 120 and one of the sufficient conditions above (or some other set of sufficient conditions) is satisfied. If such a set is found, we can usually improve the speed of convergence by using a method of

† Note that the definition of simple linear convergence in the case of a sequence of vectors is the same as that for a sequence of scalars, except that the linear convergence factor is replaced by a linear transformation.

successive displacements rather than a method of simultaneous displacements (equation 122). That is, we should iterate using

$$\begin{aligned}
x_1^{i+1} &= g_1(x_1^i, x_2^i, \ldots, x_n^i) \\
x_2^{i+1} &= g_2(x_1^{i+1}, x_2^i, \ldots, x_n^i) \\
&\vdots \\
x_n^{i+1} &= g_n(x_1^{i+1}, x_2^{i+1}, \ldots, x_{n-1}^{i+1}, x_n^i).
\end{aligned} \tag{134}$$

As with linear equations, if the simultaneous displacements iteration diverges, the corresponding successive displacements method normally diverges faster. The divergence is detectable sooner, so the method of successive displacements is still preferred. In the rest of section 3.5, if there is a choice between methods of simultaneous and successive displacements, we will choose the latter.

3.5.2 Newton's Method for Many Unknowns

Picard iteration converges only linearly, though in the limit, Aitken's δ^2 acceleration is applicable to make the convergence quadratic. However, as for one unknown, we can create a method which is itself quadratically convergent, Newton's method for many unknowns. As with one unknown, we expand each equation as a Taylor series about the root z, producing

$$\sum_{j=1}^{n} \frac{\partial f_k}{\partial x_j}(\eta^k)(x_j - z_j) = f_k(x), \quad k = 1,2,\ldots, n, \tag{135}$$

where the vector η^k has the property $\eta_j^k \in [x_j^i, z_j]$. Since we do not know η^k, we evaluate the partial derivatives instead at x^i, the approximation to z at the ith iterative step. The z_j become only approximations, x^{i+1}, so equation 135 becomes

$$\sum_{j=1}^{n} \frac{\partial f_k}{\partial x_j}(x^i)(x_j^i - x_j^{i+1}) = f_k(x^i), \quad k = 1,2,\ldots,n. \tag{136}$$

These are n linear equations in the unknowns $x_j^i - x_j^{i+1}$, $j = 1,2,\ldots,n$. Solving these equations, we can obtain a value for x^{i+1}.

The matrix specifying the coefficients of the linear equations is given by $A_{kj} = \dfrac{\partial f_k}{\partial x_j}(x^i)$. The determinant of this matrix is called the *Jacobian* of

f at x^i (f is a vector of functions), and it is written $J(f)|_{x^i}$. As is the case when $f' = 0$ near the root when solving for one unknown, if the Jacobian is near zero in a neighborhood about z of half width $|x_j^0 - z_j|$ in the jth co-ordinate (the matrix of derivatives is nearly singular), we can anticipate trouble if we apply Newton's method. In practice, this condition is very hard to test.

Equation 136, the equation for Newton's method, can also be derived by finding an n-dimensional hyperplane tangent to each $f_k(x)$ of equation 120 at x^i and finding the common zero of these hyperplanes. The method converges quadratically as long as x^0 is chosen sufficiently close to z. As we have said, finding an x^0 close to z can be very difficult.

Use of Newton's method for solving nonlinear equations with many unknowns is very time-consuming, computationally. *Every iteration* requires the evaluation of $n^2 + n$ functions, plus the solution of a set of n simultaneous linear equations. Because of this fact, it is seldom used for even medium-sized values of n. Rather, a modified version of the method is often used.

Instead of solving all of the equations simultaneously, we fit a line in the x_1 direction tangent to f_1 and move to the zero of that line. Then we fit a line in the x_2 direction tangent to f_2 and move to the zero of that line. And so on, until we have gone through the whole set of equations. Then we return to the first equation and repeat the process. This produces the iteration

$$x_j^{i+1} = x_j^i - f_j(x^i) \Big/ \frac{\partial f_j}{\partial x_j}(x^i), \quad j = 1,2,\ldots, n. \tag{137}$$

This method, which we will call *modified Newton's method* may not converge for any neighborhood of z (see exercise 3.5.3). However, it often converges if x^0 is close to z, and it requires only $2n$ function evaluations per step. When it is used, it should be used in its successive displacements form rather than in the simultaneous displacements form given in equation 137.

Another modified version of Newton's method is an n-dimensional secant method. At each step, an n-dimensional hyperplane is fit to each f_k evaluated at $n+1$ points. This method requires only n function evaluations at each step, but it also requires the solution of n simultaneous linear equations at each step as well as $n(n+1)$ function evaluations initially. The development of this method is left to the student.

PROBLEMS

3.5.1. Prove that if

a) $\underline{x} = \underline{g}(\underline{x})$ iff $\underline{f}(\underline{x}) = 0$,

b) $A = \left\{ \underline{x} \middle| \sum_{j=1}^{n} |x_j - z_j| \leq \sum_{j=1}^{n} |x_j^0 - z_j| \right\}$, where $\underline{f}(\underline{z}) = 0$,

c) $\sum_{j=1}^{n} \max_{\underline{x} \in A} \left| \frac{\partial g_j(\underline{x})}{\partial x_k} \right| < 1$ for $k = 1,2,\ldots,n$,

then the Picard iteration $\underline{x}^{i+1} = \underline{g}(\underline{x}^i)$ starting at $\underline{x}^0$ converges to $\underline{z}$.

3.5.2. Find a form of the following equations and a starting value such that Picard iteration converges. Do the iteration until two successive iterates have a relative difference less than 10^{-3} in all variables.

$log(x^3 + y) - 2x = .1$
$e^{xy} tan(x) + x^3 y = .2$

3.5.3. Consider the successive displacements form of modified Newton's method for two equations in two unknowns, $f(x,y) = 0$, $g(x,y) = 0$:

$$x_{i+1} = x_i - f(x_i, y_i) \Big/ \frac{\partial f}{\partial x}(x_i, y_i)$$

$$y_{i+1} = y_i - g(x_{i+1}, y_i) \Big/ \frac{\partial g}{\partial y}(x_{i+1}, y_i).$$

Find two functions such that their root is of multiplicity one and such that the iteration diverges for any initial guess other than the root, no matter which of the two functions is taken as f and which is taken as g. Show the mathematics you use in finding these functions.

3.5.4. a) Apply Newton's method to the equations of exercise 3.5.2.

b) Apply modified Newton's method to these equations.

3.5.5. For each of the following sets of equations, choose a method and a starting point which will lead to a solution. Justify your choices.

a) $e^x sin(y) + cos(x) cos(y) = 0$
$cos(x^2 y + xy^2) + sin(xy) = 0$

b) $xe^y + (1 + y)e^{x-1} = 2.01$
$y\, log(x + 1) + (x - 1)\, log(y + 1) = 0$

3.5.6. Each of the following systems of two nonlinear equations in two unknowns has the solution (1,1). Assuming this were not obvious but that you could guess the root as (.9,1.1), discuss precisely how you would go about finding the root in each case. Justify your answers.

a) $xy^2 + x^2 y + x^4 - 3 = 0$
$x^3 y^5 - 2x^5 y - x^2 + 2 = 0$

b) $x^7 y^5 - y^4 + xy - x = 0$
$x^6 y^8 + xy^5 - 3y + 1 = 0$

3.6 SOLUTION OF POLYNOMIAL EQUATIONS

Polynomials are common structures in applied mathematics. For example, because they are analytically easy to handle and relatively well behaved, they are used often as approximating functions in numerical analysis. Also, linear differential equations, which occur commonly in analysis of physical systems, can be analyzed to produce polynomial equations. Because polynomial equations occur commonly and have special properties which allow us to find their roots more easily than general nonlinear functions and because we much more often are interested in finding all of the roots of a polynomial than we are in finding all of the roots of some other nonlinear function, we treat the solution of this special nonlinear equation, the polynomial equation, as a separate section.

3.6.1 Properties of Polynomials and Polynomial Algorithms

Among the special properties of polynomials that make working with them easy are the following:

1. An nth-degree polynomial has exactly n complex roots, assuming a root of multiplicity m is counted as m roots.
2. Every polynomial can be written in terms of its roots ξ_i as $p(x) = K \prod_{i=1}^{n} (x - \xi_i)$, where K is a constant.
3. All roots of a polynomial have integer multiplicity.
4. Complex roots of polynomials with real coefficients come in complex conjugate pairs.
5. There exist algorithms to localize distinct roots, not necessarily of multiplicity 1, of polynomials.
6. A polynomial is completely defined by its coefficients (and thus a polynomial is stored in a computer by storing its coefficients). Two polynomials are the same if and only if, for all i, the coefficient of x^i (x to the power i) in one polynomial is the same as the coefficient of x^i in the other polynomial. This fact will be the basis of many algorithms involving polynomials.
7. Sums and products of polynomials are polynomials.
8. If $p(x)$ is a polynomial, then if $x = az + b$, then $p(x)$ evaluated as a function of z is a polynomial in z. In particular $p(x - b)$ is a polynomial in x.
9. A polynomial is differentiable as many times as desired for all finite values of x. All derivatives of a polynomial are polynomials.

10. For every finite x, a polynomial evaluated at x is finite. The magnitude of a nonconstant polynomial evaluated at $\pm\infty$ is infinite.

There are many other analytic properties of polynomials that make them useful in certain circumstances, but we need not go into them here.

We need certain basic algorithms to use in later algorithms for the solution of polynomial equations. In particular, we need algorithms for the following operations:

1. Addition or subtraction of polynomials
2. Multiplication of polynomials
3. Division of polynomials
4. Evaluation of a polynomial at some value $x = x_0$
5. Evaluation of some derivative of a polynomial at some value $x = x_0$
6. Rewriting a polynomial $p(x)$ as a polynomial $q(x - x_0)$, for some x_0 (an operation called shifting the polynomial)

All of these algorithms are based on the technique of setting equal the coefficients of corresponding terms of the polynomials on each side of an equation.

To add two polynomials, we must find a polynomial $r(x)$ which is equal to $p(x) + q(x)$, where p and q are polynomials. If we write

$$p(x) = \sum_{i=0}^{n_p} a_i x^i \tag{138}$$

and

$$q(x) = \sum_{i=0}^{n_q} b_i x^i, \tag{139}$$

then

$$r(x) = \sum_{i=0}^{max(n_p,n_q)} c_i x^i. \tag{140}$$

Assuming without loss of generality that $n_p > n_q$, we have

$$\sum_{i=0}^{n_p} c_i x^i = \sum_{i=0}^{n_p} a_i x^i + \sum_{i=0}^{n_q} b_i x^i = \sum_{i=0}^{n_p} (a_i + b_i) x^i, \tag{141}$$

where $b_i = 0$ if $i > n_q$. Setting coefficients of like terms equal, we have

$$c_i = a_i + b_i, \tag{142}$$

which when applied for $i = 0, 1, \ldots, n_p$, produces the polynomial r.

The algorithm for subtraction of polynomials is the same as that for addition except that all plus signs are replaced by minus signs.

To compute the polynomial $r(x) = p(x)q(x)$, where p and q are as before,

$$r(x) = \sum_{i=0}^{n_p+n_q} c_i x^i. \tag{143}$$

Then

$$\sum_{i=0}^{n_p+n_q} c_i x^i = \left(\sum_{k=0}^{n_p} a_k x^k\right)\left(\sum_{j=0}^{n_q} b_j x^j\right) = \sum_{k=0}^{n_p} \sum_{j=0}^{n_q} a_k b_j x^{k+j}. \tag{144}$$

The coefficient of x^i on the right side of equation 144 is $\sum_{k=0}^{i} a_k b_{i-k}$, where by convection we assume that a coefficient with a subscript less than 0 or greater than the maximum power of the polynomial of which it is a part, is equal to 0. Thus,

$$c_i = \sum_{k=0}^{i} a_k b_{i-k}, \tag{145}$$

applied for $i = 0,1,\ldots,n_p+n_q$, gives us an algorithm for finding the product of two polynomials.

When we divide the polynomial $p(x)$ by $q(x)$, we produce a quotient polynomial $\sum_{i=0}^{n_p-n_q} c_{n_q+i}\, x^i$ and the remainder polynomial $\sum_{i=0}^{n_q-1} c_i x^i$. The coefficients of these polynomials are given by the following equations:

$$\text{for} \quad k = n_p, n_p - 1, \ldots, n_q: \quad c_k = \frac{1}{b_{n_q}}\left(a_k - \sum_{i=k-n_q+1}^{min(n_p-n_q,k)} c_{n_q+i}\, b_{k-i}\right), \tag{146}$$

and

$$\text{for} \quad k = n_q-1, n_q-2, \ldots, 0: \quad c_k = a_k - \sum_{i=0}^{min(n_p-n_q,k)} c_{n_q+i}\, b_{k-i}. \tag{147}$$

The development of these equations is left to exercise 3.6.1. Note that the order of the solution for these coefficients is important because the equation for c_k involves c_i for $i > k$. Also, note the convention that if the first index value for a sum is less than the last index value, the value of the sum is 0.

To evaluate $p(x)$ at $x = x_0$, we note that if we divide $p(x)$ by $x - x_0$, we obtain a quotient $q(x)$ and a constant remainder r:

$$p(x) = q(x)(x - x_0) + r. \tag{148}$$

By property 10 of polynomials above, $q(x_0)$ is finite. Therefore, if we evaluate equation 148 at $x = x_0$, we obtain

$$p(x_0) = r. \tag{149}$$

Thus, we can evaluate p at x_0 by dividing p by $x - x_0$ and taking the remainder as the result. Writing the division algorithm for this special case, called *synthetic division* or *Horner's method,* we obtain

$$
\begin{aligned}
c_{n_p} &= a_{n_p} \\
c_{n_p-1} &= a_{n_p-1} + x_0 c_{n_p} \\
c_{n_p-2} &= a_{n_p-2} + x_0 c_{n_p-1} \\
&\vdots \\
c_i &= a_i + x_0 c_{i+1} \\
&\vdots \\
p(x_0) = r = c_0 &= a_0 + x_0 c_1,
\end{aligned}
\tag{150}
$$

which can be expressed in one equation as

$$p(x_0) = (\ldots((a_{n_p} x_0 + a_{n_p-1})x_0 + a_{n_p-2})x_0 + \cdots)x_0 + a_0, \tag{151}$$

the so-called nested form of polynomial evaluation. This algorithm (that is, synthetic division) is the most efficient way to evaluate polynomials.

Let us see how to evaluate the kth derivative of $p(x)$ at $x = x_0$. If we write p as a polynomial about x_0:

$$p(x) = \sum_{i=0}^{n_p} b_i(x - x_0)^i, \tag{152}$$

then

$$p^{(k)}(x_0) = k!b_k. \tag{153}$$

To find the kth derivative of p at x_0, we need only find b_k. To find the b_i given the a_i, the coefficients of p about 0, we first divide $p(x)$ as expressed in equation 152 by $x - x_0$ to obtain a quotient

$$p_1(x) = \sum_{i=0}^{n_p-1} b_{i+1}(x - x_0)^i \tag{154}$$

and a remainder b_0. Then we divide the quotient $p_1(x)$ by $x - x_0$ to obtain a quotient

$$p_2(x) = \sum_{i=0}^{n_p-2} b_{i+2}(x - x_0)^i \tag{155}$$

and a remainder b_1. And so on. Thus, computing the coefficients of the shifted polynomial involves (1) successively dividing the quotient polynomial resulting from the previous step by $x - x_0$, by synthetic division, and (2) taking the remainder as the next coefficient and the quotient as the polynomial to be operated on in the next step.

Restating the algorithm, we start out with $p(x)$ about $x = 0$ defined by the coefficients $a_0^{(0)}, a_1^{(0)}, a_2^{(0)}, \ldots, a_{n_p}^{(0)}$, listed in ascending order of the index. We apply synthetic division with parameter x_0 to these coefficients to produce $b_0, a_0^{(1)}, a_1^{(1)}, a_2^{(1)}, \ldots, a_{n_p-1}^{(1)}$. We apply synthetic division with the same parameter to $a_0^{(1)}, a_1^{(1)}, a_2^{(1)}, \ldots, a_{n_p-1}^{(1)}$ to produce $b_1, a_0^{(2)}, a_1^{(2)}, \ldots, a_{n_p-2}^{(2)}$. In general, at the kth step, we apply synthetic division to coefficients $a_0^{(k-1)}, a_1^{(k-1)}, \ldots, a_{n_p-(k-1)}^{(k-1)}$ to produce $b_{k-1}, a_0^{(k)}, a_1^{(k)}, \ldots, a_{n_p-k}^{(k)}$. The full shifting process takes $n_p(n_p+1)/2$ multiplications.

Returning to the question of computing derivatives, if we desire only up to the kth derivative, we need only carry out $k+1$ steps of the shifting algorithm to compute $b_0, b_1, \ldots, b_k$. To compute b_0 and b_1 alone, a more efficient method has been described by Munro and Borodin.†

3.6.2 Newton's Method Applied to Polynomials

Let us assume we can localize a real root of a polynomial so that within a known symmetric interval of the root, $p(x)$ has no near-zero derivatives. This process is aided by the algorithms mentioned above for localizing roots of polynomials. Then we can apply Newton's method to $p(x)$. Unlike the general case, we need not provide the routine with an expression for the derivative of p, because we can compute the derivative from the coefficients of p as described in section 3.6.1. That is, at any x_i, only $2n_p - 1$ multiplications are required to compute $b_0^{(i)} = p(x_i)$ and $b_1^{(i)} = p'(x_i)$. Then we iterate using

$$x_{i+1} = x_i - \frac{b_0^{(i)}}{b_1^{(i)}}, \tag{156}$$

where $b_0^{(i)}$ and $b_1^{(i)}$ are functions of x_i. As long as the root of the polynomial is of multiplicity 1, we have a simple method producing quadratic convergence to the root.

If $p(x)$ has roots of multiplicity greater than 1, we can find a polynomial which has all of the roots of $p(x)$ but of multiplicity 1. We do this by noting the following theorem.

Theorem: $p'(x)$ has as roots none of the single roots of p and all of the multiple roots of p, each with multiplicity 1 less than the multiplicity of that root in p.

† See I. Munro and A. Borodin, "Efficient Evaluation of Polynomial Forms."

Proof: Let
$$p(x) = K \prod_{i=1}^{N} (x - x_i)^{m_i}, \tag{157}$$
where the x_i are distinct. Then $p'(x)$ is also a polynomial, and for any j
$$p'(x) = K(x - x_j)^{m_j - 1} \Bigg[m_j \prod_{\substack{i=1 \\ i \neq j}}^{N} (x - x_i)^{m_i} + (x - x_j) \sum_{\substack{i=1 \\ i \neq j}}^{N} m_i (x - x_i)^{m_i - 1} \prod_{\substack{k=1 \\ k \neq i \\ k \neq j}}^{N} (x - x_k)^{m_k} \Bigg]. \tag{158}$$
At $x = x_j$ the bracketed expression in equation 158 is not equal to zero, so the multiplicity of the root x_j in $p'(x)$ is $m_j - 1$. Note also that if $m_j = 1$, $p'(x_j) \neq 0$.

The theorem above implies that the greatest common divisor of p and p' is
$$r(x) = K \prod_k (x - x_{i_k})^{m_{i_k} - 1}, \tag{159}$$
where the x_{i_k} are the roots of p with multiplicity greater than 1. Then $p(x)/r(x)$ is the polynomial with all of the roots of p but with multiplicity 1.

To find $r(x)$, we must know how to find the coefficients of $p'(x)$ and how to find the greatest common divisor of two polynomials. The former is simple, since the coefficients of p' are simply $i \cdot a_i$ for $i = 1,2,\ldots,n_p$. Finding the greatest common divisor is accomplished using the Euclidean algorithm for polynomials, which can be found in many modern algebra books.† It is outlined in exercise 3.6.2. We will not discuss this algorithm here in detail, but we note that it can require as many as $n_p - 1$ polynomial divisions and that, at the end, it involves determining whether or not a resulting value is 0. This procedure can only be accomplished approximately on a computer. Because of the work required to get a polynomial, all of whose roots are of multiplicity 1, and the difficulty in achieving the correct answer, we usually put up with the slow convergence of Newton's method in the case of multiple roots.

Having found a root ξ of p by Newton's method, the method also provides us with the coefficients of the polynomial, $q(x) \equiv$
$$p(x)/(x - \xi) = \sum_{i=0}^{n_p - 1} c_{i+1} x^i. \tag{160}$$
We say that $q(x)$ is $p(x)$ *reduced* by the root at ξ. $q(x)$ has all of the roots of

† For example, see S. MacLane and G. Birkhoff, *Algebra*, chapter 4, section 12.

$p(x)$ except ξ, unless ξ has multiplicity greater than 1 in p, in which case q has a root ξ with multiplicity one less than its multiplicity in p.

We can avoid some computation by finding the next root of p by applying Newton's method to q rather than p, since q has degree one less than p. However, since q was computed from p, it has some computational error. So a wise approach is to find a root of q within a desired accuracy and use that result, which due to an error in calculating q is only approximately a root of p, as a starting value for Newton's method applied to p. Only a few iterations will be required to converge to the true root of p. Similarly, for each new root, we find a root of an equation of one degree less than the equation for the root before, and we use that root as a starting value for Newton's method applied to p.

3.6.3 Complex Roots of Polynomials

Every polynomial of degree $n > 1$ with real coefficients has $\lfloor n/2 \rfloor$ factors of the form $x^2 - cx - d$, where c and d are real. This follows from the fact that complex roots come in conjugate pairs. So, for any pair of complex roots $\eta_k + i\zeta_k, \eta_k - i\zeta_k$ (η_k and ζ_k real), $p(x)$ is divisible by the factor $(x - (\eta_k + i\zeta_k))(x - (\eta_k - i\zeta_k)) = x^2 - 2\eta_k x + (\eta_k^2 + \zeta_k^2)$. Therefore, $c = 2\eta_k$ and $d = -(\eta_k^2 + \zeta_k^2)$ satisfy our requirement. Similarly, pairs of real roots produce a quadratic factor.

We find complex roots of a polynomial by finding a quadratic factor of the polynomial and finding the roots of that factor by the quadratic formula. We find the quadratic factor in a way similar to that used to find a linear factor as discussed in section 3.6.2. For any α and β, we can write

$$p(x) = (x^2 - \alpha x - \beta)q(x) + b_1(x - \alpha) + b_0, \tag{161}$$

where b_1 and b_0 are functions of α and β and $q(x)$ is a polynomial of degree $n_p - 2$ which is also a function of α and β. To find $\alpha = c$ and $\beta = d$, we must solve the simultaneous nonlinear equations

$$\begin{aligned} b_1(\alpha,\beta) &= 0 \\ b_0(\alpha,\beta) &= 0. \end{aligned} \tag{162}$$

As with the linear factor, we solve the equations by Newton's method. First, we make a guess α_0 and β_0. At each iterative step, if

$$\alpha_{i+1} = \alpha_i + \delta\alpha_i \tag{163}$$

and

$$\beta_{i+1} = \beta_i + \delta\beta_i, \tag{164}$$

then $\delta\alpha_i$ and $\delta\beta_i$ are the solutions of

$$\frac{\partial b_1}{\partial \alpha}(\alpha_i, \beta_i)\delta\alpha_i + \frac{\partial b_1}{\partial \beta}(\alpha_i, \beta_i)\delta\beta_i = -b_1(\alpha_i, \beta_i)$$

$$\frac{\partial b_0}{\partial \alpha}(\alpha_i, \beta_i)\delta\alpha_i + \frac{\partial b_0}{\partial \beta}(\alpha_i, \beta_i)\delta\beta_i = -b_0(\alpha_i, \beta_i). \tag{165}$$

To use this iteration, we must know how to evaluate the partial derivatives of b_1 and b_0 with respect to α and β at the point (α_i, β_i). First let us discuss how to compute b_1 and b_0. Using the division algorithm and

$$q(x) = \sum_{i=0}^{n_p - 2} b_{i+2}x^i, \tag{166}$$

we have

$$\begin{aligned}
b_n &= a_n \\
b_{n-1} &= a_{n-1} + \alpha b_n \\
b_{n-2} &= a_{n-2} + \alpha b_{n-1} + \beta b_n \\
b_{n-3} &= a_{n-3} + \alpha b_{n-2} + \beta b_{n-1} \\
&\vdots \\
b_k &= a_k + \alpha b_{k+1} + \beta b_{k+2} \\
&\vdots \\
b_1 &= a_1 + \alpha b_2 + \beta b_3 \\
b_0 &= a_0 + \alpha b_1 + \beta b_2.
\end{aligned} \tag{167}$$

Note that the form of equation 161 was chosen especially so that the last two equations of equations 167 will be of the same form as the others.

For any given α and β, the system of equations 167 defines an algorithm for computing b_1 and b_0. Furthermore, if we take the partial derivatives of equations 167, noting that $b_0, b_1, \ldots, b_{n-1}$ are functions of α, and $b_0, b_1, \ldots, b_{n-2}$ are functions of β, we obtain

$$\frac{\partial b_n}{\partial \alpha} = 0$$

$$\frac{\partial b_{n-1}}{\partial \alpha} = b_n$$

$$\frac{\partial b_{n-2}}{\partial \alpha} = b_{n-1} + \alpha \frac{\partial b_{n-1}}{\partial \alpha}$$

$$\vdots$$

$$\frac{\partial b_k}{\partial \alpha} = b_{k+1} + \alpha \frac{\partial b_{k+1}}{\partial \alpha} + \beta \frac{\partial b_{k+2}}{\partial \alpha}$$

$$\vdots$$

$$\frac{\partial b_1}{\partial \alpha} = b_2 + \alpha \frac{\partial b_2}{\partial \alpha} + \beta \frac{\partial b_3}{\partial \alpha}$$

$$\frac{\partial b_0}{\partial \alpha} = b_1 + \alpha \frac{\partial b_1}{\partial \alpha} + \beta \frac{\partial b_2}{\partial \alpha} \tag{168}$$

and

$$\frac{\partial b_n}{\partial \beta} = 0$$

$$\frac{\partial b_{n-1}}{\partial \beta} = 0$$

$$\frac{\partial b_{n-2}}{\partial \beta} = b_n$$

$$\frac{\partial b_{n-3}}{\partial \beta} = b_{n-1} + \alpha \frac{\partial b_{n-2}}{\partial \beta}$$

$$\vdots$$

$$\frac{\partial b_k}{\partial \beta} = b_{k+2} + \alpha \frac{\partial b_{k+1}}{\partial \beta} + \beta \frac{\partial b_{k+2}}{\partial \beta}$$

$$\vdots$$

$$\frac{\partial b_1}{\partial \beta} = b_3 + \alpha \frac{\partial b_2}{\partial \beta} + \beta \frac{\partial b_3}{\partial \beta}$$

$$\frac{\partial b_0}{\partial \beta} = b_2 + \alpha \frac{\partial b_1}{\partial \beta} + \beta \frac{\partial b_2}{\partial \beta}. \tag{169}$$

We see that equations 168 are of precisely the same form as equations 167 with a_i replaced by b_i, and b_i replaced by $\frac{\partial b_{i-1}}{\partial \alpha}$. We also see that equations 169 are of the same form except that a_i is replaced by b_i, and b_i is replaced by $\frac{\partial b_{i-2}}{\partial \beta}$. Since $\frac{\partial b_{i-1}}{\partial \alpha}$ and $\frac{\partial b_{i-2}}{\partial \beta}$ have the same recursive form, they are the same value. In particular,

$$\frac{\partial b_1}{\partial \alpha} = \frac{\partial b_0}{\partial \beta} \tag{170}$$

and

$$\frac{\partial b_2}{\partial \alpha} = \frac{\partial b_1}{\partial \beta}. \tag{171}$$

Therefore, if we apply the algorithm given by equations 167 to the a_i, $i = 0,1,\ldots,n$, producing the b_i, and reapply that algorithm to the b_i, $i = 0,1,\ldots,n$, producing the c_i, then

$$\frac{\partial b_0}{\partial \alpha} = c_1, \tag{172}$$

$$\frac{\partial b_1}{\partial \alpha} = \frac{\partial b_0}{\partial \beta} = c_2, \tag{173}$$

$$\frac{\partial b_1}{\partial \beta} = c_3. \tag{174}$$

We can use these values together with the values of b_1 and b_0 from the first application of the form in equation 167, to solve equation 165.

We use the results of equation 165 in equations 163 and 164 to compute new values, α_{i+1} and β_{i+1}. We then repeat the process of two evaluations using the form of equation 167 with these new values. If α_0 and β_0 have been chosen close enough to the correct answer, this iteration will converge quadratically to the coefficients of the desired quadratic factor. This process is called the *Newton-Bairstow method.*

As with Newton's method for a linear factor, when convergence to $\alpha = c$ and $\beta = d$ occurs, the method automatically produces the coefficients of $q(x)$, namely, $b_2, b_3, \ldots, b_n$, where $q(x)$ is the polynomial with all of the roots of $p(x)$ except those of the quadratic factor we have found, unless that quadratic factor appears with multiplicity greater than 1, in which case $q(x)$ has it as a factor with multiplicity 1 less than its multiplicity in $p(x)$. As with real roots, if further quadratic factors are desired, we should probably apply our algorithm to $q(x)$, and, when we find the parameters of a quadratic factor of $q(x)$ to within a desired accuracy, use those parameters as starting values of a Newton-Bairstow iteration applied to $p(x)$.

Finally, there is no requirement that the coefficients c and d produce complex roots. We may choose α_0 and β_0 so that we converge to a quadratic factor with real roots.

3.6.4 Graeffe's Root Squaring

Not all methods for finding the roots of polynomial equations are special cases of methods for solving general equations. For example, a method called *root squaring* depends on special properties of polynomials. This method is another illustration of the approach we saw in section 2.7.2 of designing an iteration that produces the domination of the largest member of a set, in this case the set of roots of a polynomial. It is sketched below and developed and analyzed in detail in exercise 3.6.7.

Let $p(x)$ be a *monic* polynomial (that is, have 1 as its high-order coefficient) such that its root of largest magnitude is strictly greater than its root of second largest magnitude. This largest root will be real. (Why?) Graeffe's root squaring method depends on the fact that the polynomial $q_0(x)$ defined by

$$q_0(x) = p_0(x)p_0(-x) \tag{175}$$

is a polynomial with all odd coefficients equal to zero, so that if we write

$$p_1(x^2) = q_0(x), \tag{176}$$

then p_1 is a polynomial of the same degree as p_0 with roots equal to the squares of all roots of p. We can iterate this process, defining

$$p_{i+1}(x^2) \equiv p_i(x)p_i(-x). \tag{177}$$

The negative of the coefficient of x^{n_p-1} in $p_k(x)$ is the sum of the roots of p_k. Therefore, because of the squaring of the roots at each iteration, the largest root will dominate this sum for large k. If $a_{k,i}$, $i = 0,1,\ldots,n$ are the coefficients of p_k, then

$$\lim_{k\to\infty} (-a_{k,n_p-1})^{1/2^k} = |\text{largest root in magnitude of } p_0(x)|. \tag{178}$$

Graeffe's root squaring does not give us a root itself but only the magnitude of a root. However, since the root is real, we can easily find which sign applies by evaluating p at the two possibilities.

Graeffe's root squaring method will produce only one root of p, namely, the root largest in magnitude. However, we can reduce the polynomial as discussed in section 3.6.2, and repeat the root squaring on the result to find its root of largest magnitude, as long as it is the only root of that magnitude.

Root squaring is an attractive approach to finding the largest root because it has nice error properties. This may be somewhat surprising because we ought to be concerned with the propagation of error in coefficients from step to step. However, the fact that we take the 2^kth root of the final result washes out this error propagation.

3.6.5 Errors in Polynomial Equations

Often the coefficients of a polynomial p have come from previous computation or measurements and thus have error. Even if the coefficients are accurately known, roundoff error will be introduced when they are represented in a computer. We must ask: How much does a small error in coefficients of p affect the roots of p?

Let $p^*(x)$ be a polynomial with inexact coefficients:

$$a_i^* = a_i + \delta a_i. \tag{179}$$

Let

$$\Delta(x) \equiv p^*(x) - p(x) = \sum_{i=0}^{n_p} \delta a_i x^i. \tag{180}$$

Let ξ be a root of $p(x)$ and $\xi + \delta\xi$ be the corresponding root of $p^*(x)$. Then

$$\begin{aligned} 0 = p^*(\xi + \delta\xi) &= p^*(\xi) + p^{*\prime}(\xi)\delta\xi + O(\delta\xi^2) \\ &= p(\xi) + \Delta(\xi) + p^{*\prime}(\xi)\delta\xi + O(\delta\xi^2) \\ &= \Delta(\xi) + p^{*\prime}(\xi)\delta\xi + O(\delta\xi^2). \end{aligned} \tag{181}$$

Thus

$$\delta\xi = -\Delta(\xi)/p^{*\prime}(\xi) + O(\delta\xi^2). \tag{182}$$

To illustrate this error propagation from coefficients of a polynomial to its roots, consider the root at $x = 20$ of the polynomial

$$p^*(x) = (x-1)(x-2)(x-3)\ldots(x-20). \tag{183}$$

We see that $a_0^* = 20!$.

First assume a_0^* has relative error 10^{-7} and all other coefficients have no error. Then $\delta a_0 = 10^{-7}(20!)$, so for $\xi + \delta\xi =$ the root at 20, $\Delta(20) = \delta a_0 20^0 = 10^{-7}(20!)$. Since $p^{*\prime}(20) = 19!$, from equation 182 the absolute error in the root at 20 is approximately $-10^{-7}(20!/19!) = -2 \times 10^{-6}$. We see that the relative error in the root at 20 is in magnitude the same as the relative error in the coefficient.

Now assume that the relative error in a_{19} is 10^{-7} and all other coefficients have no error. Since $a_{19}^* = \sum_{i=1}^{20} (-i) = -210$, $\delta a_{19} = (-210)10^{-7}$

and $\Delta(20) = (-210)10^{-7}20^{19}$, by equation 182 the error in the root at 20 is approximately 900. In this case the relative error in the coefficient has been propagated into the root with a factor of 45×10^7.

Why is the effect of coefficient error so much greater in the second case? The difficulty is caused by the fact that error in the high-order coefficients is propagated very badly for roots of large magnitude. In general, we are much better off if we can search for roots of equations with small roots rather than of those with large roots. Unfortunately, once error has been introduced, scaling or shifting of the polynomial cannot improve matters. The error in the root of the shifted polynomial will be the same as in the unshifted polynomial, if the shifting is done accurately. However, we can try to arrange our physical model so that the polynomial, the coefficients of which we must compute or measure, has small roots. Also, if we are looking for all of the roots of a polynomial, we can find the smallest roots first. This helps to ensure that, in the reduction, the propagation due to error in the root by which we are reducing is as small as possible. That is, as we do by pivoting in Gaussian elimination, we put off the major error until the end of the process to reduce propagation.

PROBLEMS

3.6.1. Let $p(x)$ and $q(x)$ be polynomials, let $r(x)$ be the polynomial which is the result without remainder of dividing $p(x)$ by $q(x)$, and let $s(x)$ be the polynomial which is the remainder of the division. Show that $r(x)$ and $s(x)$ are produced by the algorithm specified by equations 146 and 147.

3.6.2. The Euclidean algorithm to find the greatest common divisor (gcd) of the polynomials $p(x)$ and $q(x)$, where the degree of $q <$ the degree of p, is as follows:

1. Let $p_0(x) = q(x)$.
2. Let $p_1(x) = p(x)$.
3. Let $i = 1$.
4. Divide $p_{i-1}(x)$ by $p_i(x)$ to produce remainder $p_{i+1}(x)$.
5. If $p_{i+1}(x) = 0$, stop with the answer $gcd(p(x),q(x)) = p_i(x)$; otherwise increment i by 1 and go to step 4.

Apply this algorithm and the division algorithm for polynomials to find the polynomial with all of the roots of

$$x^6 + x^5 - 25x^4 - 25x^3 + 180x^2 + 108x - 432$$

but in which all roots are simple.

3.6.3. Construct an algorithm for computing $q(x) = (p(x))^n$ for a given n and polynomial $p(x) = \sum_{i=0}^{n} a_i x^i$.

3.6.4. Let $q(x) = \sum_{i=0}^{2n} b_i x^i$ be a polynomial known to be a perfect square. Let the b_i, $0 \le i \le 2n$, be given. Construct an algorithm for finding the polynomial $p = q^{1/2}$.

3.6.5. a) Apply Newton's method for polynomials to find the real root of $x^3 - 3x^2 + 4x - 1 = 0$.

b) Find the complex roots of the equation in part a by using the quadratic formula on the reduced equation.

3.6.6. Apply the Newton-Bairstow method to find the complex roots of the equation in exercise 3.6.5.

3.6.7. Let $p_0(x) = \sum_{i=0}^{n} a_{0,i} x^i$. Consider $q(x) = (-1)^n p_0(x) p_0(-x)$, where $q(x) = \sum_{i=0}^{2n} b_i x^i$.

a) Show that for i odd, $b_i = 0$. Therefore $q(x) = \sum_{i=0}^{n} b_{2i} (x^2)^i \equiv p_1(x^2)$. Let $p_1(y) = \sum_{i=0}^{n} a_{1,i} y^i$, that is, $a_{1,i} = b_{2i}$.

b) Give an algorithm to compute the $a_{1,i} (i = 0,1,\ldots,n)$, given the $a_{0,i} (i = 0,1,\ldots,n)$.

c) Assume the magnitude of the relative error in each $a_{0,i}$ is bounded by r. Give a bound on the relative error in the $a_{1,i}$ in terms of r and the $a_{0,i}$. Give the greatest lower bound for this relative error bound over all possible sets of $a_{0,i}$.

d) By iterating the step $p_{j+1}(x^2) = (-1)^n p_j(x) p_j(-x)$, we can get $p_k(y) = \sum_{i=0}^{n} a_{k,i} y^i$. Show that if $a_{0,n} = 1$ and ξ_1 is the largest root of $p_0(x)$, then $(-a_{k,n-1})^{1/2^k}$ converges to $|\xi_1|$ as k increases, and thus we can solve for $|\xi_1|$. As noted in section 3.6.4, this method, due to Graeffe, is called "root squaring."

e) Assume that the error in the $a_{j,i}$ propagates approximately as the greatest lower bound found in part c. What can you say about the relative error in computing $|\xi_1| \neq 0$ due to propagation of error in the $a_{0,i}$ when root squaring is applied? What about the effect of error generated in the computation at some intermediate iteration?

f) Apply Graeffe's root squaring method to find the largest root of $x^3 - 5x^2 + x + 1 = 0$.

REFERENCES

Hamming, R. W. *Introduction to Applied Numerical Analysis.* New York: McGraw-Hill, 1971.

MacLane, S., and Birkhoff, G. *Algebra.* New York: Macmillan, 1967.

Munro, I., and Borodin, A. "Efficient Evaluation of Polynomial Forms," *Journal of Computer and System Sciences* 6 (1972): 625–638.

Ortega, J. M., and Rheinboldt, W. C. *Iterative Solution of Nonlinear Equations in Several Variables.* New York: Academic Press, 1970.

Ralston, A. *A First Course in Numerical Analysis.* New York: McGraw-Hill, 1965.

Traub, J. F. *Iterative Methods for the Solution of Equations.* Englewood Cliffs, N. J.: Prentice-Hall, 1964.

4

APPROXIMATION

4.1 GOALS AND STRATEGY

Until now we have addressed ourselves to the numerical solution of equations. Another very important class of numerical problems involves computing values related to a function represented, not analytically, but as a set of data points (see figure 4-1). For example, one often wishes to evaluate a function at some argument value other than an argument at which a data value is given. Also, one may wish to compute the result of some operation such as differentiation or integration on that function.

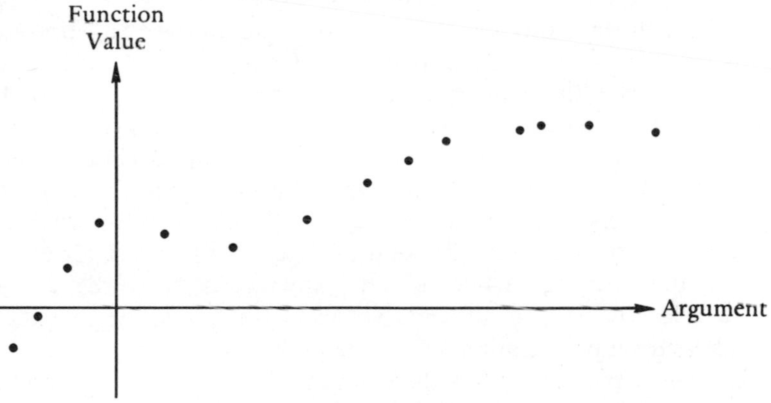

FIG. 4-1 A function represented as data points

A function is specified as a set of values at a discrete set of argument points for two reasons. First, this situation occurs when the function comes from some physical process and is only known by measuring the value of the function at various argument points or, equivalently, when the data values are the result of previous computation. Second, it arises when the computer cannot be used to efficiently evaluate an analytically definable function at arbitrary argument values in a direct way. An example of such a function is $f(x) = \int_0^x sin(1/(u^2 + 2))du$. In this case the strategy is two-part: (1) evaluate the function in some inefficient way at a set of data points to whatever accuracy is required, and (2) for routine evaluation, use techniques of approximation to calculate the function from the discrete number of data points tabulated.

The *approximation* problem can be stated as follows: Given a discrete set of argument values and a corresponding set of function values, each with some error due to computation and/or measurement, find a function defined on a desired real interval such that this function is an appropriately close approximation to the function which produced the data points. With respect to operations on the function such as differentiation, integration, and transforms, the strategy is to apply these operations to the approximating function after the approximation has been carried out.

In this book, we will direct our attention to functions of one variable. But the concepts that we will develop can be extended to functions of many variables.

What do we mean by the "best" approximation? We must exclude the answer that the approximating function should be close to the function underlying the tabulated data, because we either do not know this underlying function or cannot evaluate it for the purpose of measuring its closeness to the approximating function. Two properties suggest themselves. First, we usually assume the underlying function is *smooth,* that is, has *low average curvature,* so the smoother the approximation is, the better it is.† Second, we have measured the underlying function at the tabular argument points, so we can say that the closer the approximating function is to the data at those points, the better it is. Thus, for example, the function f_1 in figure 4-2 is a better approximation to the data than f_2 because it is closer to the data values at the tabular arguments, and f_1 is a better approximation than f_3 because it is smoother.

The above properties must be translated into a precise criterion for goodness of approximation. After this is done, one must specify the set of functions over which the criterion is to be applied to find a "best" approximating function. Most often the form of the function is constrained so that the best approximating function, according to the criterion we have established, is unique. The smoothness requirement is often met by restricting the set of functions to "appropriately smooth" functions.

† The "smoothness" of a function is defined differently here than in many mathematics books.

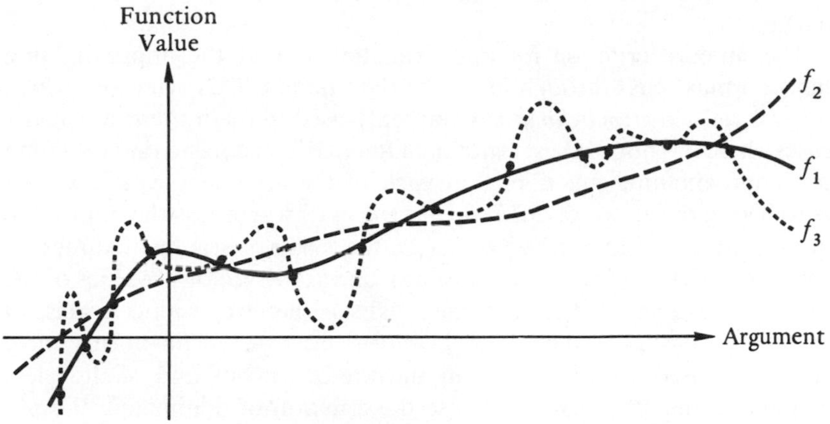

FIG. 4-2 Approximating functions fitted to data points

Let us illustrate the above by listing some possible function forms and some possible criteria. Probably the most commonly used function form is the polynomial of degree n for some fixed integer n. Another common form is the Fourier expansion: a linear combination of functions of the form $sin(k\pi x/N)$ or $cos(k\pi x/N)$, for a fixed N and with k ranging over integers between 0 and some integer less than or equal to N. Another common form for approximating functions is the rational function, defined as a quotient of two polynomials each of degree less than some fixed value n. There are an infinite number of function forms that can be used; the one that should be chosen depends first, on physical knowledge of the kind of function that produced the data, and second, on numerical matters dealt with in this chapter. We can illustrate the concepts we need using only approximating function forms which are a linear combination of basis functions, each having no variable other than the independent variable of the function. For example, the polynomials mentioned above are a linear combination of the functions x^i for $0 \leq i \leq n$; similarly, the Fourier series are linear combinations of the sinusoids indicated above. The rational functions are not of the form indicated. The student should not infer from our choice to treat only linear combinations of functions that there are not cases where an approximating form such as the rational functions can make important contributions. Rather, because of limited space, a choice to deal with only the somewhat simpler form has been made.

The criterion of fit of our approximating function specifies what we mean by the "closeness" with which the approximating function fits the data points.

The simplest criterion for approximation is that the approximating function must pass through all of the data points. This criterion, which we will call *exact matching*, is commonly used, though there are many cases where it is not the best criterion. Other criteria depend on evaluating our approximating function $\hat{f}$ at each of the argument values x_i and considering the vector of differences, $\hat{f}(x_i) - y_i$, where y_i is the data value at x_i. That is, consider three vectors: $\underline{x}$, the vector of argument values; $\underline{y}$, the vector of corresponding data values; and $\underline{\hat{f}}$, the vector of values of the function $\hat{f}$ evaluated at the argument values. Then we produce a class of criteria by trying to minimize the norm of $\underline{\hat{f}} - \underline{y}$. The norms most commonly used are the $\mathscr{L}_2$ norm and the $\mathscr{L}_\infty$ norm, and, slightly less commonly, the $\mathscr{L}_1$ norm. If we use the criterion of minimizing the $\mathscr{L}_2$ norm, the process is called *least-squares approximation*. If we minimize the $\mathscr{L}_\infty$ norm (that is, minimize the maximum deviation from the function to a data point), the process is called *Tchebycheff* or *minimax approximation*. The exact-matching criterion mentioned above simply requires that, for *any* norm,

$$\|\underline{\hat{f}} - \underline{y}\| = 0. \tag{1}$$

We have used a number of vector spaces and notations in the above paragraphs. Let us carefully specify these.

1. The infinite dimensional vector space of continuous functions over an interval on the real line. As possible basis functions, $f^k(x)$, for this vector space, we have the simple polynomials x^k, the sinusoids $sin(k\pi x/N)$ and $cos(k\pi x/N)$ where N is the half-width of the interval, and many others.
2. The vector space of one-dimensional arrays (∞-vectors) of coefficients of the basis functions, f^k, of vector space 1. That is, an element, a, of the vector space specifies the coefficients a_k of $g(x) = \sum_k a_k f^k(x)$, where g is in vector space 1. There is a different vector space for each set of basis functions.
3. The finite dimensional space of functions which are a linear combination of the first n of the basis functions of vector space 1.
4. The corresponding finite dimensional space of n-vectors of coefficients of these basis functions, one for each set of basis functions.
5. The vector space of all possible sets of m argument values. We denote vectors in this space by the vector x and specific argument values by x_i.

6. The vector space of all possible data values over these argument values. We denote by y any vector in this space, which is made up of an input value for each x_i, and by y_i some particular input value. We denote by f the m-vector of error-free function values produced by the function underlying the input data.† That is, since the input values are the function values plus error, we write

$$y_i = f_i + \varepsilon_i, \tag{2}$$

where ε_i is the error in the ith data value. We denote by $\hat{f}$ the vector of values of an approximating function evaluated over the data arguments. Since this approximating function is generally not precisely equal to the original function, we write

$$\hat{f}_i = f_i + \delta_i, \tag{3}$$

where δ_i is the error in our approximating function evaluated at x_i from the value of the underlying function evaluated there.

Approximation involves specifying a basis set in the space defined by item 3 and a norm in the space defined by item 6 and minimizing the norm.

Which criterion should we use? If we know that the data values are very accurate and that the selected form of approximating function is very good for the data we have, it makes sense to use the exact-matching criterion. If the data values have significant error, it does not make sense to force the approximating function to pass through the data points, only near them. In this case the appropriate choice of norm depends upon the assumed probability distribution of the error in the data values. (We will normally assume that the argument values have no error, a matter discussed in more detail in section 4.3.) We will see that least-squares approximation is most appropriate when the error has a normal distribution and that Tchebycheff (minimax) approximation is most appropriate when the errors are uniformly distributed. More broadly, least-squares approximation is often used when we wish to minimize the average deviation of the approximating function from the data at the expense of an occasional, rather large deviation, and Tchebycheff approximation is used when we wish to minimize the largest deviation at the expense of a slightly larger average deviation than least-squares produces. In chapter 2, we stated that the choice of norm often makes little difference in the result. Thus, the results of least-squares approximation and of Tchebycheff approximation, for example, are often not very different. Nevertheless, if our data is

† It will be clear from context whether f refers to the function f or to the m-vector f derived from the function f by evaluating it at m argument values.

more likely to have large errors than if these errors were normally distributed, the Tchebycheff approximation may give a significantly better result than least-squares approximation.

Tchebycheff approximation is also commonly used when the data points have very little error but we wish to fit the data by an approximating function that has too few degrees of freedom (too few parameters to be determined) to allow it to go through all of the data points. From the point of view of efficiency it is desirable to calculate the values of as few parameters as possible. This policy may also be desirable from the point of view of the smoothness of the approximating function. Generally speaking, we choose sets of approximating functions such that the approximating functions in the set are ordered according to smoothness. To obtain the smoothest approximation, we try to use as few of the first functions in the set as possible, because the fewer the functions we can use, that is, the smoother the functions we use to approximate with, the smoother our final approximating function will be.

4.2 POLYNOMIAL EXACT MATCHING

As noted above, the most common approximating functions are the polynomials. The most common criterion for approximation is exact matching. In both cases, the popularity is due to the mathematical tractability of the analysis and is not due to, nor indicative of, the quality of the approximation thus produced; polynomial exact matching is used far too often when it should not be. We will develop these methods first, pointing out their advantages, but we will be careful to point out their disadvantages as well so that the student will not be inclined to overuse them.

4.2.1 Interpolation Methods

Interpolation consists of the evaluation of a function at some argument values other than the ones at which it is tabulated. We accomplish this by fitting a function through some set of the tabulated data points whose argument values are close to the argument at which we desire to evaluate the function and then evaluating the fitted function at the argument in question. Fitting a polynomial which passes through $N+1$ data points depends on the following theorem.

Theorem: Given a set $\{(x_i,y_i)|i = 0,1,2,\ldots,N\}$ where the x_i are distinct, there exists a unique polynomial $p(x)$ of degree N or less for which $p(x_i) = y_i$, $0 \leq i \leq N$.

The importance of this theorem is twofold. First, given a set of data points with distinct arguments, it states that there is some polynomial passing through the data points; second, it states that no matter what method we use for computing this polynomial, we will always get the same polynomial, except for computational error. We will be developing many methods for finding this polynomial, and it is important to recognize that for a given set of data points these methods differ only in their computational error and efficiency.

Proof: We will show the existence of the desired polynomial shortly by a construction. For the moment, assume this existence, and let us prove uniqueness. Assume there exist two polynomials $p(x)$ and $q(x)$, both of degree N or less and both agreeing at all $N+1$ data points:

$$p(x_i) = y_i, \quad i = 0,1,\ldots,N \tag{4}$$

and

$$q(x_i) = y_i, \quad i = 0,1,\ldots,N. \tag{5}$$

Consider

$$s(x) = p(x) - q(x). \tag{6}$$

We know that $s(x)$ is a polynomial of degree N or less, because it is the difference of two such polynomials. But note that

$$s(x_i) = 0, \quad i = 0,1,\ldots,N. \tag{7}$$

The fundamental theorem of algebra states that an Nth-degree polynomial has only N distinct zeros unless the polynomial is identically equal to zero. Here we have the polynomial $s(x)$ with $N+1$ zeros, so

$$s(x) \equiv 0, \tag{8}$$

which implies

$$p(x) \equiv q(x). \tag{9}$$

Equation 9 contradicts our assumption that p and q are distinct polynomials, so we know there can exist only one polynomial with the required properties.

To complete the proof, we will construct the required polynomial. Note that an Nth-degree polynomial has $N+1$ parameters (its coefficients), and we have $N+1$ independent constraints given by the exact-matching criterion applied at each data point. Thus, we should not be surprised at the existence and uniqueness of the solution. Put more precisely, the constraints are of the form

$$y_i = \sum_{j=0}^{N} a_j x_i^j, \quad i = 0,1,\ldots,N. \tag{10}$$

Note that each equation is a linear equation in the a_j. For the existence and uniqueness of the solution, we would require that the matrix whose ijth element is $(x_i)^j$ is nonsingular. We could prove this, but it is easier to construct the required polynomial, and, at the same time, produce a computational algorithm for finding the polynomial.

a) Lagrange Polynomials.

Proof (continued): The basis of the construction of the exact-matching polynomial is to write it in the form

$$p(x) = \sum_{j=0}^{N} y_j L_j(x), \tag{11}$$

where each $L_j(x)$ is a polynomial of degree N. Equation 4 will be satisfied if $L_j(x)$ is 0 at every tabular argument except x_j and is 1 at x_j. If each $L_j(x)$ has the above properties, $p(x_i)$ will have the value y_i, because the fact that $L_j(x_i) = 0$ for $j \neq i$ will cause the y_j for $j \neq i$ to be multiplied by 0 and y_i to be multiplied by 1. Thus, we have reduced our problem to finding a set of polynomials $L_j(x)$, called the *Lagrange polynomials,* such that for all j

$$L_j(x_i) = \begin{cases} 0, & i \neq j \\ 1, & i = j. \end{cases} \tag{12}$$

Since $L_j(x)$ is a polynomial, to have a zero at $x = x_i$, it must have a factor $(x - x_i)$. Thus, we have

$$L_j(x) = q(x) \prod_{\substack{i=0 \\ i \neq j}}^{N} (x - x_i), \tag{13}$$

where $q(x)$ is some polynomial. But $L_j(x)$ is a polynomial of degree N or less, so $q(x)$ must be a constant. The constant is determined by the one constraint remaining to be satisfied, namely,

$$L_j(x_j) = 1, \tag{14}$$

producing
$$1 = q \prod_{\substack{i=0 \\ i \neq j}}^{N} (x_j - x_i), \tag{15}$$

from which follows

$$L_j(x) = \left(\prod_{\substack{i=0 \\ i \neq j}}^{N} (x - x_i) \right) \Bigg/ \prod_{\substack{i=0 \\ i \neq j}}^{N} (x_j - x_i). \tag{16}$$

We have now constructed the polynomial $p(x)$ with the required properties.

Note that the $L_j(x)$ depend only on the x_i, not on the y_i. A method like that above, of writing the fitting polynomial as a linear combination of polynomials depending only on the x_i values, can be used to find polynomials satisfying more general constraints than exact matching at function values. In particular, the method can be used to find a polynomial $p(x)$ which agrees with the input function f in value and in the first k_i of its derivatives at x_i, for $0 \leq i \leq N$, where k_i may vary with i (from point to point):

$$p^{(m)}(x_i) = y_i^{(m)}, \quad m = 0,1,\ldots,k_i; \quad i = 0,1,\ldots,N, \tag{17}$$

where $p(x)$ is a polynomial of degree $1 + \sum_{i=0}^{N} (k_i + 1)$ (see exercises 4.2.1 and 4.2.2).

Let us illustrate the use of Lagrange polynomials by fitting a first-degree polynomial, a straight line, through the two points (x_0, y_0) and (x_1, y_1). The Lagrange polynomials are

$$L_0(x) = (x - x_1)/(x_0 - x_1) \tag{18}$$

and
$$L_1(x) = (x - x_0)/(x_1 - x_0). \tag{19}$$

Thus, the approximating line is

$$p(x) = y_0 \frac{x - x_1}{x_0 - x_1} + y_1 \frac{x - x_0}{x_1 - x_0}, \tag{20}$$

which can be written in its more customary form as

$$p(x) = \frac{x_1 - x}{x_1 - x_0} y_0 + \frac{x - x_0}{x_1 - x_0} y_1, \tag{21}$$

a weighted average of y_0 and y_1, where if x is between x_0 and x_1, the weight of y_i is the complement of the proportion of the distance between x_0 and x_1 that is between x and x_i. Equation 20 can also be written

$$p(x) = y_0 + \frac{x - x_0}{x_1 - x_0}(y_1 - y_0), \tag{22}$$

where the interpolated value is written as the value at the beginning of the interval plus the product of the tabular y interval and the fraction of the full tabular x interval taken by $x - x_0$.

b) Truncation Error Analysis. Given a numerical method, the question we must immediately ask is: What is the error? The error can be divided into computational error and approximational or *truncation error* due to using the truncated infinite series which is the approximating function rather than the underlying function. We must analyze both. The truncation error is a function of the polynomial we have chosen to fit and thus, of the tabular arguments we have chosen to use. For a given set of tabular arguments, the truncation error is fixed. We may be able to reduce the computational error by modifying the algorithm for computing the exact-matching polynomial.

Let us analyze the truncation error first. That is, assume $p(x)$ precisely passes through each tabular point and let us evaluate the error in the interpolation, $p(x) - f(x)$. We know the error is zero at $x_0, x_1, \ldots, x_N$. Thus, we are motivated to write

$$p(x) - f(x) = K(x) \prod_{i=0}^{N} (x - x_i), \tag{23}$$

where $K(x)$ is a function to be determined. Note that, as far as we know, the root of $p(x) - f(x)$ at x_i could be of multiplicity less than 1, in which case $K(x_i)$ would be infinite.

Assume f is $N+1$ times differentiable. If K were a constant, we could solve for K by differentiating equation 23 $N+1$ times with respect to x, producing

$$-f^{(N+1)}(x) = (N+1)!K. \tag{24}$$

Since K is not a constant, differentiating equation 23 $N+1$ times produces an equation involving derivatives of $K(x)$ as well as $K(x)$ itself, so this procedure is not directly helpful. However, if we choose z as any argument value not equal to a tabular argument value, we have

$$p(x) - f(x) = K(z) \prod_{i=0}^{N} (x - x_i) \quad \text{at} \quad x = z, \tag{25}$$

or equivalently,

$$\phi(x) \equiv p(x) - f(x) - K(z) \prod_{i=0}^{N} (x - x_i) = 0 \quad \text{at} \quad x = z. \tag{26}$$

Now $K(z)$ is a constant with respect to x, so differentiating $\phi(x)$ $N+1$ times produces

$$\phi^{(N+1)}(x) = -f^{(N+1)}(x) - (N+1)!K(z). \tag{27}$$

Now we know that $\phi(x)$ has $N+2$ zeros: at each x_i and at z. We apply *Rolle's theorem,* which states that a differentiable function must have a relative maximum or relative minimum between two successive zeros, that is, if the function leaves the x axis, and then produces another zero, it must have turned around somewhere to come back. Since we are assuming that ϕ is differentiable, we conclude that in the interval bounded by the x_i and z (which we write $[x_0,x_1,\ldots,x_N,z]$), $\phi'(x)$ has at least $N+1$ zeros. Reapplying Rolle's theorem, we see $\phi''(x)$ must have N zeros in that interval,..., so $\phi^{(N+1)}(x)$ must have at least one zero in that interval. Let the value of x where that zero occurs be ξ. Then

$$0 = \phi^{(N+1)}(\xi) = -f^{(N+1)}(\xi) - (N+1)!K(z), \tag{28}$$

so
$$K(z) = -f^{(N+1)}(\xi)/(N+1)!, \quad \xi \in [x_0,x_1,\ldots,x_N,z]. \tag{29}$$

Note that ξ is a function of z. Thus, substituting x for z in equation 29 and the result in equation 23, we have for x not equal to any x_i,

$$\varepsilon(x) \equiv p(x) - f(x) = -\left(\prod_{i=0}^{N}(x - x_i)\right) f^{(N+1)}(\xi(x))/(N+1)!,$$
$$\xi(x) \in [x_0,x_1,\ldots,x_N,x]. \tag{30}$$

We have assumed $f^{(N+1)}$ exists for all values in the interval in question. Therefore, equation 30 holds for $x = x_i$, since there the error is 0. Thus, the restriction $x \neq x_i$ may be dropped, and equation 30 holds for all x.

Because we will often use the product polynomial in equation 30, we give it a name:

$$\psi(x) \equiv \prod_{i=0}^{N}(x - x_i). \tag{31}$$

It will also be helpful to give the name I to the interval for ξ in equation 30, namely, $[x_0,x_1,x_2,\ldots,x_N,x]$.

For a specific problem, we can bound the truncation error by using the largest value of $f^{(N+1)}$ over I, since we do not know what ξ is:

$$|\varepsilon(x)| \leq \frac{|\psi(x)|}{(N+1)!} \max_{\xi \in I} |f^{(N+1)}(\xi)|. \tag{32}$$

For example, consider approximating $sin(\pi/4)$ by fitting a polynomial p through the tabular points 0, $\pi/6$, $\pi/3$, $\pi/2$. Then

$$\varepsilon(\pi/4) = -\frac{(\pi/4 - 0)(\pi/4 - \pi/6)(\pi/4 - \pi/3)(\pi/4 - \pi/2)}{4!}\left(\frac{d^4}{dx^4} sin(x)\right)\bigg|_{x=\xi}, \tag{33}$$

where $\xi \in [0,\pi/2]$. Thus,

$$|\varepsilon(\pi/4)| \le (\pi/4)^4 \tfrac{1}{216} \max_{\xi \in [0,\pi/2]} |sin(\xi)| = (\pi/4)^4 \tfrac{1}{216} \approx 2 \times 10^{-3}. \tag{34}$$

Actually fitting the polynomial produces $p(\pi/4) = .706$ as compared to $sin(\pi/4) = .707$, so the error is -1×10^{-3}; our bound is reasonable. Note that in this case the sign of the fourth derivative is constant over I, so we know that the error will be negative: $-2 \times 10^{-3} \le \varepsilon(\pi/4) \le 0$. We could reduce the error bound by subtracting half of its value from the polynomial approximation, that is, by subtracting $-.001$ from .706, producing $.707 \pm .001$.

The truncation error formula, equation 30, can be used more generally to evaluate approximation strategies. First, it tells us that the farther the evaluation point x is from the tabular argument values, the larger $\psi(x)$ will be and thus the larger the anticipated error. We can conclude that (1) the $N+1$ data points used for a particular approximation should ordinarily be those in the table whose argument values are closest to the argument values at which we are evaluating; and (2) *extrapolation*, which is approximation at a value x not in the interval bounded by the tabular arguments used, is a very error-prone procedure. Compared with interpolation, the error in extrapolation may be larger not only because of the large value of $\psi(x)$ but also because of the increased interval for ξ, which produces the potential for a larger value for the derivative factor in the error (equation 30). Extrapolation at any appreciable distance from our tabular arguments is very dangerous unless we know in some detail the form of the function being approximated. If we do not, the approximating function may go in one direction and the function being approximated may go in the opposite direction (as do $p(x)$ and $f(x)$, respectively, in figure 4-3).

Another issue that equation 30 allows us to deal with is the determination of the optimum value for $N+1$, the number of data points that should be used in a particular approximation. The denominator of equation 30 would lead us to believe we should use as many points as possible if computational error considerations do not make this approach unsatisfactory (but they do). However, in many cases, truncation error considerations alone indicate that increasing the number of points is not desirable. Introducing a tabular point, x_{N+1}, has three effects on the error (equation 30).

1. It is multiplied by $x - x_{N+1}$ due to the additional term in the product $\psi(x)$.

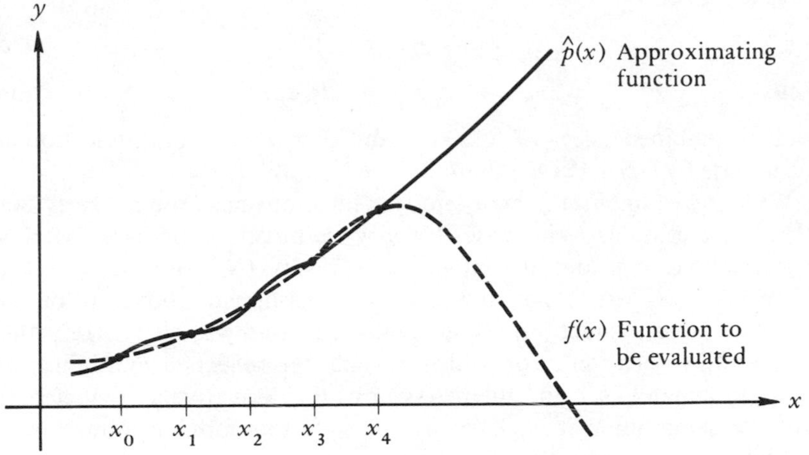

FIG. 4-3 A danger of extrapolation

2. It is divided by $N+2$ due to the additional term in the factorial in the denominator.
3. The $(N+1)$th derivative is replaced by an $(N+2)$th derivative, and the interval in which the argument of the derivative may fall is slightly enlarged.

With regard to effect 1, we assume that the best tabular arguments of those available are used, that is, x_i is the $(i+1)$th closest tabular argument to x (see figure 4-4). If the tabular arguments are approximately equally

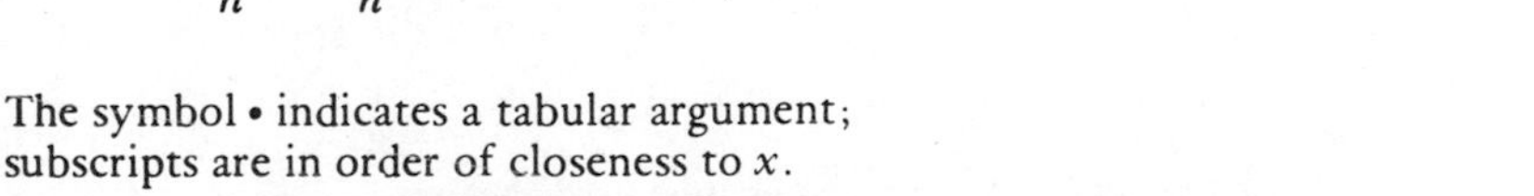

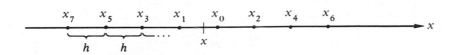

The symbol • indicates a tabular argument;
subscripts are in order of closeness to x.

FIG. 4-4 Selecting the closest of equally spaced tabular arguments to the evaluation argument

spaced with average spacing h, the x_i will be alternately on opposite sides of x and

$$x_{N+1} \approx x \pm ((N+1)/2)h. \tag{35}$$

Thus

$$|x_{N+1} - x| \approx ((N+1)/2)h, \tag{36}$$

so the combined effect of effects 1 and 2 will be a multiplication in magnitude by $((N+1)/(N+2))(h/2) \approx h/2$.

With regard to effect 3, many smooth functions have the property that $f^{(N+1)}$ decreases as N increases for a while but then increases with N. For example, consider the function $ln(x)$. Its $(N+1)$th derivative is $(-1)^N N!/x^{N+1}$, which has the property mentioned above. If one is approximating a function f that behaves in this way, it is likely that the overall truncation error, which at each step reflects a multiplication by approximately $h/2$ and the behavior of the derivative as N increases, will have a minimum at some value of N, so further tabular points should not be used (see figure 4-5).

Even if the above truncation error considerations do not lead to a limit on the number of tabular points, computational error considerations do. Computational error usually increases with the number of points used

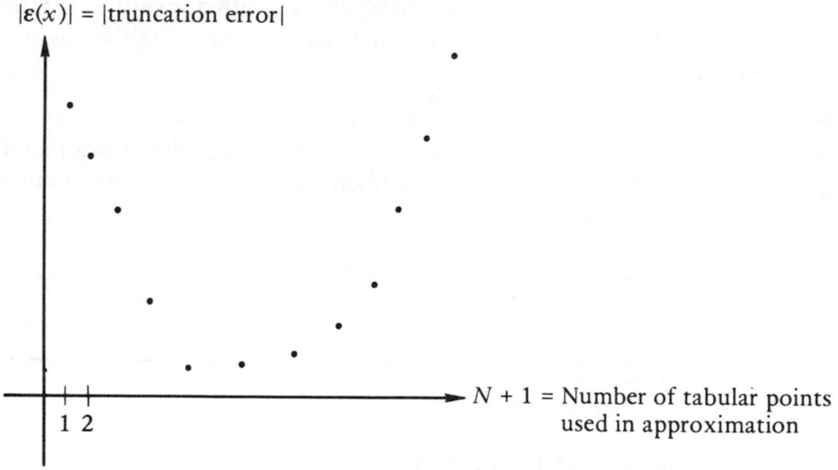

FIG. 4-5 Common truncation error behavior

(computations required). Therefore, it is almost always the case that the overall error reaches a minimum for some relatively small N, so further tabular points should not be used. We shall discuss this matter in more detail after we have further discussed computational error.

c) Barycentric Form of Lagrange Interpolation. Lagrange interpolation applied directly requires $2(N-1)$ multiplications and a division to evaluate each $L_i(x)$ and thus $2N(N+1)$ multiplications (and divisions) to evaluate the approximating polynomial through $N+1$ points by equation 11. By some clever algebra, this number can be reduced considerably. The reduction depends upon noting that the polynomial $p(x) = 1$ fits exactly through a table which has $y_i = 1$ for all x_i, $0 \leq i \leq N$. Since there is only one polynomial of degree N or less passing through these points,

$$1 = \sum_{j=0}^{N} 1 \cdot L_j(x) = \sum_{j=0}^{N} \left[\left(\prod_{\substack{i=0 \\ i \neq j}}^{N} (x - x_i) \right) \Big/ \prod_{\substack{i=0 \\ i \neq j}}^{N} (x_j - x_i) \right]. \tag{37}$$

Let $\hat{p}(x)$ be the Lagrange approximating polynomial through the data points y_i of some given problem. Then we can divide $\hat{p}(x)$ by 1 without changing its value, producing

$$\hat{p}(x) = \hat{p}(x)/1 = \sum_{j=0}^{N} y_j L_j(x) \Big/ \sum_{j=0}^{N} L_j(x). \tag{38}$$

If we divide both the numerator and the denominator of equation 38 by $\psi(x)$ and let

$$A_j = 1 \Big/ \prod_{\substack{i=0 \\ i \neq j}}^{N} (x_j - x_i), \tag{39}$$

we produce
$$\hat{p}(x) = \left(\sum_{j=0}^{N} y_j A_j/(x - x_j) \right) \Big/ \sum_{j=0}^{N} A_j/(x - x_j). \tag{40}$$

Using this so-called *barycentric form of Lagrange interpolation* (see program 4-1), computing each A_j requires N multiplications and divisions, so the whole computation requires $(N+2)(N+1)$ multiplications, approximately half that required by the direct Lagrange method. Furthermore, if we are computing the approximation for many different x values, we can compute the A_j once and for all (probably, in double precision because they will be used so much); then each evaluation will require only $2(N+1)$ multiplications and divisions.

```
/* BARYCENTRIC FORM OF LAGRANGE METHOD FOR POLYNOMIAL INTERPOLATION */
/* GIVEN IS N, THE DEGREE OF THE APPROXIMATING POLYNOMIAL; */
/* THE DATA POINTS, (X(J),Y(J)), Ø <= J <= N; */
/* AND THE EVALUATION ARGUMENT, XEVAL */

   DECLARE (X(Ø:N),          /* N+1 DATA ARGUMENTS */
            Y(Ø:N),          /* N+1 DATA VALUES */
            A(Ø:N),          /* N+1 MULTIPLYING CONSTANTS */
            XEVAL,           /* EVALUATION ARGUMENT */
            APPROX,          /* RESULTING APPROXIMATE VALUE AT X = XEVAL */
            NUM, DENOM,      /* NUMERATOR AND DENOMINATOR OF APPROXIMATION */
            TERM) FLOAT;     /* TERM OF SUM */

/* COMPUTE CONSTANT FOR EACH TERM */
   CONST: DO I = Ø TO N;
          A(I) = 1;
          PROD: DO J = Ø TO I - 1, I+1 TO N;     /* COMPUTE PRODUCT OF (X(J) - X(I)) */
                A(I) = A(I) / (X(J) - X(I));     /* OVER ALL J EXCEPT J = I */
          END PROD;
   END CONST;

/* COMPUTE APPROXIMATION */
   NUM = Ø;
   DENOM = Ø;
   SUMS: DO J = Ø TO N;        /* COMPUTE NUMERATOR AND DENOMINATOR SUMS */
         TERM = A(J) / (XEVAL - X(J));
         DENOM = DENOM + TERM;
         NUM = NUM + Y(J) * TERM;
   END SUMS;
   APPROX = NUM / DENOM;   /* COMPUTE FINAL RESULT */
```

PROGRAM 4-1 Barycentric form of Lagrange interpolation

d) Newton Divided-Difference Formula. If one knows beforehand how many data points are to be used in an approximation, the barycentric form of Lagrange interpolation provides a reasonable method for computation. However, in most cases, one does not know how many points will provide either the best approximation or an approximation of desired accuracy, because the function f being approximated is either not analytically known or too complicated to use to compute and evaluate derivatives of high order. We need a formulation that produces approximations using, successively, one data point, two data points, and so on, and that allows us to determine the optimum number of points during the computation. The Lagrange method is not satisfactory for this purpose, because if we have used it to compute the interpolation through $N+1$ points, a great deal of new work is required to evaluate the polynomial with a single point added. Each old Lagrange polynomial must be multiplied by a factor in both the numerator and the denominator, and a new Lagrange polynomial term corresponding to the new data point must be computed. An alternative approach, the *Newton divided-difference formulation,* avoids this difficulty. The approach is based on attempting to write a series like the Taylor series that will permit us to increase the degree of the approximating polynomial by simply adding a term to the preceding approximation, where that term depends on a derivative of the function being approximated. If we have only a table of arguments and function values, we do not have available values for the derivative, but these can be approximated by finite differences of the data values.

We define the first *divided-difference* of f at the points x_0 and x_1 as

$$f[x_0,x_1] \equiv (f(x_1) - f(x_0))/(x_1 - x_0). \tag{41}$$

Note that this divided-difference is an approximation to the derivative; in fact if f and f' are continuous between x_1 and x_0, the first divided-difference must be equal to the derivative at some point in the interval $[x_0,x_1]$.

We define higher divided-differences by a straightforward extension of equation 41:

$$f[x_0,x_1,x_2,\ldots,x_n] \equiv \frac{f[x_1,x_2,\ldots,x_n] - f[x_0,x_1,x_2,\ldots,x_{n-1}]}{x_n - x_0}. \tag{42}$$

We generate an interpolation formula at an arbitrary value x by noting that

$$f[x,x_0] = (f(x_0) - f(x))/(x_0 - x), \tag{43}$$

so

$$f(x) = f(x_0) + (x - x_0)f[x,x_0]. \tag{44}$$

We cannot compute $f[x,x_0]$, because doing so requires knowing $f(x)$, but we can continue the process started with equations 43 and 44 by writing

$$f[x,x_0,x_1] = (f[x_0,x_1] - f[x,x_0])/(x_1 - x), \tag{45}$$

so

$$f[x,x_0] = f[x_0,x_1] + (x - x_1)f[x,x_0,x_1]; \tag{46}$$

therefore

$$f(x) = f(x_0) + (x - x_0)f[x_0,x_1] + (x - x_0)(x - x_1)f[x,x_0,x_1]. \tag{47}$$

Applying the same technique again and again, we can show

$$\begin{aligned} f(x) = f(x_0) &+ (x - x_0)f[x_0,x_1] + (x - x_0)(x - x_1)f[x_0,x_1,x_2] \\ &+ \cdots + (x - x_0)(x - x_1)\ldots(x - x_{n-1})f[x_0,x_1,\ldots,x_n] \\ &+ (x - x_0)(x - x_1)\ldots(x - x_n)f[x,x_0,x_1,\ldots,x_n]. \end{aligned} \tag{48}$$

Theorem: Consider the polynomial $p(x)$ which is equal to the sum of all but the last term of equation 48. Then $p(x_i) = f(x_i)$ for $0 \leq i \leq n$.

Proof: The proof proceeds by induction on the terms of equation 48. As a basis step, the first term of equation 48, $f(x_0)$, agrees with f at x_0. Assume that the polynomial made up of the first $m+1$ terms of equation 48 agrees with f at x_i, $0 \leq i \leq m$. Since the $(m+2)$th term has a factor $x - x_i$ for $0 \leq i \leq m$, the first $m+2$ terms agree with f at x_i, $0 \leq i \leq m$. We have only to show that the first $m+2$ terms of equation 48 agree with f at x_{m+1}. This claim is proved using the fact that the divided-difference is a symmetric function, that is, the order of the argument points does not affect the value (see exercise 4.2.5).

We note that the identity given by equation 48 applies for $n = m$ and $x = x_{m+1}$:

$$\begin{aligned} f(x_{m+1}) = f(x_0) &+ (x_{m+1} - x_0)f[x_0,x_1] \\ &+ (x_{m+1} - x_0)(x_{m+1} - x_1)f[x_0,x_1,x_2] + \cdots \\ &+ (x_{m+1} - x_0)(x_{m+1} - x_1)\ldots(x_{m+1} - x_{m-1}) \\ &\qquad \times f[x_0,x_1,\ldots,x_m] \\ &+ (x_{m+1} - x_0)(x_{m+1} - x_1)\ldots(x_{m+1} - x_m) \\ &\qquad \times f[x_{m+1},x_0,x_1,\ldots,x_m]. \end{aligned} \tag{49}$$

But, by the symmetry of the divided-difference,

$$f[x_{m+1},x_0,x_1,\ldots,x_m] = f[x_0,x_1,\ldots,x_{m+1}]. \tag{50}$$

> Therefore, equation 49 is the same as the first $m+2$ terms of equation 48 evaluated at x_{m+1}; the first $m+2$ terms of equation 48 equal $f(x_{m+1})$.

We have shown that the nth-degree polynomial consisting of all but the last term of equation 48 agrees with f at x_i, $0 \le i \le n$. Since there is only one nth-degree polynomial with this property, this polynomial must be the same as that produced by Lagrange approximation using those points. Therefore, the last term of equation 48 must be the difference between $f(x)$ and this fitting polynomial, that is,

$$\left(\prod_{i=0}^{n} (x - x_i)\right) f[x,x_0,x_1,\ldots,x_n] = \left(\prod_{i=0}^{n} (x - x_i)\right) f^{(n+1)}(\xi(x))/(n+1)! \quad (51)$$

for some $\xi \in [x,x_0,x_1,x_2,\ldots,x_n]$. Therefore,

$$f[x,x_0,x_1,\ldots,x_n] = f^{(n+1)}(\xi(x))/(n+1)! \quad (52)$$

for some $\xi \in [x,x_0,x_1,\ldots,x_n]$.

Computing polynomial exact-matching approximations using the Newton divided-difference method proceeds as follows:

1. $p(x) = f(x_0)$.
2. $factor = 1$.
3. for $i = 1$ by 1 until stopping criterion is met:
   a) $factor = factor * (x - x_{i-1})$.
   b) $p(x) = p(x) + factor * f[x_0,x_1,\ldots,x_i]$.

The method assumes that the divided differences are available or are computed as needed. (See also program 4-2.)

One of the advantages of the divided-difference method is that a good deal of the computing can be done beforehand, independent of the x value to be used. Usually, the differences are computed in a table with the format shown in table 4-1. The x_i in the table are listed in numerical order. The subscripts in the table do not correspond directly to the subscripts in the divided-difference formula, but there exists a correspondence between the tabular arguments and the x_i of the divided-difference formula as discussed below.

We have argued that tabular points should be added into the approximation in order of the closeness of the tabular argument values to the evaluation argument. Adding the points in this order produces a notion of a *path* through the divided-difference table (see table 4-2). This path begins at a function value (0th difference) in the first column of the divided-difference table and moves from that 0th difference diagonally to an adjacent first difference; then diagonally again, but not necessarily in

TABLE 4-1 A DIVIDED-DIFFERENCE TABLE

x_0	$f(x_0) = y_0$		
		$f[x_0,x_1]$	
x_1	$f(x_1) = y_1$		$f[x_0,x_1,x_2]$
		$f[x_1,x_2]$	
x_2	$f(x_2) = y_2$		$f[x_1,x_2,x_3]$
		$f[x_2,x_3]$	
x_3	$f(x_3) = y_3$		$f[x_2,x_3,x_4]$
		$f[x_3,x_4]$	
x_4	$f(x_4) = y_4$		.
.	.	.	.
.	.	.	.
.	.	.	.
x_{n-2}	$f(x_{n-2}) = y_{n-2}$		.
		$f[x_{n-2},x_{n-1}]$	
x_{n-1}	$f(x_{n-1}) = y_{n-1}$		$f[x_{n-2},x_{n-1},x_n]$
		$f[x_{n-1},x_n]$	
x_n	$f(x_n) = y_n$		

the same direction, to an adjacent second difference; and so on, stopping at the last difference used. The data values used in the approximation corresponding to any given path-ending can be determined by drawing a triangle from the path-ending to the 0th-difference column along the diagonals of the table (look again at table 4-2). The y_i values so enclosed, and their corresponding x_i values, are the data values used. For any temporary path-ending, the difference in the next column to which the path should go is chosen from the two differences adjacent to the present path-ending. This difference is the one for which the additional argument value enclosed by the new triangle is closest to the evaluation argument.

TABLE 4-2 Building a Path in a Difference Table

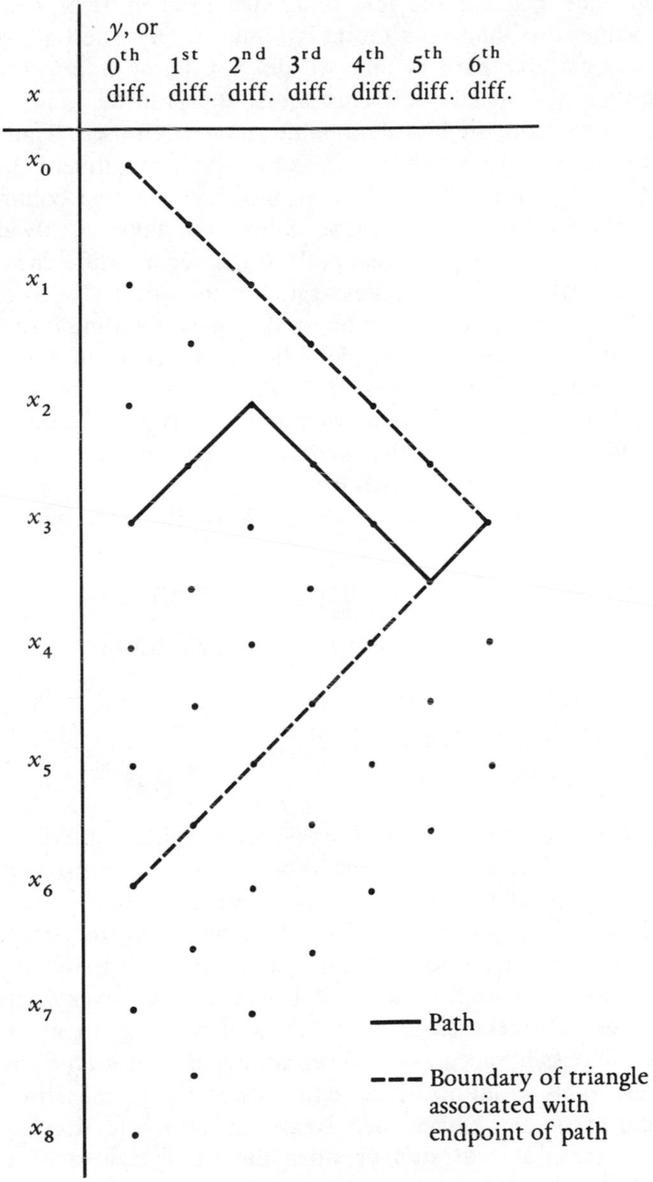

That is, at any step the approximation should involve the closest tabular argument values to the evaluation argument.

It should be emphasized that a path-ending at any given difference produces the same approximating formula as any other path-ending at that difference, because the last difference used entirely specifies the tabular values used and thus the polynomial fitted. The only reason to choose a particular path is that we do not know *a priori* where to stop, and we wish at any difference along the path to be in a position such that if we stop, the best data values have been used. Thus, for any evaluation argument x, we choose x_0 as the tabular argument closest to x and begin our path at $f(x_0)$. Then we move to the next column in the direction specified by the rule above, calling the newly involved tabular argument value x_1. We build our path to the second-difference column, calling the newly involved tabular argument point x_2, and so on. These points—the points x_0, x_1, and so on, in the approximating formula given by equation 48—depend on the path chosen, which in turn depends on the evaluation argument. The divided-difference table is ordered so that the tabular arguments increase in order, but for any given x, the subscripts corresponding to tabular values increase in order of the distance of the corresponding tabular arguments from x.

For example, consider approximating $f(2.5)$ to the third difference using table 4-3. Then

$$\begin{aligned}\hat{p}(2.5) &= f(2) + (2.5-2)f[2,3] + (2.5-2)(2.5-3)f[2,3,1.5] \\ &\quad + (2.5-2)(2.5-3)(2.5-1.5)f[2,3,1.5,1] \\ &= 83 + .5(-16) + (.5)(-.5)(-1.33) \\ &\quad + (.5)(-.5)(1)(.33) \\ &= 75.25^{-}. \end{aligned} \tag{53}$$

We stated that the advantage of the Newton divided-difference method is that it allows us to easily add tabular points into an approximation and to decide the point at which to stop as we go. How is this decision made? If there is a tolerance within which we desire to compute $f(x)$, we stop either when the error estimate first becomes less than the tolerance or when it will increase if we add another term, whichever happens first.

If the desired tolerance is zero, that is, if we wish to get the value with the best possible accuracy, we keep adding terms until the error begins to increase. This situation occurs either when the increase in the computational error at a given step is greater than the decrease in the truncation error at that step or when the truncation error begins to increase.

TABLE 4-3 USING A DIVIDED-DIFFERENCE TABLE TO APPROXIMATE $f(2.5)$

x	y			
-1	107			
		$-11/2 = -5.5$		
1	96		$-6.5/2.5 = -2.6$	
		$-6/.5 = -12$		$.6/3 = .2$
1.5	90		$-2/1 = -2$	
		$-7/.5 = -14$		$.67/2 = .33$
2	83		$-2/1.5 = -1.33$	
		$-16/1 = -16$		$.33/3.5 = .095$
3.0	67		$-3/3 = -1$	
		$-38/2 = -19$		
5	29			

Now assume we wish to stop adding terms when the error becomes less than a given tolerance other than zero. If the truncation error is large compared to the computational error, we can stop when the error bound for the truncation error given in equation 32 first becomes less than the tolerance or begins to grow. However, we often do not know the function f, or if we know it, we are not willing to evaluate or compute $f^{(N+1)}$, so we need an estimate for the error term. We note from equation 48 that the error can be written as $\psi(x)$ times a divided difference. We cannot evaluate the divided difference, $f[x_0,x_1,\ldots,x_n,x]$, because its evaluation requires knowledge of $f(x)$, the value we are looking for. But we know that equation 52 holds. Furthermore, if we evaluate equation 52 at $x = x_{n+1}$, we obtain

$$f[x_0,x_1,x_2,\ldots,x_{n+1}] = \frac{f^{(n+1)}(\xi(x_{n+1}))}{(n+1)!}, \tag{54}$$

where $\xi \in [x_0,x_1,x_2,\ldots,x_{n+1}]$. If x_{n+1} is the tabular argument closest to x

except for $x_0, x_1, \ldots, x_n$, then the interval for $\xi(x_{n+1})$ in equation 54 is close to the interval for $\xi(x)$ in equation 52. Therefore, if $f^{(n+1)}$ does not change value a great deal over that interval,

$$f^{(n+1)}(\xi(x)) \approx f^{(n+1)}(\xi(x_{n+1})), \tag{55}$$

so

$$f[x_0, x_1, x_2, \ldots, x_{n+1}] \approx f[x, x_0, x_1, \ldots, x_n]. \tag{56}$$

Even if $f^{(n+1)}$ varies a great deal over the interval, it is possible that $\xi(x) \approx \xi(x_{n+1})$ and therefore, approximation 56 holds. In either case,

$$\varepsilon(x) \approx -\psi(x) f[x_0, x_1, x_2, \ldots, x_{n+1}]. \tag{57}$$

If we have a computed value and know the error, we subtract the error from the computed value to get the correct value. In the same way, we can correct the approximation obtained from the first $n+1$ terms of the divided-difference formula by adding the negative of the right side of approximation 57. But this value is precisely the next term that would be added in the normal Newton divided-difference series. In other words, the error at any step is approximately equal in magnitude to the next term to be added in the series.

From the above results we conclude that if we cannot compute the error bound analytically, we should stop adding terms when the terms added are less than the error tolerance if this occurs. We should stop earlier if the terms added begin to increase.

We should stop adding terms before truncation error considerations tell us to if computational error begins to dominate, that is, if the computational error increase at some step is greater than the truncation error decrease. The increase in computational error due to adding the next term in the divided-difference series is largely the computational error in computing that term. By equation 57, the decrease in truncation error is approximately the next term added. So we should stop when the error in the term added is greater than the term itself. Since it is normally the case that the major part of the computational error is that in the divided difference, we stop adding terms when the error in the divided difference is approximately equal to the divided difference itself.

There are two ways to detect when the computational error in a divided difference is approximately equal to the divided difference. One way is by attaching an error bound to each function value (y_i) and computing the error bound in each difference due to propagation of these errors as well as to generation of errors in the computation needed to compute the difference. Such detailed computation is left to exercise 4.2.7. However, it should be noted here that an error in a given function value propagates into the differences included in the triangle formed by the diagonals beginning at that value, called *Pascal's triangle* (as shown in table 4-4). Another important point is that the numerators of succeeding differences normally become smaller, since each one is the difference of smoothly varying differences in the preceding column. On the other hand, error bounds for the numerator terms add, producing increasing error bounds in the numerators of succeeding differences. Therefore, the relative error normally grows as we compute higher differences.

The second way to detect 100% error in a divided difference depends on the fact that if computational errors are of the same magnitude as difference values, we can expect the errors to cause some positive values to become negative and vice versa, so the differences in a column with such errors will fluctuate in sign (more than once). Even if this sign fluctuation is due to accurate differences, differences in succeeding columns will increase, because of subtraction of adjacent differences with opposite signs. When differences increase, the terms to be added and thus the error estimates normally increase, so we should not add further terms to the approximation. Therefore, whatever the cause, if fluctuations in sign are common in an area of the difference column along our path in the difference table, we should stop adding terms.

To summarize, we should stop adding terms in a Newton divided-difference series if the tolerance desired is obtained, or if the truncation

TABLE 4-4 Propagation of Error in a Difference Table

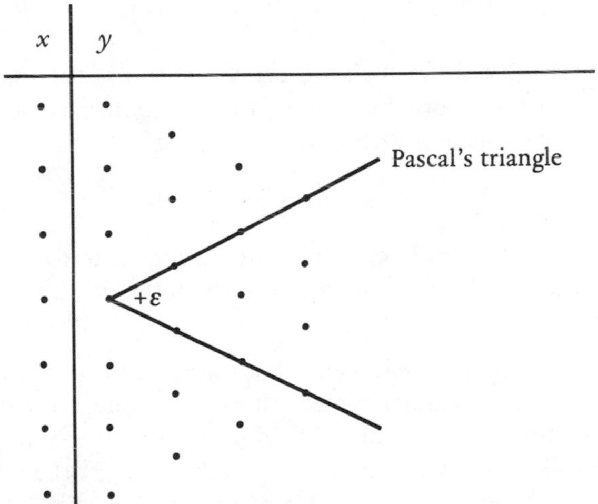

error begins to increase as terms are added, or if the computational error in a term to be added is greater than the term itself. Practically, we choose the stopping column N as follows:

Let $P+1$ be the number of the first difference column such that along our path $|\varepsilon_{comp}/diff| \approx 1$, where the y_i values are in the 0th-difference column, ε_{comp} is the propagated and generated error in a given difference near our path, and *diff* is the value of that difference. $P+1$ is either determined using numerical error estimates or taken to be the number of the first difference column such that the differences in that column fluctuate in sign near the path.

1. If $f^{(N+1)}(x)$ can be computed analytically for each N, then choose $N = min(L,M,P)$, where L is the smallest n such that

$$b_n(x) \equiv \frac{\left|\prod_{i=0}^{n} (x - x_i)\right|}{(n+1)!} \max_{\xi \in [x_0,x_1,\ldots,x_n,x]} |f^{(n+1)}(\xi)| < \beta, \tag{58}$$

where β is a desired absolute error tolerance in the approximation to $f(x)$; and M is the value of n such that

$$b_M(x) = \min_n b_n(x). \tag{59}$$

2. If $f^{(N+1)}(x)$ cannot be computed analytically, then choose $N = min(L,M,P)$, where L is the smallest n such that

$$c_n(x) \equiv \left|\prod_{i=0}^{n} (x - x_i)\right| |f[x_0,x_1,\ldots,x_{n+1}]| < \beta \tag{60}$$

(note that $c_n(x)$ without the absolute-value signs is the next term to be added to the approximation if the inequality is not satisfied); and M is the value of n such that

$$c_M(x) = \min_n c_n(x). \tag{61}$$

The Newton divided-difference method, using criterion 2 above for stopping, and assuming P is an input parameter, is fully specified in program 4-2.

e) Iterated Linear Interpolation. The interpolation process can be organized in another manner which offers the same advantage as the Newton divided-difference method: adding a tabular point is a simple procedure. Unlike the divided-difference method, it has the advantage of not requiring the preliminary computation and storage of many differences to do the interpolation. A concomitant disadvantage is that it does not allow the reuse of already computed values; the whole computation must be redone for every new evaluation argument. The method, called *iterated linear interpolation,* depends on the fact that the $(m+1)$th-degree polynomial fit can be computed as a linear interpolation between two mth-degree fits, where the argument point associated with each mth-degree fit is the tabular argument used for that mth-degree fit and not for the other. That is, if we define the 0th-degree fit,

$$p_i(x) \equiv y_i, \tag{62}$$

and then define the recurrence relation

$$p_{ij_1j_2\ldots j_mk}(x) \equiv \frac{(x_k - x)p_{ij_1j_2\ldots j_m}(x) - (x_i - x)p_{j_1j_2\ldots j_mk}(x)}{x_k - x_i} \tag{63}$$

to obtain an $(m+1)$th-degree fit from an mth-degree fit, it can be shown that

$$p_{01\ldots N}(x) = \hat{p}(x), \tag{64}$$

where $\hat{p}(x)$ is the exact-matching polynomial through the tabular points with arguments $x_0, x_1, \ldots, x_N$. The proof of this assertion, accomplished by induction, is left to exercise 4.2.8.

Since the exact-matching polynomial through a given set of points does not depend on the order of the points, the order in which the points are added in the iterated linear interpolation does not affect the final value, except for computational error. However, as with the Newton divided-difference method, we wish to add tabular points in order of the closeness of their argument values to the evaluation argument so that, whenever the iteration stops, we will have produced the interpolated value that is expected to be best among interpolations through the number of points used so far. The computation proceeds as follows.

We construct a table that has in the 0th column the x_i in order of closeness to the evaluation argument (see table 4-5). We call these argument values $x_0, x_1, \ldots$, respectively. In the first column, opposite each x_i, we place $p_i(x) = y_i$. In the second column, opposite each x_i, $i \geq 1$, we place $p_{0i}(x)$, which is computed from $p_0(x)$ and $p_i(x)$. In the third column, opposite each x_i, $i \geq 2$, we place $p_{01i}(x)$, which is computed from $p_{01}(x)$ and $p_{0i}(x)$, and so on. The order in which we do these computations follows:

1. Write x_0 and $p_0(x)$.
2. If $p_0(x)$ is an accurate enough approximation, stop.
3. If not, insert table entries x_1, $p_1(x)$, and $p_{01}(x)$.
4. If $p_{01}(x)$ is accurate enough, stop.
5. Otherwise, insert the next row: x_2, $p_2(x)$, $p_{02}(x)$, and $p_{012}(x)$.
6. Continue to build the table through steps analogous to those above until the most recently computed diagonal entry is accurate enough.

```
/* NEWTON DIVIDED-DIFFERENCE METHOD FOR POLYNOMIAL INTERPOLATION */
/* GIVEN ARE N+1 DATA ARGUMENTS, X(J), Ø <= J <= N, SUCH THAT X(J) < X(J+1); N; */
/* P, THE LAST COLUMN IN THE DIFFERENCE TABLE WHERE SIGNS DO NOT FLUCTUATE NEAR THE PATH TO BE USED; */
/* XEVAL, THE EVALUATION ARGUMENT; AND TOL, THE ERROR TOLERANCE */
/* ALSO ASSUMED AVAILABLE IS A SUBROUTINE F, WITH TWO ARGUMENT INDICES AS PARAMETERS, */
/* WHICH COMPUTES OR RETRIEVES THE DIVIDED DIFFERENCE AT THE VERTEX OF THE TRIANGLE */
/* IN THE DIFFERENCE TABLE DEFINED BY THE TWO ARGUMENT INDICES */

   DECLARE (X(Ø:N),              /* N+1 DATA ARGUMENTS */
            XEVAL,               /* EVALUATION ARGUMENT */
            APPROX,              /* RESULTING APPROXIMATE VALUE AT X = XEVAL */
            NEWTERM, NEWMAG,     /* NEW TERM IN APPROXIMATING SUM AND ITS MAGNITUDE */
            FACTOR,              /* PRODUCT OF (X - X(J)) USED */
            PREVTERM, PREVX,     /* PREVIOUS SUMMAND AND TABULAR ARGUMENT */
            TOL) FLOAT,          /* REQUIRED APPROXIMATION ACCURACY */
            (TOPNDX, BOTNDX,     /* EXTREME USED TABULAR ARGUMENT INDICES */
            P) FIXED BINARY;     /* LAST DIFFERENCE TABLE COLUMN TO BE USED */

/* FIND FIRST TABULAR ARGUMENT TO BE USED */
   PREVTERM = ABS(XEVAL - X(Ø));
   ARGFIND: DO I = 1 TO N;   /* FIND CLOSEST TABULAR ARGUMENT TO XEVAL */
            NEWTERM = ABS(XEVAL - X(I));
            IF NEWTERM < PREVTERM THEN PREVTERM = NEWTERM;
                                  ELSE GO TO INDXFOUND;
   END ARGFIND;
   INDXFOUND:  TOPNDX = I - 1;   /* I - 1 IS INDEX OF FIRST TABULAR ARGUMENT */
               BOTNDX = I - 1;

/* COMPUTE APPROXIMATION */
   APPROX = F(TOPNDX,BOTNDX);  /* Y(I - 1) */
   PREVTERM = APPROX;          /* ØTH TERM IN APPROXIMATING SUM */
   PREVX = X(TOPNDX);          /* MOST RECENTLY USED TABULAR ARGUMENT */
   FACTOR = 1;                 /* MOST RECENTLY USED MULTIPLYING FACTOR */
```

```
NEXTERM: DO I = 1 TO P;        /* COMPUTE NEXT TERM IN APPROXIMATING SUM */
         FACTOR = FACTOR *  (XEVAL - PREVX);

/* FIND NEXT CLOSEST TABULAR ARGUMENT TO XEVAL */
         IF TOPNDX = Ø THEN GO TO NEWBOT;
         IF BOTNDX = N THEN GO TO NEWTOP;
         IF ABS(XEVAL - X(TOPNDX-1)) < ABS(XEVAL - X(BOTNDX+1))
              THEN NEWTOP: DO;  /* TOPNDX - 1 IS INDEX OF NEW ARGUMENT */
                             TOPNDX = TOPNDX - 1;
                             PREVX = X(TOPNDX);
                   END NEWTOP;
              ELSE NEWBOT: DO;  /* BOTNDX + 1 IS INDEX OF NEW ARGUMENT */
                             BOTNDX = BOTNDX + 1;
                             PREVX = X(BOTNDX);
                   END NEWBOT;

/* COMPUTE TERM ASSOCIATED WITH NEXT DIFFERENCE */
         NEWTERM = FACTOR * F(TOPNDX,BOTNDX);

/* SHOULD NEW TERM BE ADDED INTO APPROXIMATION? */
         NEWMAG = ABS(NEWTERM);
         IF NEWMAG > PREVTERM THEN GO TO DONE;   /* ADD NO MORE TERMS */
         APPROX = APPROX + NEWTERM;
         PREVTERM = NEWMAG;
         IF PREVTERM < TOL THEN GO TO DONE;      /* ACCURACY ACHIEVED? */
END NEXTERM;

DONE:  /* APPROX HOLDS FINAL APPROXIMATION */
```

PROGRAM 4-2 Newton divided-difference interpolation

TABLE 4-5 ITERATED LINEAR INTERPOLATION

x_0	$p_0(x)$			
x_1	$p_1(x)$	$p_{01}(x)$		
x_2	$p_2(x)$	$p_{02}(x)$	$p_{012}(x)$	
x_3	$p_3(x)$	$p_{03}(x)$	$p_{013}(x)$	$p_{0123}(x)$
.	.	.	.	.
.	.	.	.	.
.	.	.	.	.

Note that to add a new row, that is, a new data point, we need only have stored x_0 and the diagonal of the table up to this point, because only these values are used in the computations for the next row.

We need now only answer the question of how to know when to stop. The decision is made on the same basis as it is when using the Newton divided-difference method, namely, we stop on an analytic basis, or more commonly, when we have no such basis, we stop when the difference between two successive approximations (diagonal entries) either begins to increase or becomes smaller than the error tolerance. The rules are the same as with the divided-difference method, because the polynomials in question are the same because of the uniqueness of the exact-matching polynomial through a given set of points. Note that with the Newton divided-difference method, the difference between two successive approximations is simply the term to be added whereas with iterated linear interpolation the difference must be computed directly from two successive diagonal terms.

The iterated linear interpolation method is fully specified in program 4-3.

f) Choosing Tabular Points. Thus far in our discussion of methods for general polynomial exact-matching and the error therein, we have assumed that a particular function has already been tabulated at given argument values. Now we must treat the question of choosing tabular arguments when a function has not yet been tabulated but a table is to be used for polynomial exact-matching interpolation. Such tabulation is done before one knows the value or values of the argument at which the function will be interpolated. Rather, one knows only a range for these evaluation arguments and a tolerance for the accuracy of the interpolation.

If the function variation over the argument range is very large compared to the error tolerance, it is usually necessary to have many tabular arguments over the range to achieve the desired accuracy. A single polynomial fit to all of the tabular points will be of high degree. If the accuracy of the function tabulation makes the error in higher differences large, or if the function derivatives do not die out quickly as the order of the derivative increases (as is commonly the case), a high-degree polynomial using tabular arguments across the whole range does not produce the desired accuracy. To obtain the desired accuracy, we must fit a lower-degree polynomial through only a few finely spaced tabular points with arguments near the evaluation argument. In this case, the number of points in the table must be large compared to the number of points used in any interpolation. On the other hand, if the range of possible evaluation arguments is small,† and if the number of points required to make $\psi(x)/(N+1)!$ small enough is not so large as to cause the derivative term to get large, it is best to assume, when tabulating, that every interpolation will use most or all of the tabular points.

If the number of tabular points is large compared to the number used in a given interpolation, we wish, for any evaluation argument x in the range in question, to have the tabular arguments as near as possible to x. So, assuming that all values for the evaluation argument are equally likely over the range, the tabular arguments should be equally spaced across the range. The number of such arguments required, given the degree of the interpolation polynomial to be used, can be determined from the error formula given by equation 30. Equal interval interpolation is such a common process that specialization of the formulas and techniques given in the earlier part of this section has been developed (see part g of this section).

---

† For many functions, the range of evaluation arguments can be considerably reduced by taking advantage of mathematical properties of the function. See, for example, C. T. Fike, *Computer Evaluation of Mathematical Functions,* chapter 3.

```
/* ITERATED LINEAR INTERPOLATION METHOD FOR POLYNOMIAL INTERPOLATION */
/* GIVEN ARE N+1 DATA ARGUMENTS, X(J), Ø < = J < = N, SUCH THAT X(J) < X(J+1); N; */
/* P, THE HIGHEST POSSIBLE POLYNOMIAL DEGREE (SEE DIVIDED-DIFFERENCE METHOD); */
/* XEVAL, THE EVALUATION ARGUMENT; AND TOL, THE ERROR TOLERANCE */

    DECLARE (X(Ø:N),            /* N+1 DATA ARGUMENTS */
             Y(Ø:N),            /* N+1 DATA VALUES */
             XEVAL,             /* EVALUATION ARGUMENT */
             APPROX(Ø:P),       /* ITH ELEMENT IS ITH-DEGREE APPROXIMATION */
             PREVDIF, NEWDIF,   /* SUCCESSIVE APPROXIMATION IMPROVEMENTS */
             TOL,               /* REQUIRED APPROXIMATION ACCURACY */
             NEWTAB) FLOAT,     /* NEW ENTRY IN TABLE */
             (INDEX(Ø:P),       /* ITH ELEMENT IS INDEX OF ITH TABULAR ARGUMENT USED */
             TOPNDX, BOTNDX,    /* EXTREME USED TABULAR ARGUMENT INDICES */
             P) FIXED BINARY;   /* HIGHEST ALLOWED APPROXIMATION DEGREE */

/* FIND FIRST TABULAR ARGUMENT TO BE USED */
    PREVTERM = ABS(XEVAL - X(Ø));
    ARGFIND: DO I = 1 TO N;      /* FIND CLOSEST TABULAR ARGUMENT TO XEVAL */
             NEWDIF = ABS(XEVAL - X(I));
             IF NEWDIF < PREVDIF THEN PREVDIF = NEWDIF;
                                 ELSE GO TO INDXFOUND;
    END ARGFIND;
    INDXFOUND: INDEX(Ø) = I - 1;
               TOPNDX = I - 1;
               BOTNDX = I - 1;

/* COMPUTE SUCCESSIVE APPROXIMATIONS */
    APPROX(Ø) = Y(INDEX(Ø));
    PREVDIF = 1E1Ø * APPROX(Ø);  /* ASSURE FIRST ITERATION WILL IMPROVE RESULT */

    TABCOMP: DO I = 1 TO P;      /* I IS APPROXIMATION NUMBER, THAT IS, POLYNOMIAL DEGREE */

    /* FIND  NEXT CLOSEST TABULAR ARGUMENT TO XEVAL */
             IF TOPNDX = Ø THEN GO TO NEWBOT;
             IF BOTNDX = N THEN GO TO NEWTOP;
```

```
         IF ABS(XEVAL - X(TOPNDX-1)) < ABS(XEVAL - X(BOTNDX+1))
              THEN NEWTOP: DO;  /* TOPNDX - 1 IS INDEX OF NEW ARGUMENT */
                                TOPNDX = TOPNDX - 1;
                                INDEX(I) = TOPNDX;
                   END NEWTOP;
              ELSE NEWBOT: DO;  /* BOTNDX + 1 IS INDEX OF NEW ARGUMENT */
                                BOTNDX = BOTNDX + 1;
                                INDEX(I) = BOTNDX;
                   END NEWBOT;

/* COMPUTE ITH ROW OF TABLE */
         NEWTAB = Y(INDEX(I));          /* ØTH COLUMN */
         ROWCOMP: DO J = 1 TO I;        /* J IS COLUMN NUMBER */
                  NEWTAB = ((X(INDEX(I)) - XEVAL) * APPROX(J-1)
                          - (X(INDEX(J-1)) - XEVAL) * NEWTAB))
                          / (X(INDEX(I)) - X(INDEX(J-1)));
         END ROWCOMP;

/* IS NEW APPROXIMATION MORE ACCURATE? */
         NEWDIF = ABS(NEWTAB - APPROX(I-1));
         IF NEWDIF > PREVDIF THEN OLDAPPROX: DO;  /* NO, USE OLD APPROXIMATION */
                                             I = I - 1;
                                             GO TO DONE;
                                  END OLDAPPROX;
         APPROX(I) = NEWTAB;                      /* YES, USE NEW APPROXIMATION */
         PREVDIF = NEWDIF;
         IF PREVDIF < TOL THEN GO TO DONE;        /* ACCURACY ACHIEVED? */
END TABCOMP;

I = P;    /* IF ALL P ITERATIONS WERE DONE */
DONE:  /* APPROX(I) HOLDS FINAL APPROXIMATION */
```

PROGRAM 4-3 Iterated linear interpolation

If all of the tabular points are to be used in each interpolation, a different choice of tabular arguments is indicated. Assume we have a given error bound and wish to choose tabular arguments in such a way that (1) the approximation at any point in the evaluation argument range has an error less than the error tolerance, and (2) the number of required tabular points is minimal. Our choice can be made by asking, for each given number $(N+1)$ of tabular arguments, how to choose those arguments so that the maximum approximation error magnitude across the range is as small as possible. Then for a given error tolerance, we can find the smallest number $(N+1)$ for which the maximum error is less than the bound and use the $N+1$ associated argument values.

The measure of error we have proposed to minimize is the maximum magnitude of the error over an interval. We have seen this norm of maximum magnitude before; it is called the $\mathscr{L}_\infty$ or Tchebycheff norm. To minimize this norm, we must find the tabular points which minimize $\underset{[a,b]}{max}|\varepsilon(x)|$, that is, minimize $\underset{[a,b]}{max}[|\psi(x)|\ |f^{(N+1)}(\xi(x))|]$, given values of N, a, and b. But we do not know the function $f^{(N+1)}(\xi(x))$, so we cannot truly minimize the latter form. We do the next best thing by minimizing the part of the form that we understand how to control, namely, $\underset{[a,b]}{max}|\psi(x)|$.

For $N+1$ points, $\psi(x)$ is a monic $(N+1)$th-degree polynomial with the x_i as roots. Our task then is to find the monic $(N+1)$th-degree polynomial with minimum maximum magnitude over $[a,b]$. By choosing the x_i as the roots of this "minimax" polynomial, we set $\psi(x)$ equal to the minimax polynomial, as required.

Because scaling a polynomial by a linear factor and translation of the arguments produces a polynomial, once we find the minimax polynomial for the interval $[-1,1]$, we can use that polynomial to find the minimax polynomial for any interval $[a,b]$. That is, assume we know a polynomial $p(x)$ which is the minimax monic $(N+1)$th-degree polynomial over $[-1,1]$. We can map the interval $[a,b]$ linearly into the interval $[-1,1]$ by producing for every z in $[a,b]$ a corresponding point x in $[-1,1]$:

$$x = \left(z - \frac{b+a}{2}\right) \bigg/ \frac{b-a}{2}. \tag{65}$$

Then

$$q(z) = p\left(\left(z - \frac{b+a}{2}\right) \bigg/ \frac{b-a}{2}\right) \tag{66}$$

is an $(N+1)$th-degree polynomial in z which, for every point in $[a,b]$, takes on the value of $p(x)$, where x is the value having the same relative position in $[-1,1]$ as z has in $[a,b]$. The polynomial $q(x)$ is not monic but rather has high-order coefficient $(2/(b-a))^{N+1}$, a constant for given values of a, b, and N. However,

$$q_1(z) = \left(\frac{b-a}{2}\right)^{N+1} q(z) \tag{67}$$

is a monic $(N+1)$th-degree polynomial with values that are a constant multiple of corresponding values of $p(x)$ for $x \in [-1,1]$. Therefore, if $p(x)$ is the minimax monic $(N+1)$th-degree polynomial over $[-1,1]$, then $q_1(z)$ is the minimax monic $(N+1)$th-degree polynomial in z over $[a,b]$.

How can we find the minimax monic $(N+1)$th-degree polynomial over $[-1,1]$? If we consider the problem graphically, we should expect the polynomial to have as many ripples as possible, because in reaching a maximum in magnitude and turning back towards $y = 0$, the function takes on values which are all smaller in magnitude than the already obtained maximum. If the function had not obtained the maximum (in magnitude) and turned back towards the x axis, it would have had to increase, because a monic $(N+1)$th-degree polynomial cannot remain constant for any interval. After the polynomial has reached an extremum and turned back towards the x axis, it is least wasteful of extrema if it does not take on a new extremum until it has reached the largest value obtained by the polynomial in the previous part of the interval. Thus, if there exists an equal-ripple polynomial (one for which all extrema occur at the same magnitude of y) which in the interval in question ripples as many times as possible for an $(N+1)$th-degree polynomial, it is likely to be the polynomial ($\psi(x)$) that we are looking for.

Equal-ripple behavior for $\psi(x)$ is very different from the behavior of $\psi(x)$ when the tabular arguments are equally spaced (see figure 4-6). In the latter case, ψ is symmetric in magnitude about the median of the tabular points used and the ripples strongly increase in magnitude with the distance of the ripple from that median (see exercise 4.2.10).

As soon as we mention equal-ripple functions, the sinusoids come to mind. They are particularly appropriate here, as shown by the following theorem.

Theorem: $cos((N+1)\theta)$ is an $(N+1)$th-degree polynomial in $cos(\theta)$.

Proof:
$$cos((N+1)\theta) = \mathbf{Re}[e^{i(N+1)\theta}] = \mathbf{Re}[(e^{i\theta})^{N+1}]$$

$$= \mathbf{Re}[(cos(\theta) + i\,sin(\theta))^{N+1}]$$

$$= \mathbf{Re}\left[\sum_{k=0}^{N+1} \binom{N+1}{k} i^k\, sin^k(\theta)\, cos^{N+1-k}(\theta)\right]$$

$$= \sum_{\substack{k=0 \\ k \text{ even}}}^{N+1} \binom{N+1}{k} i^k \sin^k(\theta) \cos^{N+1-k}(\theta)$$

$$= \sum_{j=0}^{\lfloor (N+1)/2 \rfloor} \binom{N+1}{2j} (-1)^j \sin^{2j}(\theta) \cos^{N+1-2j}(\theta)$$

$$= \sum_{j=0}^{\lfloor (N+1)/2 \rfloor} \binom{N+1}{2j} (-1)^j (1 - \cos^2(\theta))^j \cos^{N+1-2j}(\theta). \quad (68)$$

Since each term of the sum following the last "=" in equation 68 is an $(N+1)$th-degree polynomial in $\cos(\theta)$, the sum is such a polynomial.

Corollary: Let $\cos((N+1)\theta) = p(\cos(\theta))$. Then the coefficient of x^{N+1} in $p(x)$ is 2^N.

Proof: From the last sum in equation 68 we see that the coefficient of $\cos^{N+1}(\theta)$ is $\sum_{j=0}^{\lfloor (N+1)/2 \rfloor} \binom{N+1}{2j}$, since each term in the sum produces exactly one term in the $(N+1)$th power of $\cos(\theta)$, with the coefficient $\binom{N+1}{2j}$. We show that the high-order coefficient is 2^N by noting that the binomial theorem says that

$$0 = (1-1)^{N+1} = \sum_{\substack{k=0 \\ k \text{ even}}}^{N+1} \binom{N+1}{k} - \sum_{\substack{k=0 \\ k \text{ odd}}}^{N+1} \binom{N+1}{k}, \quad (69)$$

so

$$\sum_{\substack{k=0 \\ k \text{ even}}}^{N+1} \binom{N+1}{k} = \sum_{\substack{k=0 \\ k \text{ odd}}}^{N+1} \binom{N+1}{k}. \quad (70)$$

Also, by the binomial theorem,

$$2^{N+1} = (1+1)^{N+1} = \sum_{k=0}^{N+1} \binom{N+1}{k}$$

$$= \sum_{\substack{k=0 \\ k \text{ even}}}^{N+1} \binom{N+1}{k} + \sum_{\substack{k=0 \\ k \text{ odd}}}^{N+1} \binom{N+1}{k}. \quad (71)$$

From equations 70 and 71 we conclude that

$$\sum_{j=0}^{\lfloor (N+1)/2 \rfloor} \binom{N+1}{2j} = \sum_{\substack{k=0 \\ k \text{ even}}}^{N+1} \binom{N+1}{k} = \frac{1}{2} 2^{N+1} = 2^N. \quad (72)$$

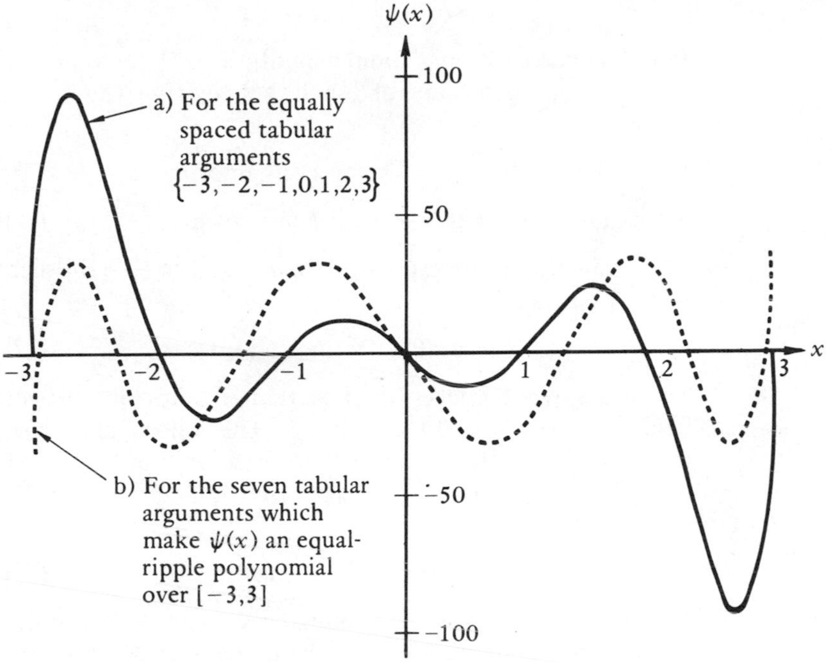

FIG. 4-6 The behavior of $\psi(x)$

We have shown that if we let $x = cos(\theta)$, then

$$cos((N+1)\theta) = cos((N+1)cos^{-1}x) \equiv T_{N+1}(x) \tag{73}$$

is an $(N+1)$th-degree polynomial in x with equal ripples (of height 1; the polynomial also takes the value 1 at the ends of $[-1,1]$). $T_{N+1}(x)$ is called the $(N+1)$th-degree *Tchebycheff polynomial.* Its high-order coefficient is 2^N, so $p(x) = T_{N+1}(x)/2^N$ is a monic $(N+1)$th-degree polynomial with equal ripples.

Theorem: $p(x) = T_{N+1}(x)/2^N$ is the minimax monic $(N+1)$th-degree polynomial over $[-1,1]$.

Proof: Assume $r(x)$ is a monic $(N+1)$th-degree polynomial with maximum magnitude over $[-1,1]$ which is less than that of $p(x)$. Then in particular at each value of x for which $p(x)$

takes on its maximum magnitude,

$$|r(x)| < |p(x)|. \tag{74}$$

But $p(x)$ takes on its maximum magnitude, $1/2^N$, for values of x corresponding to values of θ such that $cos((N+1)\theta) = \pm 1$, namely, at

$$(N+1)\theta = k\pi, \text{ for } k \text{ any integer,} \tag{75}$$

that is, at $$\theta = k\pi/(N+1), \text{ for } k \text{ any integer.} \tag{76}$$

These values of θ correspond to values of x in $[-1,1]$ such that

$$x = cos(\theta) = cos(k\pi/(N+1)). \tag{77}$$

There are $N+2$ integers k such that x takes on distinct values, namely, $k = 0,1,2,\ldots,N+1$. The values of x corresponding to these values of k decrease as k increases. At these values of x, $p(x)$ becomes alternately $1/2^N$ and $-1/2^N$. Consider

$$s(x) \equiv p(x) - r(x). \tag{78}$$

Because p and r are both monic $(N+1)$th-degree polynomials, $s(x)$ is an Nth-degree polynomial. Furthermore, we know that at the extrema of p given by equation 77, p takes on its maximum alternately positively or negatively and at each such point the magnitude of r is less than the magnitude of p. Thus, at the extrema of $p(x)$, $s(x)$ is alternately positive and negative. Since $s(x)$ is a polynomial, it is continuous, and thus between each of its $N+2$ alternate positive and negative values, it must have a zero, so $s(x)$ must have $N+1$ distinct zeros. But an Nth-degree polynomial with more than N zeros must be identically equal to 0, so

$$r(x) \equiv p(x), \tag{79}$$

a contradiction to our assumption that the magnitude of the maximum of r in $[-1,1]$ is less than that of p.

We have shown that $p(x)$ is a minimax monic $(N+1)$th-degree polynomial over $[-1,1]$. The proof that it is in fact unique follows as a corollary (see exercise 4.2.11).

Thus, we wish

$$\psi(x) = T_{N+1}(x)/2^N. \tag{80}$$

Since the x_i are the zeros of $\psi(x)$, we want the x_i to be the zeros of $T_{N+1}(x)$, that is, we want

$$x_i = cos(\theta_i) \tag{81}$$

where $$cos((N+1)\theta_i) = 0. \tag{82}$$

The points such that $$cos((N+1)\theta) = 0 \tag{83}$$

are the points such that $$(N+1)\theta = (2i+1)\pi/2. \tag{84}$$

This produces $N+1$ distinct roots of $T_{N+1}(x)$, namely,

$$x_i = cos\left(\frac{2i+1}{2(N+1)}\pi\right), \quad i = 0,1,\ldots,N. \tag{85}$$

These are the argument values at which the function should be tabulated.

It is instructive to look at the distribution of the tabular arguments in the interval $[-1,1]$ by considering the situation geometrically (see figure 4-7). The angles $(2i+1)\pi/2(N+1)$ are marked by dividing the semicircle into $2(N+1)$ equal angles. We note that the cosine of an angle is the x coordinate of the point on the unit circle where the line demarcating that angle cuts the circle. Thus, x values corresponding to these points for odd angles are the tabular arguments in question. These arguments are symmetric about zero, the middle of the interval. There is no tabular argument at either edge of the interval ($|\psi(x)|$ takes on its

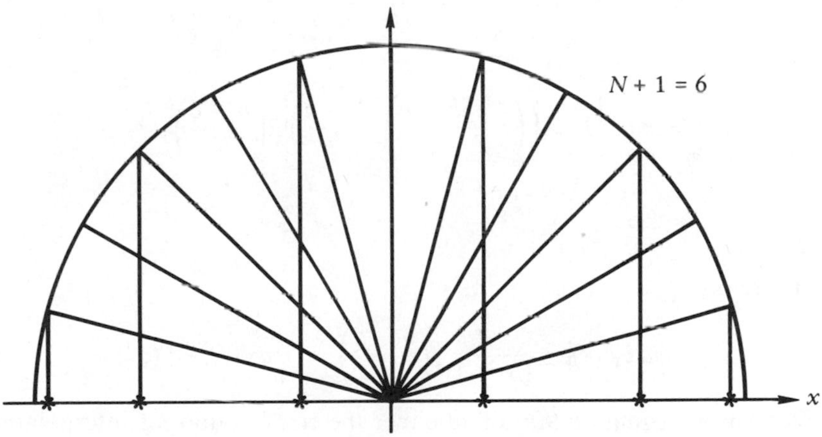

FIG. 4-7 Zeros of Tchebycheff polynomials

interval maximum at both endpoints), and the arguments are more closely packed towards the edges of the interval than towards the middle. This latter behavior compensates for the fact that the extrema of $\psi(x)$ for equally spaced arguments get larger and larger in magnitude towards the edges of the interval covered by the tabular arguments (see exercise 4.2.10). Putting a larger number of tabular arguments (zeros of $\psi(x)$) in the region of these large extrema makes $\psi(x)$ turn towards $y = 0$ sooner there and thus reduces the magnitude of the extrema.

For an interpolation range $[a,b]$ we simply map the x_i, which are the points for the interval $[-1,1]$, into the corresponding points in the interval $[a,b]$ by the inverse of the linear transformation given in equation 65, namely,

$$z_i = \left(\frac{b-a}{2}\right)x_i + \frac{b+a}{2}. \tag{86}$$

That is, the z_i are spaced in the same relative positions in $[a,b]$ that the x_i are spaced in $[-1,1]$.

The error for interpolation using the zeros of the Tchebycheff polynomial as tabular points takes a particularly simple form. From equation 32 we derive

$$\max_{[a,b]} |\varepsilon(x)| \leq \max_{x \in [a,b]} |\psi(x)| \max_{\xi \in [a,b]} |f^{(N+1)}(\xi)|/(N+1)!. \tag{87}$$

But $\psi(x)$, being a monic polynomial, is given by

$$\psi(x) = T_{N+1}\left(\left(x - \frac{b+a}{2}\right)\bigg/\frac{b-a}{2}\right)\bigg/\left(2^N\left(\frac{2}{b-a}\right)^{N+1}\right), \tag{88}$$

so

$$\max_{[a,b]} |\psi(x)| = \max_{[a,b]} \left|T_{N+1}\left(\left(x - \frac{b+a}{2}\right)\bigg/\frac{b-a}{2}\right)\right| 2\left(\frac{b-a}{4}\right)^{N+1}$$

$$= 2\left(\frac{b-a}{4}\right)^{N+1} \tag{89}$$

Therefore,

$$\max_{[a,b]} |\varepsilon(x)| \leq \frac{2}{(N+1)!}\left(\frac{b-a}{4}\right)^{N+1} \max_{[a,b]} |f^{(N+1)}(\xi)|. \tag{90}$$

We can use equation 90 to find either the error bound for interpolation in the interval $[a,b]$ for a given N or the number of points needed to achieve a desired accuracy.

For a given number of tabular points chosen as above, the tabulation must be accurate enough at each tabular point to ensure that the highest-order divided difference has relatively small computational error. Once tabulation has been done, interpolation using these points proceeds as does any interpolation where the tabular arguments are not equally spaced.

We see that choosing the optimum tabular points in the sense of minimizing the maximum magnitude of $\psi(x)$ produces interpolation with unequally spaced points. Unequal interval interpolation also occurs when the tabular points are produced from some process in which the argument values are not controllable or controlled. Another situation in which the arguments are not equally spaced arises when finding the zero of a function given only in tabulated form (see exercise 4.2.18). In this case, we treat the y values as arguments and the x values as function values and interpolate at the point $y = 0$. Since we cannot expect the y values to be equally spaced, here again we have interpolation with unequal intervals.

g) Equal-Interval Methods. As stated above, when error considerations make it necessary for the interval between tabular argument points to be very small compared to the range of tabulated arguments, so that only few points (compared to the total number of tabulated points) are needed for a good interpolation, then the tabular arguments should be equally spaced. In fact, this is the most common case. Equal-interval interpolation methods are the most commonly used.

When all intervals between adjacent tabular arguments are the same and the x_i are subscripted in increasing order of the x_i,

$$x_{i+1} - x_i = h \text{ for all } i. \tag{91}$$

It becomes convenient to rewrite the Newton divided-difference formula in terms of the interval width h. We first define an undivided difference as follows:

$$\Delta f(x) \equiv f(x+h) - f(x). \tag{92}$$

Δ is called the first *forward-difference operator*. As with divided differences, we extend the notion to higher differences by defining

$$\Delta^{i+1} f(x) \equiv \Delta(\Delta^i f(x)) = \Delta^i f(x+h) - \Delta^i f(x). \tag{93}$$

We can now rewrite divided differences in terms of forward differences as follows:

$$f[x_0,x_1] = \frac{f(x_1) - f(x_0)}{x_1 - x_0} = \frac{\Delta f(x_0)}{h}, \tag{94}$$

$$f[x_0,x_1,x_2] = \frac{f[x_1,x_2] - f[x_0,x_1]}{x_2 - x_0} = \frac{\dfrac{\Delta f(x_1)}{h} - \dfrac{\Delta f(x_0)}{h}}{2h} = \frac{\Delta^2 f(x_0)}{2h^2}, \tag{95}$$

and, by induction,

$$f[x_0,x_1,x_2,\ldots,x_{N+1}] = \frac{f[x_1,x_2,\ldots,x_{N+1}] - f[x_0,x_1,\ldots,x_N]}{x_{N+1} - x_0}$$

$$= \left(\frac{\Delta^N f(x_1)}{N!h^N} - \frac{\Delta^N f(x_0)}{N!h^N}\right)\Big/(N+1)h = \Delta^{N+1} f(x_0)/(N+1)!h^{N+1}. \tag{96}$$

We have rewritten x_i in terms of the number of intervals it is from x_0:

$$x_i = x_0 + ih. \tag{97}$$

If we do the same for x, letting s be the number of intervals from x_0 to x (where s is not generally an integer),

$$x = x_0 + sh. \tag{98}$$

If we now rewrite the Newton divided-difference formula, equation 48, entirely in terms of h, we obtain

$$\begin{aligned} f(x) = f(x_0) &+ sh\,\frac{\Delta f(x_0)}{h} + sh(s-1)h\,\frac{\Delta^2 f(x_0)}{2h^2} \\ &+ sh(s-1)h(s-2)h\,\frac{\Delta^3 f(x_0)}{3!h^3} + \cdots \\ &+ sh(s-1)h\ldots(s-(N-1))h\,\frac{\Delta^N f(x_0)}{N!h^N} \\ &+ sh(s-1)h\ldots(s-N)h\,\frac{f^{(N+1)}(\xi)}{(N+1)!},\ \xi \in [x_0,x_N,x]. \end{aligned} \tag{99}$$

With cancellation of powers of h and definition of

$$\binom{s}{i} \equiv \frac{s(s-1)\ldots(s-i+1)}{i!} \tag{100}$$

as a generalization of the binomial factor, equation 99 becomes

$$f(x) = \sum_{i=0}^{N} \binom{s}{i} \Delta^i f(x_0) + \binom{s}{N+1} h^{N+1} f^{(N+1)}(\xi),\ \xi \in [x_0,x_N,x]. \tag{101}$$

Since we are simply rewriting the Newton divided-difference formula, we know that the error term in equation 101 is approximately the next term to be added, that is,

$$\binom{s}{N+1} h^{N+1} f^{(N+1)}(\xi) \approx \binom{s}{N+1} \Delta^{N+1} f(x_0), \tag{102}$$

so
$$f^{(N+1)}(\xi) \approx \Delta^{N+1} f(x_0)/h^{N+1}. \tag{103}$$

Let it be reemphasized that nothing new is being said here; known relations are simply being rewritten in a new notation.

The formula given in equation 101 is called the *Newton forward-difference formula.* It corresponds to a path in the difference table as indicated in table 4-6, where the notation $\Delta^k y_i$ is used to mean $\Delta^k f(x_i)$. But since the interval width is constant, this path is correct (uses the closest arguments first) only if x_0 is the first element in the table and the evaluation argument x is closer to x_0 than any other tabular argument.

TABLE 4-6 NEWTON FORWARD-DIFFERENCE FORMULA

.	.			
.	.	.		
.	.	.	.	
x_{-1}	y_{-1}	.	.	.
		Δy_{-1}	.	.
x_0	y_0		$\Delta^2 y_{-1}$	.
		Δy_0		$\Delta^3 y_{-1}$
x_1	y_1		$\Delta^2 y_0$	
		Δy_1		$\Delta^3 y_0$
x_2	y_2		$\Delta^2 y_1$	
		Δy_2		$\Delta^3 y_1$
x_3	y_3		$\Delta^2 y_2$	.
		Δy_3	.	.
x_4	y_4	.	.	.
.	.	.	.	
.	.	.		
.	.			

TABLE 4-7 Gauss Backward Formula: with Forward Differences

x_0	y_0				
		Δy_0			
x_1	y_1		$\Delta^2 y_0$		
		Δy_1		$\Delta^3 y_0$	
x_2	y_2		$\Delta^2 y_1$		$\Delta^4 y_0$
		Δy_2		$\Delta^3 y_1$	
x_3	y_3		$\Delta^2 y_2$		$\Delta^4 y_1$
		Δy_3		$\Delta^3 y_2$	
x_4	y_4		$\Delta^2 y_3$		
		Δy_4			
x_5	y_5				

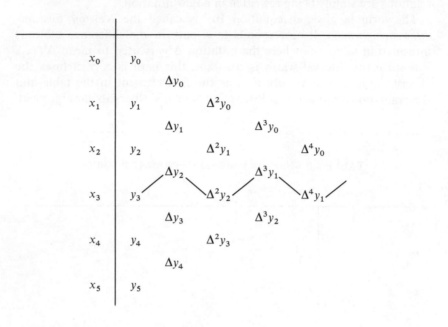

Rather, if as is more common, x is closest to some argument in the middle of the table, say x is closest to x_3, and between x_3 and x_2, then the path given in table 4-7 is most appropriate, because it adds tabular arguments in order of their closeness to x. The formula associated with this path is inconvenient to write because the subscripts on the differences used are not even approximately constant. We can avoid this difficulty by renaming the differences so that the subscript on a difference corresponds to the subscript of the argument in the middle of the interval enclosed by the arguments it uses. We define these *central differences* by

$$\delta f(x) \equiv f(x + h/2) - f(x - h/2). \tag{104}$$

Then

$$\Delta^k y_i = \delta^k y_{i+k/2}, \tag{105}$$

and table 4-7 can be rewritten as table 4-8. Note that we have not changed the values in the table, only the names of the values. An example of a difference table such as that of table 4-7 or 4-8 appears in table 4-9.

TABLE 4-8 GAUSS BACKWARD FORMULA: WITH CENTRAL DIFFERENCES

x_0	y_0				
		$\delta y_{1/2}$			
x_1	y_1		$\delta^2 y_1$		
		$\delta y_{3/2}$		$\delta^3 y_{3/2}$	
x_2	y_2		$\delta^2 y_2$		$\delta^4 y_2$
		$\delta y_{5/2}$		$\delta^3 y_{5/2}$	
x_3	y_3		$\delta^2 y_3$		$\delta^4 y_3$
		$\delta y_{7/2}$		$\delta^3 y_{7/2}$	
x_4	y_4		$\delta^2 y_4$		
		$\delta y_{9/2}$			
x_5	y_5				

As with paths in a divided-difference table, the formula includes a term for each path section. A rule for producing the term corresponding to each path section has been given by Kunz.†

Similarly, we need a rule for determining the truncation error associated with any formula. Since equal-interval difference table paths are equivalent to paths through a divided-difference table, we have, for formulas produced from the former, the same result as we had for the latter: The truncation error depends only on the place at which the path ends. We know that the formula for the path ending at $f[x_j, x_{j+1}, x_{j+2}, \ldots, x_{j+N}]$ has truncation error

$$\varepsilon(x) = -\prod_{i=j}^{j+N} (x - x_i) \frac{f^{(N+1)}(\xi)}{(N+1)!}, \quad \xi \in [x_j, x_{j+N}, x]. \tag{106}$$

† K. S. Kunz, *Numerical Analysis*, section 4.8.

TABLE 4-9 A COMPUTED DIFFERENCE TABLE

x	y	δ	δ^2	δ^3
$-.2$	1.293			
		.052		
0	1.345		.119	
		.171		$-.117$
.2	1.516		.002	
		.173		.029
.4	1.689		.031	
		.204		.002
.6	1.893		.033	
		.237		
.8	2.130			

Rewriting this formula in terms of s and central differences, we find that the error for a path ending at $\delta^k y_i$ is

$$\varepsilon(x) = -\binom{s + k/2 - i}{k+1} h^{k+1} f^{(k+1)}(\xi), \quad \xi \in [x_{i-k/2}, x_{i+k/2}, x]. \tag{107}$$

Many difference formulas have been developed and named. Each one corresponds to a path. Among the most common formulas (paths) are the Newton forward-difference formula (equation 101), which corresponds to a diagonal path with negative slope in the difference table (see table 4-6); the Newton backward-difference formula, which corresponds to a diagonal path with positive slope in the difference table (see exercise 4.2.19); the Gauss forward formula,

$$\hat{p}(x) = y_0 + \binom{s}{1}\delta y_{1/2} + \binom{s}{2}\delta^2 y_0 + \binom{s+1}{3}\delta^3 y_{1/2} + \binom{s+1}{4}\delta^4 y_0 + \binom{s+2}{5}\delta^5 y_{1/2} + \cdots, \tag{108}$$

which corresponds to a path that zigzags with alternately positive and negative slope, starting with a negative slope; and the Gauss backward formula,

$$\hat{p}(x) = y_0 + \binom{s}{1}\delta y_{-1/2} + \binom{s+1}{2}\delta^2 y_0 + \binom{s+1}{3}\delta^3 y_{-1/2} + \binom{s+2}{4}\delta^4 y_0 + \binom{s+2}{5}\delta^5 y_{-1/2} + \cdots, \qquad (109)$$

which corresponds to a path that zigzags like the Gauss forward formula except that it starts with a positive slope (see table 4-8). Either Gauss forward or Gauss backward is the best formula to use among those already covered, because in every case one of them adds the points in order of increasing distance of their arguments from the evaluation argument (see figure 4-4).

The application of the following important general point will give improved results over the Gauss formulas. If we have two estimates of a particular value, each with an error estimate, we can normally do better by taking some function of the two approximations than by taking either one. For example, given three successive iterates from an iterative method for solving nonlinear equations and information about the relation of the errors of these iterates, we were able to take a function of the iterates to get a better value. This procedure was called Aitken's δ^2 acceleration (see section 3.3.4).

With equal-interval interpolation formulas, the Gauss backward and Gauss forward formulas, each starting at the data point whose argument is closest to the evaluation argument, produce different estimates for the same value if their paths end at an odd difference. One estimate is only slightly better than the other. We can produce an improved estimate by taking an appropriate average of the two estimates produced. Note that iterated linear interpolation (equation 63) depends on the fact that taking a weighted average of these estimates produces the approximation equivalent to the one using the next difference (see section 4.2.1, part e). The approximation involving arguments x_{k-j} through x_{k+j} is a weighted sum of the approximation involving arguments x_{k-j+1} through x_{k+j} (for equal intervals, the Gauss forward formula ending at $\delta^{2j-1}x_{k+1/2}$) and the approximation involving arguments x_{k-j} through x_{k+j-1} (for equal intervals, the Gauss backward formula ending at $\delta^{2j-1}x_{k-1/2}$). The weight of the Gauss forward formula is $(x - x_{k-j})/(x_{k+j} - x_{k-j})$, and the weight of the Gauss backward formula is $(x_{k+j} - x)/(x_{k+j} - x_{k-j})$. For equal intervals, as we use higher differences in the table (j increases), the evaluation argument becomes very close to equidistant between x_{k+j} and x_{k-j}, so the weights are approximately equal. Therefore, assuming that increasing the degree of the approximating polynomial (the number of differences used) improves the approximation, we can expect to get an improved approximation by taking the unweighted average of the Gauss forward and Gauss backward formulas.

This notion of the average of a Gauss forward and Gauss backward formula allows us to extend the notion of paths in the difference table. We admit a *horizontal path section,* taking the interpolation formula term corresponding to the horizontal path section to be the average of the two immediately adjacent diagonal path sections (see Kunz).† Thus the formula corresponding to a path with horizontal sections is the average of formulas corresponding to two paths with diagonal sections only.

It is convenient to use an operator to denote the average of two adjacent differences. We define

$$\mu f(x) \equiv (f(x + h/2) + f(x - h/2))/2, \tag{110}$$

so that

$$\mu\delta^i y_j = (\delta^i y_{j+1/2} + \delta^i y_{j-1/2})/2. \tag{111}$$

With this notation, the formula corresponding to the horizontal path starting at y_0, which is the average of the Gauss forward and Gauss backward formulas starting at y_0, produces a formula called the *Stirling formula,* given by

$$\begin{aligned}\hat{p}(x) &= y_0 + \binom{s}{1}\mu\delta y_0 + \frac{\binom{s}{2} + \binom{s+1}{2}}{2}\delta^2 y_0 + \binom{s+1}{3}\mu\delta^3 y_0 \\ &\quad + \frac{\binom{s+1}{4} + \binom{s+2}{4}}{2}\delta^4 y_0 + \cdots \\ &= y_0 + s\mu\delta y_0 + \frac{s^2}{2!}\delta^2 y_0 + \frac{s(s^2-1)}{3!}\mu\delta^3 y_0 + \frac{s^2(s^2-1)}{4!}\delta^4 y_0 \\ &\quad + \frac{s(s^2-1)(s^2-4)}{5!}\mu\delta^5 y_0 + \frac{s^2(s^2-1)(s^2-4)}{6!}\delta^6 y_0 + \cdots. \end{aligned} \tag{112}$$

A horizontal path can also start between two y_i values, in which case the first interpolation formula term is the average of the two y_i values in question. This path corresponds to a formula which is the average of the Gauss forward formula beginning at the first data point in question and the Gauss backward formula beginning at the next data point. It is

† K. S. Kunz, *Numerical Analysis,* section 8.

called the *Bessel formula* and if started between y_0 and y_1 is given by

$$\hat{p}(x) = \mu y_{1/2} + \frac{\binom{s}{1} + \binom{s-1}{1}}{2} \delta y_{1/2} + \binom{s}{2} \mu \delta^2 y_{1/2} + \frac{\binom{s}{3} + \binom{s+1}{3}}{2} \delta^3 y_{1/2} + \cdots$$

$$= \mu y_{1/2} + (s - \tfrac{1}{2}) \delta y_{1/2} + \frac{s(s-1)}{2!} \mu \delta^2 y_{1/2} + \frac{(s-\frac{1}{2})(s)(s-1)}{3!} \delta^3 y_{1/2} + \frac{(s+1)(s)(s-1)(s-2)}{4!} \mu \delta^4 y_{1/2} + \frac{(s-\frac{1}{2})(s+1)(s)(s-1)(s-2)}{5!} \delta^5 y_{1/2} + \cdots. \tag{113}$$

With both the Bessel and Stirling formulas, the coefficient of an average between two differences is simply $(1/n!)\prod_i (s-i)$, where the i are the subscripts of the arguments associated with the difference in the column which preceded both differences in the average and n is the number of factors in the product. The coefficient of a difference that is not averaged is $(1/n!)\dfrac{(s-i_{max}) + (s-i_{min})}{2} \cdot \prod_{i=i_{min}+1}^{i_{max}-1} (s-i)$, where i_{max} and i_{min} are the subscripts of the elements in the 0th-difference column obtained by drawing diagonals from the difference in question to that column, and $n = i_{max} - i_{min}$.

Note that the Stirling and Bessel formulas are not strictly exact-matching formulas but rather the average of two such formulas. They match exactly at all tabular arguments but the two extreme ones used. Since these formulas are the average of two exact-matching formulas, the error in the horizontal-path formula is the average of the errors associated with the two exact-matching formulas. That is, if

$$\hat{p}(x) = (p_1(x) + p_2(x))/2 \tag{114}$$

and

$$p_1(x) = f(x) + \varepsilon_1(x) \tag{115}$$

and

$$p_2(x) = f(x) + \varepsilon_2(x), \tag{116}$$

then

$$\hat{p}(x) = f(x) + (\varepsilon_1(x) + \varepsilon_2(x))/2. \tag{117}$$

$\varepsilon_1(x)$ and $\varepsilon_2(x)$ depend only on the places where the paths corresponding to $p_1(x)$ and $p_2(x)$, respectively, end. If the two paths end at the same place (the horizontal path for $p(x)$ ends on a difference rather than between two differences),

$$\varepsilon_1(x) = \varepsilon_2(x) = (\varepsilon_1(x) + \varepsilon_2(x))/2. \tag{118}$$

Therefore, the error for any path (including a horizontal path) ending at $\delta^k y_i$ is given by equation 107.

If a path ends between two differences, the error is the average of the errors associated with the differences immediately above and below the path-ending. Thus, for example, the error in the Stirling formula approximation ending at $\mu\delta^{2n+1}y_0$ is

$$\varepsilon(x) = -\frac{1}{2}\left[\binom{s + \frac{2n+1}{2} - \left(-\frac{1}{2}\right)}{2n+2} h^{2n+2} f^{(2n+2)}(\xi_1)\right.$$

$$\left. + \binom{s + \frac{2n+1}{2} - \frac{1}{2}}{2n+2} h^{2n+2} f^{(2n+2)}(\xi_2)\right],$$

$$\text{where} \quad \xi_1 \in [x_{-(n+1)}, x_n, x] \quad \text{and} \quad \xi_2 \in [x_{-n}, x_{n+1}, x]$$

$$= \frac{-h^{2n+2}}{2}\left[\binom{s+n+1}{2n+2} f^{(2n+2)}(\xi_1) + \binom{s+n}{2n+2} f^{(2n+2)}(\xi_2)\right]$$

$$= -\frac{h^{2n+2}}{2}\binom{s+n}{2n+1}\frac{1}{2n+2}$$

$$\times [(s+n+1) f^{(2n+2)}(\xi_1) + (s-(n+1)) f^{(2n+2)}(\xi_2)]. \tag{119}$$

If we assume that

$$h^{2n+2} f^{(2n+2)}(\xi_1) \approx h^{2n+2} f^{(2n+2)}(\xi_2) \approx \delta^{2n+2} y_0, \tag{120}$$

equation 119 becomes

$$\varepsilon(x) \approx -\binom{s+n}{2n+1}\frac{1}{2n+2}\, s\delta^{2n+2} y_0. \tag{121}$$

This error term is, as with all other difference formulas, simply the next term to be added in the series. To *bound* the error in the Stirling formula ending at $\mu\delta^{2n+1}y_0$, one must take the maximum of the right side of equation 119, choosing ξ_1 and ξ_2 to make the factor in brackets (on the last line of the equation) as large as possible in magnitude. Because the Stirling formula is used only for $0 \leq s \leq 1$ (otherwise x_0, the base of

Stirling's formula, would be a different tabular argument), the functions of s in the error factor are of opposite sign, so we maximize the magnitude of the factor in brackets by choosing ξ_1 and ξ_2 so that one derivative takes its maximum over the interval associated with it and the other takes its minimum over its interval.

The error in the Bessel formula is developed in the same way as that in Stirling's, above. This development is left for the student (see exercise 4.2.20).

It has been implied that the horizontal-path formulas, which are an average between two zigzag-path formulas, are more accurate than the zigzag-path formulas. That this is in fact the case when the $(N+1)$th derivative over the range of used arguments is approximately constant is shown below.

Assume that for a given x there are enough tabular points on either side of x so that enough differences are available to allow the choice of the optimum path. Let x_0 be the largest tabular argument less than x. Then $0 \leq s \leq 1$, and the choice of formulas comes down to the Gauss forward beginning at y_0, Gauss backward beginning at y_1, Stirling beginning at y_0, Stirling beginning at y_1, and Bessel beginning at $y_{1/2}$. Assume we know that because of the behavior of the differences our approximation formula will end on an even difference, say the $(2k)$th. Then Gauss forward beginning at y_0 would end at $\delta^{2k}y_0$, Gauss backward beginning at y_1 would end at $\delta^{2k}y_1$, Stirling beginning at y_0 would end at $\delta^{2k}y_0$, Stirling at y_1 would end at $\delta^{2k}y_1$, and Bessel at $y_{1/2}$ would end at $\mu\delta^{2k}y_{1/2}$. Gauss forward and Stirling beginning at y_0 end at the same point, so they have the same error,

$$|\varepsilon_1(x)| = \left|\binom{s+k}{2k+1} h^{2k+1} f^{(2k+1)}(\xi_1)\right|, \quad \xi_1 \in [x_{-k}, x_k]. \tag{122}$$

Similarly, Gauss backward and Stirling beginning at y_1 have the same error,

$$|\varepsilon_2(x)| = \left|\binom{s+k-1}{2k+1} h^{2k+1} f^{(2k+1)}(\xi_2)\right|, \quad \xi_2 \in [x_{-k+1}, x_{k+1}]. \tag{123}$$

The Bessel formula has error

$$|\varepsilon_3(x)| = \left|\binom{s+k-1}{2k} \frac{1}{2k+1} h^{2k+1}[(s+k) f^{(2k+1)}(\xi_1) + (s-k-1) f^{(2k+1)}(\xi_2)]/2\right|. \tag{124}$$

Assuming all of the derivative terms have the same value, we can write

$$|\varepsilon_1(x)| = \left|\binom{s+k}{2k+1}h^{2k+1}f^{(2k+1)}\right|$$
$$= |(s+k)|\left|\binom{s+k-1}{2k}\frac{1}{2k+1}h^{2k+1}f^{(2k+1)}\right|, \qquad (125)$$

$$|\varepsilon_2(x)| = |(s-k-1)|\left|\binom{s+k-1}{2k}\frac{1}{2k+1}h^{2k+1}f^{(2k+1)}\right|, \quad (126)$$

$$|\varepsilon_3(x)| = \left|\left(s-\frac{1}{2}\right)\right|\left|\binom{s+k-1}{2k}\frac{1}{2k+1}h^{2k+1}f^{(2k+1)}\right|. \quad (127)$$

These errors differ only by the first factors on the right side of the equations. Since $0 \le s \le 1$, this factor in equation 127 is smallest, so we should use the Bessel formula beginning at $y_{1/2}$. This is also true if we do not assume that $f^{(2k+1)}(\xi_1) \approx f^{(2k+1)}(\xi_2)$ but choose the formula for which the error *bound* over all $0 \le s \le 1$ is least (see exercise 4.2.22).

Similarly, it can be shown that if we want to end a path on an odd difference, the Stirling formula should be used beginning at the closer of x_0 and x_1 (see exercise 4.2.21). That is, we wish to use the closest horizontal path which ends *between* two differences.

If we do not know which difference column our path will end on, it can be shown that we should use the horizontal path which starts closest to x. Thus, if $0 \le s \le \frac{1}{4}$, or $\frac{3}{4} \le s \le 1$, a Stirling formula should be used; otherwise, Bessel (see exercise 4.2.23).

It must be reemphasized that truncation error depends only on the difference at which a path ends. Thus, if the desired path-ending is known, any formula that ends at that point will do the job. The horizontal paths are helpful only in allowing us to choose a path for which every point along the path is an optimum or near-optimum position to stop, for that difference column.

We know when a path should end by the same criteria that we applied with the divided-difference formula: We stop if the terms added begin to grow or if the computational error in the differences becomes approximately equal to the differences themselves. This latter behavior is detected by noting when the computed differences fluctuate in sign in a column section near a difference to be used, or by estimating the error propagated into a difference to be used. We accomplish the estimation by estimating the error in the y_i and then noting that an error ε_i in y_i propagates into differences in the Pascal's triangle beginning at y_i, with propagated error $(-1)^k\binom{k}{j}\varepsilon_i$ in $\delta^k y_{i-k/2+j}$, $0 \le j \le k$ (see exercise 4.2.24, part a). Thus, if each y_i has an error bounded in magnitude by b, $\delta^k y_j$ has a propagated error bounded by $2^k b$ for all j (exercise 4.2.24, part b).

As an example, consider the difference table in table 4-10. The differences

TABLE 4-10 An Equal-Interval Difference Table

x	y	Δ	Δ^2	Δ^3	Δ^4	Δ^5
.2	.0034					
		103				
.4	.0137		−10			
		93		2		
.6	.0230		−8		9	
		85		11		1
.8	.0315		3		10	
		88		21		−2
1.0	.0403		24		8	
		112		29		−2
1.2	.0515		53		6	
		165		35		−2
1.4	.0680		88		4	
		253		39		−2
1.6	.0933		127		2	
		380		41		−1
1.8	.1313		168		1	
		548		42		−2
2.0	.1861		210		−1	
		758		41		3
2.2	.2619		251		2	
		1009		43		−4
2.4	.3628		294		−2	
		1303		41		3
2.6	.4931		335		1	
		1638		42		
2.8	.6569		377			
		2015				
3.0	.8584					

are not computed beyond the fifth difference because there is sign fluctuation in all areas of the Δ^5 column, so no formula should use differences beyond the fourth. By our criteria, the fourth differences in the top part of the table may be used, but those in the bottom part fluctuate in sign so should not be used.

Calculating bounds on the error in the differences supports these decisions made on the basis of sign fluctuation. The error in the y_i are bounded by 5×10^{-5}. Thus, the propagated error in the values in the Δ^3 column is bounded by $2^3 \times 5 \times 10^{-5} = 4 \times 10^{-4}$, which is less in magnitude than the values in that column. For the Δ^4 column, the propagated error is bounded by $2^4 \times 5 \times 10^{-5} = 8 \times 10^{-4}$. The probable error magnitude is less than this, approximately 2.4×10^{-4} (exercise 4.2.24, part c). Thus, the error in the values in the Δ^4 column has approximately the same magnitude as the differences in the bottom part of that column but is smaller than the differences at the top. The error in the values in the Δ^5 column, bounded by 1.6×10^{-3}, is larger than all values in that column.

We conclude that to compute $\hat{f}(2.46)$ with maximum accuracy we should stop at the third difference, whereas to compute $\hat{f}(.82)$ with maximum accuracy we should stop at the fourth difference, at least as far as computational error considerations are concerned. Thus, to compute $\hat{f}(2.46)$ with maximum accuracy, we should use the Stirling formula beginning at $x = 2.4$; to compute $\hat{f}(.82)$ to maximum accuracy, we should use the Bessel formula beginning at $x = .9$.

To compute $\hat{f}(1.42)$ with an error less than 10^{-4}, the Stirling formula beginning at $x = 1.4$ should be used, because 1.42 is closer to 1.4 than to 1.5. To compute $\hat{f}(1.48)$ with an error less than 10^{-4}, the Bessel formula beginning at $x = 1.5$ should be used, because 1.48 is closer to 1.5 than to 1.4. In these cases we do not know which difference we will end on, so we use the closest horizontal path to x.

All of the interpolation formulas above have been written in terms of differences of the y_i. They can also be written in terms of the y_i themselves (as was the Lagrange formula). Though this is not convenient with interpolation formulas, it is desirable with numerical integration formulas. The translation from differences to y_i values is accomplished using the relation

$$\Delta^k y_i = \sum_{j=0}^{k} \binom{k}{j} (-1)^j y_{i+k-j}. \tag{128}$$

This formula can be more mnemonically written as

$$\Delta^k y_i = (E - 1)^k y_i, \tag{129}$$

where $(E-1)^k$ is formally expanded as a binomial series and

$$E^j y_i \equiv y_{i+j}.\dagger \tag{130}$$

Since polynomial exact matching with equal intervals is such a common procedure, we shall comment further about the polynomial $\psi(x)$, which occurs in the error term. $\psi(x)$ is a monic $(N+1)$th-degree polynomial with equally spaced zeros. We can show that (1) ψ is symmetric in magnitude (it is either even or odd) about the center of the tabular interval enclosed by the zeros, (2) it has a single relative maximum in magnitude between each zero, and (3) the value it takes on at each such maximum increases strongly with the distance from the center of the interval in question (see figure 4-6, curve a, and exercise 4.2.10). Therefore, when more than just a few points are used for equal-interval exact-matching interpolation, except when $f^{(N+1)}(\xi(x))$ is very peculiarly behaved, the following behavior occurs:

1. The error reflects the large oscillations of $\psi(x)$, so the approximating polynomial, which is the sum of $f(x)$, a function assumed to be smooth, and this error, is nonsmooth (see figure 4-8).
2. The error for an evaluation argument near the edge of the interval enclosed by the data points is usually much larger than that for an evaluation argument near the center.

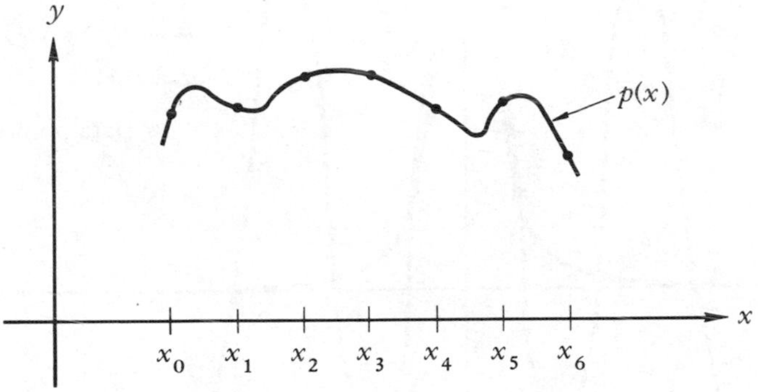

FIG. 4-8 Common behavior of high-degree polynomial fit

† The formal manipulation of the shifting operator E and other linear operators such as Δ, δ, μ, and the derivative operator D, can be made more than a mnemonic aid and can be justified algebraically. This matter will not be covered in this book.

The second statement implies the unsuitability of a single equal-interval exact-matching approximation formula through many data points for use with evaluation arguments across the whole interval enclosing the data arguments. The behavior described above also suggests that there exist smooth functions such that the equal-interval exact-matching approximation using data points with arguments over a fixed range diverges from the function in that range as the interargument interval width approaches zero. For example, $f(x) = 1/(1 + 25x^2)$ over $[-1, 1]$ is such a function (see figure 4-9). It should not surprise us that there is difficulty in fitting this function with a polynomial, because the function has the non-polynomial-like behavior of approaching an asymptote in the interval in question.

A polynomial with equally spaced zeros cannot have equal ripples. As we have seen, this behavior can cause considerable difficulty when doing exact-matching approximation over a large number of data points with equally spaced arguments. We will be interested in approximation using sinusoids, because they are equal-ripple functions with equally spaced zeros. Sinusoids are a better set of functions for approximation than the polynomials from the point of view of smoothness of fit.

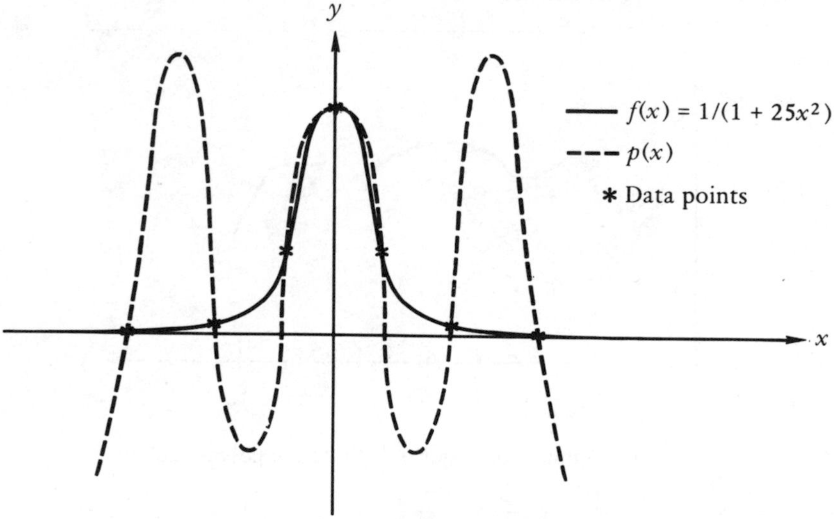

FIG. 4-9 Polynomial exact matching to non-polynomial-like function

h) Splines. At the beginning of this chapter, we recognized the desirability of approximations that are both smooth and optimum according to some criterion of closeness to the data points. Exact-matching polynomials often become unsmooth as their degree increases, because large higher-order differences corresponding to large higher-order derivatives produce large higher-order coefficients, which in turn produce large oscillations. On the other hand, if a table is fit piecewise by lower-order polynomials, the approximation at the interface between two approximations is not smooth in the sense that the slope and curvature are discontinuous at the interface (see figure 4-10). However, we can produce exact-matching piecewise fits which are smooth in slope (first derivative) and curvature (second derivative) at the interface. These piecewise fits are called *splines*. The notion of producing smooth piecewise fits can be restricted to requiring the continuity of only the first derivative at the interface or extended to require higher-order derivative continuity. We will treat the problem of first and second derivative continuity. (There also must be continuity of the approximating function due to the exact-matching requirement.)

Assume we have a table to be fit by polynomial splines in such a way that we have a different spline for each tabular interval (not necessarily of equal width). Assume further that we have fit the first m tabular intervals and wish to compute the spline for the next interval. We place four constraints upon this spline.

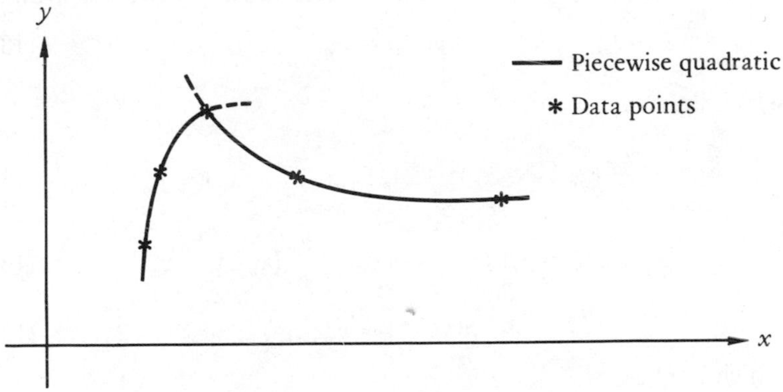

FIG. 4-10 Discontinuity of derivative and curvature in piecewise fit

1. It must agree with the data point at the left end of the interval.
2. It must agree with the data point at the right end of the interval.
3. It must agree with the previous spline in slope at the left end of the interval.
4. It must agree with the previous spline in curvature at the left end of the interval.

Since there are four constraints, a polynomial with four parameters, that is, a cubic polynomial, is required.

Let $p_{m+1}(x)$ be the cubic polynomial fit to the $(m+1)$th interval, $[x_m, x_{m+1}]$. If we let the curvature at x_m be c_m, and let c_m and c_{m+1} be parameters of $p_{m+1}(x)$, the specification of $p_{m+1}(x)$ becomes straightforward. Since $p_{m+1}(x)$ is a cubic polynomial, $p''_{m+1}(x)$ is a linear polynomial constrained to have the value c_m at x_m and c_{m+1} at x_{m+1}. Thus, $p''_{m+1}(x)$ is the exact-matching line,

$$p''_{m+1}(x) = c_m \frac{x_{m+1} - x}{x_{m+1} - x_m} + c_{m+1} \frac{x - x_m}{x_{m+1} - x_m}. \quad (131)$$

Integrating equation 131 twice produces

$$p_{m+1}(x) = \frac{c_m}{6} \frac{(x_{m+1} - x)^3}{x_{m+1} - x_m} + \frac{c_{m+1}}{6} \frac{(x - x_m)^3}{x_{m+1} - x_m} + A_{m+1}(x_{m+1} - x) + B_{m+1}(x - x_m), \quad (132)$$

where the arbitrary linear term added is written in the form $A_{m+1}(x_{m+1} - x) + B_{m+1}(x - x_m)$ for computational simplicity in the solutions for A and B. These constants are determined by two constraints:

1. $$p(x_m) = y_m, \quad (133)$$

   which implies

   $$\frac{c_m}{6}(x_{m+1} - x_m)^2 + A_{m+1}(x_{m+1} - x_m) = y_m, \quad (134)$$

   so

   $$A_{m+1} = y_m/(x_{m+1} - x_m) - \frac{c_m}{6}(x_{m+1} - x_m). \quad (135)$$

2. $$p(x_{m+1}) = y_{m+1}, \quad (136)$$

   which implies

   $$\frac{c_{m+1}}{6}(x_{m+1} - x_m)^2 + B_{m+1}(x_{m+1} - x_m) = y_{m+1}, \quad (137)$$

   so

   $$B_{m+1} = y_{m+1}/(x_{m+1} - x_m) - \frac{c_{m+1}}{6}(x_{m+1} - x_m). \quad (138)$$

The only constraint left to be satisfied, continuity of the first derivative at x_m,

$$p'_{m+1}(x_m) = p'_m(x_m), \tag{139}$$

determines a relation among c_{m+1}, c_m, and c_{m-1}. Differentiating equation 132 produces

$$p'_{m+1}(x) = -\frac{c_m}{2}\frac{(x_{m+1} - x)^2}{x_{m+1} - x_m} + \frac{c_{m+1}}{2}\frac{(x - x_m)^2}{x_{m+1} - x_m} + \frac{y_{m+1} - y_m}{x_{m+1} - x_m} - \frac{c_{m+1} - c_m}{6}(x_{m+1} - x_m). \tag{140}$$

Equation 139 becomes

$$-\frac{c_m}{2}(x_{m+1} - x_m) + \frac{y_{m+1} - y_m}{x_{m+1} - x_m} - \frac{c_{m+1} - c_m}{6}(x_{m+1} - x_m) = \frac{c_m}{2}(x_m - x_{m-1}) + \frac{y_m - y_{m-1}}{x_m - x_{m-1}} - \frac{c_m - c_{m-1}}{6}(x_m - x_{m-1}). \tag{141}$$

Simplifying equation 141 produces the following relation between the curvatures:

$$c_{m+1}(x_{m+1} - x_m) + 2c_m(x_{m+1} - x_{m-1}) + c_{m-1}(x_m - x_{m-1}) = 6\left[\frac{y_{m+1} - y_m}{x_{m+1} - x_m} - \frac{y_m - y_{m-1}}{x_m - x_{m-1}}\right]. \tag{142}$$

Note that for the whole range of the table, we are missing two constraints, namely, the curvature c_0 and slope s_0 at x_0. Given these, we can differentiate $p_1(x)$ at x_0 and set $p'_1(x_0) = s_0$ to produce

$$-\frac{c_0}{2}(x_1 - x_0) + \frac{y_1 - y_0}{x_1 - x_0} - \frac{c_1 - c_0}{6}(x_1 - x_0) = s_0, \tag{143}$$

which can be solved for c_1. Given c_0 and c_1, equation 142 can be applied recursively to produce $c_2, c_3, \ldots, c_N$. These c_i values can be substituted in equation 132 to give the formula for each spline.

Similarly, we could specify c_N and s_N and use the recurrence relation to work backwards. There are an infinite number of such pairs of boundary conditions which could be used to completely specify the splines. The most common is that the curvature is zero at both endpoints, x_0 and x_N, because the splines thus produced are the smoothest of all exact-matching fits, where maximum smoothness is defined as minimum average curvature, and average is defined according to the $\mathscr{L}_2$ norm. We prove this statement as follows:

Theorem: The cubic polynomial spline fit with $c_0 = c_N = 0$ is the function among all twice-differentiable functions $f(x)$ which exact match at x_i, $0 \le i \le N$, for which the average $\left[\frac{1}{x_N - x_0}\int_{x_0}^{x_N} (f''(x))^2 dx\right]^{1/2}$ is minimum.

Proof: The $\mathscr{L}_2$ average above is minimum when $\int_{x_0}^{x_N} [f''(x)]^2 dx$ is minimum. Let $g(x)$ be the cubic spline fit in question, and let $f(x)$ be any other twice-differentiable function over $[x_0, x_N]$. Then

$$\int_{x_0}^{x_N} [f''(x)]^2\, dx = \int_{x_0}^{x_N} [g''(x) + (f''(x) - g''(x))]^2\, dx$$

$$= \int_{x_0}^{x_N} [g''(x)]^2\, dx + 2\int_{x_0}^{x_N} g''(x)\, [f''(x) - g''(x)] dx$$

$$+ \int_{x_0}^{x_N} [f''(x) - g''(x)]^2\, dx. \qquad (144)$$

The middle integral on the right side of equation 144 can be integrated by parts to produce

$$I \equiv \int_{x_0}^{x_N} g''(x)[f''(x) - g''(x)]dx$$

$$= g''(x)[f'(x) - g'(x)]\Big|_{x_0}^{x_N} - \int_{x_0}^{x_N} [f'(x) - g'(x)]g'''(x)dx$$

$$= c_N[f'(x_N) - g'(x_N)] - c_0[f'(x_0) - g'(x_0)]$$

$$- \sum_{i=0}^{N-1} \int_{x_i}^{x_{i+1}} [f'(x) - g'(x)]K_i\, dx, \qquad (145)$$

since $g'''(x)$ is a constant (K_i) in $[x_i, x_{i+1}]$ because g is a cubic there. Since $c_N = c_0 = 0$,

$$I = -\sum_{i=0}^{N-1} K_i[f(x) - g(x)]\Big|_{x_i}^{x_{i+1}} = 0, \qquad (146)$$

because $f(x_i) = g(x_i)$ for all i by the exact-matching requirement on f and g. Thus, equation 144 becomes

$$\int_{x_0}^{x_N} [f''(x)]^2\, dx = \int_{x_0}^{x_N} [g''(x)]^2\, dx + \int_{x_0}^{x_N} [f''(x) - g''(x)]^2\, dx. \qquad (147)$$

Since the integrand of the second integral on the right side of equation 147 is always positive, we have

$$\int_{x_0}^{x_N} [f''(x)]^2\,dx \geq \int_{x_0}^{x_N} [g''(x)]^2\,dx. \qquad (148)$$

Note that in equation 148 equality holds only if $f''(x) = g''(x)$ for all $x \in [x_0, x_N]$ except for some discrete values of x.

Thus, according to the above measure of smoothness, the cubic spline fit with $c_0 = c_N = 0$ is the *smoothest exact-matching fit* through the data points. To obtain this fit, we must use these initial conditions and the recurrence relation given by equation 142 to produce the values of the c_i needed in equation 132 to produce the desired cubic splines. The unknowns are $c_1, c_2, \ldots, c_{N-1}$. The recurrence relation applied for $m = 1, 2, \ldots, N-1$ produces $N-1$ linear equations in these $N-1$ unknowns. Since the mth equation involves only c_m, c_{m-1}, and c_{m+1}, the matrix for these linear equations has only a diagonal, a superdiagonal, and a subdiagonal with nonzero elements. Such a matrix is called a *tridiagonal matrix*. The solution of linear equations involving a tridiagonal matrix is especially easy (see exercise 2.8.4).†

Once the c_i have been found, the approximation at an evaluation argument x can be computed (see program 4-4). We determine the interval $[x_m, x_{m+1}]$ in which x falls and thus determine m. Then we evaluate a form of equation 132 which has been algebraically simplified using the definitions $h \equiv x_{m+1} - x_m$, $s \equiv (x - x_m)/h$, $a = c_m h^2/6$, and $b - c_{m+1} h^2/6$.

```
/* CUBIC SPLINE EXACT-MATCHING APPROXIMATION */
/* GIVEN ARE N+1 DATA POINTS, (X(I), Y(I)), Ø <= I <= N; */
/* N; AN EVALUATION ARGUMENT, XEVAL; */
/* AND N+1 CURVATURE VALUES, C(I), AT THE X(I) */
/* MOST LIKELY, THE X(I), THE Y(I), AND TWO OF THE C(I) */
/* VALUES WERE GIVEN TO A PREVIOUS PROGRAM, */
/* AND IT COMPUTED THE REMAINING C(I) */
    DECLARE (X(Ø:N),        /* N+1 DATA ARGUMENTS */
             Y(Ø:N),        /* N+1 DATA VALUES */
             C(Ø:N),        /* N+1 CURVATURE VALUES */
             XEVAL,         /* EVALUATION ARGUMENT */
```

PROGRAM 4-4 Cubic spline exact matching

(*continued on next page*)

† See also S. D. Conte, *Elementary Numerical Analysis,* section 5.6.

```
              APPROX,     /* APPROXIMATION AT EVALUATION ARG */
              H,          /* WIDTH OF PARTICULAR INTERVAL */
              S,          /* FRACTION OF INTERVAL AND */
              SCOMP,      /* ITS COMPLEMENT */
              A, B, TEMP) FLOAT;

/* FIND INTERVAL IN WHICH XEVAL FALLS */
   FINDINT: DO I = 1 TO N-1;
              IF XEVAL < X(I) THEN GO TO INTFOUND;
   END FINDINT; /* IF XEVAL > X(N-1) THEN I = N */

/* EVALUATE POLYNOMIAL FOR INTERVAL (X(I-1),X(I)) AT XEVAL */
   INTFOUND: H = X(I) - X(I-1);
               S = (XEVAL - X(I-1)) / H;
               SCOMP = 1 - S;
               TEMP = H * H / 6;
               A = C(I-1) * TEMP;
               B = C(I) * TEMP;
   /* FINAL RESULT */
               APPROX = (A * SCOMP * SCOMP + Y(I-1) - A)
                          * SCOMP + (B * S * S + Y(I) - B) * S;
```

PROGRAM 4-4 (*continued*)

i) Economization. Sometimes, having made a polynomial exact-matching fit with a single polynomial through a set of points, we are unwilling to use this approximation because the computation of this polynomial is too time-consuming, because the degree of the polynomial is high. We need a lower-degree polynomial as close as possible to the exact-matching approximation. Clearly, this lower-degree polynomial cannot exact match, but we want it to be close to the approximating polynomial not only at the data points but in fact at every point in the interval enclosed by the tabular arguments used in the approximation. If we define closeness according to the $\mathscr{L}_\infty$ norm, that is, if we minimize the maximum difference from the original approximating polynomial, we find that the best $(N-1)$th-degree approximation to the Nth-degree exact-matching polynomial differs from the exact-matching approximation by a multiple of a polynomial which is a Tchebycheff polynomial linearly mapped from $[-1,1]$ onto the interval enclosed by the data points.

We prove the above claim by noting that the difference between the Nth-degree polynomial exact-matching approximation and any $(N-1)$th-degree polynomial is an Nth-degree polynomial, assuming the Nth-degree approximating polynomial has a nonzero coefficient of x^N. We wish this difference polynomial to be minimax over the interval in question among all Nth-degree polynomials with high-order coefficient equal to the coefficient of the exact-matching polynomial. From section 4.2.1, part f, we know this minimax polynomial is $\dfrac{a_N T_N\left(\left(x - \dfrac{x_N + x_0}{2}\right)\Big/\left(\dfrac{x_N - x_0}{2}\right)\right)}{2^{N-1}\left(\dfrac{2}{x_N - x_0}\right)^N}$,

where a_N is the coefficient of x^N in the exact-matching polynomial and x_0 and x_N are the minimum and maximum tabular arguments involved in that polynomial. The maximum value of this difference polynomial on $[x_0,x_N]$, that is, the bound on the magnitude of the interpolation error due to the economization, is $|a_N|(x_N - x_0)^N/2^{2N-1}$.

If we wish a yet more economical polynomial, one of degree $N-2$, we can reapply the above procedure, subtracting a multiple of a mapped version of T_{N-1} from the $(N-1)$th-degree polynomial produced from the first economization step. Thus, the $(N-2)$th-degree polynomial is the one which agrees most closely in the $\mathscr{L}_\infty$ sense in the interval $[x_0,x_N]$ to the $(N-1)$th-degree fit. It is normally not the case that this $(N-2)$th-degree polynomial is the closest to the Nth-degree exact-matching polynomial in the Tchebycheff sense, but it is very close to the polynomial which has the property—a polynomial that is very difficult to find. The process of subtracting multiples of Tchebycheff polynomials of linearly mapped arguments can be repeated until the compromise between the degree of the polynomial and the accuracy of the fit is that desired.

4.2.2 Extrapolation

Extrapolation is the process of fitting a function to the last few points of a table and using that function to obtain values for the tabulated function at arguments outside the tabular argument range. In section 4.2.1, part b, we noted how untrustworthy such a procedure is; emphasis and expansion on that fact is in order.

Extrapolation is dangerous because it involves building a bridge from one bank with no support at the other. One must have a great deal of information at the supported end to extrapolate successfully. The kind of information that is useful is that pertaining to the general shape of the function. Derivative information at the supported end is somewhat useful, but it loses its usefulness fairly quickly as one gets away from the point of support. More useful is information about shape such as that the function approaches an asymptote, is monitonically increasing or decreasing, is oscillating with decreasing ripples, or whatever. Most helpful is the precise functional form. For example, we may know that the function is a linear combination of functions of the form $e^{-\lambda x}$, $\lambda > 0$; or that it is an nth-degree polynomial; or perhaps that it is the quotient of two linear combinations of sinusoids. This kind of information can be obtained only from the physical basis of the function underlying the table. Since a function can go in almost any direction, no matter what its history (recall figure 4-3), we cannot obtain enough information for extrapolation from the table itself. With the kind of information described, extrapolation can be reasonably successful. Without it, it must be avoided unless the evaluation argument is very close to the tabular arguments.

4.2.3 Numerical Differentiation

In section 4.1, we noted that, when an operation is to be carried out on a tabulated function, the strategy is to approximate the table by some function and carry out the operation on the approximating function. This is how one does *numerical differentiation,* that is, differentiation of a tabulated function. One approximates the tabulated function and differentiates the approximation.

If the approximation is done by polynomial exact matching, numerical differentiation simply consists of differentiating the interpolation formulas produced in section 4.2.1. If we have

$$f(x) = \hat{p}(x) - \varepsilon(x), \tag{149}$$

we use the approximation

$$f'(x) \approx \hat{p}'(x) \tag{150}$$

or, more generally,

$$f^{(n)}(x) \approx \hat{p}^{(n)}(x). \tag{151}$$

Differentiating equation 149 produces

$$f^{(n)}(x) = \hat{p}^{(n)}(x) - \varepsilon^{(n)}(x); \tag{152}$$

the error in the nth derivative of f approximated as in equation 151 is the nth derivative of the error in the approximation to f.

An approximation for the derivative can be produced by differentiating any interpolation formula, for example, the Lagrange formula or the Newton divided-difference formula. For equal-interval formulas, we use the relation

$$\frac{d}{dx}g(s) = \frac{d}{ds}g(s)\frac{ds}{dx}. \tag{153}$$

Since

$$x = x_0 + sh, \tag{154}$$

$$\frac{ds}{dx} = \frac{1}{h}. \tag{155}$$

Therefore, if we have an interpolation formula

$$f(x) \approx \hat{q}(s), \tag{156}$$

we use the approximation

$$f'(x) \approx (1/h)\hat{q}'(s). \tag{157}$$

More generally,

$$f^{(n)}(x) \approx (1/h^n)\hat{q}^{(n)}(s). \tag{158}$$

For example, if we fit $f(x)$ by a second-degree polynomial through three equally spaced data points, x_{-1}, x_0, and x_1, Stirling's formula gives

$$\hat{q}(s) = y_0 + s\mu\delta y_0 + (s^2/2)\delta^2 y_0, \tag{159}$$

so the approximation to the first derivative of f is

$$f'(x) \approx (1/h)(\mu\delta y_0 + s\delta^2 y_0). \tag{160}$$

At $x = x_0$, that is, $s = 0$, the approximation is

$$f'(x_0) \approx \frac{1}{h}\mu\delta y_0 = \frac{1}{h}\frac{y_1 - y_{-1}}{2} = \frac{y_1 - y_{-1}}{x_1 - x_{-1}}. \tag{161}$$

We must analyze the error in the approximation to a derivative produced by differentiating an exact-matching polynomial. We will discuss only the first derivative, but our findings will be applicable to higher derivatives by extension. From equations 152 and 30, the error in the approximation to the first derivative is given by

$$\begin{aligned} \varepsilon_{\hat{p}'(x)} &= \frac{d}{dx}\left(-\frac{\psi(x)}{(N+1)!}f^{(N+1)}(\xi(x))\right) \\ &= -\frac{1}{(N+1)!}\left[\psi'(x)f^{(N+1)}(\xi(x)) + \psi(x)\frac{d}{dx}f^{(N+1)}(\xi(x))\right] \end{aligned} \tag{162}$$

We do not know what ξ is for any x, so we might suspect that we cannot compute $\frac{d}{dx}f^{(N+1)}(\xi(x))$. But the following theorem helps us.

Theorem: If $f^{(N+1)}(x)$ is differentiable,

$$\frac{d}{dx}\frac{f^{(N+1)}(\xi(x))}{(N+1)!} = \frac{f^{(N+2)}(\eta(x))}{(N+2)!} \tag{163}$$

for some $\eta \in [x_0, x_1, \ldots, x_N, x]$.

Proof: Consider $\hat{p}_{N+1}(x)$, the Newton divided-difference formula fit through $x_0, x_1, \ldots, x_N$. The error in that formula is given by

$$\varepsilon_{N+1}(x) = -\psi(x) f^{(N+1)}(\xi(x))/(N+1)!, \tag{164}$$

$\xi \in [x_0, x_1, \ldots, x_N, x]$. Thus,

$$f(x) = \hat{p}_{N+1}(x) - \varepsilon_{N+1}(x). \tag{165}$$

But we can produce a polynomial which fits through the x_i, $0 \leq i \leq N$, and any other point x_{N+1} as well by adding another term to the Newton divided-difference formula to obtain $\hat{p}_{N+2}(x)$. Then

$$\begin{aligned} f(x) &= \hat{p}_{N+2}(x) - \varepsilon_{N+2}(x) \\ &= \hat{p}_{N+1}(x) + \psi(x) f[x_0, x_1, x_2, \ldots, x_{N+1}] - \varepsilon_{N+2}(x), \end{aligned} \tag{166}$$

where

$$\psi(x) = \prod_{i=0}^{N} (x - x_i), \tag{167}$$

the polynomial in $\varepsilon_{N+1}(x)$. From equations 165 and 166, it follows that

$$-\varepsilon_{N+1}(x) = \psi(x) f[x_0, x_1, x_2, \ldots, x_{N+1}] - \varepsilon_{N+2}(x), \tag{168}$$

that is,

$$\begin{aligned} \frac{\psi(x) f^{(N+1)}(\xi(x))}{(N+1)!} &= \psi(x) f[x_0, x_1, x_2, \ldots, x_{N+1}] \\ &\quad + \frac{\psi(x)(x - x_{N+1}) f^{(N+2)}(\zeta(x))}{(N+2)!}, \end{aligned} \tag{169}$$

$\zeta \in [x_0, x_1, \ldots, x_{N+1}, x]$.

If x is not a tabular argument, $\psi(x) \neq 0$, so we can divide equation 169 by $\psi(x)$, producing

$$\begin{aligned} \frac{f^{(N+1)}(\xi(x))}{(N+1)!} &= f[x_0, x_1, x_2, \ldots, x_{N+1}] \\ &\quad + (x - x_{N+1}) \frac{f^{(N+2)}(\zeta(x))}{(N+2)!}. \end{aligned} \tag{170}$$

Equation 170 is true even if x is a tabular point by the continuity of $f^{(N+1)}$. But we know that

$$f[x_0, x_1, x_2, \ldots, x_{N+1}] = \frac{f^{(N+1)}(\xi(x_{N+1}))}{(N+1)!}, \tag{171}$$

so we have from equation 170:

$$\frac{\dfrac{f^{(N+1)}(\xi(x))}{(N+1)!} - \dfrac{f^{(N+1)}(\xi(x_{N+1}))}{(N+1)!}}{x - x_{N+1}} = \frac{f^{(N+2)}(\zeta(x))}{(N+2)!}. \quad (172)$$

Since x_{N+1} is any point at all, we can consider the limit of equation 172 as $x_{N+1} \to x$. In the limit, the left side of equation 172 becomes $\dfrac{d}{dx}\dfrac{f^{(N+1)}(\xi(x))}{(N+1)!}$ and $\zeta \to$ some $\eta \in [x_0, x_1, x_2, \ldots, x_N, x]$.

The expression for the truncation error in the approximation to the first derivative, given by equation 162, thus becomes

$$\varepsilon_{\hat{p}'(x)} = -\left[\psi'(x)\frac{f^{(N+1)}(\xi(x))}{(N+1)!} + \frac{\psi(x)f^{(N+2)}(\eta(x))}{(N+2)!}\right], \quad (173)$$

where both ξ and η are in $[x_0, x_1, \ldots, x_N, x]$. Using

$$\psi'(x) = \sum_{j=0}^{N} \prod_{\substack{k=0 \\ k \neq j}}^{N} (x - x_k), \quad (174)$$

we can find an error bound for $\varepsilon_{\hat{p}'(x)}$. In particular, at tabular arguments, $x = x_i$, $\psi(x) = 0$, so the truncation error in the derivative approximation there is

$$\varepsilon_{\hat{p}'(x)} = -\psi'(x_i)\frac{f^{(N+1)}(\xi(x_i))}{(N+1)!} = -\left(\prod_{\substack{k=0 \\ k \neq i}}^{N} (x_i - x_k)\right)\frac{f^{(N+1)}(\xi(x_i))}{(N+1)!}. \quad (175)$$

The truncation error in numerical differentiation can be quite large. If, as is common, $f^{(N+2)}(x)/(N+2)!$ has about the same magnitude as $f^{(N+1)}(x)/(N+1)!$, the second term in the derivative approximation error given by equation 173 has about the same magnitude as the interpolation error for x, and there is an additional term in the error as well. That is, the error in the derivative approximation is commonly larger than the error in interpolation. Furthermore, when $x = x_i$, equation 175 shows that the error in the derivative at x_i is of the same form as the error in the interpolation except that the polynomial in x_i which occurs as a factor is different. We should note that this multiplier is larger than the multiplier in the interpolation error when the evaluation argument x is close to x_i. This reflects the fact that even when a function matches another at certain data points, we cannot necessarily expect good agreement in the derivatives there (see figure 4-11). This fact is particularly true for high-degree polynomial fits where the approximation may not be smooth.

If we cannot improve the quality of the approximation of the derivative by raising the degree of the interpolation polynomial, perhaps we can do so by tabulating the function more finely, that is, so that the tabular arguments are closer together. This procedure might be realistic if the function were given by a subroutine so that we could evaluate f at any argument (the function is not tabulated once and for all), but we did not have the function analytically so we could not differentiate it analytically. However, even this approach is limited in the amount of accuracy that can be obtained, because of problems of computational error. Numerical differentiation formulas have the form: a difference in y divided by a difference in x. For example, fitting a second-degree polynomial to three equally spaced points, x_{-1}, x_0, and x_1, and differentiating this polynomial produces the approximation to the derivative at x_0:

$$\hat{p}'(x_0) = (y_1 - y_{-1})/(x_1 - x_{-1}). \tag{176}$$

As x_1 approaches x_{-1}, y_1 approaches y_{-1}, and thus we are subtracting two numbers with approximately the same value and then dividing this difference by a very small number. We know that the subtraction in the numerator can cause very bad propagation of relative errors generated in the computation of the required function values, y_1 and y_{-1}.

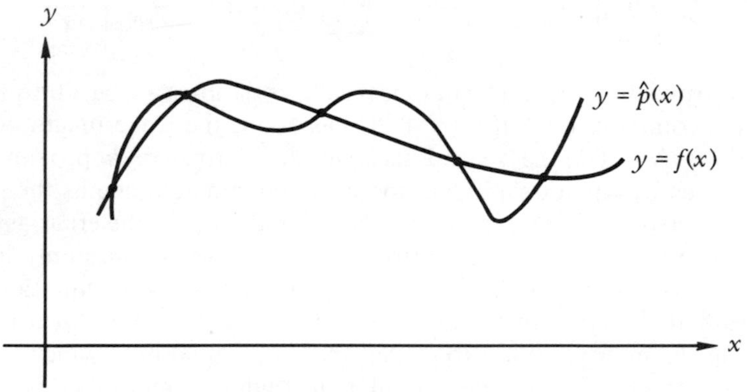

FIG. 4-11 Disagreement of derivative of function and that of exact-matching approximation at tabular arguments

This relative error will become a very large absolute error when the difference is divided by a very small number. For a given accuracy in computing the data values y_i, there is a limit on the extent to which we can make the difference between tabulated arguments small and still decrease the overall error. After some point, the computational error increases more than the truncation error decreases for any further decrease in the distance between tabular arguments. Therefore, there is a limit in the accuracy in the approximation to the derivative (see exercise 4.2.33). This limit increases quickly as the error in the y_i increases, so it is very important to keep the error in the y_i small, or conversely, to avoid computing derivatives using polynomial exact-matching methods when there is significant data error.

All of the arguments above about the inaccuracy of derivative computation hold even more strongly for higher-order derivatives. Therefore, it is very important to avoid numerical differentiation wherever possible. This can be done by doing analytic differentiation when possible and by transforming problems which produce the need for derivatives that must be taken numerically into corresponding integral problems. As we will see, numerical integration is quite an accurate process. Thus, for example, differential equations can be transformed into equivalent integral equations.

4.2.4 Numerical Integration

Numerical integration, sometimes called *quadrature,* involves the integration of an approximating function. It is more accurate than numerical differentiation because it tends to cancel errors in the approximation to the function (see figure 4-12). When one considers that integration is the inverse of differentiation, this behavior seems reasonable. Since differentiation maps small approximation errors into large errors, integration maps large approximation errors into small errors. Furthermore, whereas numerical differentiation has difficulties due to the factor $1/h$, producing a limit on the degree to which the truncation error can usefully be made small, numerical integration has a factor h, which permits us to make the truncation error as small as desired by taking small h, without producing large computational error.

Numerical integration is necessary both when the function is not available analytically and when the function is available analytically but the integral cannot be obtained analytically. In fact, the second case is the more common one. Therefore, we shall discuss methods that assume the argument values to be used can be specified by the methods. In the case of a fixed table, the applicable methods will be restricted.

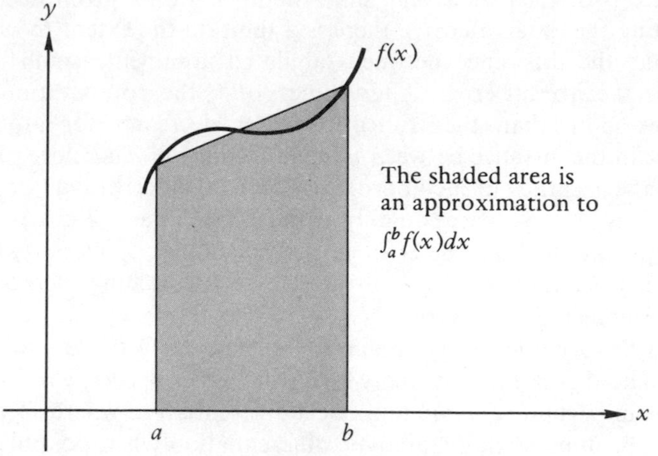

FIG. 4-12 The accuracy of numerical integration; cancellation of approximation errors

a) Newton-Cotes Methods. The Newton-Cotes methods involve the integration of exact-matching polynomial approximation formulas where the arguments are equally spaced. That is, to evaluate

$$I \equiv \int_a^b f(x)dx, \tag{177}$$

we divide the interval $[a,b]$ into m equal intervals, defining $x_0 = a$, $x_m = b$, and

$$x_i = x_0 + i\frac{b-a}{m}, \tag{178}$$

and we fit an mth-degree polynomial at these arguments and integrate that polynomial to produce the approximation $\hat{I}$.

For $m = 0$ the approximating polynomial is the constant y_0. The approximation to the integral is the area of the rectangle of width $b - a$ and height y_0, so

$$\hat{I} = (b-a)y_0. \tag{179}$$

This integration method is called the *rectangular rule*. It is seldom used because it has large error due to the fact that the approximation is based on information at only one end of the interval.

For $m = 1$ the polynomial fitted is a straight line and the approximation to the integral is the area of the trapezoid under this straight line (see figure 4-13). This integration method is called the *trapezoidal rule*. Mathematically, it involves integrating the first two terms of the Newton forward-difference formula at x_0:

$$\hat{I} = \int_{x=x_0}^{x=x_1} (y_0 + s\Delta y_0)dx, \tag{180}$$

where

$$x = x_0 + sh \tag{181}$$

and

$$h = b - a = x_1 - x_0. \tag{182}$$

Transforming equation 180 to an integral in s, we obtain

$$\hat{I} = \int_{s=0}^{s=1} (y_0 + s\Delta y_0)h\,ds = h\left(y_0 + \frac{1}{2}\Delta y_0\right) = \frac{h}{2}(y_0 + y_1). \tag{183}$$

Notice that the right side of equation 183 is precisely the area of a trapezoid with height h and parallel sides of lengths y_0 and y_1.

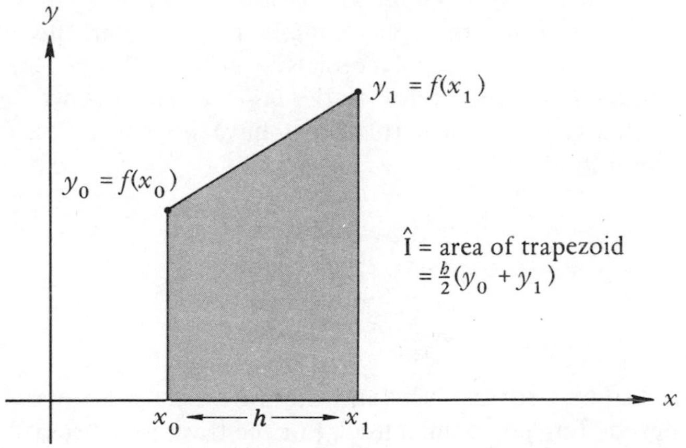

FIG. 4-13 Numerical integration using the trapezoidal rule

If $m = 2$, that is, if we divide the integration range into two intervals, fit a parabola, and integrate it, we have

$$h = (b - a)/2 = (x_2 - x_0)/2 \tag{184}$$

and produce the result

$$\begin{aligned}\hat{I} &= \int_{x_0}^{x_2} \left(y_0 + s\Delta y_0 + \frac{s(s-1)}{2} \Delta^2 y_0 \right) dx \\ &= \int_0^2 \left(y_0 + s\Delta y_0 + \frac{s(s-1)}{2} \Delta^2 y_0 \right) h\, ds = h\left(2y_0 + 2\Delta y_0 + \frac{1}{3}\Delta^2 y_0 \right) \\ &= \frac{h}{3}(y_0 + 4y_1 + y_2).\end{aligned} \tag{185}$$

The integration method given by equation 185 is called *Simpson's rule*.

Higher-order Newton-Cotes rules are developed in the same way. Generally, these rules are written in the form

$$\hat{I} = h \sum_{i=0}^{m} w_i y_i \tag{186}$$

for the mth-order rule. The weights w_i for $1 \le m \le 8$ are given in table 4-11. The w_i for the mth-order rule can also be developed by a method of undetermined coefficients, depending on the fact that the mth-order rule must have no truncation error for any mth-degree or lower polynomial because the fitted polynomial equals the underlying polynomial by the uniqueness of exact-matching polynomials. In particular, the mth-order rule must be true for the polynomials $x^0, x^1, \ldots, x^m$, and these $m+1$ constraints determine the values for the $m+1$ unknowns, w_i, $0 \le i \le m$. Moreover, the fact that each rule must have no truncation error for $f(x) = x^0 = 1$ implies that

$$\int_a^b 1\, dx = \sum_{i=0}^{m} w_i \cdot 1, \tag{187}$$

that is,

$$\sum_{i=0}^{m} w_i = b - a. \tag{188}$$

The truncation error in a Newton-Cotes rule is the integral of the error in the interpolating polynomial used. For the trapezoidal rule

$$\begin{aligned}\varepsilon_{trap} &= -\int_{x_0}^{x_1} \frac{\psi(x) f^{(2)}(\xi(x))}{2!} dx \\ &= -\int_{x_0}^{x_1} (x - x_0)(x - x_1) \frac{f^{(2)}(\xi(x))}{2} dx, \quad \xi \in [x_0, x_1].\end{aligned} \tag{189}$$

TABLE 4-11 Weights and Error Coefficients for Simple Newton-Cotes Rules

$$\hat{I} = \sum_{i=0}^{m} w_i y_i; \quad w_i = w_{m-i}, \quad i > m/2$$

$$\varepsilon_{\hat{I}} = \begin{cases} k_m h^{m+3} f^{(m+2)}(\eta), & \eta \subset [x_0, x_m], \quad \text{if } m \text{ even} \\ k_m h^{m+2} f^{(m+1)}(\eta), & \eta \in [x_0, x_m], \quad \text{if } m \text{ odd} \end{cases}$$

Polynomial Degree, or Number of Intervals m	Weights w_i i: 0	1	2	3	4	Error Coefficient k_m
1	$\frac{1}{2}$					$\frac{1}{12}$
2	$\frac{1}{3}$	$\frac{4}{3}$				$\frac{1}{90}$
3	$\frac{3}{8}$	$\frac{9}{8}$				$\frac{3}{80}$
4	$\frac{14}{45}$	$\frac{64}{45}$	$\frac{24}{45}$			$\frac{8}{945}$
5	$\frac{95}{288}$	$\frac{375}{288}$	$\frac{250}{288}$			$\frac{275}{12,096}$
6	$\frac{41}{140}$	$\frac{216}{140}$	$\frac{27}{140}$	$\frac{272}{140}$		$\frac{9}{1,400}$
7	$\frac{5,257}{17,280}$	$\frac{25,039}{17,280}$	$\frac{9,261}{17,280}$	$\frac{20,923}{17,280}$		$\frac{8,183}{518,400}$
8	$\frac{3,956}{14,175}$	$\frac{23,552}{14,175}$	$-\frac{3,712}{14,175}$	$\frac{41,984}{14,175}$	$-\frac{18,160}{14,175}$	

Source: R. W. Hamming, *Numerical Methods for Scientists and Engineers,* page 157.

$\psi(x)$ maintains constant sign over the integration interval, so the integral mean-value theorem applied to the last integral in equation 189 implies

$$\varepsilon_{trap} = -\frac{f^{(2)}(\eta)}{2}\int_{x_0}^{x_1}(x-x_0)(x-x_1)dx, \quad \eta\in[x_0,x_1]$$

$$= -\frac{f^{(2)}(\eta)}{2}\int_0^1 hs\,h(s-1)\,h\,ds = h^3 f^{(2)}(\eta)/12. \tag{190}$$

The error in Simpson's rule integration is obtained in a similar way. However, a surprising result occurs. Instead of having truncation error of order h^4 as we might expect (one h from $dx = h\,ds$ and one from each of three factors $(x - x_i) = h(s - i)$ in $\psi(x)$ in the interpolation error term), the quadrature truncation error is of order h^5, because

$$\int_{x_0}^{x_2}\psi(x)dx = 0. \tag{191}$$

This result follows from writing $f(x)$ as the sum of a third-degree interpolation formula through x_0, x_1, x_2, and $x_1 + h\varepsilon$, and its approximational error:

$$f(x) = \left(y_0 + s\Delta y_0 + \frac{s(s-1)}{2}\Delta^2 y_0\right)$$
$$+ (x-x_0)(x-x_1)(x-x_2)f[x_0,x_1,x_2,x_1+h\varepsilon]$$
$$+ \frac{(x-x_0)(x-x_1)(x-x_2)(x-(x_1+h\varepsilon))}{4!}f^{(4)}(\xi(x)),$$
$$\xi\in[x_0,x_2,x]. \tag{192}$$

Noting that the first three terms of the third-degree interpolation formula constitute the second-degree interpolation formula which is integrated to produce Simpson's rule, we conclude that the approximational error in Simpson's rule is the integral of the last two terms on the right side of equation 192:

$$-\varepsilon_{Simp} = \int_{x_0}^{x_2}(x-x_0)(x-x_1)(x-x_2)f[x_0,x_1,x_2,x_1+h\varepsilon]dx$$
$$+ \int_{x_0}^{x_2}(x-x_0)(x-x_1)(x-x_2)(x-(x_1+h\varepsilon))\frac{f^{(4)}(\xi(x))}{4!}dx,$$
$$\xi(x)\in[x_0,x_1]. \tag{193}$$

The first integral in equation 193 is 0, because, as we discovered in section 4.2.1, part g, $\psi(x)$ is an odd function about the point x_1 (that is,

$\psi(x_1 - x) = -\psi(x_1 + x)$). Therefore, equation 193 after transformation from x to s becomes

$$-\varepsilon_{Simp} = \frac{h^5}{4!}\int_0^2 s(s-1)(s-2)(s-(1+\varepsilon))f^{(4)}(\xi(x))ds. \tag{194}$$

Letting $\varepsilon \to 0$, we obtain

$$-\varepsilon_{Simp} = \frac{h^5}{4!}\int_0^2 s(s-1)^2(s-2)f^{(4)}(\xi(x))ds. \tag{195}$$

The polynomial in s in the integrand in equation 195 maintains constant sign over the integration range, so the integral mean-value theorem can be applied to produce

$$-\varepsilon_{Simp} = \frac{h^5}{4!} f^{(4)}(\eta)\int_0^2 s(s-1)^2(s-2)ds, \quad \eta \in [x_0, x_2], \tag{196}$$

from which we find that

$$\varepsilon_{Simp} = \frac{h^5}{90} f^{(4)}(\eta), \quad \eta \in [x_0, x_2]. \tag{197}$$

The fact that the integral of $\psi(x) \equiv h^3 s(s-1)(s-2)$ is zero for the quadratic interpolation means that Simpson's rule ($m = 2$) is correct not only for all second-degree polynomials, but also for all third-degree polynomials. This property is apparent when we note that the fourth derivative in equation 197 is zero for any third-degree polynomial.

More generally, the behavior exhibited above is true for all Newton-Cotes rules for an even m. That is,

$$\int_{x_0}^{x_m} \prod_{i=0}^{m} (x - x_i)dx = 0 \tag{198}$$

is true for all even integers m but not true for odd integers m, so the order of h in the error for even m is one degree higher than we might expect. The above proof of this latter result for $m = 2$ does not generalize for $m \geq 4$. Rather the required proof (see exercise 4.2.35) shows that for even m the error in the integral over $[x_0, x_m]$ of the mth-degree exact-matching polynomial fitted through $x_0, x_1, x_2, \ldots, x_m$ is the same as the error in the integral over $[x_0, x_m]$ of the $(m+1)$th-degree polynomial fitted through $x_0, x_1, x_2, \ldots, x_{m+1}$:

$$\int_0^m h^{m+2}\left(\prod_{i=0}^{m}(s-i)\right)\frac{f^{(m+1)}(\xi(x))}{(m+1)!}ds = \int_0^m h^{m+3}\left(\prod_{i=0}^{m+1}(s-i)\right)\frac{f^{(m+2)}(\zeta(x))}{(m+2)!}ds. \tag{199}$$

Furthermore, it can be shown that

$$\int_0^m h^{m+2}\left(\prod_{i=0}^{m}(s-i)\right)\frac{f^{(m+1)}(\xi(x))}{(m+1)!}\,ds$$

$$= \frac{h^{m+3}}{(m+2)!} f^{(m+2)}(\eta) \int_0^m \left(\prod_{i=0}^{m+1}(s-i)\right) ds, \quad \text{for some } \eta \in [x_0,x_m]. \qquad (200)$$

Thus, from equations 199 and 200,

$$\frac{h^{m+3}}{(m+2)!}\int_0^m \left(\prod_{i=0}^{m+1}(s-i)\right) f^{(m+2)}(\zeta(x))\,ds$$

$$= \frac{h^{m+3}}{(m+2)!} f^{(m+2)}(\eta) \int_0^m \left(\prod_{i=0}^{m+1}(s-i)\right) ds, \quad \text{for some } \eta \in [x_0,x_m]. \qquad (201)$$

In other words, although the integral mean-value theorem does not apply to $\int_0^m \left(\prod_{i=0}^{m+1}(s-i)\right) f^{(m+2)}(\zeta(x))ds$, because $\prod_{i=0}^{m+1}(s-i)$ is not of constant sign over $[0,m]$, still the result of the integral mean-value theorem if formally applied is true in this case.

To summarize, for even m, the mth-degree Newton-Cotes rule integrates $(m+1)$th-degree polynomials exactly, as does the $(m+1)$th-degree rule. Both the mth-degree and $(m+1)$th-degree rules have error terms of the form $kh^{m+3}f^{(m+2)}(\eta)$, $\eta \in [a,b]$, where the multiplying constant k for each rule is tabulated in table 4-11. It can be observed there that the multiplying constant for even m is less in magnitude than the multiplying constant for the succeeding odd m for which the rest of the error term is the same. Therefore, the even m Newton-Cotes rules should be chosen over the odd m rules.

Because high-order derivatives of most functions are large, we want to avoid the Newton-Cotes rules for large m. But to make the errors small, we must use a small value for h, producing many tabular points over $[a,b]$. For any given value of m, we can accomplish both of these aims by dividing the integration range $[a,b]$ into $n = km$ intervals, rewriting our original integral as

$$I = \int_{x_0}^{x_m} f(x)dx + \int_{x_m}^{x_{2m}} f(x)dx + \cdots + \int_{x_{(k-1)m}}^{x_n} f(x)dx \qquad (202)$$

and applying the mth-order Newton-Cotes rule to each of the integrals on the right side of equation 202. The resulting algorithm is called the *composite Newton-Cotes rule* of degree m. For example, the composite trapezoidal rule (see program 4-5) is

$$\hat{I}_1 = (h/2)(y_0 + 2y_1 + 2y_2 + \cdots + 2y_{n-1} + y_n), \tag{203}$$

and the composite Simpson's rule (see program 4-6) is

$$\hat{I}_2 = (h/3)(y_0 + 4y_1 + 2y_2 + 4y_3 + 2y_4 + \cdots + 4y_{n-1} + y_n). \tag{204}$$

Note that with Simpson's rule we require that n be an even number.

For illustration, we apply the composite Simpson's rule with $n = 4$ to $I_2 = \int_0^\pi sin(x)dx$ (integrating, we see $I_2 = 2$). Then $h = (\pi - 0)/4 = \pi/4$, so $\hat{I}_2 = (\pi/12)(sin(0) + 4sin(\pi/4) + 2sin(\pi/2) + 4sin(3\pi/4) + sin(\pi)) = 2.005$, which is not a bad approximation.

The composite Newton-Cotes rules are more frequently used than the simple rules because of their better error properties. The error in the composite rules is the sum of the errors in each individual integral in the sum of integrals. Therefore, the error in the composite trapezoidal rule is

$$\varepsilon(\hat{I}_1) = \sum_{i=0}^{n-1} \frac{h^3 f^{(2)}(\eta_i)}{12}, \quad \eta_i \in [x_i, x_{i+1}]. \tag{205}$$

```
/* TRAPEZOIDAL RULE FOR NUMERICAL INTEGRATION OF F(X) */
/* OVER INTERVAL XMIN TO XMAX */
/* SUBROUTINE F IS GIVEN */
/* INTEGRATION LIMITS, XMIN AND XMAX, ARE GIVEN */
/* NUMBER OF TABULAR INTERVALS, N, IS GIVEN */

   DECLARE N FIXED BINARY,   /* NUMBER OF TABULAR INTERVALS */
           (H,               /* TABULAR INTERVAL WIDTH */
           XMIN, XMAX,       /* INTEGRATION LIMITS */
           X,                /* TABULAR ARGUMENT */
           INTEGRAL) FLOAT;  /* INTEGRAL VALUE */

/* INITIALIZE */
   H = (XMAX - XMIN) / N;
   INTEGRAL = F(XMIN) + F(XMAX);
   X = XMIN;

/* COMPUTE WEIGHTED SUM OF TABULAR VALUES */
   SUMLOOP: DO I = 1 TO N-1;
               X = X + H;
               INTEGRAL = INTEGRAL + 2 * F(X);
   END SUMLOOP;

/* NORMALIZE INTEGRAL VALUE */
   INTEGRAL = (H/2) * INTEGRAL;
```

PROGRAM 4-5 Trapezoidal rule for numerical integration

```
/* SIMPSON'S RULE FOR NUMERICAL INTEGRATION OF F(X) */
/* OVER INTERVAL XMIN TO XMAX */
/* SUBROUTINE F IS GIVEN */
/* INTEGRATION LIMITS, XMIN AND XMAX, ARE GIVEN */
/* NUMBER OF TABULAR INTERVALS, N, IS GIVEN; N IS EVEN */

     DECLARE (N,                 /* NUMBER OF TABULAR INTERVALS */
              M) FIXED BINARY,   /* NUMBER OF PANELS */
              (H,                /* TABULAR INTERVAL WIDTH */
              XMIN, XMAX,        /* INTEGRATION LIMITS */
              X,                 /* TABULAR ARGUMENT */
              INTEGRAL) FLOAT;   /* INTEGRAL VALUE */

/* INITIALIZE */
    H = (XMAX - XMIN) / N;
    M = N / 2;
    INTEGRAL = F(XMIN) + F(XMAX);
    X = XMIN;

/* COMPUTE WEIGHTED SUM OF TABULAR VALUES */
    SUMLOOP: DO I = 1 TO M-1;
                 X = X + 2 * H;
                 INTEGRAL = INTEGRAL + 4 * F(X-H) + 2 * F(X);
    END SUMLOOP;
    INTEGRAL = INTEGRAL + 4 * F(X+H);

/* NORMALIZE INTEGRAL VALUE */
    INTEGRAL = (H/3) * INTEGRAL;
```

PROGRAM 4-6 Simpson's rule for numerical integration

Since the integral mean-value theorem also applies to the weighted sums of functions with weights of the same sign, equation 205 can be written

$$\varepsilon(\hat{I}_1) = f^{(2)}(\eta) \sum_{i=0}^{n-1} \frac{h^3}{12}, \quad \eta \in [x_0, x_n], \tag{206}$$

so

$$\varepsilon(\hat{I}_1) = \frac{nh^3}{12} f^{(2)}(\eta), \quad \eta \in [x_0, x_n]. \tag{207}$$

Since by definition

$$nh = b - a, \tag{208}$$

we have finally

$$\varepsilon(\hat{I}_1) = \frac{b-a}{12} h^2 f^{(2)}(\eta), \quad \eta \in [x_0, x_n]. \tag{209}$$

Equation 209 can be used to find the interval width h needed to achieve an integration error less than a given tolerance.

By similar manipulation of the error in the simple Simpson's rule, we find that the truncation error in the composite Simpson's rule is

$$\varepsilon(\hat{I}_2) = \sum_{i=0}^{n/2-1} h^5 f^{(4)}(\eta_i)/90 = \frac{n}{2} h^5 f^{(4)}(\eta)/90 = \frac{b-a}{180} h^4 f^{(4)}(\eta), \quad (210)$$

where $\eta_i \in [x_{2i}, x_{2i+2}]$ and $\eta \in [x_0, x_n]$. Note that the order of h in the error term of a composite rule is one less than its order in the corresponding simple rule.

To fully analyze the error, we must analyze the computational error as well as the truncation error. We assume that the generated error in developing the sum for any integration rule is negligible compared to the errors due to evaluating the y_i. When using any rule of the form

$$\hat{I} = h \sum_{i=0}^{n} w_i y_i, \quad (211)$$

we deal with inaccurate values

$$y_i^* = y_i + \varepsilon_i. \quad (212)$$

Then the computational error in $\hat{I}$ is

$$\varepsilon_{comp}(\hat{I}) = \sum_{i=0}^{n} w_i \varepsilon_i. \quad (213)$$

If we assume that each ε_i is of approximately the same magnitude and bounded by B, we can show

$$|\varepsilon_{comp}(\hat{I})| \le B \sum_{i=0}^{n} |w_i|. \quad (214)$$

But, using equation 188,

$$\sum_{i=0}^{n} |w_i| = \sum_{i=0}^{n} w_i + \sum_{i=0}^{n} (|w_i| - w_i) = b - a + 2 \sum_{\substack{i \ni: \\ w_i < 0}} |w_i|. \quad (215)$$

Therefore, if all of the weights are positive, the computational error bound is as small as possible and equal to $(b - a)B$. Negative weights increase the computational error. Since the sum of the weights is constant, the negative weights must be counterbalanced by corresponding positive weights, producing a larger sum of weight magnitudes than would be necessary if all of the weights were positive. Since the greater the sum of the weight magnitudes, the greater the possible propagated error, we do not want to use rules with negative weights.

The Newton-Cotes rules for $m \geq 8$ have negative weights, so we wish $m < 8$. Since we have rejected the odd m rules, we are left with $m = 2$, 4, and 6. Most commonly, Simpson's rule ($m = 2$) is used because of possible large values of the derivatives of order 6 and 8 which appear in the $m = 4$ and $m = 6$ rules respectively, and because of the simple multipliers used in Simpson's rule.

b) Richardson Extrapolation; Romberg Integration. We can improve the results from the Newton-Cotes rules by applying the general notion that a weighted average of two different estimates of the same value is a more accurate value than either of the initial estimates if the weights are chosen according to a known relation between the errors in the individual estimates. Let us apply this notion to the trapezoidal rule (such an application will be seen to be equivalent to application to Simpson's rule, but the procedure is simpler for the trapezoidal rule). In particular, the trapezoidal rule applied with interval width h and the trapezoidal rule applied with interval width $h/2$ give two different estimates for I, so a linear combination of these two estimates can give a value for I of improved accuracy. That is, if we write the application of the composite trapezoidal rule with interval width h as $T^0(h)$, then

$$T^0(h) = I + \frac{b-a}{12} h^2 f''(\eta_1), \quad \eta_1 \in [a,b], \tag{216}$$

and

$$T^0\left(\frac{h}{2}\right) = I + \frac{b-a}{12}\left(\frac{h}{2}\right)^2 f''(\eta_2), \quad \eta_2 \in [a,b]. \tag{217}$$

If $f''(\eta_1)$ equals $f''(\eta_2)$, the error in $T^0(h)$ is four times that in $T^0(h/2)$; therefore,

$$T^1(h/2) \equiv \frac{4T^0(h/2) - T^0(h)}{3} = I. \tag{218}$$

Since the equality of the second derivative cannot be assumed, $T^1(h/2)$ is only an approximation for I with error $((b-a)/36)h^2(f''(\eta_2) - f''(\eta_1))$.

If the error in trapezoidal integration is written in a different form, we can show that the weighted average given by equation 218 produces an error with order greater than 2 in h. As a first step, we write I as a sum of single-interval integrals:

$$\int_{x_i}^{x_{i+1}} f(x)dx = \int_0^1 \left[(y_i + s\Delta y_i) + \sum_{j=1}^{k} \binom{s}{j} \Delta^j y_i + O(h^{k+1})\right] h\, ds. \tag{219}$$

The first two terms in the bracket which is the integrand in equation 219 produce the trapezoidal rule, and the remaining terms produce the error. If we sum over all the integration intervals and expand all y_i in Taylor series about either x_0 or x_n, we produce $T^0(h)$ plus a polynomial in h

plus a final term with as high an order of h as we like (see exercise 4.2.38). The coefficients of the polynomial are functions of x_0 and x_n, because they involve derivatives at x_0 and x_n, but they are not functions of h. Furthermore, since we know $I = T^0(h) + O(h^2)$, the coefficients of h^0 and h^1 are zero in the polynomial in question. Thus, we have

$$I = T^0(h) + \sum_{i=2}^{k} c_i h^i + O(h^{k+1}). \tag{220}$$

If we evaluate equation 220 with interval $h/2$, we obtain

$$I = T^0(h/2) + \sum_{i=2}^{k} c_i (h/2)^i + O(h^{k+1}). \tag{221}$$

Multiplying equation 221 by 4, subtracting equation 220, and dividing the result by 3 eliminates the h^2 term. This produces

$$I = (4T^0(h/2) - T^0(h))/3 + \sum_{i=3}^{k} d_i h^i + O(h^{k+1}), \tag{222}$$

where

$$d_i = (4/2^i - 1)c_i/3. \tag{223}$$

By taking a linear combination of two trapezoidal results, we have increased the order of h in the error term from 2 to 3. Let

$$T^1(h/2) \equiv (4T^0(h/2) - T^0(h))/3. \tag{224}$$

Then equation 222 is of the same form as equation 220 with $T^1(h/2)$ instead of $T^0(h)$ and coefficients d_i instead of c_i, and starting at one higher degree of h. Thus, we can repeat the process of weighted averaging, taking an appropriate weighted average of $T^1(h/2)$ and $T^1(h/4)$ to eliminate the h^3 term in equation 222. This notion of repeatedly halving the interval and using the results to eliminate the lowest-order error term is called *Richardson extrapolation* (extrapolation of h to zero). The particular application of Richardson extrapolation to the trapezoidal rule is called *Romberg integration*.

Due to a reason that is much the same as that causing the even m Newton-Cotes rules to have an order of error that is one degree higher than might be expected, the c_i in equation 220 are zero for odd i. Therefore, the error in $T^1(h/2)$ is $O(h^4)$, not $O(h^3)$; the error in T^2, the weighted average using the T^1 results, is $O(h^6)$; and so forth. This behavior is not necessary to the application of Richardson extrapolation, but it speeds up the convergence of that extrapolation. We see, then, that the ith step of the Romberg integration should be

$$T^i(h/2^i) = \frac{4^i T^{i-1}(h/2^i) - T^{i-1}(h/2^{i-1})}{4^i - 1}. \tag{225}$$

It can be shown that

$$T^i(h/2^i) = I + k_i h^{2i+2} f^{(2i+2)}(\eta), \quad \eta \in [a,b] \tag{226}$$

for some constant k_i. We should be concerned about the possibility that the high-order derivatives in the approximational error in the result of the ith step of Romberg integration will cause this error to be large for large i. But it can be shown that the k_i approach zero so quickly as i gets large that for all functions where all required derivatives exist, the approximational error in the limit approaches zero. Intuitively, since the error in the trapezoidal rule approaches zero as $h \to 0$ and the error in the result of Romberg integration is smaller than the error in the trapezoidal rule for any h, its error approaches zero also.

It can also be shown that the computational error in Romberg integration is bounded as $i \to \infty$ (see exercise 4.2.39). In fact, for most functions, the convergence to zero of the approximational error is so fast that the amount of computational error is considerably less than the limit; only a small amount of computational error is generated in the few steps needed to obtain the desired tolerance for the truncation error.

One of the reasons for the good behavior of Romberg integration with respect to computational error is that all of the weights w_{ij} in

$$T^i(h) = h \sum_{j=0}^{(b-a)/h} w_{ij} y_j \tag{227}$$

are positive for any i, a situation which we have seen is desirable. We can also show that the weights w_{1j} are precisely the weights of Simpson's rule; the first step of Romberg integration produces the same result as Simpson's rule. This result justifies the statement made above that we would not lose anything by starting with the trapezoidal rule. Had we started with Simpson's rule in our Romberg integration, we would simply have begun with the T^1 values. For $i > 1$, Romberg results are not the same as those of any Newton-Cotes rule.

Thus, Romberg integration is desirable from the point of view both of approximational error and of computational error. Therefore, it is a method of choice among equal-interval methods. The method is summarized as follows (see also program 4-7). We start the algorithm by evaluating

$$T^0(b-a) = \frac{b-a}{2}(f(b) + f(a)). \tag{228}$$

We then evaluate

$$T^0\left(\frac{b-a}{2}\right) = \frac{1}{2}\left(T^0(b-a) + (b-a) f\left(\frac{b+a}{2}\right)\right). \tag{229}$$

```
/* ROMBERG INTEGRATION TO INTEGRATE F(X) FROM XMIN TO XMAX */
/* GIVEN ARE XMIN; XMAX; AN ERROR TOLERANCE, TOL; A MAX # OF */
/* ITERATIONS, MAXIT; AND SUBROUTINE F */

   DECLARE (XMIN, XMAX,            /* INTEGRATION LIMITS */
            H, ARG,                /* INTERVAL WIDTH, EVAL ARG */
            ROMBROW(Ø:MAXIT),      /* LAST ROW OF TABLE */
            NEWTAB, OLD1, OLD2,    /* TEMPORARY TABLE ENTRIES */
            DENOM, TOL) FLOAT,     /* 4**I, ERROR TOLERANCE */
            (N,                    /* (XMAX - XMIN) / H */
            MAXIT)                 /* MAX # OF ITERATIONS */
            FIXED BINARY;

/* INITIALIZE */
   H = XMAX - XMIN;
   /* SIMPLE TRAPEZOIDAL RULE */
   ROMBROW(Ø) = (H/2) * (F(XMAX) + F(XMIN));
   N = 1;
/* COMPUTE AS MANY TABLE ROWS AS NECESSARY */
   TABCOMP: DO K = 1 TO MAXIT;  /* K IS ROW NUMBER */
               OLD1 = ROMBROW(Ø);
   /* COMP NEW TRAPEZOIDAL RULE RESULT FOR INTERVAL H/2 */
               NEWTAB = Ø;
               ARG = XMIN + H/2;
               SUMNEW: DO J = 1 TO N;  /* SUM NEW FUNC VALUES */
                          NEWTAB = NEWTAB + F(ARG);
                          ARG = ARG + H;
               END SUMNEW;
               ROMBROW(Ø) = (OLD1 + H * NEWTAB) / 2;
   /* COMPUTE REMAINDER OF KTH ROW */
               DENOM = 4;
               ROWCOMP: DO I = 1 TO K; /* I IS COLUMN NUMBER */
                           /* SAVE VALUE IN PREVIOUS ROW */
                           OLD2 = ROMBROW(I);
                           ROMBROW(I) = ROMBROW(I-1)
                              + (ROMBROW(I-1) - OLD1) / (DENOM - 1);
                           OLD1 = OLD2;
                           DENOM = DENOM * 4;
               END ROWCOMP;
   /* SET UP NEXT ROW COMPUTATION */
               H = H/2;
               N = 2 * N;
   /* TEST IF ACCURACY ACHIEVED */
               IF ABS(ROMBROW(K) - ROMBROW(K-1)) < TOL
                  THEN GO TO DONE;
   END TABCOMP;
/* IF MAXIT ITERATIONS HAVE BEEN DONE AND ACCURACY NOT ACHIEVED */
   K = MAXIT;
   DONE:  /* ROMBROW(K) HOLDS INTEGRAL VALUE */
```

PROGRAM 4-7 Romberg integration

Note that the evaluation of the trapezoidal rule at half the interval does not require reusing any of the previously used function values. It can be computed as an average of the value of the trapezoidal rule for the previous interval and the constant h times the sum of the data values with arguments halfway between all of the arguments of the previously used values:

$$T^0\left(\frac{h}{2}\right) = \frac{1}{2}\left[T^0(h) + h \sum_{i=1}^{(b-a)/h} f(a + (i - \tfrac{1}{2})h)\right]. \tag{230}$$

Having computed $T^0(b-a)$ and $T^0((b-a)/2)$, we compute

$$T^1\left(\frac{b-a}{2}\right) = T^0\left(\frac{b-a}{2}\right) + \frac{T^0\left(\frac{b-a}{2}\right) - T^0(b-a)}{3}. \tag{231}$$

Note the form used to compute T^1:

$$T^i(h/2) = T^{i-1}(h/2) + \frac{T^{i-1}(h/2) - T^{i-1}(h)}{4^i - 1}. \tag{232}$$

This form is better computationally than the form given by equation 225 (see exercise 4.2.39). If the value $T^1((b-a)/2)$ is accurate enough, we stop the process. If not, we halve the interval once more, computing $T^0((b-a)/4)$ from equation 230, applying equation 232 for $h = (b-a)/2$ and $i = 1$ to produce $T^1((b-a)/4)$. Reapplying equation 232 with $h = (b-a)/2$ and $i = 2$, we produce $T^2((b-a)/4)$. The overall process can be visualized as developing elements of table 4-12 row by row. Saving only the previous row, we compute a new row and decide whether the diagonal element is accurate enough. If so, we stop, and if not, we add another row.

How do we know whether a given value is accurate enough? Since we know that the process converges in both the horizontal and vertical directions, we can simply continue the process until two adjacent vertical or horizontal values agree to the number of significant digits required; then we stop.

As an illustration of the above algorithm, let us apply it to evaluate $\int_0^\pi sin(x)dx$ (which has the value 2) with an error tolerance of .01. We start by computing $T^0(\pi) = (\pi/2)(0 + 0) = 0$. We then compute $T^0(\pi/2) = \frac{1}{2}(T^0(\pi) + \pi sin(\pi/2)) = 1.571$, and $T^1(\pi/2) = 1.571 + (1.571 - 0)/3 = 2.194$. Since $|2.194 - 1.571| > .01$, we continue our computation process: $T^0(\pi/4) = \frac{1}{2}(T^0(\pi/2) + (\pi/2)(sin(\pi/4) + sin(3\pi/4))) = 1.895$, $T^1(\pi/4) = 1.895 + (1.895 - 1.571)/3 = 2.004$ (note that this result is the

TABLE 4-12 ROMBERG INTEGRATION

h	$T^m(h)$ m: 0	1	2	3	$\cdots$
$b-a$	•				
$\frac{b-a}{2}$	•	•			
$\frac{b-a}{4}$	•	•	•		
$\frac{b-a}{8}$	•	•	•	•	
$\vdots$					

same as the result we obtained for $h = \pi/4$ in our preceding illustration of Simpson's rule), and $T^2(\pi/4) = 2.004 + (2.004 - 2.194)/15 = 1.999$. Since $|1.999 - 2.004| < .01$, we stop with the approximation for the integral, 1.999. Note how much better this approximation is than the trapezoidal rule or Simpson's rule using the same number of evaluation arguments.

c) Gaussian Integration; Orthogonal Functions. Both the Newton-Cotes rules and the more desirable Romberg integration require that the integrand be evaluated at equal intervals. We can ask the same question we asked with respect to interpolation: To minimize the error, how should we choose the argument points used? With interpolation we chose the zeros of a Tchebycheff polynomial. For integration we will see that we should choose the zeros of other special polynomials.

As with Romberg integration we will try to produce an error term with the highest possible order in h. When discussing Newton-Cotes integration, we noted that this was equivalent to making the rule exact for a polynomial of as high a degree as possible. In use of the method of undetermined coefficients, each simple polynomial x^i for which the method was exact imposed one constraint on the method. If we use m data points and use a formula of the form

$$\hat{I} = \sum_{i=1}^{m} w_i y(x_i), \tag{233}$$

we are free to choose the m weights w_i and the m arguments x_i. Thus, we should be able to satisfy $2m$ constraints, so we should expect the formula to be exact for all polynomials of degree $2m-1$ or less.

Our strategy with integration formulas has been to find an exact-matching interpolation formula and integrate it. We will continue this strategy. Let us take the arguments at which the integrand will be evaluated, the x_i, as parameters. To find an interpolating polynomial that is exact for all polynomials of degree less than $2m$, we need $2m$ constraints. We obtain m constraints by requiring exact matching at each argument. We can obtain m more by requiring that the interpolation function agree in its derivative at each argument point as well. We will not let the fact that we have not assumed that the derivatives are available bother us right now; we will deal with this problem later.

Interpolation requiring exact matching of both function values and derivative values at all arguments is called *Hermite interpolation*. The resulting interpolation polynomial is given by

$$\hat{p}(x) = \sum_{j=1}^{m} (H_j(x)y_j + h_j(x)y'_j), \tag{234}$$

where

$$h_j(x) = (x - x_j)[L_j(x)]^2 \tag{235}$$

and

$$H_j(x) = d_j h_j(x) + [L_j(x)]^2, \tag{236}$$

where $L_j(x)$ is the jth Lagrange polynomial and d_j is the constant, $-2 \sum_{\substack{i=1 \\ i \neq j}}^{m} (1/(x_j - x_i))$ (see exercise 4.2.1, part a). It can be shown that the error in Hermite interpolation is

$$\varepsilon(x) = -[\psi(x)]^2 f^{(2m)}(\xi)/(2m)!, \quad \xi \in [x_1, x_2, \ldots, x_m, x] \tag{237}$$

(see part b of exercise 4.2.1).

We integrate the polynomial in equation 234 to obtain the desired approximation to the integral I:

$$\hat{I} = \sum_{j=1}^{m} \left[y_j \int_a^b H_j(x)dx + y'_j \int_a^b h_j(x)dx \right]. \tag{238}$$

Since it is assumed that we do not necessarily know the y'_j, we wish to eliminate the y'_j from equation 238. We can do so by requiring that the integral multiplying y'_j be zero for each j. The arguments x_i are parameters of $h_j(x)$, and we wish to choose the x_i so that

$$\int_a^b h_j(x)dx = 0, \quad j = 1,2,\ldots,m. \tag{239}$$

To summarize, we do a Hermite interpolation, so the fit of $\hat{p}(x)$ will be correct for all polynomials of degree less than $2m$, but we do not have to know the m values y'_j. Rather, by choosing the x_i we have to satisfy a different set of m constraints, namely, that equation 239 holds.

At this point, it is useful to linearly map the interval $[a,b]$ to the interval $[-1,1]$ as we did when we were considering the choice of points for interpolation. We let

$$z = \frac{b-a}{2}x + \frac{b+a}{2}. \tag{240}$$

If $z \in [a,b]$, then $x \in [-1,1]$; and if

$$f(z) \equiv g(x), \tag{241}$$

then
$$I \equiv \int_a^b f(z)dz = \frac{b-a}{2}\int_{-1}^1 g(x)dx. \tag{242}$$

Therefore, we need only consider the problem of integration over $[-1,1]$. For the rest of this section, we will assume that the integration is for a function $f(x)$ over $[-1,1]$, that the x_i to be chosen are for this interval, and that the Lagrange polynomials are for these x_i. We require

$$0 = \int_{-1}^1 h_j(x)dx = \int_{-1}^1 (x - x_j)[L_j(x)]^2\, dx, \quad j = 1,2,\ldots,m. \tag{243}$$

Since
$$(x - x_j)L_j(x) = K\psi(x) \tag{244}$$

for some constant K, equation 243 can be rewritten

$$\int_{-1}^1 \psi(x)L_j(x)dx = 0, \quad j = 1,2,\ldots,m. \tag{245}$$

Equation 245 is equivalent to the requirement that for any constants c_j,

$$\int_{-1}^1 \psi(x) \sum_{j=1}^m c_jL_j(x)dx = 0. \tag{246}$$

Any polynomial $p(x)$ of degree $m-1$ or less can be written as $\sum_{j=1}^m c_jL_j(x)$, because it can be tabulated at the x_i and then, by the uniqueness of polynomial exact-matching approximation, $p(x)$ is equal to the resulting Lagrange approximation. Therefore, equation 246 is equivalent to the requirement that for any polynomial $p(x)$ of degree less than m,

$$\int_{-1}^{1} \psi(x)p(x)dx = 0. \tag{247}$$

In section 2.5.2, two n-vectors u and v were defined to be orthogonal if

$$\sum_{i=1}^{n} u_i v_i = 0. \tag{248}$$

If we think of a function over an interval as the list of the function values for all points in the interval (there are an uncountable number of such points, but that does not bother us), the sum in equation 248 becomes an integral; we say two functions $u(x)$ and $v(x)$ are orthogonal over an interval $[a,b]$ if

$$\int_{a}^{b} u(x)v(x)dx = 0. \tag{249}$$

More correctly, we say the functions u and v are orthogonal according to the inner product

$$(u,v) \equiv \int_{a}^{b} u(x)v(x)dx. \tag{250}$$

Then equation 247 can be expressed as follows: The x_i must be chosen such that $\psi(x)$ is orthogonal over $[-1,1]$ to every polynomial $p(x)$ of degree less than m.

Consider the vector space of polynomials, with an inner product defined as above over $[-1,1]$. Then $\psi(x)$ is an mth-degree polynomial in this vector space. We wish this polynomial to be orthogonal to every polynomial of degree less than m. Assume we can find a set of orthogonal polynomials $P_i(x)$, such that P_i is a monic polynomial of ith-degree and P_i and P_j are orthogonal for $i \neq j$. Then we can show that the P_i, $i = 0,1,\ldots,m-1$ span the space of all polynomials of degree less than m (see exercise 4.2.41); in fact, they are an orthogonal basis for that space. So if $\psi(x)$ is orthogonal to each P_i, $0 \leq i < m$, it will be orthogonal to every polynomial of degree less than m, because each such polynomial can be written as a linear combination of the P_i, $0 \leq i < m$. But $\psi(x)$ is monic, so what we desire as $\psi(x)$ is the monic mth-degree polynomial orthogonal to each P_i, $0 \leq i < m$. That is, we want

$$\psi(x) = P_m(x). \tag{251}$$

We find the orthogonal polynomials, $P_i(x)$, by Gram-Schmidt orthogonalization using the basis vectors $x^0, x^1, x^2, \ldots, x^m$. The monic orthogonal polynomials thus produced are called the *Legendre poly-*

nomials, $\psi(x)$ is the mth Legendre polynomial, so the zeros of the mth Legendre polynomial are the zeros of $\psi(x)$, the x_i. We have found the argument points which do the job desired. These Legendre polynomial zeros have been tabulated for all small integers m, and all are in the interval $[-1,1]$. The weights w_i have also been tabulated for these m.

Having taken the x_i as the zeros of the mth Legendre polynomial, the second integral in equation 238 is zero for all j, so

$$\hat{I} = \sum_{j=1}^{m} w_j y_j \tag{252}$$

with

$$w_j = \int_{-1}^{1} H_j(x)dx, \tag{253}$$

where $H_j(x)$ depends on the x_i and $\hat{I}$ is the approximation to the integral which agrees with all polynomials of degree less than $2m$. This integration formula is called *Gaussian integration*, or more specifically *Gauss-Legendre integration* (see program 4-8).

```
/* GAUSSIAN INTEGRATION - INTEGRAND IS F(X) */
/* GIVEN ARE INTEGRATION LIMITS, XMIN AND XMAX; */
/* NUMBER OF EVALUATION POINTS, N; AND SUBROUTINE F */
/* SUBROUTINES X AND W, EACH FUNCTIONS OF TWO PARAMETERS, */
/* ARE ASSUMED TO BE PROVIDED AS PART OF THIS PROGRAM. */
/* FOR EACH, THE FIRST PARAM IS THE NUMBER OF POINTS USED */
/* IN THIS QUADRATURE, AND THE SECOND IS THE POINT INDEX. */
/* X PROVIDES ARG POINTS IN (-1,1) AND W PROVIDES WEIGHTS */

   DECLARE (XMIN, XMAX,       /* INTEGRATION LIMITS */
            INTEGRAL,         /* APPROXIMATE INTEGRAL */
            STRETCH, OFFSET)  /* CONSTANTS TO TRANSFORM */
                    FLOAT,    /* (-1,1) TO (XMIN,XMAX) */
            N FIXED BINARY;   /* NUMBER OF POINTS */
                              /* TO BE USED IN QUADRATURE */

/* COMPUTE INTERVAL TRANSFORMATION CONSTANTS */
   STRETCH = (XMAX - XMIN) / 2;
   OFFSET  = (XMAX + XMIN) / 2;

/* COMPUTE APPROXIMATE INTEGRAL */
   INTEGRAL = Ø;
   INTEGCMP: DO I = 1 TO N;
               INTEGRAL = INTEGRAL +
                     W(N,I) * F(STRETCH * X(N,I) + OFFSET);
   END INTEGCMP;
   INTEGRAL = STRETCH * INTEGRAL;
```

PROGRAM 4-8 Gaussian integration

If $[a,b] \neq [-1,1]$, the value of $\hat{I}$ from equation 252 must be multiplied by $(b-a)/2$ to produce the correct value of the approximation to the integral $\int_a^b f(z)dz$ (see equation 242):

$$\hat{I} = \frac{b-a}{2} \sum_{j=1}^{m} w_j f(z_j), \tag{254}$$

where

$$z_j = \frac{b-a}{2} x_j + \frac{b+a}{2}. \tag{255}$$

Thus, carrying out a Gaussian integration of $I = \int_a^b f(z)dz$ involves choosing a value of m for which the truncation error, as developed below, is appropriately small, finding from a table the values of w_j and x_j corresponding to m, transforming the x_j ($\in [-1,1]$) to z_j ($\in [a,b]$) using equation 255, and then evaluating $\hat{I}$ using equation 254.

For example, let $m = 3$ and $I = \int_0^\pi sin(x)dx$ (so $I = 2$). Then the table will give $-x_1 = x_3 = .7746$ and $x_2 = 0$, and $w_1 = w_3 = \frac{5}{9}$ and $w_2 = \frac{8}{9}$. Thus $z_1 = (-.7746\pi/2 + \pi/2) = .1127\pi$, $z_2 = (0\pi/2 + \pi/2) = \pi/2$, and $z_3 = (.7746\pi/2 + \pi/2) = \pi - .1127\pi$. Thus $\hat{I} = (\pi/2)(\frac{5}{9}sin(.1127\pi) + \frac{8}{9}sin(\pi/2) + \frac{5}{9}sin(\pi - .1127\pi)) = 2.001$; with only three points, we have obtained more accuracy than we did using Simpson's rule with five points and as much accuracy as when using Romberg integration with five points.

Let us discuss the w_j values further. Substituting equation 236 into 253, we obtain

$$w_j = d_j \int_{-1}^{1} h_j(x)dx + \int_{-1}^{1} [L_j(x)]^2 dx = \int_{-1}^{1} [L_j(x)]^2 dx \tag{256}$$

by equation 239. Since each w_j is an integral of a perfect square, each w_j is positive—a desirable situation from the point of view of computational error.

The error in Gauss-Legendre integration is the integral of the Hermite interpolation error given by equation 237:

$$\varepsilon_{Gauss} = -\int_a^b \frac{[\psi_{a,b}(z)]^2 f^{(2m)}(\xi(z))}{(2m)!} dz, \quad \xi \in [a,b], \tag{257}$$

where $\psi_{a,b}$ is the error polynomial involving the z_i as transformed back to $[a,b]$ from the x_i in $[-1,1]$. But the integral mean-value theorem is applicable and

$$\psi_{a,b}(z) = \left(\frac{b-a}{2}\right)^m \psi_{-1,1}(x) = \left(\frac{b-a}{2}\right)^m P_m(x), \tag{258}$$

so

$$\varepsilon_{Gauss} = -\frac{f^{(2m)}(\eta)}{(2m)!}\left(\frac{b-a}{2}\right)^{2m+1}\int_{-1}^{1}[P_m(x)]^2 dx$$
$$= -\frac{(b-a)^{2m+1}(m!)^4}{(2m+1)[(2m)!]^3} f^{(2m)}(\eta), \quad \eta \in [a,b]. \tag{259}$$

For smooth functions $f(x)$, Gaussian integration requires fewer data points to meet a given error tolerance in the approximation than Romberg integration does. Fewer points, of course, means smaller computational error, assuming the x_i are tabulated accurately enough. However, because these values must be tabulated and stored in high precision, Gaussian integration is not entirely desirable from the point of view of computer storage requirements. Furthermore, for functions f with large high-order derivatives, Romberg integration can in fact give better results for the same number of data points than Gaussian integration does. These two methods are common methods of choice; either one of them or a method called Clenshaw-Curtis integration (described in the next part of this section) is usually the best method to use for any given numerical integration.

d) Weighted Integration. In our work thus far, we have produced approximations to an integral by fitting the integrand by a polynomial and integrating the approximating polynomial. As long as one applies composite rules with a small enough interval width, this approach produces adequate truncation error properties. If the integrand is very un-polynomial-like, however, the required interval width may be so small that the computational error is undesirably large and the application of the method is inefficient. In such a case, one does better to fit the integrand by a function with properties more like that of the integrand itself. This is just another instance of the principle that polynomials are not always the best functions for approximation; rather, approximational functions should be chosen on the basis of the shape of the function being approximated.

If the function being approximated is known to be oscillatory, a polynomial fit is not ideal, especially if there are a large number of oscillations over the integration interval. In this case, it is more profitable to approximate f by $p(x)sin(\omega x)$, $p(x)cos(\omega x)$, or a linear combination of these two. Since for any polynomial p the integral of either of these approximating functions is analytically computable, we can produce integration formulas just like those for pure polynomial approximating functions. The resulting method, which is produced in a way directly analogous to the Newton-Cotes rules, is called *Filon integration* if a quadratic polynomial is fitted (see exercise 4.2.44).

Similarly, if the integrand rises exponentially, we can approximate with $p(x)e^{ax}$, or if the integrand falls exponentially, by $p(x)e^{-ax}$. Both of these approximating functions are analytically integrable.

In general we wish to approximate the integrand by

$$f(x) \approx w(x)p(x), \tag{260}$$

where $w(x)p(x)$ is analytically integrable, $w(x)$ characterizes the predominant shape of $f(x)$, and $p(x)$ characterizes the slight variations of the function from $w(x)$ and thus is smooth.

Weighted integration formulas can be developed for Gaussian integration, that is, the argument points can be chosen to maximize the degree of $p(x)$ for which the formula is exact. The analysis for Gauss-Legendre integration ($w(x) = 1$) given above holds for arbitrary $w(x)$, except that the inner product in question is

$$(u,v) \equiv \int_a^b w(x)u(x)v(x)dx. \tag{261}$$

For m points, the argument points to be chosen are the zeros of the mth-degree polynomial of the orthogonal polynomials according to this inner product. For $w(x) = e^{-x}$, $a = 0$, and $b = \infty$, the orthogonal polynomials are called the *Laguerre polynomials* and the method is called *Laguerre-Gauss quadrature* (see exercise 4.2.45). For $w(x) = e^{-x^2}$, $a = -\infty$, and $b = \infty$, the orthogonal polynomials are called the *Hermite polynomials* and the method is called *Hermite-Gauss quadrature*. For $w(x) = 1/\sqrt{1-x^2}$, $a = -1$, and $b = 1$, the orthogonal polynomials are called the *Tchebycheff polynomials* and the method is called *Tchebycheff-Gauss quadrature* (see exercise 4.2.46).† Tchebycheff-Gauss quadrature has the interesting property that the integration weights w_j are equal (see exercise 4.2.46, part b).

In Tchebycheff-Gauss quadrature, a factor in the integrand is approximated by a polynomial which exact-matches at the zeros of the Tchebycheff polynomial. We know that this polynomial is according to the Tchebycheff norm the best single approximating polynomial over the interval. In the unweighted case, we can also produce an integration formula based on doing such an approximation of the integrand and

† For these and other Gaussian integration formulas, see F. B. Hildebrand, *Introduction to Numerical Analysis*, chapter 8.

integrating the approximating polynomial. Such a procedure is equivalent to transforming the integrand by changing variables from x to θ using $x = cos(\theta)$ and approximating the resulting function by a sum of cosine functions. (We will see in section 4.3.7 that such an approximation is not only accurate but also efficient, if an algorithm called the Fast Fourier transform is used.) The quadrature method based on the above approach is called *Clenshaw-Curtis integration.* It has been reported by Gentleman that this method competes well with both Romberg and Gaussian integration in terms of efficiency and approximational error and is better in terms of propagated error, so it is a method of choice.†

PROBLEMS

4.2.1. Assume we are given $\{(x_i,y_i,y_i')|i = 0,1,\ldots,N\}$, a set of argument values and, for each argument, the value of a function f and a value for its derivative at that argument.

a) Find a $(2N+1)$th-degree polynomial $p(x)$ such that for each i, $p(x_i) = y_i$ and $p'(x_i) = y_i'$. As mentioned in section 4.2.4, part c, interpolation using this polynomial is called Hermite interpolation.

b) Find the error in this approximation at an arbitrary value of x.

c) For what values of x would you expect Hermite interpolation using the tabular arguments $x_1,x_3,x_5,\ldots,x_{2N+1}$ to be superior to Lagrange interpolation using the tabular arguments $x_0,x_1,x_2,\ldots,x_{2N+1}$?

4.2.2. Assume we are given the value of $f(x)$ at n values of x $(x = x_1,x_2,\ldots,x_n)$ and $f'(x_k)$ for some k between 1 and n. Develop an interpolation polynomial of minimum degree which agrees with these $n+1$ values.

4.2.3. Assume we wish to tabulate the function $f(x) = (2/\sqrt{\pi})\int_0^x u^{1/2}e^{-u}\,du$ at points equally spaced in the interval given below, in order to use linear exact-matching interpolation to approximate $f(x)$ for any x in the interval with an approximational error of no more than 10^{-4}.

a) For the interval [0,2], how many tabular points need there be?

b) Repeat part a for the interval [1,3].

4.2.4. Assume the function $f(x) = e^{-x^2/4}$ is tabulated with equal spacing $h = .2$ on the interval [0,2]. Bound the maximum approximational error magnitude over all approximation arguments in the interval if quadratic exact-matching interpolation is used.

† W. M. Gentleman, "Implementing Clenshaw-Curtis Quadrature."

4.2.5. Show that the divided difference is a symmetric function of its arguments, that is, $f[x_0,x_1,x_2,x_3,\ldots,x_N] = f[x_{i_0},x_{i_1},x_{i_2},\ldots,x_{i_N}]$, where $(i_0,i_1,i_2,\ldots,i_N)^T$ is any permutation of the vector $(0,1,2,\ldots,N)^T$.

4.2.6. Assume you have only the top diagonal row of a divided-difference table and the tabular argument points, x_i, in storage in a computer.

a) How can you add one point at the bottom of the table and compute the next divided difference in the row?

b) How can you add a point at the top and find the new top row?

4.2.7. Consider the data (x,y) in table 4-3. Assume an error of 1 is made in $y(1.5)$, that is, $y^*(1.5) = 91$. The relative error in this value is $\frac{1}{90} = 1.1\%$ (in this problem, use ε/y as the definition of relative error in y^*, not ε/y^*). Assume there is no error in any y_i other than $y(1.5)$. What is the relative error in each difference in the divided-difference table? Notice the tendency of the relative error as the order of the difference increases.

4.2.8. Prove by induction that the polynomial given by iterated linear interpolation using a given set of points is the same as the polynomial given by the Lagrange method using that set of points. That is, let $p_i(x) = y_i$ and $p_{ij_1j_2\ldots j_nk}(x) = [(x_k - x)p_{ij_1j_2\ldots j_n}(x) - (x_i - x)p_{j_1j_2\ldots j_nk}(x)]/(x_k - x_i)$ and prove by induction that $p_{01\ldots N}(x) = \hat{p}(x)$, where $\hat{p}(x)$ is the polynomial determined by the Lagrange method passing through $(x_0,y_0),(x_1,y_1),\ldots,(x_N,y_N)$.

4.2.9. Carry out the following problem, first using Lagrange interpolation (in barycentric form), then using the divided-difference method, and then using iterated linear interpolation: Compute $f(2.16)$ using the table for f below, first using a quadratic polynomial and then using a cubic polynomial. Compare your answers to the correct value of this function e^x: $f(2.16) = 8.6711$.

x	$f(x)$
2.00	7.3891
2.10	8.1662
2.25	9.4877
2.30	9.9742

4.2.10. Let $x_i = x_0 + ih$, $i = 0,1,\ldots,n$, where $n \geq 1$. Let $\psi(x) = \prod_{i=0}^{n} (x - x_i)$.

a) Show that if n is even, $\psi(x_{n/2} + x) = -\psi(x_{n/2} - x)$, that is, ψ is odd about $x_{n/2}$.

b) Show that if n is odd and $w = (x_0 + x_n)/2$, $\psi(w + x) = \psi(w - x)$, that is, ψ is even about w.

c) Show that $\psi(x)$ has exactly one extremum between x_i and x_{i+1}, $0 \leq i \leq n-1$.

d) Show that the magnitude of the value of ψ at its extrema is a strictly increasing function of the magnitude of the distance of the extremum from $(x_0 + x_n)/2$.

e) Show by example that there exists an unequally spaced set of x_i, $0 \le i \le n$, such that $x_0 < x_1 < \cdots < x_n$ and the maximum magnitude of ψ in the range $[x_0, x_n]$ does not occur in $[x_0, x_1]$ or in $[x_{n-1}, x_n]$.

4.2.11. Let $p_n(x)$ be a monic polynomial of degree n whose maximum magnitude over $[-1,1]$ is minimum over all monic polynomials of degree n. Show that $p_n(x)$ is unique.

4.2.12. Say we wish to fit the function $log(x)$ for $1 \le x \le 2$ with as few tabular points as possible but ensure, in doing so, that polynomial exact-matching interpolation using these points will give an approximational error of magnitude less than 10^{-6}.

a) At which values of x should we tabulate the function?

b) What would be the maximum error if the interval were covered by the same number of equally spaced points? (You will need to solve a polynomial equation numerically.)

4.2.13. Assume $f(x) = e^{-x2/2}$ is to be approximated on $[0,1]$ using polynomial exact-matching.

a) At how many arguments need we tabulate to ensure that the approximational error magnitude is less than 10^{-5}?

b) What will be the tabular arguments if a single polynomial will be fit to all of the points?

c) What will the arguments be if they must be equally spaced in $[-h, 1+h]$ (where h is the interval you choose), and a polynomial will be fit to four points?

d) Same as part c except the arguments are in $[0,1]$.

4.2.14. Let $f(x) = x^{2/3}$, tabulated on $[1,2]$ with an interargument interval of .01. How many points must we use to interpolate $f(x)$, $1 \le x \le 2$, if the approximational error magnitude in an exact-matching polynomial is to be no more than 10^{-6}?

4.2.15. Consider the function $f(x) = xe^x$ over the interval $[1,2]$.

a) How many tabular points do we need if the magnitude of the relative error in polynomial exact-matching interpolation must be less than 10^{-3}?

b) Choose the tabular arguments which will minimize the maximum value of the easily controllable part of the error term and make a table using these points.

c) Using this table, compute an estimate for $f(1.6)$ using iterated linear interpolation.

4.2.16. Assume we wish to evaluate the function e^{-x} on the interval $[0,10]$.

a) How many tabular points are needed for polynomial exact matching to produce a relative truncation error less than 10^{-4}?

b) The problem could be solved by storing an accurate value of e^{-1} and then computing $e^{-x} = e^{-\lfloor x \rfloor} e^{-(x - \lfloor x \rfloor)}$ where $e^{-\lfloor x \rfloor}$ is computed as $e^{-1} \cdot e^{-1} \cdot e^{-1} \ldots e^{-1}$, with $\lfloor x \rfloor$ factors, and $e^{-(x-\lfloor x \rfloor)}$ is computed by interpolation in $[0,1]$. Assuming the error in the computation of $e^{-\lfloor x \rfloor}$ is negligible, how many tabular points are needed in

$[0,1]$ to interpolate $e^{-(x-\lfloor x \rfloor)}$ by polynomial exact matching so that the relative error in the computed value of e^{-x} is less than 10^{-4}?

4.2.17. Consider the function $f(x) = (3 - x)ln(x + 1)$ over the interval $[1,2]$.

a) How many tabular points do we need if the magnitude of the *relative* error in polynomial exact-matching interpolation must be less than 5×10^{-3}?

b) Choose the tabular arguments which will minimize the maximum value of the easily controllable part of the error term and make a table using these points.

c) Using this table, compute an estimate for $f(1.6)$ using Newton divided-difference interpolation with just enough terms to obtain the required accuracy.

4.2.18. Consider the following tabulated function.

x	$f(x)$
.5	−4.125
1.0	−4.000
1.5	−1.875
2.0	3.000

a) Using iterated linear interpolation, find the best approximation for the root of f between 1.5 and 2.

b) If you are told that $f(x) = (x^2 - 3)(x + 1)$ [the root is therefore at $x = 1.732$], can you explain the behavior that you found in part a?

4.2.19. Define the backward difference $\nabla f(x) \equiv f(x) - f(x - h)$, so $\nabla y_i = y_i - y_{i-1}$. Define ∇^k by extension.

a) In terms of $\nabla^k y_n$, $k = 0,1,\ldots$, write the "Newton backward-difference formula" which corresponds to the diagonal path with positive slope beginning at y_n by writing the divided-difference formula and translating to backward differences.

b) When would you use this formula?

4.2.20. Show that the error in the Bessel formula ending at $\mu\delta^{2n}y_{1/2}$ is as given in equation 124.

4.2.21. Show that if $f^{(2n+2)}(x)$ is constant, the Stirling formula ending at $\mu\delta^{2n+1}y_0$ gives a smaller truncation error for $-\frac{1}{2} < s < \frac{1}{2}$ than a formula ending at $\delta^{2n+1}y_{-1/2}$ or one ending at $\delta^{2n+1}y_{1/2}$.

4.2.22. Assume that we wish to do equal-interval polynomial exact matching, ending on an even difference. As in equations 122–124, let $\varepsilon_1(x)$ be the error at x in the Gauss forward and Stirling formulas beginning at y_0, $\varepsilon_2(x)$ be the error at x in the Gauss backward and Stirling formulas beginning at y_1, and $\varepsilon_3(x)$ be the error at x in the Bessel formula beginning at $y_{1/2}$. Let $A_i = \max_{\substack{0 \le s \le 1 \\ x_{-k} \le \xi_1 \le x_k \\ x_{-k+1} \le \xi_2 \le x_{k+1}}} |\varepsilon_i(x)|$, where $x = x_0 + sh$ and the maxima over s, ξ_1, and ξ_2 are taken independently (even though ξ_1 and ξ_2 are functions of s). Show $A_3 < A_1$ and $A_3 < A_2$.

4.2.23. For $0 < s < \frac{1}{4}$ and assuming the derivative term is approximately constant over an appropriate interval, we know that the Stirling formula beginning at y_0 gives a better approximation than the Bessel formula beginning at $y_{1/2}$ if the path ends on an odd column and vice versa if the path ends on an even column. Show, however, that the ratio of the Stirling error for an even column-ending to the Bessel error for that column-ending is less in magnitude than the ratio of the Bessel error for the succeeding odd column-ending to the Stirling error for that column-ending.

4.2.24. a) Show that if y_i has an error ε_i and all other y_n have no error, the propagated error in $\delta^k y_{i-k/2+j}$ is $(-1)^k\binom{k}{j}\varepsilon_i$ for $0 \le j \le k$.

b) Show that if each y_i has an error bounded in magnitude by b, the error in $\delta^k y_j$ is bounded in magnitude by $2^k b$ for any j.

c) Show that if the ε_i are uncorrelated and have equal variance σ, then $var(y_i) = \sigma^2$. If the error in ε is uniformly distributed in $[-b,b]$, show that the standard deviation in $\delta^4 y_i$ is $\sqrt{\frac{70}{3}}b \approx 4.83b$.

4.2.25. Assume you are given the table below for the function $y(x)$.

x	y
0	4
1	7
2	12
3	20
4	33
5	54

a) Using difference techniques, find the polynomial of lowest degree passing through the points of the table above. Identify the numerical method used.

b) Which standard polynomial exact-matching interpolation method should be used to find $y(2.4)$ with an approximational error of magnitude less than 2? Why?

4.2.26. Assume you are given the table below for the function $y(x)$.

x	y
0	1
1	9
2	31
3	61
4	69
5	1

a) Using difference techniques, find the polynomial of lowest degree passing through the points of the table above. Identify the numerical methods used.

b) Which standard polynomial exact-matching interpolation method should be used to find $y(2.4)$ with an approximational error of magnitude less than 2? Why?

4.2.27. Consider the following tabulated function where the x_i are equally spaced:
$y_0 = 150$, $y_1 = 122$, $y_2 = 102$, $y_3 = 76$, $y_4 = 58$, $y_5 = 35$,
$y_6 = 12$, $y_7 = -5$, $y_8 = -12$, $y_9 = -3$, $y_{10} = 27$.

a) At which difference column should you end to interpolate with greatest accuracy for $y_{2.5}$?

b) For $y_{7.5}$?

c) What would be the path ending for the interpolation in part a?

d) For the interpolation in part b?

4.2.28. Consider the following section of a difference table (for which the Gauss forward formula could be used for interpolation):

x	y	δ	δ^2	δ^3	δ^4	δ^5	δ^6	δ^7
0	0		2		120		720	
		1		60		360		0
1								

Say we wish to compute $f(1.2)$ ending on the fifth difference using Stirling's formula.

a) Where in the table should we start?

b) Fill in the tabular values needed.

c) Execute the interpolation.

4.2.29. Consider the following table with differences indicated.

x	y	Δ	Δ^2	Δ^3	Δ^4	Δ^5
.20	.20134					
		•				
.21	.21155		•			
		•		•		
.22	.22178		•		•	
		•		•		•
.23	.23203		•		•	
		•		•		•
.24	.24231		•		•	
		•		•		•
.25	.25261		•		•	
		•		•		•
.26	.26294		•		•	
		•		•		
.27	.27329		•			
		•				
.28	.28367					

a) For what values of x would you use this specific path, assuming you do not know beforehand at what difference you wish to stop?

b) Give an expression (in terms of $s = (x - .26)/.01$ and derivatives of f) for the error when this path is used out to the last difference indicated.

c) Which formula would you use to compute $f(.234)$ to within some error bound to be given?

d) Which formula would you use to compute $f(.234)$ using up to third differences?

e) Bound the error in part d, given that $f(x) = sinh(x)$.

4.2.30. Assume we have a set of data points (x_i, y_i), $i = 0,1,2,\ldots,N$, where the x_i are equally spaced with interval h, and we wish to fit the data by cubic splines, $p_i(x)$, as in section 4.2.1, part h. Assume that instead of using the boundary conditions $c_0 = c_N = 0$, where c_i is the curvature at x_i, we are given measured values for c_0 and c_1. We can use the recurrence relation given by equation 142 to compute $c_2, c_3, c_4, \ldots$, and thus the splines $p_i(x)$.

a) Show that in this case equation 142 can be rewritten $hc_{m+1} + 4hc_m + hc_{m-1} = (6/h)\delta^2 y_m$.

b) Assume there is measurement error in our values for c_0 and c_1, but the y_i are error-free. Give a recurrence relation for γ_m, the error in c_m.

c) By applying the techniques discussed in section 1.3.2, analyze the absolute stability of the operator defined by the above recurrence relation, assuming the y_i are error-free. How does this stability depend upon h?

4.2.31. Assume that we have tabulated the (un-polynomial-like) function $f(x) = 1/(1 + 9x^2)$ at $x = 0,1,2,3,4$.

a) Approximate $f(1.5)$ using the exact-matching cubic polynomial through the four nearest tabular points to 1.5.

b) Approximate $f(1.5)$ using a cubic spline, assuming f is tabulated at only the points used in part a.

c) Approximate $f(1.5)$ using a cubic spline, assuming f is tabulated at all five points.

d) Explain the behavior you observe.

4.2.32. Consider the function $f(x) = x^3 - 5x^2 + 6x$ on the interval $[-1,4]$. Economize this function to a quadratic polynomial, and then economize that quadratic to a linear polynomial. Graph the three polynomials over the interval in question.

4.2.33. Consider the numerical differentiation formula $f'(x_0) \approx (y_1 - y_{-1})/2h$.

a) In terms of h and derivatives of f, what is the truncation error of this approximation?

b) Assuming $|\varepsilon_{y_1}| < B$ and $|\varepsilon_{y_{-1}}| < B$, h has only relative representation error less than R, relative error less than R is generated at each arithmetic step, and products of error bounds are negligible (for example, neglect RB), what is a bound on the magnitude of the computational error in this approximation?

c) Bound the magnitude of the overall error of the approximation.

d) Assume $B = 2.5 \times 10^{-3}$, $R = 10^{-4}$, $(y_1 - y_{-1})/h = .1$ for all h, and
$|f'(x)| < 1 \times 10^{-2}, x \in [x_{-1}, x_1]$
$|f''(x)| < 2 \times 10^{-2}, x \in [x_{-1}, x_1]$
$|f'''(x)| < 6 \times 10^{-2}, x \in [x_{-1}, x_1]$
$|f''''(x)| < 2.4 \times 10^{-1}, x \in [x_{-1}, x_1]$.
What value of h gives the smallest bound on the overall error magnitude?

e) What is the resulting error bound if we use the value of h obtained in part d?

f) What is the relative error bound corresponding to that absolute error bound?

4.2.34. Stirling's formula can be written

$$y = y_0 + s\mu\delta y_0 + \sum_{j=1}^{\infty} (g_j(x)\delta^{2j}y_0 + f_j(x)\mu\delta^{2j+1}y_0),$$

where
$$g_j(x) = \frac{s^2(s^2-1)(s^2-4)\dots(s^2-(j-1)^2)}{(2j)!}$$

and
$$f_j(x) = \frac{s(s^2-1)(s^2-4)\dots(s^2-j^2)}{(2j+1)!}.$$

a) Show by differentiation that

$$y_0' = \frac{1}{h}\left(\mu\delta - \frac{\mu\delta^3}{3!} + \frac{1(4)}{5!}\mu\delta^5 - \frac{1(4)(9)}{7!}\mu\delta^7 + \cdots\right)y_0.$$

b) Use this formula with the table for $f(x) = sinh(x)$ below to compute $f'(.4)$.
(1) With $h = .002$ using up to first differences.
(2) With $h = .001$ using up to first differences.
(3) With $h = .001$ using up to third differences.
Which is the most accurate? Note $f'(.4) = cosh(.4) = 1.081072$.

x	$sinh(x)$
.398	.408591
.399	.409671
.400	.410752
.401	.411834
.402	.412915

c) Show that truncating at the second difference gives $y_0'' \approx (\delta^2 y_0)/h^2$.

d) Show by using Taylor's theorem that the error in the formula of part c is given by $(h^2/12)f^{(4)}(\xi)$, where $\xi \in [x_{-1}, x_1]$.

4.2.35. Let
$$\phi_j(s) \equiv \prod_{i=0}^{j} (s - i).$$

Let $x_i = x_0 + ih$ and $x = x_0 + sh$, and let $p_j(x)$ be the exact-matching polynomial through $f(x_i)$, $i = 0,1,\dots,j$. We know that

$$\varepsilon_j(x) \equiv p_j(x) - f(x) = -h^{j+1}\frac{\phi_j(s)f^{(j+1)}(\xi_j(x))}{(j+1)!},$$

where $\xi_j(x) \in [x_0, x_1, \ldots, x_j]$. Furthermore, we know that

$$\varepsilon_j(x) = \frac{-\phi_j(s)\Delta^{j+1}f(x_0)}{(j+1)!} + \varepsilon_{j+1}(x).$$

Finally, if $\hat{I} = \int_{x_0}^{x_{2k}} p_{2k}(x)dx$, we know that

$$\varepsilon_{\hat{I}} \equiv \int_{x_0}^{x_{2k}} p_{2k}(x)dx - \int_{x_0}^{x_{2k}} f(x)dx = \int_{x_0}^{x_{2k}} \varepsilon_{2k}(x)dx.$$

a) Show that

$$\varepsilon_{\hat{I}} = \int_{x_0}^{x_{2k}} \varepsilon_{2k}(x)dx = \int_{x_0}^{x_{2k}} \varepsilon_{2k+1}(x)dx,$$

that is, $$-\varepsilon_{\hat{I}} = \int_0^{2k} h^{2k+2}\frac{\phi_{2k}(s)f^{(2k+1)}(\xi_{2k}(x))}{(2k+1)!}ds$$

$$= \int_0^{2k} h^{2k+3}\frac{\phi_{2k+1}(s)f^{(2k+2)}(\xi_{2k+1}(x))}{(2k+2)!}ds.$$

b) $$\varepsilon_{\hat{I}} \equiv \int_{x_0}^{x_{2k}} \varepsilon_{2k}(x)dx = -\int_0^{2k} h^{2k+1}\frac{\phi_{2k}(s)f^{(2k+1)}(\xi_{2k}(x))}{(2k+1)!}ds.$$

(1) Apply integration by parts to the right side of this equation using the definition $g(s) = \int_0^s \phi_{2k}(s)ds$.

(2) Using $$\frac{df^{(2k+1)}(\xi_{2k}(x))}{ds} = \frac{df^{(2k+1)}(\xi_{2k}(x))}{dx}\frac{dx}{ds}$$

and the result of equation 163,

show that $$\varepsilon_{\hat{I}} = -\int_0^{2k} h^{2k+3}\frac{g(s)f^{(2k+2)}(\zeta(x))}{(2k+2)!}dx,$$

where $\zeta(x) \in [x_0, x_{2k}, x]$.

(3) Show that $g(s)$ is of constant sign.

(4) By part 3 above, the integral mean-value theorem is applicable to the integral in part 2. Apply it, and then carry out an integration by parts to produce

$$\varepsilon_{\hat{I}} = \frac{-h^{2k+3}f^{(2k+2)}(\eta)}{(2k+2)!}\int_0^{2k} \phi_{2k+1}(s)ds, \quad \eta \in [x_0, x_{2k}].$$

Note that parts a and b imply that

$$\int_0^{2k} \phi_{2k+1}(s) f^{(2k+2)}(\xi_{2k+1}(x))ds = f^{(2k+2)}(\eta) \int_0^{2k} \phi_{2k+1}(s)ds, \ \eta \in [x_0, x_{2k}];$$

the integral mean-value theorem applied formally gives the correct result despite the fact that its assumptions are not satisfied.

4.2.36. Suppose the function tabulated in exercise 4.2.25 were to be integrated between $x = 1$ and $x = 5$ by applying Simpson's rule with $h = 1$.

a) Would you expect a discrepancy between the result and that obtained by integrating the interpolating polynomial that you found in exercise 4.2.25, part a, between the same limits?

b) Why or why not?

c) Estimate the error of the Simpson's rule integration.

4.2.37. Consider the integral $\int_0^1 sin(\pi x^2/2)dx$. Say we wish to integrate this numerically, with an approximational error of magnitude less than 10^{-3}.

a) What interval width, h, will we need if we wish to use the trapezoidal rule?

b) Simpson's rule?

c) Do the Simpson's rule integration with the interval you find.

4.2.38. From equation 219 we know that

$$\varepsilon_{T^0(h)} \equiv T^0(h) - \int_{x_0}^{x_m} f(x)dx = -h \sum_{i=0}^{m-1} \int_0^1 \sum_{j=1}^{k} \binom{s}{j} \Delta^j y_i ds + O(h^{2k+1}).$$

a) Move the integral sign so that only functions of s are in the integrand, and change the order of summation to produce

$$\varepsilon_{T^0(h)} = -h \sum_{j=1}^{k} S_j \sum_{i=0}^{m-1} \Delta^j y_i + O(h^{2k+1}), \text{ where } S_j \equiv \int_0^1 \binom{s}{j} ds.$$

b) Using the definition that $\Delta^j y_i \equiv \Delta^{j-1} y_{i+1} - \Delta^{j-1} y_i$, show that

$$\sum_{i=0}^{m-1} \Delta^j y_i = \Delta^{j-1} y_m - \Delta^{j-1} y_0.$$

Note that this result states that the difference operator Δ is the inverse of the summation operator $\sum$ in the same sense that the differential operator d is the inverse of the integral operator $\int$. Thus evaluating $\sum_{i=i_1}^{i_2} f_i$ involves finding the function g such that $f_i = \Delta g_i$, and then evaluating the sum as $\sum_{i=i_1}^{i_2} f_i = g_{i_2+1} - g_{i_1}$.

c) From parts a and b, we have

$$\varepsilon_{T^0(h)} = -h \sum_{j=1}^{k} S_j (\Delta^{j-1} y_m - \Delta^{j-1} y_0) + O(h^{2k+1}).$$

By equation 128, we can expand $\Delta^{j-1}y_m$ in terms of y_i, $m \le i \le m+j-1$ and $\Delta^{j-1}y_0$ in terms of y_i, $0 \le i \le j-1$:

$$\Delta^{j-1}y_m = \sum_{i=0}^{j-1} k_i y(x_{m+i})$$

and

$$\Delta^{j-1}y_0 = \sum_{i=0}^{j-1} k_i y(x_i).$$

Expand each $y(x_{m+i})$ in the first sum as a Taylor series about x_m and each $y(x_i)$ in the second sum as a Taylor series about x_0, in all cases with the error term $O(h^k)$. If $\Delta^{j-1}y_m$ is written as $\sum_{n=0}^{k-1} q_{jn}h^n$ and $\Delta^{j-1}y_0$ as $\sum_{n=0}^{k-1} r_{jn}h^n$, each with error term $O(h^k)$, give an expression for the q_{jn} in terms of the k_i and derivatives of y about x_m and for the r_{jn} in terms of the k_i and derivatives of y about x_0.

d) Using the first equation in part c, show that equation 220 holds and give an expression for the c_i in terms of the q_{jn}, the r_{jn}, and the S_j. You may assume $c_0 = c_1 = 0$ by the argument immediately before equation 220.

4.2.39. Romberg integration for $I = \int_a^b f(x)dx$ is specified by the following relations:

(1) $T^0(b-a) = \dfrac{b-a}{2}(f(b)+f(a));$

(2) $T^0(h/2) = \dfrac{1}{2}\left[T^0(h) + h\sum_{i=1}^{N} f(a+(i-\tfrac{1}{2})h)\right]$, where $h = (b-a)/N$;

(3) $T^{i+1}(h/2) = T^i(h/2) + \dfrac{T^i(h/2) - T^i(h)}{4^{i+1}-1}.$

Assume that the absolute error generated in evaluating $f(x)$ is less than C and that the error generated elsewhere is negligible compared to the errors propagated from errors in $f(x)$.

a) Show that the propagated error in $T^0(h/2)$ is bounded by the same constant for all h used in a Romberg integration (all values in the T^0 column of the Romberg table have the same error bound). What is the constant?

b) Show that the error bound for $T^{i+1}(h/2)$ is bounded by a function of i only (not of h). What is the function? Show that this error bound approaches a limit as $i \to \infty$.

Hints: $\prod_{i=1}^{\infty} a_i = e^{\sum_{i=1}^{\infty} log(a_i)}$, and $log(1 + x/(y-1)) < x/y$ if $(x-2)y > -2$.

c) Assume the following:

(1) $T^i(h) = I + \varepsilon_1$ and $T^i(h/2) = I + \varepsilon_2$ where $|\varepsilon_1| < b_1$ and $|\varepsilon_2| < b_2$, where b_1 and b_2 are known.

(2) Generated error is *not* negligible. Rather, the relative generated error for all arithmetic operations is bounded by R.

(3) Products of error bounds are negligible (for example, neglect Rb_1).

(4) $4^{i+1} - 1$ can be evaluated exactly.

Give a bound on the error in $T^{i+1}(h/2)$ as computed in equation 3 above.

d) How does a generated error E in $T^i(h)$ at step i of the Romberg integration propagate into the result $T^j(h/2^k)$ of step $j > i$? (Obtain an approximate answer only.) Note that the bound on the contribution of E into the jth step is greater than E. Therefore, an error E at each step would eventually cause the overall error bound (the sum of the contributions due to each step) to $\rightarrow\infty$; the Romberg integration process might diverge.

e) What property of digital computer arithmetic will nullify the behavior described in part d, that is, will cause the error propagation due to using equation 3 to stop?

4.2.40. Consider the function $g(x) = e^{-x}f(x)$, where $g(x)$ is not polynomial-like, but $f(x)$ is, over the interval $[0,50]$.

a) Assume we wish a subroutine to evaluate $g(x)$ for any x in $[0,50]$, but $f(x)$ is too complicated to evaluate every time we need a value of $g(x)$. A good solution is to tabulate $f(x)$ and for the given input argument, x_{inp}, for our subroutine, interpolate a value y from the table by polynomial exact-matching interpolation or a variant, and estimate $g(x_{inp})$ as $e^{-x_{inp}}y$. Briefly and without mathematical analysis, why is this a better scheme than tabulating $g(x)$ and doing polynomial exact matching in that table?

b) Assume we tabulate f at $x = 0,1,2,\ldots,50$ as shown below.

x	$f(x)$
0	−5.1
1	−1.0
2	4.9
3	12.9
4	22.8
5	34.7
.	.
.	.
.	.

(1) By examining the difference table, decide which difference to stop on.
(2) Carry out the approximation for $g(2.8)$, explaining what you are doing.
(3) Estimate the approximational error magnitude in your answer.

c) Assume we wish to tabulate f at points x_1 and x_2 and to fit to g through these points a single minimum-degree weighted (by e^{-x}) exact-matching polynomial over the interval. Set up (but do not solve) explicit equations to find x_1 and x_2 if we wish the magnitude of the controllable part of the error in $\hat{g}$ (the polynomial $\times\, e^{-x}$) to be as small as possible over $[0,50]$ according to the $\mathscr{L}_\infty$ norm.

d) Assume we wish to evaluate $\int_0^{50} g(x)dx$. One possibility for computing this value is, for each tabular interval in the table of part b, to fit f by an exact-matching line and integrate the approximation to g produced by weighting the resulting line by e^{-x}, and then sum the results for all of the intervals. Show that the error in the integral over a single tabular interval is $O(h^3)$, where h is the table interval-width.

e) We can further show that the error in the approximation for the integral over $[0,50]$ is $O(h^2)$. Better still, we can show that it is $kh^2 + O(h^3)$ for some constant k independent of h. A variation on Romberg integration is applicable here.

(1) Set up this algorithm for two steps.

(2) Compare the improvement from step to step to that achieved in normal Romberg integration. (Discuss; do not do numerical calculations.)

f) Would you expect a nonzero difference between the value for $\int_0^{50} g(x)dx$ achieved in part d and the value obtained by integrating the function obtained in part c? Why or why not?

4.2.41. Let $q_i(x)$ be a monic polynomial of degree i. Show that $q_i(x)$, $0 \leq i \leq m$, span the space of all polynomials of degree m, that is, any polynomial $p(x)$ in the space can be written $p(x) = \sum_{i=0}^{m} a_i q_i(x)$.

4.2.42. Consider the integral $\int_0^1 e^{-x}\,dx$. Say we wish to integrate this numerically with an approximational error of magnitude less than 2×10^{-5}.

a) What interval width and thus how many data points will we need if we wish to use the trapezoidal rule?

b) To use Simpson's rule?

c) Do the Simpson's rule integration using the interval width that you find.

d) Do Romberg integration (beginning with $h = 1$) until two successive vertical or horizontal entries differ by less than 2×10^{-5}. How many data points did you need?

e) How many points would be required by Gauss-Legendre integration?

f) What would the points be?

g) What would the weights be?

h) Do the Gauss-Legendre integration.

Note that the correct value of the integral is $1 - e^{-1} = .632121$.

4.2.43. Assume we wish to find a set of $n+1$ arguments at which we can tabulate $f(x)$ so that the mean-square error over $[a,b]$ in interpolation, $(1/(b-a))\int_a^b[\varepsilon(x)]^2dx$, is minimum, where $f(x) = p(x) + \varepsilon(x)$ and $p(x)$ is the polynomial which matches $f(x)$ at all of the arguments. Assume that the term $f^{(n+1)}(\xi(x))$ is not known and cannot be controlled. At which values of x should we tabulate $f(x)$? *Hint:* Use the integral mean-value theorem to take $f^{(n+1)}$ outside the integral, and minimize the remaining integral. Watch for a condition encountered in section 4.2.4, part c.

4.2.44. We wish to approximate

$$I \equiv \int_a^b f(x)sin(\omega x)dx \text{ by } \hat{I} \equiv \sum_{i=1}^{m} \int_{x_{2i-2}}^{x_{2i}} p_i(x)sin(\omega x)dx,$$

where $x_k = a + kh$ with $h = (b-a)/2m$, and where $p_i(x)$ is a quadratic polynomial that matches $f(x)$ at x_{2i-2}, x_{2i-1}, and x_{2i}. Let $j = 2i - 1$. Then

$$\hat{I} = \sum_{i=1}^{m} \int_{x_{j-1}}^{x_{j+1}} p_i(x) sin(\omega x) dx,$$

where $p_i(x)$ matches $f(x)$ at x_{j-1}, x_j, and x_{j+1}.

a) Write $p_i(x)$ as a divided-difference formula where the points used are x_{j-1}, x_{j+1}, and x_j in the order specified.

b) Integration by parts of $I_i \equiv \int_{x_{j-1}}^{x_{j+1}} p_i(x) sin(\omega x) dx$ produces

$$I_i = \left[-(1/\omega)p_i(x)cos(\omega x) + (1/\omega^2)p_i'(x)sin(\omega x) + (1/\omega^3)p_i''(x)cos(\omega x)\right]\Big|_{x_{j-1}}^{x_{j+1}},$$

since $p_i''(x)$ is a constant. Use this formula to show that if

$$\begin{aligned}\theta = \omega h, \quad I_i/h = &-(1/\theta)(y_{j+1}cos(\omega x_{j+1}) - y_{i-1}cos(\omega x_{j+1}))\\ &+ (1/\theta^2)(\tfrac{3}{2}y_{j+1}sin(\omega x_{j+1}) + \tfrac{3}{2}y_{j-1}sin(\omega x_{j-1})\\ &+ \tfrac{1}{2}y_{j+1}sin(\omega x_{j-1}) + \tfrac{1}{2}y_{j-1}sin(\omega x_{j+1}) - 2y_j(sin(\omega x_{j+1}) + sin(\omega x_{j-1})))\\ &+ (1/\theta^3)(y_{j+1}cos(\omega x_{j+1}) - y_{j-1}cos(\omega x_{j-1})\\ &- y_{j+1}cos(\omega x_{j-1}) + y_{j-1}cos(\omega x_{j+1}) - 2y_j(cos(\omega x_{j+1}) - cos(\omega x_{j-1}))).\end{aligned}$$

c) Each of the terms above is of the form $Cy_k sin(\omega x_n)$ or $Cy_k cos(\omega x_n)$. For each term such that $k \neq n$, rewrite the term as $Cy_k\{sin \text{ or } cos\}(\omega(x_k + (n-k)h))$. Then use the formula for sines and cosines of sums of angles to show that

$$\begin{aligned}I_i/h = &A(y_{j+1}\,cos(\omega x_{j+1}) - y_{j-1}\,cos(\omega x_{j-1}))\\ &+ B(y_{j+1}\,sin(\omega x_{j+1}) + y_{j-1}\,sin(\omega x_{j-1}))\\ &+ C(y_j\,sin(\omega x_j)),\end{aligned}$$

where

$$\begin{aligned}A &= (1/\theta^3)(1 - cos(2\theta)) - (1/2\theta^2)sin(2\theta) - 1/\theta,\\ B &= (1/2\theta^2)(3 + cos(2\theta)) - (1/\theta^3)sin(2\theta), \quad \text{and}\\ C &= (4/\theta^3)sin(\theta) - (4/\theta^2)cos(\theta).\end{aligned}$$

These three constants can be rewritten as

$$\begin{aligned}A &= (-1/\theta^3)(\theta^2 + \theta sin(\theta)cos(\theta) - 2sin^2(\theta)),\\ B &= (1/\theta^3)(\theta(1 + cos^2(\theta)) - 2sin(\theta)cos(\theta)), \quad \text{and}\\ C &= (4/\theta^3)(sin(\theta) - \theta cos(\theta)).\end{aligned}$$

d) Using $\hat{I} = \sum_{i=1}^{m} I_i$, show that

$$\begin{aligned}\hat{I} = h\Bigg[&A(f(b)cos(\omega b) - f(a)cos(\omega a)\\ &+ B\left(f(a)sin(\omega a) + 2\sum_{i=1}^{m-1} f(a + 2ih)sin(\omega(a + 2ih)) + f(b)sin(\omega b)\right)\\ &+ C\left(\sum_{i=1}^{m} f(a + (2i-1)h)sin(\omega(a + (2i-1)h))\right)\Bigg].\end{aligned}$$

4.2.45. Consider the integral $I = \int_0^\infty e^{-x} f(x)dx$. Assume we wish to find a quadrature formula $\hat{I}$ for I such that $\hat{I} = \sum_{k=1}^{m} w_k f(x_k)$ and $\hat{I} = I$ if $f(x)$ is any polynomial of degree $2m-1$ or less. The x_k will be the zeros of the mth Laguerre polynomial. Find the monic Laguerre polynomials, their zeros (x_k), and the associated weights (w_k), for $m = 1,2,3$. The error associated with this Laguerre-Gauss quadrature is

$$\frac{(m!)^2}{(2m)!} f^{(2m)}(\xi) \text{ where } 0 < \xi < \infty.$$

4.2.46. Consider the integral $I = \int_{-1}^{1} \frac{f(x)}{\sqrt{1-x^2}} dx$. Assume we wish to find a quadrature formula $\hat{I}$ for I such that $\hat{I} = \sum_{k=1}^{m} w_k f(x_k)$ and $\hat{I} = I$ if $f(x)$ is any polynomial of degree $2m-1$ or less.

a) Show that the x_k will be the zeros of the mth Tchebycheff polynomial.

b) Show that the weights are equal. *Hint:*

(1) Write the jth Lagrange polynomial, $L_j(x)$, in terms of the mth-degree Tchebycheff polynomial, $T_m(x)$.

(2) Expand $T_m(x)/(x - x_j)^2$ as a polynomial plus a remainder to get

$$w_j = \frac{1}{T'_m(x_j)} \int_{-1}^{1} \frac{1}{\sqrt{1-x^2}} \frac{T_m(x)}{x - x_j} dx.$$

(3) Show that if θ_j is a root of $cos(m\theta)$, then $cos(m\theta)cos((m+1)\theta_j)$

$$= -\sum_{k=0}^{m} \Delta_{(k)} [C_k(\theta)C_{k-1}(\theta_j) - C_{k-1}(\theta)C_k(\theta_j)],$$

where $C_k(\theta) = \begin{cases} cos(k\theta) & \text{if } k \geq 0 \\ 0 & \text{if } k < 0. \end{cases}$

(4) Using part 3 of this hint and the trigonometric identity

$$cos((k+1)\theta) + cos((k-1)\theta) = 2cos(k\theta)cos(\theta),$$

show $$\frac{cos(m\theta)}{cos(\theta) - cos(\theta_j)} = \frac{-2}{cos((m+1)\theta_j)} \left(\sum_{k=0}^{m} cos(k\theta)cos(k\theta_j) - \frac{1}{2} \right).$$

You can then easily integrate the formula for w_j in part 2 of the hint.

c) Show that the error associated with this Tchebycheff-Gauss integration is $(2\pi/(2^{2m}(2m)!)) f^{(2m)}(\xi)$ where $|\xi| < 1$.

4.3 LEAST-SQUARES APPROXIMATION

Exact matching as a criterion for approximation makes no sense if the errors in the data values y_i are not negligible. There is no reason for the approximating function to pass precisely through the data points if the data points themselves are in error. We want the approximating function to pass near the data points, but requiring it to go through them would cause it to include ripples that fit the error rather than the underlying error-free function we are trying to approximate.

If we do not do exact-matching approximation, we choose a norm which measures the difference between the approximating function and the data, and we approximate by minimizing that norm. Which norm should we choose? We will show that this depends on the probability distribution assumed to characterize the error in the data. However, as we discussed in chapter 2, most often the result is not sensitive to the choice of norm; therefore, we will sometimes choose one norm over the other on the basis of its mathematical tractability or the simplicity of the algorithm it produces rather than because it is the correct norm for the error distribution in question.

We assume in the following discussion that the measured function values y_i are in error but the argument values x_i are error free. In other words, the y_i are random variables while the x_i are deterministic; each x_i is determined by the experimenter as the value of the independent variable (for example, time or distance) at which he chooses to measure the corresponding y_i.

4.3.1 Estimation Theory

We need to support the statement that for a given error distribution the best approximation is that which minimizes the norm of $\hat{f} - y$. Such support requires a careful definition of the "best approximation." Such definitions and the mathematics on which they are based fall into an area of the theory of probability called *estimation theory*. We think of $\hat{f}$ as an estimator for f, and we define what we mean by the best estimate $\hat{f}$. There are a number of possible definitions. One of the most reasonable and most common says to choose as $\hat{f}$ that function which is most probably the one that produced the data values, y, given the data values. In other words, we wish to maximize $p_{f|y^*}(\hat{f}|y)$, where all of the arguments are vectors. This probability is called the *a posteriori probability* of $\hat{f}$ given y, the probability after the experiment, that is, knowing the data values y, of the function which produced the data being $\hat{f}$. The resulting best estimate is called the *maximum a posteriori probability estimate,* or just the *MAP estimate.*

Applying Bayes' rule to the above *a posteriori* probability, we conclude

$$p_{f|y^*}(\hat{f}|y) = p_{y^*|f}(y|\hat{f})p_f(\hat{f})/p_{y^*}(y). \tag{262}$$

Since the denominator of the right side of equation 262 does not involve $\hat{f}$, to maximize the left side of equation 262 with respect to $\hat{f}$, we need only maximize the numerator. That is, the MAP estimate $\hat{f}$ is that $\hat{f}$ which maximizes $p_{y^*|f}(y|\hat{f})p_f(\hat{f})$. The first term in the above expression is called the *conditional probability* of the data given the underlying function. It expresses the distribution of error in the data. The second factor is called the *a priori probability* of $\hat{f}$. It tells the likelihood before any data is measured of the given function occurring. That is, on some physical basis we may know, for example, that third-degree polynomials with high-order coefficient between 1 and 2 are more likely than some other function type, and the *a priori* probability expresses this fact.

If we have no *a priori* knowledge of functions' relative likelihood of underlying the data, we wish to choose as the *a priori* probability that probability distribution for which there is the maximum uncertainty. A theorem in information theory states that the uncertainty is maximum when all possibilities have equal probability.† Thus, in this case, if nothing is known about the underlying function, we wish $p_f(\hat{f})$ to be a uniform distribution over all possibilities $\hat{f}$, that is, a constant function of $\hat{f}$. In this situation, to maximize the *a posteriori* probability, that is, to maximize the numerator on the right side of equation 262, is to maximize over all $\hat{f}$ the conditional probability of y given $\hat{f}$. The $\hat{f}$ which results, namely, the MAP estimate assuming the greatest *a priori* uncertainty for f, is called the *maximum likelihood estimate,* or the *ML estimate.* It is this maximum likelihood estimate that we will use.

Normally, the error in one y_i is independent of the errors in other y_i. It is also most often the case that the errors,

$$\varepsilon_i = y_i - f_i, \tag{263}$$

have mean zero. That is,

$$E(y_i) = f_i; \tag{264}$$

over many experiments the average measured value is the underlying

† See N. Abramson, *Information Theory and Coding.*

function value. Finally, it is common that the errors ε_i are distributed normally with variance σ_i^2:

$$p_{y_i^*|f_i}(y_i|\hat{f}_i) = \frac{1}{\sqrt{2\pi}\,\sigma_i} \; e^{-\frac{1}{2}\left(\frac{\hat{f}_i - y_i}{\sigma_i}\right)^2}. \tag{265}$$

In this case the conditional probability we are maximizing has the following form:

$$p_{y^*|f}(y|\hat{f}) = \frac{1}{(2\pi)^{N/2} \prod_{i=1}^{N} \sigma_i} \; e^{-\frac{1}{2}\sum_{i=1}^{N}\left(\frac{\hat{f}_i - y_i}{\sigma_i}\right)^2}, \tag{266}$$

where N is the number of measured values. Maximizing this probability over all vectors $\hat{f}$ requires that we maximize the exponent of e, that is, that we minimize the sum in the exponent,

$$S = \sum_{i=1}^{N} \left(\frac{\hat{f}_i - y_i}{\sigma_i}\right)^2. \tag{267}$$

In effect, we choose the function $\hat{f}$ to minimize the weighted sum of the squares of the differences between the function values at x_i and the data values at x_i, where the weights are proportional to the reciprocal of the variance at the respective points.

What we have shown is that for a set of data with independent zero-mean normally distributed errors, the maximum likelihood estimate among a set of functions is that function for which the weighted sum of squares S is minimum. This approximation is called a *weighted least-squares approximation.* We can restate the conclusion in terms of norms as follows: The maximum likelihood estimate under the conditions above is that function which minimizes $\|\hat{f} - y\|$, where

$$\|z\| \equiv |Wz|, \tag{268}$$

where W is the diagonal matrix with

$$W_{ii} = 1/\sigma_i. \tag{269}$$

If all of the variances σ_i^2 are equal (if the error distribution does not vary from point to point), the maximum likelihood estimate under the above assumption is the simple least-squares approximation: the function $\hat{f}$ which minimizes $|\hat{f} - y|$.

Let us consider the more general situation where the errors in the y_i are independent and zero-mean and come from the same distribution:

$$p_{y_i^*|f_i}(y_i|\hat{f}_i) = C_n \, e^{-K_n\left|\frac{\hat{f}_i - y_i}{k_n}\right|^n}. \tag{270}$$

The argument above for $n = 2$ applies here so that if $\hat{f}_i \equiv \hat{f}(x_i)$, the ML estimate $\hat{f}$ is the function from the set of functions considered which minimizes $\sum_{i=1}^{N} |\hat{f}_i - y_i|^n$, that is, minimizes $\|\hat{f} - y\|_n$.

As $n \to \infty$ the ML estimate is the function which minimizes the Tchebycheff norm, and the probability distribution given by equation 270 approaches a constant C_∞ for $|\hat{f}_i - y_i| < |k_n|$ and zero for $|\hat{f}_i - y_i| > |k_n|$. In other words, if the error distribution is uniform, Tchebycheff approximation is indicated.

To summarize, given data points which have independent zero-mean error with a distribution given by equation 270: (1) if $n = 2$ (the error is normally distributed), the maximum likelihood (ML) estimate among a set of functions is the function which minimizes the Euclidean norm of $\hat{f} - y$, and (2) if $n = \infty$ (the error is uniformly distributed), the ML estimate is the function which minimizes the Tchebycheff norm of $\hat{f} - y$. These two values of n are the most common. The only other one that arises with any frequency is $n = 1$, in which case the error is said to have an exponential distribution, and $\hat{f}$ is chosen to minimize the $\mathscr{L}_1$ norm of $\hat{f} - y$.

Since the normal distribution is the most common error distribution, least-squares approximation is the most common of the family of approximational methods described above. Another reason for its common use is that it is easily analyzed mathematically and easily implemented. For the latter reasons, it is often used even when the assumption of the normality of the error is either incorrect or unsupported.

Still Tchebycheff approximation is relatively well understood and should be used if the error distribution is closer to uniform than normal, that is, if the probability of an error does not diminish quickly with the size of the error. The Tchebycheff norm of a vector is the maximum element in magnitude of that vector. Therefore, Tchebycheff approximation involves minimizing the maximum element magnitude. The process of minimizing and maximizing falls under the subject of linear and nonlinear programming. If $\hat{f}$ is a linear combination of basis functions, the problem can be solved by linear programming techniques. Otherwise, nonlinear programming techniques are required.† We will not discuss either case further in this text. Rather we will illustrate by least-squares approximation many of the notions of approximation through minimization of a norm.

† See D. F. McAllister, "Algorithms for Chebychev Approximation over Finite Sets."

4.3.2 Linear Least-Squares Approximation

All of the above analysis applies whether the function $\hat{f}$ is a linear combination of basis functions or not. Assume now that $\hat{f}$ is a linear combination of basis functions. Furthermore, assume that the basis functions, f^j, are linearly independent in the sense that the vectors f^j defined by

$$f_i^j \equiv f^j(x_i), \quad i = 1,2,\ldots,N, \tag{271}$$

are linearly independent. Since the f^j are linearly independent, there can be no more than N functions because we know that N linearly independent N-vectors span N-space. If there are N f^j vectors, they form a basis for the space, so there exists a linear combination of the f^j which equals any vector in the space. In particular, if we have N linearly independent basis functions, there exists a linear combination g of these basis functions which precisely produces y. Thus the vector $g - y = 0$, so $\hat{f} = g$ is the least-squares approximation. In other words, if there are as many basis functions as data points and the basis functions are linearly independent, the exact-matching approximation is the least-squares approximation. Furthermore, it is the best approximation according to any norm we might choose, since for all norms

$$\|0\| = 0. \tag{272}$$

If there are fewer than N basis functions, f^j, it may be that y cannot be written as a linear combination of the f^j, so we may arrive at an approximation which is not an exact-matching approximation. It is this situation which normally exists when we talk about least-squares approximation.

We wish to approximate a set of data y by a function

$$f(x) = \sum_{j=1}^{m} a_j f^j(x) \tag{273}$$

according to the least-squares criterion. A special case of this approximation problem is polynomial least-squares approximation, where

$$f^j(x) = x^{j-1}. \tag{274}$$

We choose the parameters a_j to minimize the sum of squares

$$S = \sum_{i=1}^{N} \frac{1}{\sigma_i^2}\left(\sum_{j=1}^{m} a_j f^j(x_i) - y_i\right)^2. \tag{275}$$

This minimization is accomplished by setting the partial derivative of S

with respect to each a_j equal to 0. Since from equation 275

$$\frac{\partial S}{\partial a_k} = 2 \sum_{i=1}^{N} \frac{1}{\sigma_i^2} \left(\sum_{j=1}^{m} a_j f_i^j - y_i \right) f_i^k, \tag{276}$$

we choose the $\hat{a}_j$, the best estimates of the a_j, so that

$$\sum_{i=1}^{N} \frac{1}{\sigma_i^2} \sum_{j=1}^{m} \hat{a}_j f_i^j f_i^k = \sum_{i=1}^{N} \frac{1}{\sigma_i^2} y_i f_i^k, \quad k = 1,2,\ldots,m. \tag{277}$$

Reversing the order of summation on the left side of equation 277, we obtain

$$\sum_{j=1}^{m} \hat{a}_j \left(\sum_{i=1}^{N} \frac{1}{\sigma_i^2} f_i^j f_i^k \right) = \sum_{i=1}^{N} \frac{1}{\sigma_i^2} y_i f_i^k, \quad k = 1,2,\ldots,m. \tag{278}$$

Equation 278 is a set of m linear equations in the m unknowns $\hat{a}_j$. It can be written

$$F\hat{a} = d, \tag{279}$$

where

$$F_{kj} = \sum_{i=1}^{N} \frac{1}{\sigma_i^2} f_i^j f_i^k, \quad 1 \le k \le m, \quad 1 \le j \le m, \tag{280}$$

and

$$d_k = \sum_{i=1}^{N} \frac{1}{\sigma_i^2} y_i f_i^k, \quad 1 \le k \le m. \tag{281}$$

If all variances are equal, the weights,

$$w_i = 1/\sigma_i^2, \tag{282}$$

can be removed from equations 280 and 281. The equations given by equation 278 (or equivalently, equation 279), are called the *normal equations.* Solving the normal equations is solving a set of linear equations, a matter covered in chapter 2.

4.3.3 Error Propagation; Estimation of Parameters

Assuming we have solved equation 279, we wish to know the accuracy of the parameters $\hat{a}_j$ as compared to the "correct" values a_j in the expansion of the underlying function f written in terms of the basis functions f^j:

$$f(x) = \sum_{j=1}^{m} a_j f^j(x), \tag{283}$$

assuming f can be exactly expressed in this form. This assumption must be true for

$$\hat{f}(x) = \sum_{j=1}^{m} \hat{a}_j f^j(x) \tag{284}$$

to be a maximum likelihood estimate for f.

Since the errors in the y_i are given probabilistically, we should expect to determine probabilistic measures of the errors in the $\hat{a}_j$. We can accomplish this by taking an *expected value* of the vector $\hat{a}$, where the expected value of a vector is defined as the vector of expected values of the elements. Then we have

$$E(\hat{a}) = E(F^{-1}d). \tag{285}$$

Since F is a deterministic matrix and expected value is a linear operator,

$$E(\hat{a}) = F^{-1}E(d). \tag{286}$$

From equation 281,

$$\begin{aligned} E(d_k) &= E\left(\sum_{i=1}^{N} \frac{1}{\sigma_i^2} y_i f_i^k\right) \\ &= \sum_{i=1}^{N} \frac{1}{\sigma_i^2} f_i^k E(y_i) \\ &= \sum_{i=1}^{N} \frac{1}{\sigma_i^2} f_i^k f_i \\ &= \sum_{i=1}^{N} \frac{1}{\sigma_i^2} f_i^k \sum_{j=1}^{m} a_j f_i^j \\ &= \sum_{j=1}^{m} a_j \sum_{i=1}^{N} \frac{1}{\sigma_i^2} f_i^k f_i^j \\ &= (Fa)_k. \end{aligned} \tag{287}$$

Therefore,

$$E(d) = Fa, \tag{288}$$

so

$$E(\hat{a}) = F^{-1}E(d) = a. \tag{289}$$

We have shown that the parameters $\hat{a}_j$ are unbiased; their mean over many experiments is their underlying value.

We must now determine the variances and covariances of the parameters $\hat{a}_j$:

$$cov(\hat{a}_j, \hat{a}_k) \equiv E(\hat{a}_j\hat{a}_k) - E(\hat{a}_j)E(\hat{a}_k) = E(\hat{a}_j\hat{a}_k) - a_j a_k. \tag{290}$$

$\hat{a}_j\hat{a}_k$ is the jkth element of the matrix $\hat{a}\hat{a}^T$, so the variances and covariances can be determined from the expected values of that matrix, producing the *covariance matrix*. Again using equation 279, we obtain

$$E(\hat{a}\hat{a}^T) = E(F^{-1}dd^TF^{-1T}). \tag{291}$$

Since

$$F^{-1T} = F^{T-1} \tag{292}$$

$(F^{-1T}F^T = (FF^{-1})^T = I)$, and since F is symmetric (see equation 280),

$$E(\hat{a}\hat{a}^T) = F^{-1}E(dd^T)F^{-1}. \tag{293}$$

Now

$$\begin{aligned} E(dd^T)_{kl} &= E(d_kd_l) \\ &= E\left(\sum_{i=1}^{N} \frac{1}{\sigma_i^2} y_i f_i^k \sum_{j=1}^{N} \frac{1}{\sigma_j^2} y_j f_j^l\right) \\ &= \sum_{i=1}^{N} \sum_{j=1}^{N} \frac{1}{\sigma_i^2\sigma_j^2} f_i^k f_j^l E(y_i y_j). \end{aligned} \tag{294}$$

Since the y_i are independent (actually, they need only be uncorrelated) and

$$E(y_i^2) = f_i^2 + \sigma_i^2, \tag{295}$$

$$E(dd^T)_{kl} = \sum_{i=1}^{N} \frac{1}{\sigma_i^2} f_i^k f_i \sum_{j=1}^{N} \frac{1}{\sigma_j^2} f_j^l f_j + \sum_{i=1}^{N} \frac{1}{\sigma_i^2} f_i^k f_i^l. \tag{296}$$

By equation 287,

$$\sum_{i=1}^{N} \frac{1}{\sigma_i^2} f_i^k f_i = (Fa)_k, \tag{297}$$

and by equation 280,

$$\sum_{i=1}^{N} \frac{1}{\sigma_i^2} f_i^k f_i^l = F_{kl}, \tag{298}$$

so equation 296 produces

$$E(dd^T)_{kl} = (Fa)_k(Fa)_l + F_{kl}. \tag{299}$$

Writing equation 299 in matrix form, we have

$$E(dd^T) = (Fa)(Fa)^T + F = Faa^TF + F. \tag{300}$$

Using equations 293 and 300, we arrive at

$$\begin{aligned} E(\hat{a}\hat{a}^T) &= F^{-1}Faa^TFF^{-1} + F^{-1}FF^{-1} \\ &= aa^T + F^{-1}, \end{aligned} \tag{301}$$

so the covariance matrix of the $\hat{a}_j$ is

$$E(\hat{a}\hat{a}^T) - aa^T = F^{-1}. \tag{302}$$

We note here that the $\hat{a}_j$ are uncorrelated if and only if F^{-1} is diagonal, a situation which is true if and only if F is diagonal, that is, if and only if the functions f^j are orthogonal according to the following inner product:

$$(g,h) \equiv \sum_{i=1}^{N} \frac{1}{\sigma_i^2} g_i h_i. \tag{303}$$

Since it is desirable for the parameters to be uncorrelated (the errors in them uncoupled in the mean), it is desirable for the functions f^j to be orthogonal. (We will see even stronger reasons for this orthogonality in section 4.3.4.)

We are also often interested in the mean and variance of the value of the approximating function at a given point, x. The mean is given by

$$E(\hat{f}(x)) = E\left(\sum_{j=1}^{m} \hat{a}_j f^j(x)\right) = \sum_{j=1}^{m} E(\hat{a}_j) f^j(x)$$

$$= \sum_{j=1}^{m} a_j f^j(x) = f(x). \tag{304}$$

The variance is given by

$$\begin{aligned} var(\hat{f}(x)) \equiv E(\hat{f}(x) - E(\hat{f}(x)))^2 &= E(\hat{f}(x) - f(x))^2 \\ &= E\left(\sum_{j=1}^{m} (\hat{a}_j - a_j) f^j(x)\right)^2 \\ &= \sum_{j=1}^{m} \sum_{k=1}^{m} cov(\hat{a}_j,\hat{a}_k) f^j(x) f^k(x) \\ &= \sum_{j=1}^{m} \sum_{k=1}^{m} F_{jk}^{-1} f^j(x) f^k(x) \\ &= \phi^{xT} F^{-1} \phi^x, \end{aligned} \tag{305}$$

where ϕ^x is the m-vector with elements defined by

$$\phi_j^x \equiv f^j(x). \tag{306}$$

If F^{-1} is diagonal, that is, if the f^j are orthogonal, then

$$F_{jj}^{-1} = 1/(f^j,f^j) \tag{307}$$

and

$$var(\hat{f}(x)) = \sum_{j=1}^{m} \frac{1}{(f^j,f^j)} (f^j(x))^2. \tag{308}$$

Finally, least-squares approximation is sometimes used to estimate the variance of the error in the measured data values. Assume that we know the ratio between the variances in the data values y_i:

$$\sigma_i = k_i\sigma_1, \quad 2 \leq i \leq N \tag{309}$$

(for example, we may know that all of the variances are the same). We wish to estimate

$$\begin{aligned} \sigma_1^2 &= E[(y_1 - E(y_1))^2] \\ &= E[(y_1 - f(x_1))^2]. \end{aligned} \tag{310}$$

If we could find $f(x_1)$ and had many measurements for y_1, the maximum likelihood estimate for σ_1^2 would be

$$\hat{\sigma}_1^2 = \frac{1}{N}\sum_{j=1}^{N}((y_1)_j - f(x_1))^2 \tag{311}$$

(see exercise 4.3.2). We do not have many measurements for y_1, but we are assuming the y_i are normally distributed, so $(y_i - f(x_i))/k_i$ is distributed with the same distribution as $y_1 - f(x_1)$: Therefore, we could use as our estimator

$$\hat{\sigma}_1^2 = \frac{1}{N}\sum_{i=1}^{N}\left(\frac{y_i - f(x_i)}{k_i}\right)^2. \tag{312}$$

The difficulty with equation 312 is that we do not have the values $f(x_i)$ but only estimates of them, $\hat{f}(x_i)$, from the least-squares fit. We want to use these estimates in place of the values of the underlying function f. When we do this for the case where $f(x)$ is a constant and for all i, $k_i = 1$ (that is, when we are estimating the variance of a single random variable), we know that the maximum likelihood estimate is given by

$$\hat{\sigma}^2 = \frac{1}{N-1}\sum_{i=1}^{N}(y_i - \hat{f})^2 = \frac{1}{N-1}\sum_{i=1}^{N}(y_i - \bar{y})^2, \tag{313}$$

where

$$\bar{y} \equiv \frac{1}{N}\sum_{i=1}^{N} y_i \tag{314}$$

(see exercise 4.3.3). The denominator, $N-1$, reflects the fact that given $\hat{f}$ and $N-1$ of the y_i, the remaining y_i value can be determined. That is, there are only $N-1$ free (independently specifiable) y_i, given $\hat{f}$. We say that the problem has only $N-1$ *degrees of freedom*. Another way of looking at the situation is that we cannot estimate a variance if $N = 1$ and we do not know the mean. The denominator of equation 313 reflects this fact. If the denominator were N instead, an estimate of the variance would be produced, an obviously ridiculous result.

With the case at hand, where $\hat{f}$ involves m parameters determined from the y_i, only $N-m$ y_i are independently specifiable, given $\hat{f}$. Therefore, we should expect an estimation formula,

$$\hat{\sigma}_1^2 = \frac{1}{N-m} \sum_{i=1}^{N} \left(\frac{y_i - \hat{f}(x_i)}{k_i} \right)^2. \tag{315}$$

This is in fact the maximum likelihood estimator (see exercise 4.3.4). Here we will show only that it is an unbiased estimator for σ_1^2. We investigate the value

$$E(\hat{\sigma}_1^2) = \frac{1}{N-m} \sum_{i=1}^{N} \frac{1}{k_i^2} E(y_i - \hat{f}(x_i))^2. \tag{316}$$

The expected value in the sum can be rewritten

$$\begin{aligned} E(y_i - \hat{f}(x_i))^2 &= E((y_i - f_i) - (\hat{f}_i - f_i))^2 \\ &= E(y_i - f_i)^2 - 2E((y_i - f_i)(\hat{f}_i - f_i)) + E(\hat{f}_i - f_i)^2. \end{aligned} \tag{317}$$

The first term on the right side of equation 317 is simply

$$E(y_i - f_i)^2 = \sigma_i^2 = k_i^2 \sigma_1^2. \tag{318}$$

The last term on the right side of equation 317 is given by equation 305 as

$$E(\hat{f}_i - f_i)^2 = \phi^{i^T} F^{-1} \phi^i, \tag{319}$$

where

$$\phi_j^i \equiv f^j(x_i). \tag{320}$$

The expected value in the middle term in equation 317 is

$$E((y_i - f_i)(\hat{f}_i - f_i)) = E\left((y_i - f_i) \sum_{j=1}^{m} (\hat{a}_j - a_j) f_i^j \right). \tag{321}$$

By equations 279, 281, and 320,

$$\hat{a} = F^{-1} d = F^{-1} \sum_{k=1}^{N} \frac{1}{\sigma_k^2} y_k \phi^k. \tag{322}$$

By equation 289,

$$a = F^{-1} \sum_{k=1}^{N} \frac{1}{\sigma_k^2} f_k \phi^k, \tag{323}$$

and since

$$y_i - f_i \equiv \varepsilon_i, \tag{324}$$

$$\begin{aligned} \hat{a}_j - a_j &= \left(F^{-1} \sum_{k=1}^{N} \frac{1}{\sigma_k^2} (y_k - f_k) \phi^k \right)_j \\ &= \sum_{k=1}^{N} \frac{1}{\sigma_k^2} \varepsilon_k (F^{-1} \phi^k)_j. \end{aligned} \tag{325}$$

Using equations 321, 324, and 325, we have

$$E((y_i - f_i)(\hat{f}_i - f_i)) = \sum_{j=1}^{m} f_i^j E\left(\varepsilon_i \sum_{k=1}^{N} \frac{1}{\sigma_k^2} \varepsilon_k (F^{-1}\phi^k)_j\right)$$

$$= \sum_{j=1}^{m} f_i^j \sum_{k=1}^{N} \frac{1}{\sigma_k^2} (F^{-1}\phi^k)_j E(\varepsilon_i \varepsilon_k). \tag{326}$$

Since

$$E(\varepsilon_i \varepsilon_k) = \begin{cases} 0, & i \neq k \\ \sigma_i^2, & i = k, \end{cases} \tag{327}$$

equation 326 becomes

$$E((y_i - f_i)(\hat{f}_i - f_i)) = \sum_{j=1}^{m} f_i^j (F^{-1}\phi^i)_j = \phi^{i^T} F^{-1} \phi^i. \tag{328}$$

Using equations 318, 319, and 328 in equation 317, we have

$$E(y_i - \hat{f}(x_i))^2 = k_i \sigma_1^2 - \phi^{i^T} F^{-1} \phi^i. \tag{329}$$

Using this result in equation 316,

$$E(\hat{\sigma}_1^2) = \frac{1}{N - m} \sum_{i=1}^{N} \left(\sigma_1^2 - \frac{\sigma_1^2}{\sigma_i^2} \phi^{i^T} F^{-1} \phi^i\right)$$

$$= \frac{\sigma_1^2}{N - m}\left(N - \sum_{i=1}^{N} \sum_{j=1}^{m} \sum_{k=1}^{m} \frac{1}{\sigma_l^2} f_i^j F_{jk}^{-1} f_l^k\right). \tag{330}$$

The triple sum in equation 330 can be rewritten as

$$\sum_{j=1}^{m} \sum_{k=1}^{m} F_{jk}^{-1} \sum_{i=1}^{N} \frac{1}{\sigma_i^2} f_i^j f_i^k = \sum_{j=1}^{m} \sum_{k=1}^{m} F_{jk}^{-1} F_{kj}$$

$$= \sum_{j=1}^{m} I_{jj} = m. \tag{331}$$

Using this result in equation 330, we have

$$E(\hat{\sigma}_1^2) = \frac{\sigma_1^2}{N - m} (N - m) = \sigma_1^2, \tag{332}$$

our desired result.

If the basis functions are orthogonal according to the inner product defined by equation 303, a computationally simpler form for $\hat{\sigma}_1^2$ than that given by equation 315 is

$$\hat{\sigma}_1^2 = \frac{1}{N - m}\left(\sum_{i=1}^{N} (y_i/k_i)^2 - \sum_{j=1}^{m} \left(\hat{a}_j^2 \sum_{i=1}^{N} (f_i^j/k_i)^2\right)\right) \tag{333}$$

(see exercise 4.3.7).

4.3.4 Polynomial Least-Squares Approximation; Linear Least-Squares Solution Algorithm

As an example of least-squares approximation and its error analysis, let us consider polynomial least-squares approximation, where the basis functions are

$$f^j(x) = x^{j-1}, \tag{334}$$

resulting in an approximation by an $(m-1)$th-degree polynomial. To simplify our analysis, we assume that all of the variances σ_i^2 are constant. This assumption allows us to drop out all factors involving σ_i in the F matrix and the d vector. The inclusion of different σ_i values would change the following results only in that the factor $1/\sigma_i^2$ would appear in every sum over the data points.

The normal equations for polynomial least-squares approximation with an $(m-1)$th-degree polynomial are

$$F\hat{a} = d, \tag{335}$$

where

$$F_{kj} = \sum_{i=1}^{N} x_i^{j+k-2}, \quad 1 \le k \le m, \quad 1 \le j \le m, \tag{336}$$

and

$$d_k = \sum_{i=1}^{N} y_i x_i^{k-1}, \quad 1 \le k \le m. \tag{337}$$

The problem of fitting a least-squares line (in which case, $m = 2$) occurs commonly. In this case

$$\hat{f}(x) = \hat{a}x + \hat{b}, \tag{338}$$

where

$$\hat{a} = \frac{N \sum_{i=1}^{N} x_i y_i - \sum_{i=1}^{N} x_i \sum_{i=1}^{N} y_i}{N \sum_{i=1}^{N} x_i^2 - \left(\sum_{i=1}^{N} x_i \right)^2} \tag{339}$$

and

$$\hat{b} = \frac{1}{N} \sum_{i=1}^{N} y_i - \hat{a} \frac{1}{N} \sum_{i=1}^{N} x_i \tag{340}$$

(see exercise 4.3.8, part a). The variances and covariance of these coefficients are given by

$$var(\hat{a}) = \frac{\sigma^2}{\sum_{i=1}^{N} x_i^2 - \frac{1}{N} \left(\sum_{i=1}^{N} x_i \right)^2}, \tag{341}$$

$$var(\hat{b}) = \sigma^2 \frac{\dfrac{1}{N}\sum_{i=1}^{N} x_i^2}{\sum_{i=1}^{N} x_i^2 - \dfrac{1}{N}\left(\sum_{i=1}^{N} x_i\right)^2}, \tag{342}$$

$$cov(\hat{a},\hat{b}) = -\sigma^2 \frac{\dfrac{1}{N}\sum_{i=1}^{N} x_i}{\sum_{i=1}^{N} x_i^2 - \dfrac{1}{N}\left(\sum_{i=1}^{N} x_i\right)^2} \tag{343}$$

(see exercise 4.3.8, part b). Analyzing these three equations, we see that the variances and covariance increase with the variance σ^2 of the y_i. This should not be surprising. Also of interest is the sign of the covariance of $\hat{a}$ and $\hat{b}$. Because the denominator is positive (it can be rewritten $\sum_{i=1}^{N}\left(x_i - \frac{1}{N}\sum_{j=1}^{N} x_j\right)^2$), the sign of the covariance is opposite that of $\sum_{i=1}^{N} x_i$. This can be interpreted graphically as in figure 4-14. If the x_i values are predominantly positive, then if the slope of $\hat{f}$ is less (greater) than it should be, the y-intercept will tend to be greater (less) than it should be and vice versa. Similarly, if the x_i are predominantly negative, if the slope of $\hat{f}$ is less (greater) than it should be, the y-intercept will tend to be less (greater) than it should be and vice versa.

In statistics one sometimes has two random variables (for example, weight and height) that one has reason to believe are related. Often the parameters of this relationship are obtained by least-squares fitting. In statistics the least-squares fitting is called finding the *regression of y on x*. In particular, if a line is being fitted by least-squares techniques, the method is called *linear regression*. The least-squares solution techniques that we have developed in this section are applicable to regression. In particular, equations 339 and 340 can be rewritten using probabilistic notation as

$$\hat{a} = \frac{\dfrac{N}{N-1}\left(\dfrac{1}{N}\sum_{i=1}^{N} x_i y_i - \left(\dfrac{1}{N}\sum_{i=1}^{N} x_i\right)\left(\dfrac{1}{N}\sum_{i=1}^{N} y_i\right)\right)}{\dfrac{N}{N-1}\left(\dfrac{1}{N}\sum_{i=1}^{N} x_i^2 - \left(\dfrac{1}{N}\sum_{i=1}^{N} x_i\right)^2\right)}$$

$$= \frac{s_{xy}}{s_x^2} = \frac{s_{xy}}{s_x s_y}\frac{s_y}{s_x} = r_{xy}\frac{s_y}{s_x}, \tag{344}$$

and

$$\hat{b} = \bar{y} - \hat{a}\bar{x}. \tag{345}$$

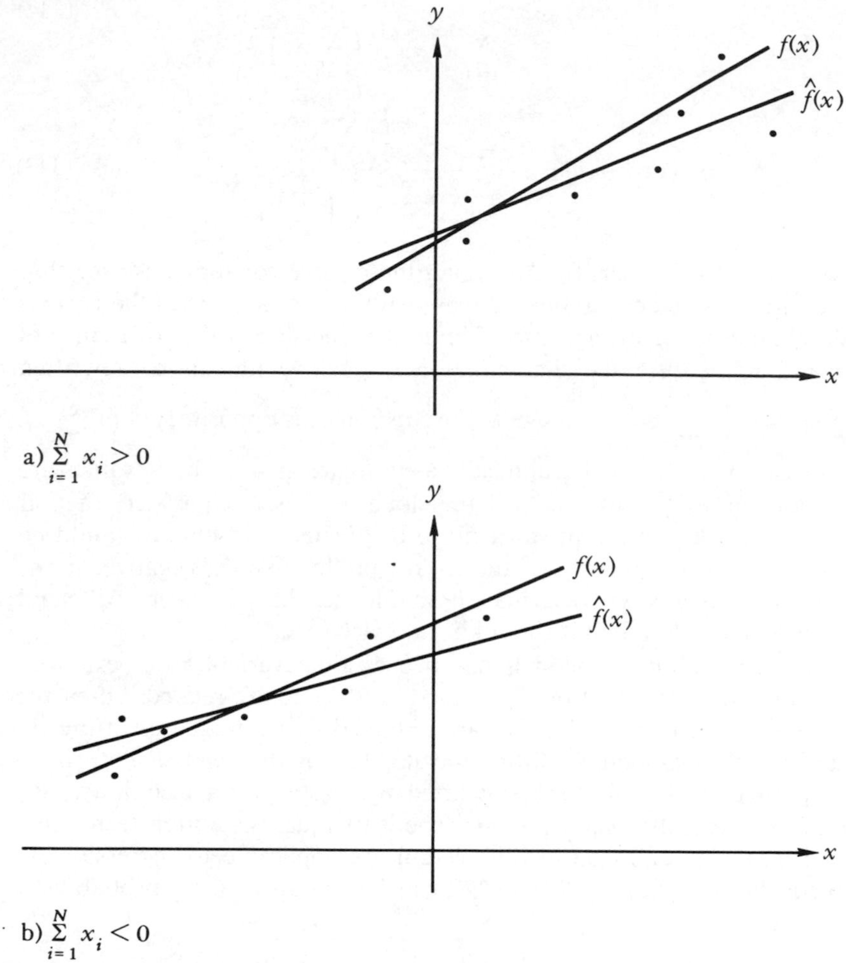

FIG. 4-14 Relation between covariance of the coefficients of the least-squares line and $\sum_{i=1}^{N} x_i$

Differences between regression and least-squares approximation stem from the fact that in regression the x_i are random variables whereas in least-squares approximation the x_i are deterministic. Thus, in regression we can talk about the regression line of x on y; we minimize the sum of squares of deviations in x rather than of deviations in y. But this notion is not particularly reasonable in the context of numerical analysis least-squares approximation. In statistical regression, if the errors in x_i and y_i are approximately equal, it may also be reasonable to minimize the sum of squares of perpendicular distances from the data points to the regression line rather than of the deviations in x or y (see exercise 4.3.11).

Returning to the least-squares approximation problem, we must now ask how to computationally solve the normal equations,

$$F\hat{a} = d. \tag{346}$$

We can solve the equations directly, for example by Gaussian elimination, if the matrix F is reasonably well-conditioned. However, as we will see, most often, this is very much not the case.

A matrix is well-conditioned to the extent that its row vectors are not close to one another in direction. Consider the odd rows of the F matrix. The jth element of the kth row is

$$F_{kj} = \sum_{i=1}^{N} x_i^{k+j-2}, \tag{347}$$

and the jth element of the $(k+2)$th row is

$$F_{k+2,j} = \sum_{i=1}^{N} x_i^{k+j}. \tag{348}$$

We can see that for most sets of x_i and values of k, the sign of F_{kj} will be the same as the sign of $F_{k+2,j}$. That is to say, the kth and $(k+2)$th row vectors will be very much in the same direction. It is likely all of the odd rows will have a direction in the same orthant (a generalization of quadrant to N-space) and all of the even rows will be in the same orthant. Thus, we should suspect that the conditioning of the F matrix will be quite poor for any reasonably large m.

As an example of how bad the conditioning is, consider the problem where the x_i values are approximately uniformly distributed in the interval $[0,1]$. Then

$$\sum_{i=1}^{N} x_i^{k+j} \approx N \int_0^1 x^{k+j}\,dx = \frac{N}{k+j+1}, \tag{349}$$

so

$$F_{kj} \approx \frac{N}{k+j-1}. \tag{350}$$

The matrix F is N times the matrix called the $m \times m$ *Hilbert matrix*, the normalized determinant of which is listed for $m \leq 6$ in table 4-13. As we can see, the normalized determinant of the $m \times m$ Hilbert matrix decreases quickly as m increases.

TABLE 4-13 NORMALIZED DETERMINANT FOR $m \times m$ POLYNOMIAL LEAST-SQUARES NORMAL EQUATIONS

m	Normalized Determinant
1	1
2	1.2×10^{-1}
3	1.3×10^{-3}
4	1.1×10^{-6}
5	6.1×10^{-11}
6	2.5×10^{-16}

It can be seen that direct solution of the normal equations for $m \geq 3$ is a very error-prone process. We must reformulate the problem so that the normal equations are more well-conditioned. Ideally, we would like to reformulate the problem so that the matrix F is diagonal. Such a matrix not only is very well-conditioned but also results in a very simple solution procedure. We notice that the matrix F is defined as

$$F_{kj} = \sum_{i=1}^{N} f_i^j f_i^k = (f^j, f^k). \tag{351}$$

Therefore, for the matrix F to be diagonal, we would like

$$(f^j, f^k) = 0, \quad k \neq j; \tag{352}$$

that is, we want the vectors corresponding to the basis functions to be orthogonal. We can produce such basis vectors by applying Gram-Schmidt orthogonalization (see section 2.5.2) to our initial basis vectors, in the case of polynomials, x^j. That is, we define

$$f^1(x) \equiv x^0 = 1, \tag{353}$$

and define the vector f^1 from the function f^1. We find the orthogonal function f^2 by applying the Gram-Schmidt method, producing

$$f^2(x) \equiv x^1 - \frac{(x^1, x^0)}{(x^0, x^0)} x^0. \tag{354}$$

We do this successively, defining the vector f^k from the function

$$f^k(x) \equiv x^{k-1} - \sum_{j=1}^{k-1} \frac{(x^{k-1}, f^j)}{(f^j, f^j)} f^j(x). \tag{355}$$

The difficulty with the application of Gram-Schmidt orthogonalization is that, as we have noted above, the vectors associated with the functions x^{k-1} and x^{k-1-2n}, for n an integer, are very close in direction. Therefore, x^{k-1} can be very closely represented as a linear combination of the f^j, $1 \le j \le k-1$. That is, the difference between x^{k-1} and the sum in equation 355 will be small. We know that computing such a difference can cause a large propagated relative error in f^k, especially considering all the error which can be generated and propagated in the large amount of computation needed to compute the sum. This large error is in turn propagated into the f^j for $j > k$. We have traded a poorly conditioned set of normal equations for a poorly conditioned Gram-Schmidt orthogonalization. Still, at least the f^k for small k are reasonably accurate and we can carry out the process of least-squares approximation for $m \le 5$ using the above method.

We need a better conditioned and more efficient method for finding the orthogonal vectors f^k. Such a method can be developed by taking advantage of the fact that the f^n, $1 \le n \le k$, are k orthogonal vectors in the space of polynomials of degree $k-1$ or less. Since that space is of dimension k (it is spanned by the vectors corresponding to the functions $x^0, x^1, \ldots, x^{k-1}$), the f^n are an orthogonal basis for the set. Therefore, any polynomial of degree $k-1$ or less can be written in terms of the f^n. In particular, let the $f^n(x)$ be monic polynomials, as we have above. Then

$$f^k(x) - xf^{k-1}(x) = \sum_{j=1}^{k-1} a_{kj}x^{j-1} = \sum_{j=1}^{k-1} b_{kj}f^j(x), \tag{356}$$

because the left-hand side must be a $(k-2)$th-degree polynomial. Finding the b_{kj} is quite straightforward because of the orthogonality of the f^j. We rewrite equation 356 as

$$f^k(x) = xf^{k-1}(x) + \sum_{j=1}^{k-1} b_{kj}f^j(x). \tag{357}$$

We have $k-1$ orthogonality conditions to be met:

$$(f^k, f^n) = 0, \quad 1 \le n \le k-1. \tag{358}$$

These conditions determine the $k-1$ b_{kj} values. The effect of the first $k-3$ of the constraints in equation 358 is as follows.

$$\begin{aligned} 0 = (f^k, f^n) &= (xf^{k-1}, f^n) + \sum_{j=1}^{k-1} b_{kj}(f^j, f^n) \\ &= (f^{k-1}, xf^n) + b_{kn}(f^n, f^n), \quad 1 \le n \le k-3. \end{aligned} \tag{359}$$

(Note that we making use of the fact that inner products are linear functions of each entry.) But xf^n is a polynomial of degree n and therefore can be written

$$xf^n(x) = \sum_{j=1}^{n+1} c_{nj} f^j(x), \tag{360}$$

so the first product in the right side of equation 359 can be written

$$(f^{k-1}, xf^n) = \sum_{j=1}^{n+1} c_{nj}(f^{k-1}, f^j). \tag{361}$$

But by assumption $k-1 > n+1$, so all inner products in the sum in equation 361 are zero; therefore,

$$(f^{k-1}, xf^n) = 0. \tag{362}$$

Thus, from equations 359 and 362,

$$0 = b_{kn}(f^n, f^n), \tag{363}$$

and since $(f^n, f^n) \neq 0$,

$$b_{kn} = 0, \quad 1 \leq n \leq k-3. \tag{364}$$

Equation 357 becomes

$$f^k(x) = xf^{k-1}(x) + b_{k,k-1} f^{k-1}(x) + b_{k,k-2} f^{k-2}(x). \tag{365}$$

We have two remaining orthogonality conditions to be met, namely,

$$(f^k, f^{k-2}) = 0 \tag{366}$$

and

$$(f^k, f^{k-1}) = 0, \tag{367}$$

to determine the two remaining parameters. Applying equation 366, we see

$$0 = (f^k, f^{k-2}) = (xf^{k-1}, f^{k-2}) + b_{k,k-1}(f^{k-1}, f^{k-2}) + b_{k,k-2}(f^{k-2}, f^{k-2}). \tag{368}$$

By the orthogonality relation, the second term on the right side of equation 368 is zero. The first term can be simplified as follows:

$$\begin{aligned}(xf^{k-1}, f^{k-2}) &= (f^{k-1}, xf^{k-2}) \\ &= \left(f^{k-1}, f^{k-1} + \sum_{j=1}^{k-2} c_{k-2,j} f^j\right) \\ &= (f^{k-1}, f^{k-1}) + \sum_{j=1}^{k-2} c_{k-2,j}(f^{k-1}, f^j) \\ &= (f^{k-1}, f^{k-1}).\end{aligned} \tag{369}$$

Therefore, we have

$$0 = (f^{k-1}, f^{k-1}) + b_{k,k-2}(f^{k-2}, f^{k-2}) \tag{370}$$

or

$$b_{k,k-2} = -(f^{k-1}, f^{k-1})/(f^{k-2}, f^{k-2}). \tag{371}$$

Applying condition 367, we arrive similarly to the equation

$$b_{k,k-1} = -(xf^{k-1}, f^{k-1})/(f^{k-1}, f^{k-1}). \tag{372}$$

Therefore, given

$$f^1(x) \equiv 1 \tag{373}$$

and by Gram-Schmidt orthogonalization

$$f^2(x) \equiv x - \frac{1}{N}\sum_{i=1}^{N} x_i, \tag{374}$$

by recursive use of equation 365 we can find in order $f^3, f^4, \ldots, f^m$. We still have the propagation problem due to the application of the recurrence relation, but the recurrence relation itself is not badly conditioned. Furthermore, note that to compute the parameters to compute f^k, we need only compute (f^{k-1}, f^{k-1}) and (xf^{k-1}, f^{k-1}). The inner product in the denominator of equation 371 has been computed at a previous step. Therefore, we have a fairly efficient and well-conditioned algorithm for compuing the f^k.

Note that the existence of a recurrence relation is directly related to the fact that we are finding orthogonal functions. The method used above can be applied to produce recurrence relations for orthogonal functions other than the orthogonal polynomials.

We have now found a set of polynomials f^k which are orthogonal over the data points x_i. (Let us emphasize that the orthogonality is with respect to a set of data points. A set of polynomials orthogonal over one set of data points is not necessarily orthogonal over another set of data points.) The advantages of this orthogonality are many.

1. The normal equations become diagonal and thus are well-conditioned and easy to solve. They are of the form

$$(f^j, f^j)\hat{a}_j = (f^j, y), \quad 1 \leq j \leq m, \tag{375}$$

which leads to

$$\hat{a}_j = (f^j, y)/(f^j, f^j), \quad 1 \leq j \leq m. \tag{376}$$

Note that the denominator in equation 376 has already been computed when developing the orthogonal functions f^k.

2. The formula for $\hat{a}_j$ does not involve m, that is, $\hat{a}_j$ is not a function of the degree of the polynomial being fitted. Thus, if the $(m-1)$th-degree least-squares polynomial does not fit the data well enough, it is easy to find the mth-degree polynomial. We simply apply the

recurrence relation to f^m and f^{m-1} to produce f^{m+1}, then apply equation 376 to compute $\hat{a}_{m+1}$, and add $\hat{a}_{m+1}f^{m+1}$ to the $(m-1)$th-degree polynomial. This is very unlike the situation where simple nonorthogonal polynomials are used as basis functions, in which case increasing the value of m requires re-solving all of the normal equations from scratch. This ease in adding terms to the fitting polynomial makes it possible to find efficiently the polynomial of desired accuracy of fit.

3. As noted in section 4.3.3, the fact that the f^j are orthogonal means that the $\hat{a}_j$ are uncorrelated, and thus the formula for $var(\hat{f}(x))$ is much simpler than it would be if the $\hat{a}_j$ were correlated.
4. Many other formulas, such as that for the estimator $\hat{\sigma}_1^2$, for $var(y_1)$ (see equation 333), become much simpler.

The recurrence relation method allows us to find the orthogonal polynomials for any set of tabular arguments $\{x_i \mid 1 \le i \le N\}$. However, if the arguments are equally spaced, such a solution is unnecessary, because the coefficients of the orthogonal polynomials have been tabulated in the case where tabular arguments are the integers between 0 and $N-1$. It can be shown (see exercise 4.3.14) that if the tabular arguments are equally spaced, if we find the least-squares polynomial approximation $\hat{q}(z)$ for the data points $\{(i-1,y_i) \mid 1 \le i \le N\}$, then the least-squares polynomial approximation, $\hat{p}(x)$, for the data points $\{(x_i,y_i) \mid 1 \le i \le N\}$ is given by

$$\hat{p}(x) = \hat{q}((x - x_1)/h), \tag{377}$$

where

$$h = x_{i+1} - x_i. \tag{378}$$

```
/* POLY LEAST-SQUARES APPROX USING RECURRENCE RELATION */
/* GIVEN ARE NUMBER OF DATA POINTS, N; NUMBER OF POLY TERMS */
/* DESIRED, M; AND DATA POINTS (X(I),Y(I)), 1 < = I < = N. */
/* OUTPUT ARE COEFFICIENTS A(K), 1 < = K < = M, BY WHICH KTH */
/* POLY MUST BE MULTIPLIED TO PRODUCE APPROXIMATING POLY; */
/* AND FACTORS B1(K), 2 < = K < = M, AND B2(K), 3 < = K < = M, */
/* USED TO COMPUTE COEFFICIENTS OF ORTHOGONAL POLY & THUS */
/* OF APPROXIMATING POLY. COEFFICIENTS OF (M-I)TH-DEGREE */
/* APPROX POLY ARE COMPUTED FROM P(X) = SUM(A(K) * FK(X)), */
/* WHERE ORTHOGONAL POLY FK(X) ARE COMPUTED FROM F1(X) = 1; */
/* F2(X) = X - B1(2); FKPLUS1(X) = (X - B1(K+1)) * FK(X) */
/* - B2(K+1) * FKMINUS1(X), FOR K > = 2 */
```

PROGRAM 4-9 Unweighted polynomial least-squares approximation

(*continued on next page*)

```
    DCL (M, /* NUM OF POLY TERMS DESIRED, DEGREE IS M-1 */
        N) FIXED BINARY, /* NUMBER OF DATA POINTS */
        (X(N),              /* DATA ARGUMENTS */
        Y(N),               /* DATA VALUES */
        A(M),               /* APPROXIMATION COEFFICIENTS */
        B1(2:M), B2(3:M), /* RECURRENCE RELATION FACTORS */
    /* VALUES AT X(I) OF LAST TWO ORTHOGONAL POLYNOMIALS */
        FOLD(N), FNEW(N),
    /* INNER PRODS(FK,FK) AND (FKMINUS1,FKMINUS1) */
        FFPROD, OLDFPROD,
    /* INNER PRODS (X*FK,FK) AND (Y,FK) */
        XFPROD, YFPROD,
    /* FK(X(I))**2, OLDFNEW(I) TO BE NEW FOLD(I) */
        POLYSQ, FHOLD) FLOAT;

/* SET UP F1 AND F2 */
    OLDFPROD = N;
    YFPROD  = Ø; XFPROD = Ø;
    SUMXY: DO I = 1 TO N;
           XFPROD = XFPROD + X(I);
           YFPROD = YFPROD + Y(I);
    END SUMXY;
    A(1) = YFPROD / N; B1(2) = XFPROD / N;
    FEVAL:  DO I = 1 TO N;  /* EVAL F1 AND F2 AT THE X(I) */
           FOLD(I) = 1;
           FNEW(I) = X(I) - B1(2);
    END FEVAL;
/* RECUR, COMPUTING B1(K), B2(K), AND A(K) */
    RECUR: DO K = 2 TO M;
           XFPROD = Ø; YFPROD = Ø; FFPROD = Ø;
           INPROD: DO I = 1 TO N; /* COMPUTE INNER PRODUCTS */
                   YFPROD = YFPROD + Y(I)  * FNEW(I);
                   POLYSQ = FNEW(I)  * FNEW(I);
                   FFPROD = FFPROD + POLYSQ;
                   XFPROD = XFPROD + X(I)  * POLYSQ;
           END INPROD;
           A(K) = YFPROD / FFPROD;
           IF K = M THEN GO TO DONE;
    /* SET UP (K+1)TH ORTHOGONAL POLYNOMIAL */
           B1(K+1) = XFPROD / FFPROD;
           B2(K+1) = FFPROD / OLDFPROD;
           FNCEVAL: DO I = 1 TO N; /*
    /* COMPUTE NEW FOLD(I) & FNEW(I) */
           FHOLD  = FNEW(I);
           FNEW(I) = (X(I) - B1(K+1))  * FNEW(I)
                     - B2(K+1)  * FOLD(I);
           FOLD(I) = FHOLD;
           END FNCEVAL;
           OLDPROD = FFPROD;
    END RECUR;
    DONE:  /* RESULTS ARE IN ARRAYS A, B1, B2 */
```

PROGRAM 4-9 (*continued*)

The full algorithm for unweighted polynomial least-squares approximation over arbitrary arguments is given in program 4-9. As an illustration, let us apply the algorithm to the data points (0,1), (1,3), (2,3), (5,5). We compute $f^1(x) = 1$, so by equation 376,

$$\hat{a}_1 = (1 + 3 + 3 + 5)/(1 + 1 + 1 + 1) = 3.$$

That is, the 0th-degree least-squares polynomial is

$$\hat{p}_0 = \hat{a}_1 f^1 = 3.$$

By equation 374,

$$f^2(x) = x - \tfrac{1}{4}(0 + 1 + 2 + 5) = x - 2.$$

So by equation 376,

$$\hat{a}_2 = ((-2)\cdot 1 + (-1)\cdot 3 + 0 + 3\cdot 5)/(4 + 1 + 0 + 9) = \tfrac{5}{7}.$$

Thus, the least-squares line is

$$\hat{p}_1 = \hat{p}_0 + \hat{a}_2 f^2 = 3 + \tfrac{5}{7}(x - 2) = \tfrac{5}{7}x + \tfrac{11}{7}.$$

Using the recurrence relation, equation 365,

$$f^3(x) = x(x - 2) + b_{32}(x - 2) + b_{31}\cdot 1,$$

where $\quad b_{31} = -(4 + 1 + 0 + 9)/(1 + 1 + 1 + 1) = -\tfrac{7}{2}$

and $\quad b_{32} = -(0 + 1\cdot 1 + 0 + 5\cdot 9)/(4 + 1 + 0 + 9) = -\tfrac{23}{7}$

by equations 371 and 372. Therefore

$$f^3(x) = x^2 - 2x - \tfrac{23}{7}x + \tfrac{46}{7} - \tfrac{7}{2} = x^2 - \tfrac{37}{7}x + \tfrac{43}{14}.$$

By equation 376,

$$\hat{a}_3 = \frac{\tfrac{43}{14}\cdot 1 + (-\tfrac{17}{14})\cdot 3 + (-\tfrac{7}{2})\cdot 3 + \tfrac{23}{14}\cdot 5}{(\tfrac{43}{14})^2 + (-\tfrac{17}{14})^2 + (-\tfrac{7}{2})^2 + (\tfrac{23}{14})^2} = -\frac{160}{811},$$

so the quadratic least-squares approximation is

$$\hat{p}_2 = \hat{p}_1 + \hat{a}_3 f_3 = \tfrac{5}{7}x + \tfrac{11}{7} - \tfrac{160}{811}(x^2 - \tfrac{37}{7}x + \tfrac{43}{14}).$$

We now have a method of finding a least-squares polynomial for a given m. If we do not know the degree of polynomial we wish to fit, we may start with $m = 1$ and then add terms, after each addition computing $\hat{\sigma}$ by equation 333. This gives us a measure of the closeness of fit. We can stop when $\hat{\sigma}$ is as small as desired. In particular, if we have reason to believe that the underlying function is a polynomial, we often wish to know the degree of that polynomial. We note that as long as the polynomial being fit is less in degree than the underlying polynomial, an increase in m will make the shape of the fitting polynomial significantly closer to the shape of the underlying polynomial, thus significantly decreasing $\hat{\sigma}$. However, once the correct m has been reached, increasing the degree of the approximating polynomial will help to fit it to the noise but will not improve its fit to the basic data shape; only minor decrease in $\hat{\sigma}$ can be expected, assuming $m \ll N$ and the relative error in the y_i is not large. Therefore, if we plot $\hat{\sigma}$ vs. m, we choose m as the last value for which the $\hat{\sigma}$ at that m is significantly less than the $\hat{\sigma}$ at $m-1$ (see exercise 4.3.15).

4.3.5 Constraints

Sometimes we are required to find the least-squares approximation among all functions which satisfy a particular constraint or set of constraints. For example, given a set of data points, we might be asked to find the least-squares solution among all functions which have a certain value at some specific argument point. These problems can be solved by standard analytic methods of constrained minimization such as the method of Lagrange multipliers. Alternatively and more desirably, the fitting function can be written so that it already satisfies the appropriate constraints, with the result that the parameters can be determined as in the unconstrained least-squares problem. The solution then proceeds as with the unconstrained least-squares problem.

The following example serves not only to demonstrate the solution of a constrained approximation problem by the technique described above, but also to show the generality of the notions of orthogonality and inner product, and of the approximational techniques based on these notions.

Assume we are given a set of data points, $\{(x_i, y_i) | 1 \leq i \leq N\}$, and asked to find with respect to these points the least-squares $(m-1)$th-degree polynomial $\hat{p}$ such that

$$\hat{p}(x_{c_1}) = y_{c_1} \tag{379}$$

and

$$\hat{p}(x_{c_2}) = y_{c_2}, \tag{380}$$

where x_{c_1} and x_{c_2} are not among the x_i and $m \leq N-2$. Such a problem, a hybrid between an exact-matching problem and a least-squares problem, arises, for example, when some of the values of the underlying function are known because of its physical basis, but the remaining approximational parameters must be determined.

The fitting polynomial $\hat{p}(x)$ must be the sum of a first-degree polynomial which fits through the two constraining points and an $(m-1)$th-degree polynomial which is zero at the two constraining arguments:

$$\hat{p}(x) = (x - x_{c_1})(x - x_{c_2}) \sum_{j=1}^{m-2} a_j x^{j-1} + y_{c_1} + (x - x_{c_1}) y[x_{c_1}, x_{c_2}], \tag{381}$$

where $y[x_{c_1}, x_{c_2}]$ is a first divided-difference, a constant for any given (x_{c_1}, y_{c_1}) and (x_{c_2}, y_{c_2}). We will want to satisfy orthogonality relations, so we recast the problem in terms of orthogonal polynomials f^j, where the inner product according to which orthogonality is defined has not yet been determined, but where $f^j(x)$ is a $(j-1)$th-degree polynomial:

$$\hat{p}(x) = (x - x_{c_1})(x - x_{c_2}) \sum_{j=1}^{m-2} b_j f^j(x) + y_{c_1} + (x - x_{c_1}) y[x_{c_1}, x_{c_2}]. \tag{382}$$

As with the unconstrained problem, the normal equations are obtained by setting equal to zero the derivative of the sum of the squared deviations with respect to each parameter:

$$\frac{\partial}{\partial b_k}\left(\sum_{i=1}^{N} (\hat{p}(x_i) - y_i)^2 \right) = 0, \quad 1 \le k \le m-2, \tag{383}$$

which implies

$$\sum_{i=1}^{N} \hat{p}(x_i) f^k(x_i)(x_i - x_{c_1})(x_i - x_{c_2}) = \sum_{i=1}^{N} y_i f^k(x_i)(x_i - x_{c_1})(x_i - x_{c_2}), \quad 1 \le k \le m-2. \tag{384}$$

Substituting equation 382 in equation 384, we produce

$$\sum_{i=1}^{N} \sum_{j=1}^{m-2} b_j f^j(x_i) f^k(x_i)(x_i - x_{c_1})^2 (x_i - x_{c_2})^2 = \sum_{i=1}^{N} ((y_i f^k(x_i)(x_i - x_{c_2}) - y[x_{c_1}, x_{c_2}])(x_i - x_{c_1}) - y_{c_1}), \quad 1 \le k \le m-2. \tag{385}$$

If we identify the right side of equation 385 by the name d_k and switch the order of summation on the left side of the equation, we find

$$\sum_{j=1}^{m-2} F_{kj} b_j = d_k, \quad 1 \le k \le m-2, \tag{386}$$

where

$$F_{kj} \equiv \sum_{i=1}^{N} f^j(x_i) f^k(x_i)(x_i - x_{c_1})^2 (x_i - x_{c_2})^2. \tag{387}$$

To make the normal equations 386 easily and accurately solvable, we want

$$F_{kj} = 0 \quad \text{if} \quad k \neq j. \tag{388}$$

In other words, we wish to choose the polynomials f^j to be orthogonal according to the inner product

$$(f^j, f^k) = \sum_{i=1}^{N} f_i^j f_i^k (x_i - x_{c_1})^2 (x_i - x_{c_2})^2 \tag{389}$$

(see exercise 4.3.19). We can find these orthogonal polynomials by the recurrence relation method described in section 4.3.4, where the inner product used is the one described in equation 389 (see exercise 4.3.20). Once we have the f^k, the solution of the normal equations 386 is trivial.

A similar approach is applicable if derivatives of y are constrained. Such constraints appear if we attempt to produce least-squares splines, that is, if we break up the range of the data arguments into a number of subintervals and fit each subinterval with a different least-squares polynomial but insist that the polynomials agree in value and first and second derivatives at the interfaces between the intervals. Such a solution might be desirable, because it allows us to fit each interval with a low-degree and thus smooth polynomial, maintaining the smoothness across the subinterval interfaces, while still following the shape of the underlying function which may not be characterizable over a large range by a low-degree polynomial.

PROBLEMS (4.3.1 through 4.3.22)

4.3.1. Let $f^1(x) = 1$ and $f^2(x) = 1/x$. Find a least-squares fit $\hat{f}(x) = \hat{a}_1 f^1(x) + \hat{a}_2 f^2(x)$ with respect to the data points $\{(1,1), (2,2), (4,5)\}$, where it is assumed the variance of the error in each y_i is constant with respect to i.

4.3.2. Assume $\{y_i | 1 \leq i \leq N\}$ are a set of independent samples from a normal probability distribution with known mean $\mu = M$. Show that the maximum likelihood estimate of the variance of the distribution is $\hat{\sigma}^2 = \frac{1}{N} \sum_{i=1}^{N} (y_i - M)^2$. Note that the maximum likelihood estimator is the value of $\hat{\sigma}^2$ which maximizes $p_{\{y_i^*\}|\mu,\sigma^2}(\{y_i\}|M,\hat{\sigma}^2)$.

4.3.3. Assume $\{y_i | 1 \leq i \leq N\}$ are a set of independent samples from a normal distribution. Show by the following sequence of steps that the maximum likelihood estimate of the variance σ^2 of this distribution is given by

$$\hat{\sigma}^2 = \frac{1}{N-1}\sum_{i=1}^{N}(y_i - \bar{y})^2, \quad \text{where} \quad \bar{y} = \frac{1}{N}\sum_{i=1}^{N} y_i.$$

a) We wish to maximize $p_{\{y_i^*\}|\sigma^2}(\{y_i\}|\hat{\sigma}^2)$ over $\hat{\sigma}^2$. Write this probability in terms of $p_{\{y_i^*\}|\sigma^2,\mu}(\{y_i\}|\hat{\sigma}^2,M)$ and $p_{\mu|\sigma^2}(M|\hat{\sigma}^2)$, where M is the unknown distribution mean.

b) By the maximum uncertainty principle inherent in maximum likelihood estimation, we can assume $p_{\mu|\sigma^2}(M|\hat{\sigma}^2)$ is a constant. Therefore, show that $p_{\{y_i^*\}|\sigma^2}(\{y_i\}|\hat{\sigma}^2)$ is proportional to

$$\int_{-\infty}^{\infty} \frac{1}{\sqrt{2\pi}\hat{\sigma}^N} e^{\frac{-1}{2\hat{\sigma}^2}\sum_{i=1}^{N}(y_i - M)^2}\, dM.$$

c) Replace $y_i - M$ in the above expression by $(y_i - \bar{y}) - (M - \bar{y})$. Expand the square in terms of these parenthesized expressions to produce the result that $p_{\{y_i^*\}|\sigma^2}(\{y_i\}|\hat{\sigma}^2)$ is proportional to

$$e^{\frac{-1}{2\hat{\sigma}^2}\sum_{i=1}^{N}(y_i - \bar{y})^2} \int_{-\infty}^{\infty} \frac{1}{\sqrt{2\pi}\hat{\sigma}^N} e^{\frac{-N}{2\hat{\sigma}^2}(M-\bar{y})^2}\, dM.$$

d) Take $1/\sqrt{N}\hat{\sigma}^{N-1}$ out of the integral in part c and note that the resulting integrand is the probability density for a random variable M normally distributed with mean $\bar{y}$ and variance $\hat{\sigma}^2$. Using this fact, integrate to show that $p_{\{y_i^*\}|\sigma^2}(\{y_i\}|\hat{\sigma}^2)$ is proportional to

$$\frac{1}{\hat{\sigma}^{N-1}} e^{\frac{-1}{2\hat{\sigma}^2}\sum_{i=1}^{N}(y_i - \bar{y})^2}$$

e) Maximize the expression in part d over $\hat{\sigma}^2$ to produce the required result.

4.3.4. Assume $\{y_i|1 \le i \le N\}$ are a set of independent samples, each from a different normal distribution with the same variance σ^2 but with a different mean $\mu_i = M_i$, $1 \le i \le N$. Assume $M_i = \sum_{j=1}^{m} a_j f^j(x_i)$, where $f^1(x) \equiv 1$ and $m < N$. Show that the maximum likelihood estimator for σ^2 is $\hat{\sigma}^2 = \frac{1}{N-m}\sum_{i=1}^{N}\left(y_i - \sum_{j=1}^{m}\hat{a}_j f^j(x_i)\right)^2$, where the $\hat{a}_j$ are the coefficients of the least-squares fit to the y_i.

The proof of the above result should proceed very much like that of exercise 4.3.3, except that instead of one unknown parameter, the mean value M, which is assumed to have equal *a priori* probability over $[-\infty,\infty]$, there are m parameters $\hat{a}_1, \hat{a}_2, \ldots, \hat{a}_m$, each of which has this property.

4.3.5. Let $\{(x_i, y_i)|i = 1,2,\ldots,N\}$ be a set of arguments and data values. Further, let $\{f^k(x)|k = 1,2,\ldots,N\}$ be orthonormal functions over the argument values.

a) Show we can fit this data in the least-squares sense by $\hat{f}(x) = \sum_{k=1}^{m}\hat{a}_k f^k(x)$, where $m \le N$ and $\hat{a}_k = \sum_{i=1}^{N} y_i f^k(x_i)$.

b) Assume that in computing $f^k(x_i)$ we make error δ_{ki}. Assume that δ_{ki} and δ_{kj} are uncorrelated for $i \ne j$, $E(\delta_{ki}) = 0$, and $var(\delta_{ki}) = \sigma^2_{comp}$. Further assume that each

y_i has error ε_i, ε_i and ε_j are uncorrelated for $i \neq j$, $E(\varepsilon_i) = 0$, and $var(\varepsilon_i) = \sigma_{inp}^2$. Finally, assume ε_i and δ_{kj} are independent for all i, k, and j. What is the variance in the computed value of $\hat{a}_k$, where expected values are taken over probability distributions corresponding to both sources of error?

4.3.6. Consider the general least-squares problem (variable σ_i) discussed in this section. Assume the vectors f^j defined by equation 271 are orthonormal according to the inner product given by equation 303. If $\hat{a}$ is the vector of least-squares coefficients, show that $var(\hat{a}_k)$ is independent of k. What is the value of $var(\hat{a}_k)$?

4.3.7. Assume that for $j \neq n$,

$$\sum_{i=1}^{N} (f_i^j f_i^n / k_i^2) = 0, \quad \text{where} \quad k_i \equiv \sigma_{y_i}/\sigma_{y_1}.$$

a) Show that equation 315 can be rewritten as

$$\hat{\sigma}_1^2 = \frac{1}{N-m}\left(\sum_{i=1}^{N} (y_i/k_i)^2 - \sum_{j=1}^{m}\left(\hat{a}_j^2 \sum_{i=1}^{N} (f_i^j/k_i)^2\right)\right).$$

Hint: Write an equation for y_i in terms of all N of the f^j, and use that form in the cross term when expanding the square in equation 315.

b) Prove Parseval's theorem, which states that

$$\sum_{i=1}^{N} (y_i/k_i)^2 = \sum_{j=1}^{N}\left(\hat{a}_j^2 \sum_{i=1}^{N} (f_i^j/k_i)^2\right).$$

4.3.8. a) Show that given a set of data points $\{(x_i, y_i) | 1 \leq i \leq N\}$, where each y_i has error with variance σ^2, the coefficients of the least-squares line, $\hat{f}(x) = \hat{a}x + \hat{b}$, are given by equations 339 and 340.

b) Show that the variances and covariance of the least-squares line coefficients, $\hat{a}$ and $\hat{b}$, are given by equations 341, 342, and 343.

c) Show that $var(\hat{f}(x)) = var(\hat{a})x^2 + 2cov(\hat{a},\hat{b})x + var(\hat{b})$.

4.3.9. Given the data $\{(1,1.3), (2,1.8), (3,2.9), (5,5.1)\}$, where all of the y_i have independent errors with the same variance, σ^2, and the underlying function is assumed to be a straight line,

a) Compute the least-squares line.

b) Estimate σ^2.

c) Estimate $var(\hat{f}(2))$ by using the result of exercise 4.3.8, part c, and equations 341, 342, and 343, which in turn require the estimate of σ^2 found in part b above.

4.3.10. Repeat exercise 4.3.9 but assume the variance of the error in y_i varies over i such that $\sigma_i^2 = x_i\sigma^2$, where σ^2 is a constant (and not the variance of anything).

4.3.11. In fitting a least-squares line to the points $\{(x_i, y_i) | 1 \leq i \leq N\}$ when there are errors of the same magnitude in the x_i and the y_i, it is reasonable to minimize the sum of the squares of the perpendicular distances from the data points to the fitted line. Give equations for the slope and y-intercept of the fitted line as a function of the x_i and the y_i.

4.3.12. a) By Gram-Schmidt orthogonalization, find four orthogonal polynomials over the argument values $x_1 = -2$, $x_2 = -1$, $x_3 = 0$, and $x_4 = 3$.

b) Using the recurrence relation method, find the same four polynomials.

4.3.13. Use the recurrence relation method to find the best least-squares 0th-, 1st-, 2nd-, and 3rd-degree polynomials for the data $\{(-3,6), (-2,12), (0,1), (1,0), (4,85)\}$, assuming the variance of the error at each point is the same.

4.3.14. Assume we are given a set of values $\{(x_i, y_i, \sigma_i^2) | 1 \le i \le N\}$, where σ_i^2 is the variance of the error in $y_i \equiv y(x_i)$. Let $z_i = ax_i + b$, where a and b are constants.

Consider solving the linear least-squares problem with the basis functions f^j, $1 \le j \le m$, using the data $\{(z_i, y_i, \sigma_i^2) | 1 \le i \le N\}$. Let $\hat{f}(z)$ be the result of that solution.

Define $g^j(x) \equiv f^j(ax + b)$. Let $\hat{g}(x)$ be the result of solving the linear least-squares problem with the basis functions $g^j(x)$ using the original data.

a) Show that $\hat{g}(x) = \hat{f}(ax + b)$.

b) Show that if f^j is a $(j-1)$th-degree polynomial, then so is g^j.

4.3.15. Pick $p(x)$ as any 5th-degree polynomial with five real roots between -10 and 10.

a) Write and run a program to do the following:
Compute $p(x)$ at the 31 points: $x = -15(1)15$. To each point, add normally distributed error with the same standard deviation $= .05\overline{|p(x_i)|}$, where the bar indicates an average taken over the x_i inside the outside zeros of $p(x)$. For $m = 1,2,3,\ldots,11$, compute $\hat{p}_{(m)}$, the least-squares fitting polynomial of degree $m-1$, and $\hat{\sigma}_{(m)}^2$, the estimate of the variance of the data, assuming it comes from a polynomial of degree m.

b) Plot $\hat{\sigma}_{(m)}^2$ vs. m and note the behavior of the plotted points.

4.3.16. Consider fitting a data set over the arguments $x_i = 0,1,2,\ldots,N$ by a linear combination of the functions $f^k(x) = 2^{-kx}$, $k = 0,1,\ldots,N$, using the least-squares criterion: $\hat{f}(x) = \sum_{k=0}^{m} \hat{a}_k f^k(x)$.

a) If the data represent samples from a function $h(x)$ with error, what properties would you expect $h(x)$ to have if the data is to be well fit by the linear combination of the $f^k(x)$?

b) What property of the $f^k(x)$ would lead you to believe that the solution of the normal equations to find $\{\hat{a}_k | k = 0,1,\ldots,m\}$ will be ill-conditioned?

c) We wish to build a new set of functions $g^k(x) = \sum_{j=0}^{k} c_{kj} f^j(x)$ such that $c_{kk} = 1$ and if $\hat{f}(x) = \sum_{k=0}^{m} \hat{b}_k g^k(x)$, the difficulty encountered in part b will not occur in finding $\{\hat{b}_k | k = 0,1,\ldots,m\}$. What other advantages will this formulation of the problem have over the formulation in part b?

d) Find $g^0(x)$, $g^1(x)$. Note $\sum_{x=0}^{N} 2^{-kx} = (1 - 2^{-k(N+1)})/(1 - 2^{-k})$.

e) Find a recurrence relation for $g^{i+1}(x)$. First show a recurrence relation exists, and then find the appropriate constants.

f) As $N \to \infty$, the largest root of $g^k(x)$ gets large for large k. How would you find the largest root of $g^k(x)$ for some given $k \gg 0$ (without using roots of $g^j(x)$ for $j < k$)? Note the difficulties involved in using methods other than that chosen for finding the root. *Note:* This is a tricky question in the area of the solution of nonlinear equations, not in the area of approximation.

4.3.17. Consider the set of functions $f^k(x) = e^{-x}p^{k-1}(x)$, $k = 1,2,\ldots,m$, where $p^k(x)$ is a polynomial of degree k.

a) Using the facts that $\Delta e^{-ax} = e^{-a(x+1)} - e^{-ax} = (e^{-a} - 1)e^{-ax}$ and $\sum_{i=j}^{k} f_i \Delta g_i$ $= f_{k+1}g_{k+1} - f_j g_j - \sum_{i=j}^{k} g_{i+1}\Delta f_i$ (summation by parts), find $p^0(x)$ and $p^1(x)$ such that $f^1(x)$ and $f^2(x)$ are orthonormal over $\{x_i\} = \{0,1,2,\ldots,N-1\}$.

b) How many orthonormal f^k can there be, that is, what is the maximum possible value of m?

c) What are the characteristics of data to which such a set of approximating functions should be applied? Compare the effectiveness of these functions for approximating such data with that of orthogonal polynomials.

d) Given $\{y_i | 1 \le i \le N\}$, describe in detail how you would find the least-squares approximation $\hat{f}(x) = \sum_{k=1}^{m} \ddot{a}_k f^k(x)$.

e) If the function underlying the data is believed to be of the form $e^{-x}q(x)$, where $q(x)$ is a polynomial of degree less than m, how would you find the degree of q?

4.3.18. Let $\{f^j(x) | 1 \le j \le m\}$ be a set of functions such that the vectors f^j defined by $f_i^j \equiv f^j(x_i)$ are linearly independent over the set $\{x_i | i = 1,2,\ldots,N\}$. Define the $(m \times m)$ matrix F as in equation 280. Since F is symmetric, it has m orthonormal eigenvectors, v^k, $1 \le k \le m$, with corresponding eigenvalues λ_k.

a) Show all of the λ_k are strictly positive, that is, F is positive definite.

b) Let $g^k(x)$, $1 \le k \le m$, be linear combinations of the $f^j(x)$, that is, let $g^k(x) = \sum_{j=1}^{m} a_{kj} f^j(x)$, such that the $g^k(x)$ are orthonormal over the x_i according to the inner product given by equation 303. Show that $a_{kj} = v_j^k/\lambda_k^{1/2}$ produces such a set of g^k.

4.3.19. a) Show that the operator defined by equation 389 is an inner product.

b) Show that in the least-squares problem where the approximating function is constrained to pass through (x_{c_1}, y_{c_1}) and (x_{c_2}, y_{c_2}), in order for the normal equations to be uncoupled, the basis functions should be orthogonal according to the inner product defined by equation 389.

4.3.20. Find the least-squares polynomial of degree 3 which passes through (0,0) and (1,1), where the remaining data points are $\{(-2,-7), (-1,-1), (2,7), (3,25)\}$, assuming the variance of the error in the y_i is constant.

4.3.21. Find the least-squares line for the following set of (x_i,y_i) pairs, where $\sigma_i^2 = x_i$: $\{(0,1), (1,3), (3,2), (5,7)\}$.

4.3.22. Assume we wish to construct a least-squares cubic spline over $\{x_j, x_{j+1}, \ldots, x_{j+k-1}\}$ which agrees at x_j in value, slope, and curvature with a spline over a previous interval. Assume the previous spline has been calculated, so that $\hat{f}(x_j)$, $\hat{f}'(x_j)$, and $\hat{f}''(x_j)$ can be assumed to be given. Find expressions for the coefficients of the cubic spline, $\hat{f}(x)$, to be fit over the above argument points.

4.3.6 Orthogonal Functions

Given an inner product and a set of argument points, there are an infinite number of sets of functions orthogonal over that set of points according to that inner product. Any of these sets can be used for least-squares approximation. Which set should be used depends on (1) assumptions about the function underlying the data, (2) the smoothness of approximation obtainable with the set of functions, and (3) the computational efficiency when using the set of functions.

To make the comparison of sets of orthogonal functions easier, we will consider only normalized functions. In doing so, we lose nothing since

$$\hat{f}_{(m)}(x) = \sum_{j=1}^{m} \hat{a}_j f^j(x) \tag{390}$$

can be equivalently written

$$\hat{f}_{(m)}(x) = \sum_{j=1}^{m} \hat{b}_j g^j(x), \tag{391}$$

where

$$g^j(x) = f^j(x)/(f^j,f^j)^{1/2}, \tag{392}$$

the normalized form of $f^j(x)$, and

$$\hat{b}_j = \hat{a}_j(f^j,f^j)^{1/2}. \tag{393}$$

Then from equation 376,

$$\hat{b}_j = (g^j,y). \tag{394}$$

For simplicity, let us assume $var(y_i)$ is constant. Then from equation 333,

$$\hat{\sigma}^2_{(m)} = \frac{1}{N-m}\left(\sum_{i=1}^{N} y_i^2 - \sum_{j=1}^{m} \hat{b}_j^2\right), \tag{395}$$

where $\hat{\sigma}^2_{(m)}$ is the estimator of $var(y_i)$, assuming the underlying function values $f(x_i)$ can be exactly matched by $\sum_{j=1}^{m} b_j g^j(x)$ for some set of b_j. We have previously discussed the fact that $\hat{\sigma}^2_{(m)} \approx var(y_i)$ when the value of m is large enough so that the above exact-matching assumption can be expected to hold, that is, when the approximation given by $\hat{f}(x)$ cannot be improved any further.

Thus from equation 395, m is large enough when

$$\sum_{j=1}^{m} \hat{b}_j^2 \approx \left(\sum_{i=1}^{N} y_i^2 \right) - (N - m)var(y_i). \tag{396}$$

Assuming the relative error in the y_i is small, that is, $var(y_i) \ll y_i^2$, $1 \leq i \leq N$, the last term in equation 396 is negligible compared to $\sum_{i=1}^{N} y_i^2$. Therefore, we can say m is large enough when

$$\sum_{j=1}^{m} \hat{b}_j^2 \approx \sum_{i=1}^{N} y_i^2 \equiv Y, \tag{397}$$

where we note that Y is independent of the set of basis functions. Parseval showed that

$$\sum_{j=1}^{N} \hat{b}_j^2 = Y \tag{398}$$

(recall exercise 4.3.7) so we conclude that $\sum_{j=1}^{m} \hat{b}_j^2$ is an increasing bounded function of m.

If the basis functions g^j are such that $\hat{b}_j^2$ drops quickly as a function of j, equation 398 implies that relation 397 must hold for a small value of m; only a few basis functions will be necessary for an adequate fit. We will see that such a situation is desirable from the points of view of both the efficiency and the smoothness of the approximation. We will also see that using certain sinusoids as basis functions results in this desirable behavior of the $\hat{b}_j^2$.

All sets of orthogonal functions over a set of arguments have some geometric characteristics in common. Sets of continuous orthogonal functions have the property that if the functions are ordered in increasing order of the number of zero crossings in the interval bounded by the data arguments, the number of such zeros increases linearly with the function number. That is, generally the ith function in this ordering has $i-1$ or i zeros in the interval (for orthogonal polynomials according to any inner product, the ith-degree polynomial has exactly i distinct zeros in the interval†). Thus, orthogonal functions are inherently oscillatory with greater and greater frequency as i increases.

Let us give an argument that intuitively supports the statement above. We assume that the first function, $f^1(x)$, in a set of orthogonal functions is the constant function. Other functions in the set are orthogonal to this constant function only if, on the average, they are positive half the time and negative half the time. In particular, a function with a single zero crossing must be positive on one side of the zero crossing and negative on the other side of the zero crossing, in such a way that the average of all function values at data arguments to the left of the zero crossing have the same magnitude as the average of all function values at data arguments to the right of the zero crossing (see $f^2(x)$ in figure 4-15).

The third function must be orthogonal to both of the first two functions; not only must it be on the average half positive and half negative, but also its product with the second function, which is itself half negative and half positive, must be half negative and half positive. Therefore, during the range in which the second function is positive, the third function must be on the average half positive and half negative; it must behave similarly in the part of the range in which the second function is negative. Therefore, we obtain the third function $f^3(x)$ as in figure 4-15.

By the same sort of argument, we can show that the fourth function must be like $f^4(x)$ in figure 4-15. In summary, we know that, for a given function to be orthogonal to another function, its product with that other function must have zero as an average value over the data points. For this to occur with respect to functions which do not oscillate very much, the new function must oscillate faster than any of the other functions, so that its product with those functions is sometimes positive and sometimes negative. Thus we see, for example, that the function f^i in our example has $i-1$ zero crossings over the interval defined by the data arguments. This sort of behavior is generally the case. That is, a set of orthogonal functions can be ranked in order of the number of oscillations in the data argument interval, and increasing the number of orthogonal functions means adding functions which are more and more oscillatory.

Assume we have ordered our set of orthogonal functions in increasing order of number of oscillations, as is the case, for example, with orthogonal polynomials ordered by degree. Furthermore, assume that all of the basis functions are normalized (have norm 1) according to the $\mathscr{L}_2$ norm.

---

† See G. Szegö, *Orthogonal Polynomials*, page 46, with

$$\alpha(x) = \begin{cases} 0, & x < x_0 \\ i, & x_i \le x < x_{i+1}, \quad 1 \le i < N-1 \\ N, & x \ge x_N. \end{cases}$$

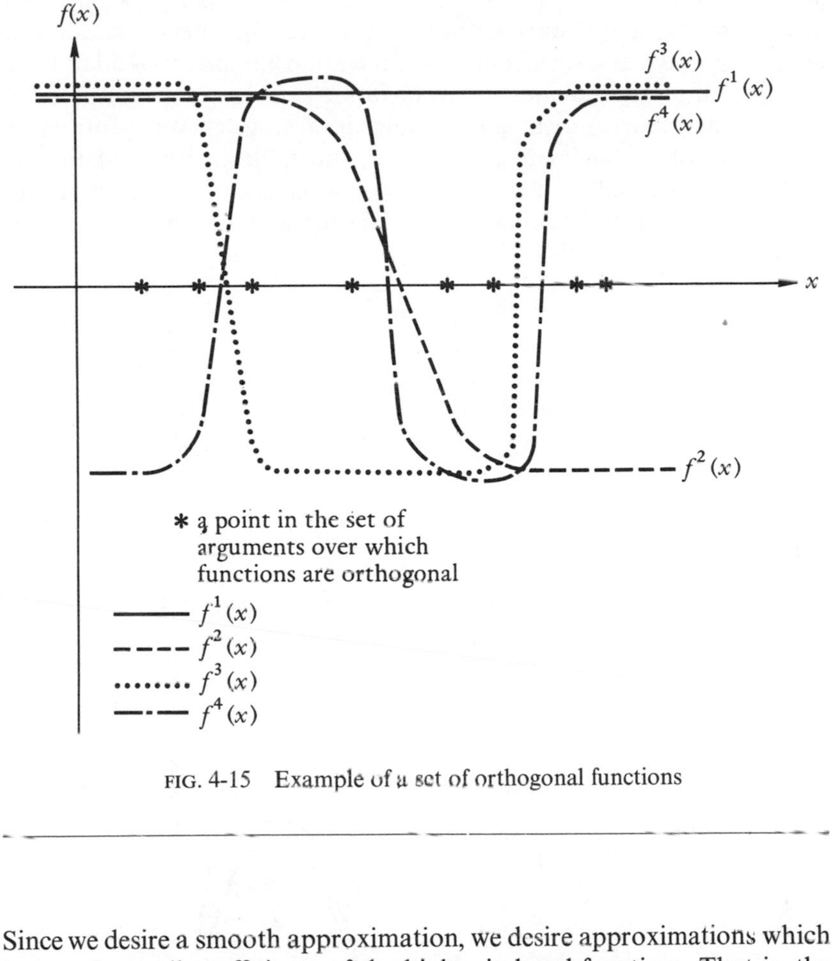

FIG. 4-15 Example of a set of orthogonal functions

Since we desire a smooth approximation, we desire approximations which have only small coefficients of the higher-indexed functions. That is, the set of orthogonal basis functions that we use in approximating should have the property that a good approximation can be obtained using only the first few members of the set. This property results in both smooth fit and efficient computation. Furthermore, the basis functions should be as smooth as possible for the given number of zero crossings. That is, the average curvature of each function should be as small as possible. This behavior occurs when all peaks and valleys of the function are of approximately the same magnitude. Relatively large excursions from zero require a large curvature to return to near zero quickly enough to make the $\mathscr{L}_2$ norm 1. For example, consider over the interval $[0,3\pi]$ the function $f(x) = (2/3\pi)^{1/2}\, sin(x)$, which has equal ripples, and the function $g(x) = -(70/\pi)^{1/2}(1/9\pi^4)x(x-\pi)(x-2\pi)(x-3\pi)$, which does not (see figure 4-16). Both functions are normalized over the interval in question according to the $\mathscr{L}_2$ norm, both have the same number of oscillations in the interval in question, and both have the same zeros in the interval. The maximum curvature magnitude and the average curvature (using the $\mathscr{L}_2$ norm) of f and of g are given in the table in figure 4-16. The function with equal ripples has a much smaller maximum curvature magnitude and average curvature than the function which does not have equal ripples.

	$f(x) = \left(\frac{2}{3\pi}\right)^{1/2} sin(x)$	$g(x) = -\left(\frac{70}{\pi}\right)^{1/2} \frac{1}{9\pi^4} x(x-\pi)(x-2\pi)(x-3\pi)$
Maximum curvature magnitude over $[0,3\pi]$	.46	1.17
Average curvature over $[0,3\pi]$	.33	.44

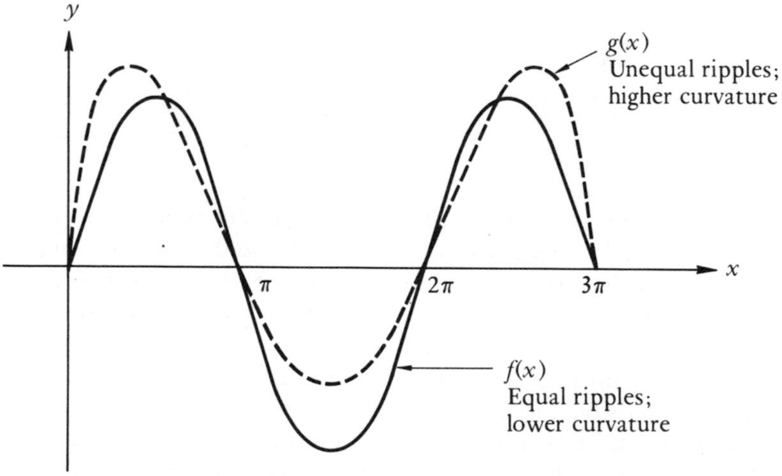

FIG. 4-16 The relation of ripple equality to curvature

Therefore, we desire a set of basis functions such that smooth functions are fit with relatively few of the set of basis functions and such that the basis functions, which we know must oscillate, have ripples which are not far from equal.

If we assume that the data arguments are not far from equally spaced, we know that at least the highest-degree orthogonal polynomial must have a zero between every pair of data arguments, and thus have nearly equally spaced zeros. It can be shown that other of the high-degree polynomials have their zeros approximately equally spaced. But we know from our discussion of the function $\psi(x)$ (see section 4.2.1, part f) that polynomials with equally spaced zeros are functions with ripples that are far from equal. To put it another way, if the data arguments are equally spaced, the orthogonal polynomials are quite nonsmooth for even low-degree polynomials. This early lack of smoothness results in having to make a choice between (a) using only the first members of the orthogonal set, producing a smooth but poor fit, (b) using the first and middle members of the set, producing a nonsmooth fit, or (c) using very small amounts of the high-degree polynomials to try to correct for non-smoothness of the previous fit, a strategy which may result either in continued lack of smoothness or in a smooth fit but a very inefficient one. Therefore, for equally spaced data argument values, polynomial least-squares approximation is useful only for the lowest-degree polynomials over a relatively small argument range (or as splines) unless the underlying function is itself a polynomial. Otherwise, we should use a set of functions with more equal ripples and for which fewer basis functions are necessary to produce an adequate approximation at the data arguments. The *sinusoids,* sometimes called the *Fourier functions,* make up such a set of functions. The problems of approximation with these functions will take our attention in the next section.

4.3.7 Fourier Approximation

a) Concepts and Methods. Given N data points with equally spaced data arguments, it can be shown that there exist N functions orthogonal over these data arguments which are each either sine or cosine of a multiple of x. We shall discuss only the case when N is even and the data arguments are consecutive integers, but the above statement is correct for any interval and for any number of data points. For the case of an odd number of consecutive integers, the development is analogous to that below for an even number of integers (see exercise 4.3.27). For equally spaced data arguments with noninteger spacing, the least-squares approximation problem can be solved by linearly mapping the data argument variable (which we call z) such that the data arguments are mapped onto consecutive integers in the approximating argument (which we call x), solving the least-squares problem for the integer arguments, and mapping the resulting approximation in x back to the original variable z by the linear argument mapping which is the inverse of the original mapping (see exercise 4.3.25).

Consider the set of data arguments $\{x_i\} = \{0,1,2,\ldots,2n-1\}$. We can show that the following functions are orthogonal over these argument values:

$$\begin{aligned} f^1(x) &= cos(0\pi x/n) = 1, \\ f^{2j}(x) &= cos(j\pi x/n), \quad j = 1,2,\ldots,n, \\ f^{2j+1}(x) &= sin(j\pi x/n), \quad j = 1,2,\ldots,n-1 \end{aligned} \tag{399}$$

(see exercise 4.3.26). Notice that $sin(j\pi x/n)$ is not included for $j = 0$ or $j = n$ because the function is identically zero at all of the data arguments. Furthermore, notice that for $n < j \leq 2n$ if x is a data argument,

$$\begin{aligned} cos(j\pi x/n) &= cos\left(2\pi x - \frac{(2n-j)\pi x}{n}\right) \\ &= cos(2\pi x)cos((2n-j)\pi x/n) \\ &= cos((2n-j)\pi x/n) \end{aligned} \tag{400}$$

and

$$\begin{aligned} sin(j\pi x/n) &= sin\left(2\pi x - \frac{(2n-j)\pi x}{n}\right) \\ &= -cos(2\pi x)sin((2n-j)\pi x/n) \\ &= -sin((2n-j)\pi x/n). \end{aligned} \tag{401}$$

A similar result is obtained for $sin(j\pi x/n)$ and $cos(j\pi x/n)$ for $j > 2n$. Thus, for j outside the range given in equation 399, we simply get vectors already in the set or their negative. The high-frequency sinusoids look like lower-frequency sinusoids as far as their values at the data points are concerned.

To use this set of functions in least-squares approximation, we need to use equation 376, which requires the value of

$$(f^j,f^j) = \sum_{x=0}^{2n-1} [f^j(x)]^2 \tag{402}$$

for each j. It can be proved that

$$\sum_{x=0}^{2n-1} [f^1(x)]^2 = \sum_{x=0}^{2n-1} [f^{2n}(x)]^2 = 2n \tag{403}$$

and

$$\sum_{x=0}^{2n-1} [f^j(x)]^2 = n, \quad 2 \leq j \leq 2n-1 \tag{404}$$

(see exercise 4.3.26, part c).

Solving the least-squares approximation problem with this set of orthogonal functions,

$$\hat{f}(x) = \sum_{j=1}^{m} \hat{a}_j f^j(x), \tag{405}$$

by equations 376, 403, and 404 produces

$$\hat{a}_j = \begin{cases} \sum_{i=1}^{N} y_i f^j(x_i)/2n, & j = 1,2n \\ \sum_{i=1}^{N} y_i f^j(x_i)/n, & 2 \leq j \leq 2n-1, \end{cases} \tag{406}$$

where $N = 2n$ and $x_i = i - 1$. The approximation coefficients $\hat{a}_j$ are sometimes called the *discrete Fourier transform* of the discrete function y.

We will show that this sinusoidal approximation is quite efficient in the sense that we can obtain a rather good approximation using a very small m. But we may object that this efficiency is impaired by the necessity to compute the values $f^j(x_i)$, that is, to apply many times a subroutine which evaluates sinusoids. We can overcome this objection by noting that there exists a recurrence relation by which these sinusoids can be evaluated quite efficiently. We first compute $cos(\pi/n)$ and $sin(\pi/n)$. Then, using the formulas for the sine and cosine of sums of angles,

$$cos\left(\frac{\pi x}{n}\right) = cos\left(\frac{\pi(x-1)}{n}\right)cos\left(\frac{\pi}{n}\right) - sin\left(\frac{\pi(x-1)}{n}\right)sin\left(\frac{\pi}{n}\right) \tag{407}$$

and

$$sin\left(\frac{\pi x}{n}\right) = sin\left(\frac{\pi(x-1)}{n}\right)cos\left(\frac{\pi}{n}\right) + cos\left(\frac{\pi(x-1)}{n}\right)sin\left(\frac{\pi}{n}\right), \tag{408}$$

we have recurrence relations for computing f^2 and f^3. Finally, using

$$cos\left(\frac{j\pi x}{n}\right) = cos\left(\frac{(j-1)\pi x}{n}\right)cos\left(\frac{\pi x}{n}\right) - sin\left(\frac{(j-1)\pi x}{n}\right)sin\left(\frac{\pi x}{n}\right) \tag{409}$$

and

$$sin\left(\frac{j\pi x}{n}\right) = sin\left(\frac{(j-1)\pi x}{n}\right)cos\left(\frac{\pi x}{n}\right) + cos\left(\frac{(j-1)\pi x}{n}\right)sin\left(\frac{\pi x}{n}\right), \tag{410}$$

we have a recurrence relation to compute f^j for $j > 3$. Therefore, the only f values that need to be computed by other than a recurrence relation are $cos(\pi/n)$ and $sin(\pi/n)$.

Let us note some properties of our approximation using sinusoids.

1. Each of the f^j are cyclic functions with period $2n$. Therefore $\hat{f}(x)$ is a cyclic function with period $2n$. Effectively, we are fitting the data extended cyclically with period $2n$ by a cyclic function $\hat{f}(x)$ with the same period.

   Because the $f^j(x)$ are cyclic, the orthogonality relationships hold over any $2n$ consecutive integers. Therefore, the theory above holds for $x_i = i - M$ for any integer M. Note that if we have a problem with equal spacing and map the original argument range so that data arguments z_i are mapped onto the consecutive integers x_i, the mapping is a function of M, and the basis functions in the original argument z are a function of M. Therefore, different least-squares fits are obtained for different values of M. A very commonly used value of M is $M = n$. In this case the approximation arguments are symmetrically placed with regard to the origin except for the additional argument at n:

$$\{x_i\} = \{-n+1, -n+2, \ldots, -1, 0, 1, 2, \ldots, n\}. \qquad (411)$$

2. Let the x_i be the symmetric integers, with the addition of n, given above. Assume the data is from an even function $y(x)$, that is, one symmetric about the y axis:

$$y(x) = y(-x). \qquad (412)$$

   Then by equation 406 the coefficients of the sine function, $\hat{a}_j$ for $j = 3,5,7,\ldots,2n-1$, are zero by the following argument:

$$\begin{aligned} n\hat{a}_{2j+1} &= \sum_{x=-n+1}^{n} y(x)sin(j\pi x/n), \quad j = 1,2,\ldots,n-1 \\ &= y(0)\,sin(0) + y(n)\,sin(j\pi) \\ &\qquad + \sum_{x=1}^{n-1} (y(x)sin(j\pi x/n) + y(-x)sin(-j\pi x/n)) \\ &= 0 + 0 + \sum_{x=1}^{n-1} (y(x) - y(-x))sin(j\pi x/n) \\ &= 0, \text{ since } y(x) \text{ is even.} \end{aligned} \qquad (413)$$

   Thus if $y(x)$ is even, the least-squares fit is a *pure cosine series*. This result is useful not only when the data is actually even, but more often, if N is the number of given data points, we let $n = N - 1$, map the N data arguments linearly onto $0,1,\ldots,n$, reflect the resulting discrete function about the origin, and then fit the resulting discrete function by a pure cosine series using the values at the symmetric set of data arguments used for the pure cosine series (see figure 4-17, part b). This process, in which the basis functions are cosine functions only, produces yet a different least-squares fit from the ones above.

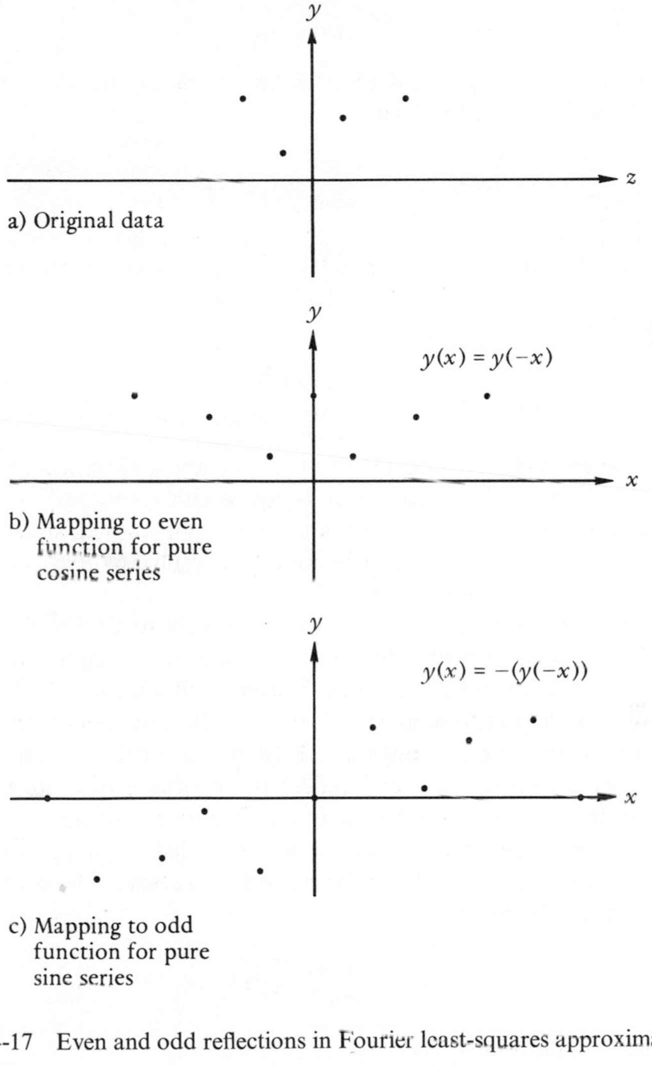

FIG. 4-17 Even and odd reflections in Fourier least-squares approximations

3. Similarly, if $y(x)$ is odd (that is, its values on the left side of the y axis are the negative of its corresponding values on the right side of the axis:

$$y(x) = -y(-x)), \tag{414}$$

we can show that the least-squares Fourier approximation over the same symmetric set of data arguments produces close to a *pure sine series*. First note that equation 414 implies that

$$y(0) = 0. \tag{415}$$

Therefore every coefficient of a cosine term in the Fourier series approximation is of the form

$$\begin{aligned}
&(1/C_j) \sum_{x=-n+1}^{n} y(x)cos(j\pi x/n), \quad j = 0,1,\ldots,n, \\
&= (1/C_j)(y(n)cos\, j\pi + y(0))cos\, 0 + \sum_{x=1}^{n-1} (y(x) + y(-x))cos(j\pi x/n) \\
&= (-1)^j y(n)/C_j,
\end{aligned} \tag{416}$$

where
$$C_j = \begin{cases} 2n, & j = 0 \text{ or } j = 2n \\ n, & \text{otherwise.} \end{cases} \tag{417}$$

Thus the coefficients of cosine terms in the approximation are zero if $y(n) = 0$. Note that this condition is equivalent to a condition that a cyclic continuation of y with period $2n$ is single-valued, since y being odd implies $y(n) = -y(-n)$, y being cyclic implies $y(n) = y(-n)$, and together, these imply $y(n) = 0$.

As with the pure cosine series, it is unlikely to find a set of data which is in fact odd and has a zero value at the right end point. However, we can take the given N arguments, let $n = N + 1$, map the smallest data argument onto 1, the next largest onto 2, and so on, with the largest mapped onto $n - 1$, then arbitrarily set the value at $x = 0$ and $x = n$ to zero, and reflect the resulting discrete function negatively about the y axis (see figure 4-17, part c). Fitting this discrete function produces a pure sine series as a least-squares fit, yet a different least-squares fit from those produced above. The coefficients, $\hat{a}_j$, of this least-squares fit,

$$\hat{f}(x) = \sum_{j=1}^{m} \hat{a}_j\, sin(j\pi x/n), \tag{418}$$

are from equation 406 given by

$$\hat{a}_j = (1/n) \sum_{x=-n+1}^{n} y(x)sin(j\pi x/n)$$

$$= (1/n)\left(y(0)\,sin(0) + y(n)\,sin(j\pi) + \sum_{x=1}^{n-1} (y(x) - y(-x))sin(j\pi x/n)\right)$$

$$= (2/n) \sum_{x=1}^{n-1} y(x)sin(j\pi x/n), \quad j = 1,2,\ldots,m. \tag{419}$$

4. Since sines and cosines can be written as exponential functions of imaginary arguments, the whole of the foregoing can be written in terms of these exponentials. We write

$$\hat{f}(x) = \sum_{j=m_1}^{m_2} \hat{b}_j \, e^{-i\frac{j\pi x}{n}}, \tag{420}$$

where i is the square root of -1 and normally $m_1 = -m_2$ if $m_2 < n$ and $m_1 = -n+1$ and $m_2 = n$ if we are using the full set of basis functions. Equation 420 is equivalent to equation 405 if one of these conditions holds and

$$\hat{b}_j = \begin{cases} \hat{a}_{2j}, & j = 0 \text{ or } j = n \\ \frac{1}{2}(\hat{a}_{2j} + i\hat{a}_{2j+1}), & 0 < j < n \\ \frac{1}{2}(\hat{a}_{2|j|} - i\hat{a}_{2|j|+1}), & j < 0. \end{cases} \tag{421}$$

It is left to the student (1) to check this fact; (2) to show that the complex exponential functions which are the basis functions in equation 420 are orthogonal according to the inner product $(f^j, f^k) = \sum_{i=1}^{2n} f^j(x_i) f^{k*}(x_i)$, where "*" indicates complex conjugate, and thus the procedure for finding their coefficients (equation 376) is applicable, even in this case of complex functions and coefficients, and even if the y_i are complex; and (3) to show that finding the $\hat{b}_j$ directly by this method results in precisely the answer given by equation 421 if the y_i are real. (See exercise 4.3.28.)

It is sometimes useful to think of the Fourier approximation problem according to the complex characterization given by equation 420. This approach permits one to take advantage of some of the characteristics of the exponential function in applying Fourier theory.

We have discussed the above characteristics of Fourier approximation in order to develop the most efficient of the forms above and also to understand the applications of Fourier approximation and its extensions. The efficiency of Fourier approximation depends on the behavior of the approximation coefficients $\hat{a}_j$ as j increases. As soon as $\hat{a}_j$ becomes small, we can neglect the corresponding terms while obtaining a good fit; we can stop adding terms into our approximation.

If $y(x)$ is a smooth function, the $\hat{a}_j$ can be approximated by integrals:

$$\sum_{x=-n+1}^{n} y(x)sin(j\pi x/n) \approx \int_{-n}^{n} y(x)sin(j\pi x/n)dx \tag{422}$$

and

$$\sum_{x=-n+1}^{n} y(x)cos(j\pi x/n) \approx \int_{-n}^{n} y(x)cos(j\pi x/n)dx. \tag{423}$$

Therefore, we are interested in the following theorem, which describes the behavior of these integrals as j gets large.

Theorem: If $g(x)$ is piecewise continuous and bounded over $[-n,n]$ and j is an integer, as $j \to \infty, |j \int_{-n}^{n} g(x)sin(j\pi x/n)dx| \leq$ a constant $< \infty$, that is, the integral above is $O(1/j)$.

Proof: Since $g(x)$ is piecewise continuous, approximate $g(x)$ by a finite number of straight lines: $a_i x + b_i$, $x_i \leq x \leq x_{i+1}$. Then

$$I \equiv \int_{-n}^{n} g(x)sin(j\pi x/n)dx \approx \sum_i \int_{x_i}^{x_{i+1}} (a_i x + b_i)sin(j\pi x/n)dx$$

$$= \sum_i \frac{n}{j\pi}\Bigg[(b_i + a_i x_i)cos(j\pi x_i/n) - (b_i + a_i x_{i+1})\,cos(j\pi x_{i+1}/n)$$

$$+ \frac{n}{j\pi} a_i(sin(j\pi x_{i+1}/n) - sin(j\pi x_i/n))\Bigg]. \tag{424}$$

Thus, as $j \to \infty$,

$$jI \to \frac{n}{\pi}\sum_i [g(x_i^+)cos(j\pi x_i/n) - g(x_{i+1}^-)cos(j\pi x_{i+1}/n)]. \tag{425}$$

If x_i is not a discontinuity of f, successive terms cancel, so

$$jI \to \frac{n}{\pi}\Bigg(g(-n)\,cos(j\pi) - g(n)cos(j\pi)$$

$$+ \sum_{\substack{x \in [-n,n] \text{ and} \\ g \text{ discont. at } x}} (g(x^+) - g(x^-))cos(j\pi x/n)\Bigg). \tag{426}$$

By hypothesis, the number of discontinuities is finite and $g(x^+)$ and $g(x^-)$ are bounded, so

$$|jI| \leq \frac{n}{\pi}\Bigg(|g(n) - g(-n)| + \sum_{\substack{x \in [-n,n] \text{ and} \\ g \text{ discont. at } x}} |g(x^+) - g(x^-)|\Bigg). \tag{427}$$

A similar theorem holds when the sine function is replaced by the cosine function. For

$$I \equiv \int_{-n}^{n} g(x)cos(j\pi x/n)dx \tag{428}$$

we can show that as $j \to \infty$,

$$|jI| \le \frac{n}{\pi} \sum_{\substack{x \in [-n,n] \text{ and} \\ g \text{ discont. at } x}} |g(x^+) - g(x^-)|. \tag{429}$$

Thus, for smooth functions y,

$$\hat{a}_j = O(1/j); \tag{430}$$

that is, $\hat{a}_j$ gets small like $1/j$. This rate of convergence for these coefficients is not very fast. However, as shown below, the pure sine series and pure cosine series approximations developed above have coefficients which converge to zero much more quickly.

Assume that we have mapped our data argument values z_i onto the integers between 0 and n, where the function underlying the data has piecewise continuous first derivatives. Then if we let $y(i)$, $0 \le i \le n$, be the mapped data values and reflect this discrete function about the y axis and fit the result by a pure cosine series as described above, we see that the coefficient of $cos(j\pi x/n)$ in the cosine series is

$$\begin{aligned}\hat{a}_j &= \frac{1}{C_j} \sum_{x=-n+1}^{n} y(x)\, cos(j\pi x/n) \\ &\approx \frac{1}{C_j} \int_{-n}^{n} y(x)\, cos(j\pi x/n)dx,\end{aligned} \tag{431}$$

where C_j is equal to either n or $2n$, depending on the value of j. If we integrate the integral in equation 431 by parts, we produce

$$\int_{-n}^{n} y(x)cos(j\pi x/n)dx = (-n/j\pi) \int_{-n}^{n} y'(x)sin(j\pi x/n)dx. \tag{432}$$

Since by assumption y' is piecewise continuous, the integral on the right side of equation 432 is $O(1/j)$, so

$$\hat{a}_j = O(1/j^2). \tag{433}$$

If we integrate the right side of equation 432 once more, we produce

$$\begin{aligned}&\int_{-n}^{n} y(x)cos(j\pi x/n)dx \\ &\quad = (n/j\pi)^2 \left[(y'(n) - y'(-n))(-1)^j - \int_{-n}^{n} y''(x)cos(j\pi x/n)dx \right],\end{aligned} \tag{434}$$

assuming y'' exists. We note that to make the term in brackets $O(1/j)$, we require that

$$y'(n) = y'(-n), \tag{435}$$

a condition which is not true in general (see figure 4-18, part a) and which we cannot make true in any obvious way. Therefore the coefficients for the least-squares fit using a pure cosine series diminish like $1/j^2$ but no faster.

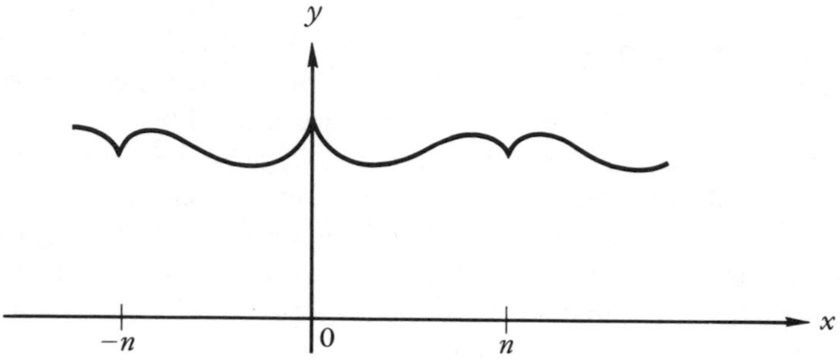

a) $y(x)$ even:
$y'(n^-) \neq y'(-n^+)$, $y'(0^+) \neq y'(0^-)$,
that is, $y'(x)$ is discontinuous
at $x = n$ and $x = 0$

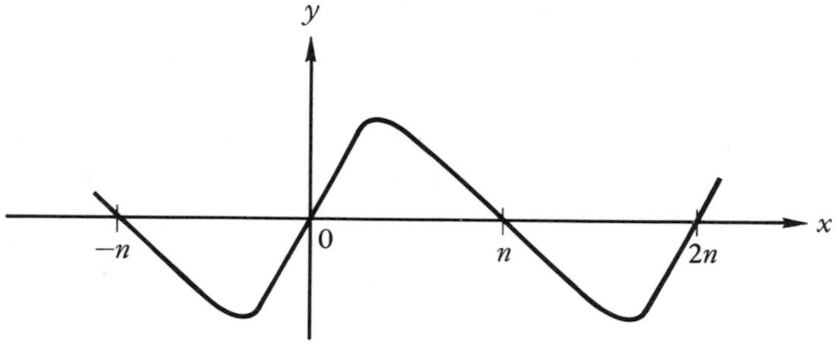

b) $y(x)$ odd:
$y'(x)$ is continuous at
$x = n$ and $x = 0$

FIG. 4-18 The continuity of $y(x)$ and $y'(x)$

Let us now consider the pure sine series approximation where $y(x)$ is an odd function. Then the coefficient of $sin(j\pi x/n)$ is given by

$$\hat{a}_j = (1/n) \sum_{x=-n+1}^{n} y(x)sin(j\pi x/n)$$

$$\approx (1/n) \int_{-n}^{n} y(x)sin(j\pi x/n)dx. \tag{436}$$

Remember that $y(x) = -y(-x)$ and $y(n) = y(-n) = 0$. If we integrate the integral in equation 436 by parts, we obtain

$$\int_{-n}^{n} y(x)sin(j\pi x/n)dx$$

$$= (-n/j\pi)\left[(y(n) - y(-n))(-1)^j - \int_{-n}^{n} y'(x)cos(j\pi x/n)dx\right]$$

$$= (n/j\pi) \int_{-n}^{n} y'(x)cos(j\pi x/n)dx. \tag{437}$$

Note that equation 437 depends on the fact that $y(n) = y(-n)$, or in other words, that y as continued cyclically, which is in fact the function being fitted, is continuous at $x = n$.

If we can show that $y''(x)$ exists and is piecewise continuous and bounded on $[-n,n]$, then we can integrate again by parts, producing

$$\int_{-n}^{n} y(x)sin(j\pi x/n)dx = -(n/j\pi)^2 \int_{-n}^{n} y''(x)sin(j\pi x/n)dx$$

$$= O(1/j^3). \tag{438}$$

We prove the required property of y'' as follows. The function $y(x)$ is produced by (1) mapping the function underlying the data onto $[1,n-1]$, (2) extending the result to zero at $x=0$ and $x=n$, (3) reflecting the result negatively about $x=0$, and (4) extending the result of the reflection cyclically with period $2n$. Since by assumption the function underlying the data is smooth over the interval included by the data arguments z_i, it is reasonable to assume that it is twice differentiable over that interval. Thus, $y(x)$ will be twice differentiable on $[-n+1,-1]$ and $[1,n-1]$. Furthermore, it is correct to assume that the extension to zero at $x=0$ and $x=n$ can be done in a twice-differentiable way, so the only question of twice differentiability occurs at the reflection point $x=0$ and the cyclic extension points $x = \pm n$. If we can show $y'(x)$ is continuous at $x=0$ and $x = \pm n$ (see figure 4-18, part b), $y''(x)$ will be at least piecewise continuous and bounded on $[-n,n]$.

The derivative at $x = 0$ from the right is given by

$$y'(0^+) = \lim_{x\downarrow 0} \frac{y(x) - y(0)}{x - 0} = \lim_{x\downarrow 0} \frac{y(x)}{x}. \tag{439}$$

The derivative from the left at $x = 0$ is given by

$$\begin{aligned} y'(0^-) &= \lim_{x\downarrow 0} \frac{y(-x) - y(0)}{-x - 0} = \lim_{x\downarrow 0} \frac{y(-x)}{-x} \\ &= \lim_{x\downarrow 0} \frac{-y(x)}{-x} = \lim_{x\downarrow 0} \frac{y(x)}{x}. \end{aligned} \tag{440}$$

Since we have assumed that both the derivative from the right and the derivative from the left exist and have shown they equal the same value, y' is continuous at 0. Similarly, using the fact that y is cyclic with period $2n$ and odd,

$$\begin{aligned} y'(n^-) &= \lim_{x\downarrow 0} \frac{y(n - x) - y(n)}{-x} \\ &= \lim_{x\downarrow 0} \frac{-y(n - x)}{x}, \end{aligned} \tag{441}$$

and

$$\begin{aligned} y'(n^+) &= \lim_{x\downarrow 0} \frac{y(n + x) - y(n)}{x} \\ &= \lim_{x\downarrow 0} \frac{y(n + x - 2n)}{x} \\ &= \lim_{x\downarrow 0} \frac{y(-n + x)}{x} = \lim_{x\downarrow 0} \frac{-y(n - x)}{x}. \end{aligned} \tag{442}$$

Therefore y' is continuous at $x = n$, and since y is cyclic, y' is continuous at $x = -n$.

We have shown above that by appropriate mapping and approximating with a pure sine series, we can make the approximating coefficients converge in magnitude like $1/j^3$. We can go no further with this approach. The second derivative of the function underlying the original data may not be continuous.† Even if it is, there is no data transformation that assures the continuity of the second derivative of the cyclic transformed function, y.

Because the coefficients of the pure sine series fit get small faster than

† See C. Lanczos, *Applied Analysis,* chapter 5, section 11.

those of other least-squares fitting techniques, this least-squares procedure is most efficient when doing Fourier least-squares approximation. Note, however, that the continuity, or at least the smoothness, of y', and thus the $O(1/j^3)$ convergence of the $\hat{a}_j$, depends upon the fact that adding zeros at either end of the original function $y(x)$ does not produce a sharp change from the values at the extreme data arguments. This is not necessarily true; there is nothing that prevents the first and last data values from being large compared to the average difference between adjacent data values. We can get around this difficulty by mapping the original data arguments, not onto $\{1,2,\ldots,n-1\}$, but onto $\{0,1,2,\ldots,n\}$. That is, we let $n = N - 1$. Then we subtract from this mapped function the straight line drawn between $y(0)$ and $y(n)$. That is, we define

$$w(x) = y(x) - \left(y(0) + \frac{y(n) - y(0)}{n}x\right). \tag{443}$$

By this *subtraction of the linear tendency* of y from y, we produce a function w which is smooth and appropriately differentiable if y is and which has the property

$$w(0) = w(n) = 0. \tag{444}$$

We can then reflect w negatively about $x = 0$ and find the sine series fit to w.

Since the coefficients of this sine series become small quickly, we need only use the first few sine functions, so the approximation will be smooth. If we add to this smooth approximation the line subtracted in equation 443, we add a line (a smooth function) to a smooth approximation, producing a smooth approximation to y. This smooth approximation can then be mapped back to the original argument range to get the approximation to the original data. This final approximation will not be a least-squares Fourier approximation but will be a smooth, low-error approximation to y, which is all that we normally desire.

The recommended Fourier approximation procedure is given in programs 4-10 and 4-11. Given a set of data, $\{(z_i, y_i) | i = 1,2,\ldots,N\}$, where $z_{i+1} - z_i = h$, we define

$$w_i = y_i - \left(y_1 + \frac{y_N - y_1}{N-1}(i-1)\right). \tag{445}$$

We compute successive coefficients of the sine series fit to the w_i:

$$\hat{a}_j = \frac{2}{N-1}\sum_{x=1}^{N-2} w_{x+1} \sin\left(\frac{j\pi x}{N-1}\right). \tag{446}$$

```
/* FOURIER LEAST-SQUARES APPROXIMATION WITH LINEAR TENDENCY SUBTRACTION */
/* COEFFICIENT CALCULATION */
/* GIVEN ARE NUMBER OF DATA POINTS, N; DATA VALUES, Y(I), 1 < = I < = N, AT */
/* EQUALLY SPACED ARGUMENTS; AND NUMBER OF SINE TERMS, M, IN APPROXIMATION */
/* OUTPUT IS M SINUSOID COEFFICIENTS A(J), 1 < = J < = M; */
/* AND LINE COEFFICIENTS, LINEØ AND LINE1 */
/* IT IS ASSUMED SINTAB(I,J) CONTAINS SIN(J*PI*I/(N-1)) */

    DECLARE (N,                  /* NUMBER OF DATA POINTS */
             M) FIXED BINARY,    /* NUMBER OF SINE TERMS IN APPROXIMATION */
             (Y(N),              /* DATA VALUES */
             A(M),               /* SINUSOID COEFFICIENTS IN APPROXIMATION */
             SINTAB(N,M),        /* SINUSOID VALUES */
             LINEØ, LINE1,       /* LINE COEFFICIENTS IN APPROXIMATION */
             AFACTOR) FLOAT;

/* COMPUTE LINE COEFFICIENTS */
    LINEØ = Y(1);
    LINE1 = (Y(N) - Y(1)) / (N - 1);
/* SUBTRACT LINEAR TENDENCY FROM DATA VALUES */
    LINSUB: DO I = 2 TO N-1;
            Y(I) = Y(I) - (LINEØ + LINE1 * (I-1));
    END LINSUB;
/* COMPUTE SINUSOID COEFFICIENTS */
    AFACTOR = 2 / (N-1);
    COEF: DO J = 1 TO M;
          A(J) = Ø;
          COEFSUM: DO I = 1 TO N-2;
                   A(J) = A(J) + Y(I+1) * SINTAB(I,J);
          END COEFSUM;
          A(J) = AFACTOR * A(J);
    END COEF;
```

PROGRAM 4-10 Fourier least-squares approximation with linear tendency subtraction; coefficient calculation

```
/* FOURIER LEAST-SQUARES APPROXIMATION WITH LINEAR TENDENCY SUBTRACTION */
/* EVALUATION OF APPROXIMATION AT ARGUMENT */
/* GIVEN ARE NUMBER OF DATA POINTS, N; NUMBER OF SINE TERMS, M, IN APPROXIMATION; */
/* APPROXIMATION COEFFICIENTS FROM COEFFICIENT CALCULATION PROGRAM: */
/* LINEØ AND LINE1 OF LINE TERMS, AND A(J), 1 < = J < = M, OF SINE TERMS; */
/* Z1, THE FIRST TABULAR ARGUMENT; Z, THE EVALUATION ARGUMENT; */
/* AND THE TABULAR ARGUMENT INTERVAL WIDTH, H */
/* OUTPUT IS VALUE OF APPROXIMATION, APPROX */
   DECLARE (N,                  /* NUMBER OF DATA POINTS */
            M) FIXED BINARY,    /* NUMBER OF SINE TERMS */
            (LINEØ, LINE1, A(M), /* APPROXIMATION COEFFICIENTS */
            Z1, Z,              /* FIRST ARG, EVALUATION ARG */
            H, S,               /* ARG INTERVAL, ARG IN UNITS OF H */
            SINTABLE(M),        /* TABLE OF SIN(J*PI*(Z-Z1) / ((N-1)*H) */
            COSTABLE(M),        /* TABLE OF COS(J*PI*(Z-Z1) / ((N-1)*H) */
            APPROX) FLOAT;      /* RESULTING APPROX AT Z */
/* COMPUTE LINE TERMS */
   S = (Z - Z1) / H;
   APPROX = LINE1 * S + LINEØ;
/* COMPUTE TABLE OF SINES AT S */
   S = 3.141592 * S / (N - 1);
   SINTABLE(1) = SIN(S);
   COSTABLE(1) = COS(S);
   TABCOMP: DO J = 2 TO M;
            SINTABLE(J) = SINTABLE(J-1) * COSTABLE(1) + COSTABLE(J-1) * SINTABLE(1);
            COSTABLE(J) = COSTABLE(J-1) * COSTABLE(1) - SINTABLE(J-1) * SINTABLE(1);
   END TABCOMP;
/* COMPUTE SINE TERMS */
   SINETERM: DO J = 1 TO M;
            APPROX = APPROX + A(J) * SINTABLE(J);
   END SINETERM;
```

PROGRAM 4-11 Fourier least-squares approximation with linear tendency subtraction; evaluation of approximation at argument

We wish to stop computing the $\hat{a}_j$ when they reflect largely error. It can be shown that when the $\hat{a}_j$ reflect largely error, assuming the error is independent, the smoothness assumption made in the above development no longer holds and the $\hat{a}_j$ no longer falls in magnitude like $1/j^3$. Therefore, we need only detect the point at which this fall-off stops occurring, that is, where the $\hat{a}_j$ flatten out as a function of j, to determine where to stop. This situation is much preferable to that with polynomials, where one must compute an estimate of the variance at each step and watch its behavior. Here we need compute nothing new; we need only watch the behavior of the coefficients themselves.

Assume m is the number of terms we have decided to include in our series. Then $\hat{a}_m$ is the last coefficient used. Our approximation is

$$\hat{w}(x+1) = \sum_{j=1}^{m} \hat{a}_j \sin\left(\frac{j\pi x}{N-1}\right), \tag{447}$$

so

$$\hat{y}(x+1) = y_1 + \frac{y_N - y_1}{N-1}x + \sum_{j=1}^{m} \hat{a}_j \sin\left(\frac{j\pi x}{N-1}\right), \tag{448}$$

where x need not be an integer. Writing equation 448 in terms of the original argument variable z, we obtain

$$\hat{f}(z) = y_1 + \frac{y_N - y_1}{N-1}\left(\frac{z - z_1}{h}\right) + \sum_{j=1}^{m} \hat{a}_j \sin\left(\frac{j\pi(z - z_1)}{(N-1)h}\right). \tag{449}$$

Consider the following example for five data points (z_i, y_i), that is, $N = 5$. Let the data points be $\{(.4,1.0), (.6,1.2), (.8,1.8), (1.0,2.0), (1.2,1.8)\}$. First we map the data arguments linearly onto $x_i = 0,1,2,3,4$, producing $\{(0,1.0), (1,1.2), (2,1.8), (3,2.0), (4,1.8)\}$. Then we subtract the linear tendency $1.0 + ((1.8 - 1.0)/4)x = 1.0 + .2x$ from the data, producing the following set of $(i-1, w_i)$ pairs: $\{(0,0), (1,0), (2,.4), (3,.4), (4,0)\}$. We fit this set by a pure sine series

$$\hat{w}(x+1) = \sum_{j=1}^{m} \hat{a}_j \sin\left(\frac{j\pi x}{4}\right),$$

where $m \leq 3$. We compute the $\hat{a}_j$ by equation 446 as

$$\hat{a}_j = \frac{2}{4}\sum_{x=1}^{3} w_{x+1} \sin\left(\frac{j\pi x}{4}\right),$$

so

$$\hat{a}_1 = \frac{1}{2}\left(0 \sin\left(\frac{\pi}{4}\right) + .4 \sin\left(\frac{2\pi}{4}\right) + .4 \sin\left(\frac{3\pi}{4}\right)\right) = .2(1 + \sqrt{2}),$$

$$\hat{a}_2 = \frac{1}{2}\left(0 \sin\left(\frac{2\pi}{4}\right) + .4 \sin\left(\frac{4\pi}{4}\right) + .4 \sin\left(\frac{6\pi}{4}\right)\right) = -.2,$$

$$\hat{a}_3 = \frac{1}{2}\left(0\ sin\left(\frac{3\pi}{4}\right) + .4\ sin\left(\frac{6\pi}{4}\right) + .4\ sin\left(\frac{9\pi}{4}\right)\right) = .2(-1 + \sqrt{2}).$$

Since the number of data points in this example is small, we do not see the $O(1/j^3)$ behavior of the $\hat{a}_j$, so we must use some other criterion to stop the series. In the above, we simply computed the coefficients for all of the terms.

The final approximation, then, is

$$\begin{aligned}\hat{f}(z) &= 1.0 + .2\left(\frac{z - .4}{.2}\right) + \sum_{j=1}^{3} \hat{a}_j\ sin\left(\frac{j\pi(z - .4)}{4(.2)}\right)\\ &= .6 + z + .2(1 + \sqrt{2})sin\left(\frac{\pi(z - .4)}{.8}\right) - .2\ sin\left(\frac{\pi(z - .4)}{.4}\right)\\ &\quad + .2(-1 + \sqrt{2})sin\left(\frac{3\pi(z - .4)}{.8}\right).\end{aligned}$$

Let us now discuss the Fourier approximation problem with N data points in its complex form. Then the approximation is given by equation 420, where the Fourier coefficients $\hat{b}_j$ (discrete Fourier transform of y) are given by

$$\hat{b}_j = \frac{1}{N}\sum_{x=0}^{N-1} y_x\, e^{i\frac{2j\pi x}{N}}, \quad m_1 \leq j \leq m_2. \tag{450}$$

We note that to compute each $\hat{b}_j$ requires N complex multiplications, and therefore to compute M $\hat{b}_j$ values requires $M \times N$ complex multiplications. If the y_x are real, the requirement is for $M \times N$ multiplications of a real number and a complex number, in effect, $2(M \times N)$ real multiplications.

We see that it would require $2N^2$ real multiplications to calculate the coefficients of the exact-matching approximation $\hat{f}$ which is produced when all N Fourier functions are used:

$$\hat{f}(x) = \sum_{j=-N/2+1}^{N/2} \hat{b}_j\, e^{-i\frac{2\pi jx}{N}}. \tag{451}$$

This function is useful in areas other than just approximation (see part b of this section). As a result, much effort has been put into the development of efficient algorithms for computing the $\hat{b}_j$, where j ranges over all integers between $-N/2 + 1$ and $N/2$, that is, algorithms which require fewer than the $2N^2$ multiplications required by the straightforward method of computation. In particular, an algorithm called the *Fast Fourier transform*

(*FFT*) has been developed by Cooley and Tukey† and is described particularly well by Gold and Rader.‡ This algorithm requires $N\ log_2(N)$ real multiplications to compute all the $\hat{b}_j$ if the data values y_x are real. Therefore, it is also useful in the approximation problem when $2M > log_2(N)$.

The FFT is based on the fact that we can use the properties of the exponential that

$$e^{i(x+y)} = e^{ix}\, e^{iy}, \tag{452}$$

$$\left(e^{i\frac{2\pi}{N}}\right)^{N} = 1, \tag{453}$$

and

$$\left(e^{i\frac{2\pi}{N}}\right)^{N/2} = -1 \tag{454}$$

to convert N^2 multiplications into two parts, each requiring $(N/2)^2$ multiplications, with an overhead of $N/2$ multiplications. By applying the same tactic to each of the two parts requiring $(N/2)^2$ multiplications, again applying the tactic to each of the four resulting parts requiring $(N/4)^2$ multiplications, and so on, we produce the FFT algorithm. The details of the development follow.

It is convenient to reorder the $\hat{b}_j$. Since for x an integer

$$e^{i\frac{2(j+N)\pi x}{N}} = e^{i\frac{2\pi jx}{N}}\, e^{i2\pi x} = e^{i\frac{2\pi jx}{N}}, \tag{455}$$

if we formally compute $\hat{b}_{N+j}$, we see that

$$\hat{b}_{N+j} = \hat{b}_j; \tag{456}$$

if we need all of the coefficients, we can just as well compute $\hat{b}_j$, for $j=0,1,\ldots,N-1$, as $\hat{b}_j$, for $j=-N/2+1,-N/2+2,\ldots,-1,0,1,2,\ldots,N/2-1$. When we finish the computation, $\hat{b}_{N/2+1}$ can be identified with $\hat{b}_{-N/2+1}$; $\hat{b}_{N/2+2}$ can be identified with $\hat{b}_{-N/2+2}$; and so on.

Our problem then is to calculate

$$\hat{b}_j = \frac{1}{N}\sum_{x=0}^{N-1} y_x e^{i\frac{2j\pi x}{N}}, \quad j = 0,1,\ldots,N-1. \tag{457}$$

Note that equation 457 can be written in matrix form:

$$\hat{b} = Gy, \tag{458}$$

where all subscripts are zero origin and where

$$G_{jk} = \frac{1}{N} e^{i\frac{2\pi jx}{N}}. \tag{459}$$

---

† J. W. Cooley and J. W. Tukey, "An Algorithm for the Machine Calculation of Complex Fourier Series."

‡ B. Gold and C. M. Rader, *Digital Processing of Signals,* chapter 6.

Evaluating equation 457 for all j requires N^2 multiplications of a real number by a complex number. By taking advantage of the exponential properties of the matrix elements, however, we can reduce the number of multiplications required.

If we divide the data points into those with odd and even subscripts, we can rewrite equation 457 as

$$N\hat{b}_j = \sum_{k=0}^{N/2-1} y_{2k}\, e^{i\frac{2j\pi 2k}{N}} + \sum_{k=0}^{N/2-1} y_{2k+1}\, e^{i\frac{2j\pi(2k+1)}{N}}$$

$$= \left(\sum_{k=0}^{N/2-1} y_{2k}\, e^{i\frac{2j\pi k}{N/2}}\right) + e^{i\frac{2j\pi}{N}}\left(\sum_{k=0}^{N/2-1} y_{2k+1}\, e^{i\frac{2j\pi k}{N/2}}\right). \tag{460}$$

If we only consider the integers j between 0 and $N/2-1$, the last expression given by equation 460 shows $\hat{b}_j$ as a weighted sum of terms where except for a constant factor the first term is the jth Fourier coefficient for the data made up of the even y values alone and the second term is the jth Fourier coefficient for the odd values alone. That is, the calculation of the $\hat{b}_j$, for $j = 0,1,\ldots,N/2-1$, would require $(N/2)^2$ complex multiplications to calculate the first sum for all these values of j, $(N/2)^2$ multiplications to calculate the second sum for all these values of j, and $N/2$ multiplications to take the weighted sum of the appropriate terms to produce the $\hat{b}_j$.

What about the $\hat{b}_j$ for $N/2 \le j \le N-1$ or equivalently $\hat{b}_{j+N/2}$ for $0 \le j \le N/2-1$? We note that from equation 460,

$$N\hat{b}_{j+N/2} = \sum_{k=0}^{N/2-1} y_{2k}\, e^{i\frac{2(j+N/2)\pi k}{N/2}} + e^{i\frac{2(j+N/2)\pi}{N}} \sum_{k=0}^{N/2-1} y_{2k+1}\, e^{i\frac{2(j+N/2)\pi k}{N/2}}$$

$$= \sum_{k=0}^{N/2-1} y_{2k}\, e^{i\frac{2j\pi k}{N/2}} + e^{i\pi} e^{i\frac{2j\pi}{N}} \sum_{k=0}^{N/2-1} y_{2k+1}\, e^{i\frac{2j\pi k}{N/2}}$$

$$= \left(\sum_{k=0}^{N/2-1} y_{2k}\, e^{i\frac{2j\pi k}{N/2}}\right) - e^{i\frac{2j\pi}{N}}\left(\sum_{k=0}^{N/2-1} y_{2k+1}\, e^{i\frac{2j\pi k}{N/2}}\right). \tag{461}$$

That is, the $\hat{b}_j$ for $N/2 \le j \le N-1$ involve precisely the same terms as the $\hat{b}_j$ for $0 \le j \le N/2-1$. Only the sign of the weight in the sum changes. Therefore, no multiplications other than those required for $0 \le j \le N/2-1$ are required for $N/2 \le j \le N-1$. By writing the discrete Fourier transform of the y_x as the weighted sum of the discrete Fourier transform of the even y_x and that of the odd y_x, we have reduced the number of multiplications by almost a half.

Consider each one of the discrete Fourier transforms with $N/2$ terms, namely, the discrete Fourier transform of the odd y_x and the discrete Fourier transform of the even y_x. If $N/2$ is divisible by 2, rather than compute them directly we can use the above tactic to compute their values. Each $N/2$-element discrete Fourier transform will require two discrete Fourier transforms of $N/4$ terms each, plus $N/4$ multiplications to produce the weighted sum. Thus, we can recast the original N-element discrete Fourier transform into a computation which requires four $N/4$-element discrete Fourier transforms, plus two sets of weighted sums, each requiring $N/4$ multiplications, plus one set of weighted sums requiring $N/2$ multiplications. That is, the number of multiplications required by the new process is

$$4(N/4)^2 + 2(N/4) + N/2 = N^2/4 + 2(N/2). \tag{462}$$

If $N/4$ is divisible by two, the tactic can be used again; the total number of multiplications required will be $N^2/8 + 3(N/2)$. In general, if $N = 2^n$, the tactic can be used n times, and the total number of multiplications required will be

$$N^2/2^n + n(N/2) = N + n(N/2). \tag{463}$$

However, the N multiplications represented by the first N on the right side of equation 463 are the multiplications required by N one-element discrete Fourier transforms. But a one-element discrete Fourier transform is a multiplication of the one data value by 1, so these N multiplications need not be carried out. Therefore, the total number of multiplications (of complex numbers times complex numbers) required is

$$n(N/2) = (N/2)\, log_2(N). \tag{464}$$

The above analysis applies in general for complex data values y_x. We have shown that if all the y_x are complex, the Fast Fourier transform requires $(N/2)\, log_2(N)$ complex multiplications. It can be shown that if the y_x are real, the Fast Fourier transform requires only $N\ log_2(N)$ real multiplications (see exercise 4.3.33). In the complex case, $N\ log_2(N)$ complex additions are required, and in the real case, $(3N/2)\, log_2(N)$ real additions are required (see exercise 4.3.33).

Although the above description of the Fast Fourier transform is definitive, it gives little indication of exactly how the algorithm proceeds. Table 4-14, called a flow graph, shows how to compute $N\hat{b}_j$, $0 \leq j \leq N-1$, for $N = 8$. It can be generalized to produce the method for any $N = 2^n$. Each node in this diagram corresponds to a value to be computed. We proceed from left to right, computing all values in one column before going on to the next column. Each node has two branches entering from the left, proceeding from two nodes in the previous column.

TABLE 4-14 FAST FOURIER TRANSFORM FOR $N = 8$

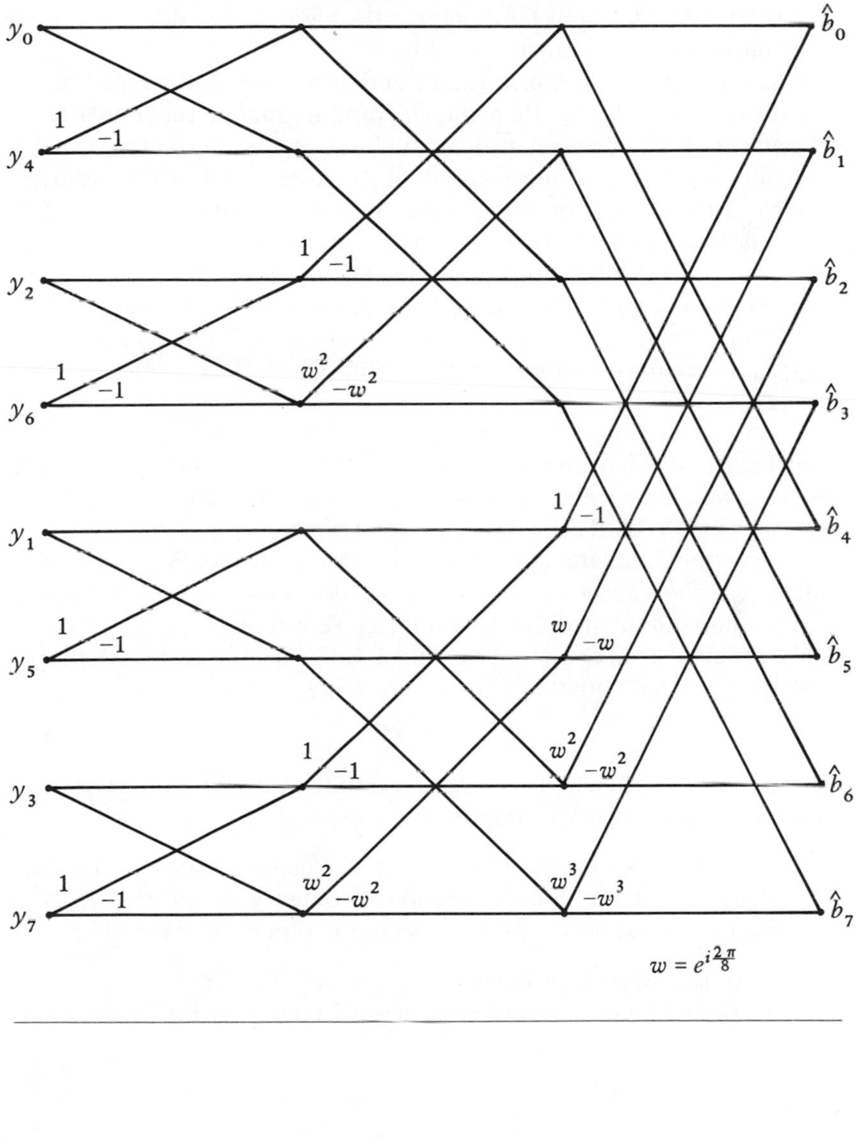

The two nodes represent the two summands in the weighted sum which make up the value represented by the node entered. One branch is weighted and one is unweighted (that is, weighted by 1). In table 4-14, the unweighted branches are unlabeled. A weight is written next to each weighted branch. The weight is the number by which the value of the input node must be multiplied. Note that from each node from which a weighted branch proceeds, there proceeds another weighted branch with the negative of the weight of the first branch.

The Fast Fourier transform indicated by the flow graph of table 4-14 can be rediagramed in many ways, depending upon the order of the nodes in each column. Such reorderings, which can be seen in Gold and Rader, provide different methods for designing the algorithm to determine what the multiplier on any branch should be.†

It has been shown by Kaneko and Liu that not only is the Fast Fourier transform more efficient than direct computation of the Fourier coefficients, but also its error propagation properties are better than the straightforward methods for computing the discrete Fourier transform.‡

What if the number of data points, N, is not an integer power of 2? An algorithm similar to that described above exists for $N = 2 \cdot 3^n$, but most often the FFT is applied when $N = 2^n$. This property can be achieved by designing the data-gathering procedure with the need for $N = 2^n$ in mind. Another approach is to pad the data with an appropriate number of zeros or appropriately extrapolated data values so that $N = 2^n$.

b) Fourier Analysis and Filtering. There are applications of Fourier analysis, alluded to above, in which the decomposition of functions into sinusoids is used for more than least-squares approximation. This decomposition facilitates the analysis of a special class of linear operators called *stationary linear operators* (defined below). Such transformations of input into output are quite common, as we will see.

We conceive of an operator s as a black box which transforms an input function f into an output function g. We write

$$g = s(f), \tag{465}$$

where g and f are functions of some continuous variable x (or $\underline{x}$) or some discrete variable x_i (or $\underline{x}_i$). Here are some examples:

1. A camera is an operator on an input function which is the light intensity as a function of position, producing an output function which is the exposure of the film as a function of position.

† B. Gold and C. M. Rader, *Digital Processing of Signals,* chapter 6.

‡ T. Kaneko and B. Liu, "Accumulation of Round-Off Error in Fast Fourier Transforms."

2. An electrical circuit has an input voltage which is a function of time that it transforms into an output voltage which is also a function of time.
3. Suppose we have the function f tabulated at N equally spaced arguments, x_i, with interval h, and we want to know at each x_i the integral of f over a range of width $2h$ centered at x_i. We might approximate this integral about x_i by applying Simpson's rule to $\int_{x_{i-1}}^{x_{i+1}} f(x)dx$. The resulting transformation of an input, x_i, to an output, the approximation to the integral about x_i, is a discrete variable operator defined by

$$g(x_i) \equiv [s(f)](x_i) = (h/3)(f(x_{i-1}) + 4f(x_i) + f(x_{i+1})). \quad (466)$$

Note that $[s(f)](x)$ means the function $s(f)$ evaluated at x.

As we know, an operator is said to be *linear* if, for any input functions f and g and constants α and β,

$$s(\alpha f + \beta g) = \alpha s(f) + \beta s(g). \quad (467)$$

With a linear operator, if the input function can be analyzed as a weighted sum of members of some set of basis functions, the output can be synthesized as the same weighted sum of the set of functions produced by applying the operator to each member of the basis set:

$$s\left(\sum_{i=1}^{n} a_i f^i\right) = \sum_{i=1}^{n} a_i s(f^i). \quad (468)$$

We will be using the sinusoids as such a basis set because of some very convenient properties they have with respect to stationary linear operators (still to be defined).

We can show that for the discrete variable case, a linear operator can be written as a matrix:

$$[s(f)](x_i) = \sum_{j=j_1}^{j_2} A_{ij} f(x_j). \quad (469)$$

For the continuous variable case, the sum becomes an integral and linear operators can generally be written

$$g(x) = [s(f)](x) = \int_c^d a(x,y) f(y) dy. \quad (470)$$

This statement must be modified slightly to take into account linear operators like the differential operator, which can be written as the limit of such an integral. Also, we can set $c = -\infty$ and $d = \infty$ if we assume the convention that $f(y) = 0$ outside of the original interval $[c,d]$.

Assume $a(x,y)$ is a function only of the difference between x and y:

$$a(x,y) = w(x - y). \tag{471}$$

In this case, if the input f is shifted right by z to create a new input,

$$f_z(y) = f(y - z), \tag{472}$$

then the corresponding output $g_z(x)$, from equation 470, is

$$g_z(x) = [s(f)](x) = \int_{-\infty}^{\infty} w(x - y)f(y - z)dy. \tag{473}$$

Letting $u = y - z$, we see

$$g_z(x) = \int_{-\infty}^{\infty} w((x - z) - u)f(u)du = g(x - z); \tag{474}$$

if the input is shifted right by z, the only effect is to shift the output by the same amount. An operator with this property is called *shift invariant* or *stationary*. Thus, we have defined stationary linear operators.

Many common physical systems can be characterized or approximated by stationary linear operators. Both the electrical circuit and the camera given as examples above have this property. For example, consider the electrical circuit with no current flowing. Its reaction to a given input voltage applied beginning at time $t = t_0$ is linear and independent of t_0 except that the output voltage will be shifted to be relative to t_0. In the same manner, with a camera, the blurring due to imaging is approximately linear and independent of position.

It can be proved that only operators such that $a(x,y) = w(x - y)$, or in the discrete case $A_{ij} = w_{i-j}$, are stationary (see exercise 4.3.34). Thus a stationary linear operator in a continuous variable can be written

$$g(x) = [s(f)](x) = \int_{-\infty}^{\infty} w(x - y)f(y)dy, \tag{475}$$

and a stationary linear operator in a discrete variable can be written

$$g(x_i) = \sum_{j=-\infty}^{\infty} w_{i-j}f(x_j) \tag{476}$$

(for example, see the Simpson's rule operator defined in equation 466). Such an operator is called a *convolution* of w with f. If we let $u = x - y$, equation 475 can be rewritten

$$g(x) = \int_{-\infty}^{\infty} w(u)f(x - u)du = \int_{-\infty}^{\infty} w(y)f(x - y)dy; \tag{477}$$

convolution is commutative in w and f (see exercise 4.3.35).

Equation 475 (and also 477) states that the value of g at x is a weighted sum of surrounding values of f, where the weights depend only on the distance from x of the argument of the f value in question. For each output value x, $f(x)$ is weighted by $w(0)$, $f(x+.5)$ is weighted by $w(-.5)$, $f(x-.5)$ is weighted by $w(.5)$, and so on, and the results are summed to produce $g(x)$. We can think of convolution as, for any output argument x, centering a weighting function $w(-u)$ at the value x (reflecting $w(u)$ about $u=0$ and shifting the result so that $w(0)$ is at $y=x$), weighting all values of $f(y)$ by the corresponding values of this result of reflection and shifting, and summing these weighted $f(y)$ values. (See figure 4-19.) For a new value of x, $w(-u)$ is shifted to the new x value, and the weighting and summing is repeated. Thus, in the case where $\int_{-\infty}^{\infty} w(u)du = 1$, convolution is sometimes called *running averaging.*

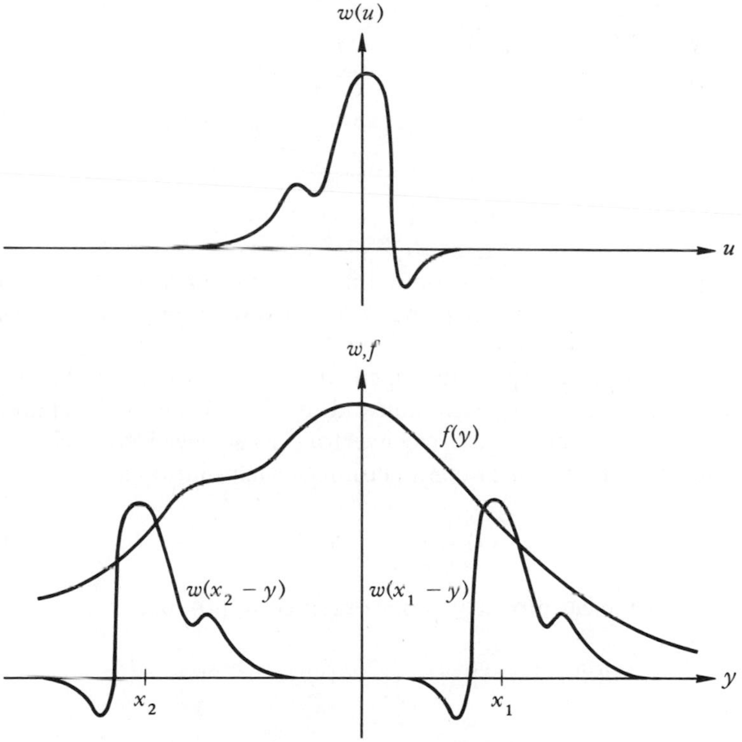

FIG. 4-19 Reflection and shifting in convolution

The sinusoids (complex exponentials) are useful in the analysis of stationary linear operators because, in the vector space of absolutely integrable functions on $[-\infty,\infty]$, the sinusoids, $e^{-i2\pi vx}$, $-\infty < v < \infty$, are the eigenvectors of all stationary linear operators. Thus, their use has an uncoupling effect, as discussed in section 2.6.

To show that the sinusoids are eigenvectors, we must show that convolution of any function with a sinusoid at a given frequency produces a constant times that sinusoid. This behavior depends on the property of the exponential ($e^{-i2\pi vx}$) that shifting (x by z) multiplies the exponential by a constant ($e^{i2\pi vz}$):

$$e^{-i2\pi v(x-z)} = e^{-i2\pi vx}e^{i2\pi vz}. \tag{478}$$

The predicted result follows immediately:

$$g(x) = \int_{-\infty}^{\infty} w(x-y)e^{-i2\pi vy}dy \tag{479}$$

with the change of variable $z = x - y$ produces

$$g(x) = \int_{-\infty}^{\infty} w(z)e^{-i2\pi v(x-z)}dz$$

$$= e^{-i2\pi vx}\int_{-\infty}^{\infty} w(z)e^{i2\pi vz}dz \equiv e^{-i2\pi vx}W(v), \tag{480}$$

where $$W(v) \equiv \int_{-\infty}^{\infty} w(z)e^{i2\pi vz}dz. \tag{481}$$

$W(v)$ is the eigenvalue corresponding to the eigenvector $e^{-i2\pi vx}$. $W(v)$ taken as a function of the frequency v is called the *transfer function* of the stationary linear operator.

Assume we can compute the eigenvalue $W(v)$ for each v, that is, the effect of the operator on each sinusoid, $e^{-i2\pi vx}$. Then, as in chapter 2, the uncoupling effect of the eigenvectors is achieved by writing any input function f as a linear combination of such sinusoids,

$$f(x) = \int_{-\infty}^{\infty} F(v)e^{-i2\pi vx}dv, \tag{482}$$

and using the linear property of our operator to produce

$$g(x) = [s(f)](x) = \int_{-\infty}^{\infty} F(v)s(e^{-i2\pi vx})dv$$

$$= \int_{-\infty}^{\infty} W(v)F(v)e^{-i2\pi vx}dv. \tag{483}$$

That is, we can synthesize $g = s(f)$ as a linear combination of sinusoids, where the coefficient in g of $e^{-i2\pi vx}$ is its coefficient in the decomposition of f times the transfer function value $W(v)$:

$$G(v) = W(v)F(v), \tag{484}$$

where $G(v)dv$ is the coefficient of $e^{-i2\pi vx}$ in the sinusoidal decomposition of $g(x)$:

$$g(x) = \int_{-\infty}^{\infty} G(v)e^{-i2\pi vx}dv. \tag{485}$$

Note the simplification obtained by this approach as compared with that of viewing a stationary linear operator as convolution. Convolution consists of shifting the reflected weighting function past the input function and, for each such shift, computing a sum of products. In contrast, after decomposition of the input into sinusoidal components, our new operation consists only of a point-by-point multiplication of $W(v)$ by $F(v)$. Whereas many values of $f(x)$ affect $g(x)$ (the $f(x)$ are coupled), only one value of $F(v)$ affects $G(v)$ (the $F(v)$ are uncoupled).

The same statements with only minor changes apply for the discrete variable case. We will discuss this matter later in this part of section 4.3.7. First, because problem analysis is most commonly done in terms of functions of a continuous variable, we must discuss the extension to the continuous variable case of the theory of the decomposition of functions into sinusoids, for which the discrete variable case was developed in part a of this section. In part c of this section, we will discuss the effect of approximating these continuous functions and processes by finite discrete processes.

Consider the discrete Fourier transform where the data points are taken to be between $-N/2+1$ and $N/2$:

$$\hat{b}_j = \frac{1}{N} \sum_{k=-N/2+1}^{N/2} y_k e^{i2\pi jk/N}, \tag{486}$$

and the corresponding inverse Fourier transform which expresses the exact-matching function $\hat{g}(z)$ as a linear combination of complex exponentials:

$$\hat{g}(z) = \sum_{j=-N/2+1}^{N/2} \hat{b}_j e^{-i2\pi jz/N}. \tag{487}$$

Let the y_k be produced as a result of mapping the function f defined at N equally spaced arguments, x_k, between a and d onto the integers between $-N/2+1$ and $N/2$. That is, let

$$\Delta x = (d-a)/N, \tag{488}$$

$$x_k = (a+d)/2 + k\,\Delta x, \quad -N/2+1 \le k \le N/2, \tag{489}$$

$$y_k = f(x_k), \quad -N/2+1 \le k \le N/2, \tag{490}$$

and
$$x = (a+d)/2 + z\,\Delta x. \tag{491}$$

Then equations 486 and 487 can be rewritten

$$\begin{aligned}\hat{b}_j &= \frac{1}{N} \sum_{k=-N/2+1}^{N/2} f(x_k) e^{i2\pi jk/N} \\ &= \frac{1}{d-a}\,\Delta x \sum_{k=-N/2+1}^{N/2} f(x_k)\, e^{i\frac{2\pi jk\,\Delta x}{d-a}}\end{aligned} \tag{492}$$

and
$$\begin{aligned}\hat{g}(z) \equiv \hat{f}(x) &= \sum_{j=-N/2+1}^{N/2} \hat{b}_j\, e^{-i\frac{2\pi j}{N}\left(\frac{x-(a+d)/2}{\Delta x}\right)} \\ &= \sum_{j=-N/2+1}^{N/2} \hat{b}_j\, e^{-i2\pi j\frac{x-(a+d)/2}{d-a}}.\end{aligned} \tag{493}$$

The approximating function $\hat{f}(x)$ given by equation 493 is a fit which exactly matches f at the N data arguments equally spaced between a and d.

Now if we let $N \to \infty$ (let the number of (equally spaced) data arguments in $[a,d]$ become infinite), $\hat{f}$ becomes an exact-matching function at all data points: $\hat{f}(x) \to f(x)$ on $[a,d]$. It can be shown that this convergence is uniform; that is, for all points in the interval $[a,d]$, for any ε, there exists an integer N_ε such that, for any $N > N_\varepsilon$,

$$|f(x) - \hat{f}(x)| < \varepsilon \quad \text{if} \quad x \in [a,d] \tag{494}$$

(see exercise 4.3.38). Remember that this behavior does not take place when expanding in terms of polynomials orthogonal over equally spaced data arguments.

As $N \to \infty$, equation 493 becomes

$$\hat{f}(x) = \sum_{j=-\infty}^{\infty} \hat{b}_j\, e^{-i2\pi j\frac{x-(a+d)/2}{d-a}}. \tag{495}$$

To produce the corresponding form for equation 492, we first use equation 489 to obtain

$$k\,\Delta x = x_k - (a+d)/2. \tag{496}$$

We substitute this result in equation 492 to obtain

$$\hat{b}_j = \frac{1}{d-a} \sum_{\substack{x_k = a+\Delta x \\ \text{with step } \Delta x}}^{d} f(x_k)\, e^{i2\pi j\frac{x_k-(a+d)/2}{d-a}}\,\Delta x,$$

$$-N/2+1 \le j \le N/2. \tag{497}$$

As $N \to \infty$, $\Delta x \to 0$ and equation 497 becomes an integral:

$$\hat{b}_j = \frac{1}{d-a} \int_a^d f(x)\, e^{i2\pi j \frac{x-(a+d)/2}{d-a}}\, dx, \quad -\infty \leq j \leq \infty. \tag{498}$$

Equation 495 is called a *Fourier series* for f and equation 498 gives the coefficients of the Fourier series by the so-called *finite Fourier transform* of f. Note that the finite Fourier transform is defined for any function f which is given *on an interval,* where the integral given by equation 498 exists, in particular at least where f is piecewise continuous and bounded on $[a,d]$. Furthermore, note that the fit given by $\hat{f}$ in equation 495 exact-matches with f throughout the interval $[a,d]$ and its values outside that interval are determined by the fact that it is periodic with period $(d-a)$.

A common situation is when the interval $[a,d]$ is symmetric about zero. If $a = -d$, equation 495 becomes

$$\hat{f}(x) = \sum_{j=-\infty}^{\infty} \hat{b}_j e^{-ij\pi x/d}, \tag{499}$$

and equation 498 becomes

$$\hat{b}_j = \frac{1}{2d} \int_{-d}^{d} f(x) e^{ij\pi x/d} dx. \tag{500}$$

If we define

$$v_j \equiv j/2d, \tag{501}$$

equation 499 can be rewritten

$$\hat{f}(x) = \sum_{j=-\infty}^{\infty} \hat{b}_j e^{-i2\pi v_j x} \tag{502}$$

and equation 500 can be rewritten

$$\hat{b}_j = \frac{1}{2d} \int_{-d}^{d} f(x) e^{i2\pi v_j x} dx, \quad -\infty \leq j \leq \infty. \tag{503}$$

Note that since j ranges between $-\infty$ and ∞, v_j also ranges between $-\infty$ and ∞ with step

$$\Delta v = 1/2d. \tag{504}$$

This result can be substituted in equations 502 and 503 to produce

$$\hat{f}(x) = \sum_{\substack{v_j=-\infty \\ \text{with step } \Delta v}}^{\infty} \frac{\hat{b}_j}{\Delta v} e^{-i2\pi v_j x} \Delta v \tag{505}$$

and

$$\hat{b}_j = \Delta v \int_{-d}^{d} f(x) e^{i2\pi v_j x} dx, \quad -\infty \leq j \leq \infty. \tag{506}$$

Finally, if we define the *coefficient density* of $e^{-i2\pi vx}$ over a range $[u,v]$ in v as the sum of the coefficient ($\hat{b}_j$) values over all j such that $v \in [u,v]$, divided by $v-u$, and we call $F(v_j)$ the coefficient density in $[v_j - \Delta v/2, v_j + \Delta v/2]$, we have

$$F(v_j) = \hat{b}_j/\Delta v, \tag{507}$$

so equations 505 and 506 can be rewritten

$$\hat{f}(x) = \sum_{\substack{v_j=-\infty \\ \text{with step } \Delta v}}^{\infty} F(v_j)e^{-i2\pi v_j x}\Delta v \tag{508}$$

and

$$F(v_j) = \int_{-d}^{d} f(x)e^{i2\pi v_j x}dx. \tag{509}$$

We are now in a position to discuss the Fourier analysis (sinusoidal decomposition) of a continuous function over $[-\infty,\infty]$. Having done the above analysis with the interval $[-d,d]$, we let $d\rightarrow\infty$, and thus by equation 504, $\Delta v\rightarrow 0$. Then the sum in equation 508 becomes an integral:

$$\hat{f}(x) = \int_{-\infty}^{\infty} F(v)e^{-i2\pi vx}dv, \tag{510}$$

and equation 509 becomes

$$F(v) = \int_{-\infty}^{\infty} f(x)e^{i2\pi vx}dx. \tag{511}$$

Equation 511 gives $F(v)$ as the *infinite continuous Fourier transform* of $f(x)$. Often the phrase "infinite continuous" is omitted. Equation 510 gives $\hat{f}(x)$ as the *inverse Fourier transform* of $F(v)$.

Let us make some observations about the above development used to produce this transform. First, note that in the process of going from the case of discrete values of the "frequency" v_j to continuous frequency v, we changed from discrete coefficients $\hat{b}_j$ to a coefficient *density* $F(v)$. In general, the relation between a discrete function and the continuous function which is its limit is that the continuous function is the limit of the density of the values of the discrete function. That is, the continuous function is given by the limit of the sum of the discrete function values in an interval width Δv, divided by Δv, where v is the variable being changed from discrete to continuous. Equivalently, if Δv is a very small interval, the coefficient value in Δv is given by the product of the density in Δv and the width of the interval, Δv. Thus, the "dv" appearing in an integral should not be thought of as simply a notational device to tell what the variable of integration is, but as $\lim_{\Delta v\rightarrow 0} \Delta v$. More completely, when we take the limit as $\Delta v\rightarrow 0$ of coefficient values, $\sum_j (G(v_j)\Delta v)$, where

G gives coefficient density, we should think that the finite sum becomes an integral ($\Sigma \to \int$), the discrete density becomes a continuous density ($G(v_j) \to G(v)$), and the interval width becomes infinitesimal ($\Delta v \to dv$), so the sum becomes $\int_v G(v)dv$.

The second observation we should make about the above development of the Fourier transform is more specific to this particular development. The observation is that the transform $F(v)$ was produced as the limit as $d \to \infty$ of $\int_{-d}^{d} f(x)e^{i2\pi v_j x}dx$ (see equation 503), where $v_j = j\frac{1}{2d}$, an integer multiple of $\Delta v = 1/2d$. This fact must be remembered, because it is necessary in a number of places in the further analysis and use of the transform.

Let us now comment on the Fourier transform itself. We shall assume that the integrals in equations 510 and 511 exist. Then $\hat{f}$ is an exact-matching function over $\lim_{d\to\infty}[-d,d] = [-\infty,\infty]$, so we can replace "$\hat{f}$" by "$f$" in equation 510:

$$f(x) = \int_{-\infty}^{\infty} F(v)e^{-i2\pi vx}dv. \tag{512}$$

Furthermore, it can be shown that both the Fourier transform given by equation 511 and the inverse Fourier transform given by equation 512 are 1 to 1.† That is, to each function f there corresponds only one function F which under the inverse Fourier transform maps into f and vice versa. We will generally use the notation of capital letters to stand for the function which is the Fourier transform of the function named by the corresponding small letter.

Let us restate our findings about the Fourier transform in terms that are familiar from the finite discrete case. The complex exponentials $e^{-i2\pi vx}$ over $[-\infty \le v \le \infty]$ are orthogonal functions of x according to the inner product

$$(f_{v_1}, f_{v_2}) \equiv \lim_{d\to\infty} \int_{-d}^{d} f_{v_1}(x) f_{v_2}^*(x)dx, \tag{513}$$

where $$v_j = j/2d \quad \text{for } j \text{ an integer.} \tag{514}$$

The Fourier transform given by equation 512 writes f as a linear combination of sinusoidal components $e^{-i2\pi vx}$, where the coefficient in the linear combination at frequency v is $F(v)dv$. Just as in the discrete case, the coefficient density function F can be computed by taking the inner product of f with the exponential function at the frequency in question.

---

† The result follows directly from the Fourier Integral Theorem, which is proved, for example, in C. Lanczos, *Discourse on Fourier Series*, section 28.

The Fourier transform has some useful properties.

1. The Fourier transform and inverse Fourier transform are the same with the exception of the sign in the exponent. That is, the Fourier transform of the Fourier transform of $f(x)$ is $f(-x)$. Furthermore, if the Fourier transform of $f(x)$ is $F(\nu)$, the inverse Fourier transform of $f(\nu)$ is $F(-x)$, and thus the Fourier transform of $F(-x)$ is $f(\nu)$.
2. The Fourier transform of the impulse at the origin (the Dirac delta function at $x = 0$) is unity:

$$\int_{-\infty}^{\infty} \delta(x)e^{i2\pi\nu x}dx = e^{i2\pi\nu 0} = 1. \tag{515}$$

This means that the impulse at zero contains an equal amount of the sinusoid at each frequency. More generally, the Fourier transform of an arbitrarily positioned impulse is

$$\int_{-\infty}^{\infty} \delta(x - x_0)e^{i2\pi\nu x}dx = e^{i2\pi\nu x_0}, \tag{516}$$

a sinusoid.
3. By applying the logic of item 1 above to equation 516, we see that

$$\int_{-\infty}^{\infty} e^{-i2\pi\nu_0 x}e^{i2\pi\nu x}dx = \delta(\nu - \nu_0). \tag{517}$$

Equation 517 can be interpreted as a statement of the orthogonality of two sinusoids, on which we have already remarked. That is, if $\nu_0 \neq \nu$, $e^{-i2\pi\nu x}$ and $e^{-i2\pi\nu_0 x}$ are orthogonal and the result is zero; if $\nu_0 = \nu$, in the discrete case we produce a coefficient with value 1, that is, the coefficient density at ν_0 is $\delta(\nu - \nu_0)$.

Summarizing the above, the Fourier transform of the impulse function is a sinusoid, and the Fourier transform of a sinusoid is an impulse function. Likewise, the inverse Fourier transform of an impulse function is a sinusoid and the inverse Fourier transform of a sinusoid is an impulse function.

4. Functions defined on a finite range, whether continuous or discrete, have a Fourier transform which is discrete, and functions defined on an infinite range have a Fourier transform which is continuous. (We have not covered the case of a discrete function on an infinite range, but the result holds for such a function.) We can understand this result if we remember that a function defined over a finite range is thought of as one period of a periodic function. Therefore, in the above, we can substitute "periodic" and "nonperiodic" for "finite range" and "infinite range," respectively. The resulting statement, that

the transform of a periodic function is discrete, is explained when we note that the sinusoids with nonzero coefficients in the expression of a periodic function as a linear combination of sinusoids (the Fourier transform) must share the period of the periodic function, that is, they occur at discrete multiples of a basic frequency.

5. Discrete functions have a Fourier transform defined over a finite frequency range, and continuous functions have a Fourier transform defined over an infinite frequency range. We can understand this result by remembering that any sinusoid outside of some finite frequency range agrees with some sinusoid inside the range at the integer argument values (see section 4.3.7, part a). We will see in section 4.3.7, part c, that the statement holds for any discrete equal spacing. Therefore, the sinusoids outside the appropriate frequency range add nothing at the discrete argument points, and the discrete function can be fully defined as a linear combination of sinusoids inside the range.

To summarize, discrete and continuous functions correspond respectively to finite range and infinite range for the Fourier transform or inverse Fourier transform.

We are now ready to return to the analysis of stationary linear operators in terms of their effect on sinusoids. First, let us summarize the findings we made earlier about stationary linear operators in the continuous variable case (see equations 475–485). We found that any such operator consisted of a convolution:

$$g(x) = \int_{-\infty}^{\infty} w(x-y)f(y)dy. \tag{518}$$

If the input function, f, was taken to be a sinusoid, $e^{-i2\pi vx}$, the output was given by

$$g(x) = W(v)e^{-i2\pi vx}, \tag{519}$$

where the transfer function, $W(v)$, was given by

$$W(v) = \int_{-\infty}^{\infty} w(x)e^{i2\pi vx}dx; \tag{520}$$

that is, $W(v)$ is the Fourier transform of $w(x)$.

By the linearity of convolution, if $f(x)$ is a weighted sum of such sinusoids, that is, if

$$f(x) = \int_{-\infty}^{\infty} F(v)e^{-i2\pi vx}dv, \tag{521}$$

then $g(x)$ is also a weighted sum of such sinusoids, where, since the weight of $e^{-i2\pi\nu x}$ in $f(x)$ is $F(\nu)d\nu$, its weight in $g(x)$ is $W(\nu)F(\nu)d\nu$:

$$g(x) = \int_{-\infty}^{\infty} W(\nu)F(\nu)e^{-i2\pi\nu x}d\nu. \tag{522}$$

Equations 518 and 522 together state that the inverse Fourier transform of a product of two functions is equal to the convolution of inverse transforms of the two functions. Equivalently, the Fourier transform of $g(x)$ is $W(\nu)F(\nu)$:

$$G(\nu) = W(\nu)F(\nu); \tag{523}$$

the Fourier transform of a convolution of two functions is equal to the product of the Fourier transforms of the respective functions.

Because the Fourier transform and inverse Fourier transform are similar, it can also be shown that the Fourier transform of the product of two functions is equal to the convolution of the Fourier transforms of the functions (see exercise 4.3.40). Equivalently, the inverse Fourier transform of a convolution of two functions is equal to the product of the inverse Fourier transforms of the functions.

If we let the input function $f(y)$ be the impulse function $\delta(y)$, we produce

$$g(x) = \int_{-\infty}^{\infty} w(x - y)\delta(y)dy = w(x). \tag{524}$$

That is, the weighting function associated with a given stationary linear operator is given by the response of that system to an impulse. As a result, the weighting function is called the *impulse response function* of the operator.

Equation 520 shows that the transfer function, $W(\nu)$, of a stationary linear operator is the Fourier transform of the impulse response function, $w(x)$, of the operator. Thus, convolution of an input function $f(x)$ with the impulse response function $w(x)$ is equivalent to the combination of (1) sinusoidal decomposition (Fourier transform) of $f(x)$, (2) for each resulting component (at frequency ν), multiplication of its weight $F(\nu)$ by a transfer function value $W(\nu)$, where the function $W(\nu)$ is the Fourier transform of $w(x)$, and (3) synthesis of the resulting sinusoidal components (inverse Fourier transform).

When we discuss the effect of a system in terms of its effect on sinusoidal components, we say we are considering this effect in the *frequency domain.* Analyzing the effect of the operator in the frequency domain is useful because the relatively complicated process of convolution has been transformed into the relatively simple process of multiplication. The simplicity helps us to understand the effect of the operator and as a result to design operators with appropriate behavior. Analysis in the frequency domain is also helpful from a computational point of view; if we desire to compute the result of a stationary linear operator on a given input function, rather than compute the convolution, it is often more efficient to compute the Fourier transform, multiply by the transfer function in the frequency domain, and then compute the inverse Fourier transform. We will discuss each of these advantages of Fourier analysis.

The analysis and synthesis of stationary linear operators is a procedure which cuts across many different areas of application of the subfield of mathematics called *analysis*. Both the analysis and design of such an operator depend on understanding the effect or the desired effect of the operator in the frequency domain. In the design of a stationary linear operator, we synthesize a transfer function that has the desired effect on the input data. This transfer function is often called a *filter*, a name which comes from the first filters that were used. These filters were unity at some frequencies and zero at the rest, resulting in an operator which "filtered out" the signal components at frequencies for which the filter was zero and "passed" the signal components at frequencies for which the filter was 1.

Frequency domain analysis of operators and filter design cover a very broad, deep area.† We shall discuss only a few general aspects of the matter. Then we shall consider two examples of filter design, which hopefully will give some of the spirit of the process.

In general, low-frequency components of a function, that is, components consisting of low-frequency sinusoids, correspond to smooth, slowly varying components. High-frequency components correspond to high oscillation or, more generally, to sharp changes. Thus, smooth functions have relatively larger amounts of low-frequency components (small $|\nu|$) than they have of high-frequency components (large $|\nu|$), and functions that are very edgy or jittery have relatively larger amounts of high-frequency components. Therefore, filters which emphasize low frequencies (have larger values at low frequencies) tend to have a smoothing effect on the input; filters which increase in magnitude with frequency magnitude tend to increase both the definition of edges and the amount of high-frequency oscillations in the input. Figure 4-20 illustrates the effects of various filters on a given input function. In this figure, the magnitudes of the filters are graphed on the ordinate, because the filters themselves are, in general, complex functions, and so are difficult to graph. The relative amount of the real and imaginary parts of a filter at any frequency determines only the relative amount by which the sine and cosine terms, at that frequency, of the input are multiplied. We will ignore this matter in our discussion.

† See B. Gold and C. M. Rader, *Digital Processing of Signals*.

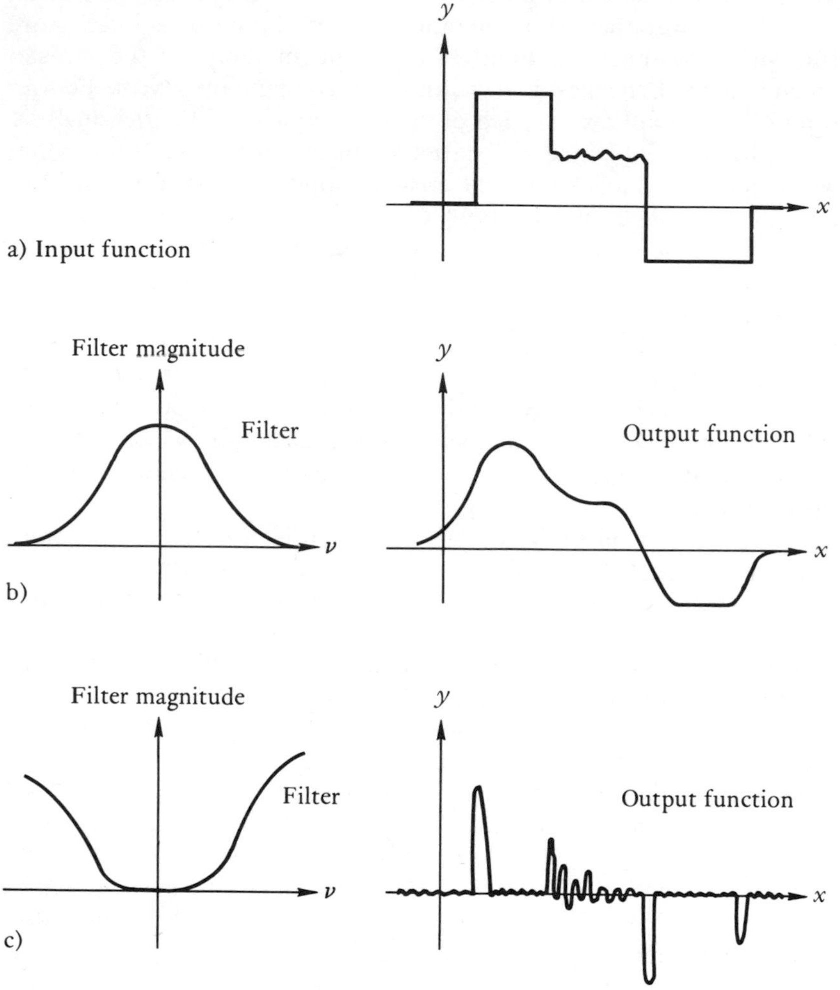

FIG. 4-20 The effects of filters on an input function

As one example of filter synthesis, consider a situation in which we have an input function $f(x)$ made up of a signal $s(x)$ plus noise (error) $n(x)$. Assume that the noise has a random value at each value of the input argument x. We do not know $n(x)$, but we often can say that it is a member from an ensemble of noise functions with certain distribution characteristics. As a result, we can make certain statements about the ensemble of the Fourier transforms $N(\nu)$ of the members of the original ensemble.

In particular, let us assume a situation which is quite common, namely, that the noise is very nonsmooth and thus has significant high-frequency components. More specifically, let us assume a type of noise called *white noise* (containing all frequencies, as in white light), for which the expected value (over the ensemble) of $N(\nu)$ at any frequency ν is 0 but the variance of the noise at any ν, $E[|N(\nu)|^2]$, is constant as a function of ν. (From exercise 4.3.6, we can see that this property always holds if $n(x_1)$ and $n(x_2)$ are uncorrelated if $x_1 \neq x_2$.) Let us further assume that our signal function is relatively smooth, in which case, it has larger amounts of low-frequency components than it has of high-frequency components. Its Fourier transform, or sinusoidal composition, will be of a shape like that given in figure 4-21 (again, only the magnitude of the ordinate is graphed). The transform of the noise is also plotted.

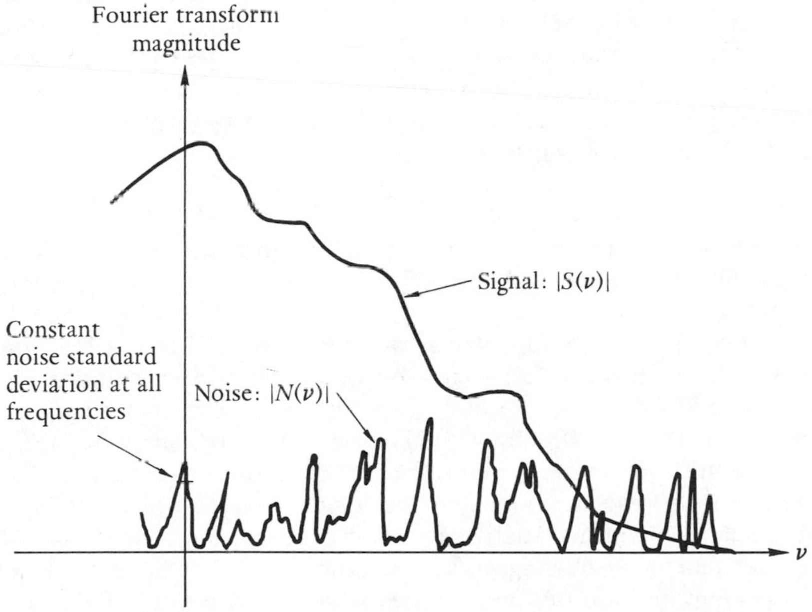

FIG. 4-21 Transforms of signal and noise of an input function

Since the input function is the sum of the signal and the noise, we can see that at low frequencies the signal dominates the noise but at high frequencies the noise dominates the signal. It should be remembered, however, that the magnitude of the sum of signal and noise transforms is not the sum of the magnitudes of the transforms, since these transforms are complex. Nevertheless, the general idea applies. If our object is to remove noise in the input function, we should emphasize the frequencies at which the ratio of the magnitude of the signal to the magnitude of the noise is high and deemphasize the frequencies at which it is very low. In particular, a filter which is near unity as long as the signal is considerably larger than the noise and falls to zero as the noise begins to become comparable in magnitude to the signal will smooth the input function.

As a second example of filter synthesis, consider the problem of inverting the effect of a stationary linear operator imposed by a physical process. For example, we may wish to invert the blurring caused by a particular camera in order to produce a sharper image. Assume that the physical process we are trying to invert is characterized by a stationary linear operator with transfer function $W(v)$. Also, assume that this physical system has input $f(x)$ and output $g(x)$, with sinusoidal compositions (Fourier transforms) $F(v)$ and $G(v)$ respectively. Then, $G(v) = F(v)W(v)$. The output $g(x)$ is the input to the linear operator we are trying to design. Clearly, if we multiply $G(v)$ by $1/W(v)$, that is, if we apply a filter consisting of the reciprocal of the physical system transfer function, we produce a result $H(v)$ which has the property

$$H(v) = G(v)(1/W(v)) = F(v). \tag{525}$$

By the uniqueness of the Fourier transform, the inverse Fourier transform of our filtered result H will be precisely the physical system input that we are trying to retrieve.

Assume that the operator which we are trying to invert is blurring. Such blurring reduces the magnitude of high-frequency components, because the positive and negative parts of a high-frequency sine wave are averaged. That is, the transfer function of the original system tends to fall with frequency. As a result, the reciprocal of the transfer function will rise with frequency, and thus the filter we desire will boost high-frequency components relative to low-frequency components. If the function that we are filtering, $g(x)$, consists not only of the blurred version of the input but also of some additive white noise, a filter of the type we have designed will boost the values at high frequencies which consist largely of noise, and thus will boost the noise in the result. Because this effect is normally unacceptable, most probably the filter we wish will rise like the reciprocal of $W(v)$ until v reaches a frequency such that the noise becomes too large a part of the function being processed. Then the filter will fall to zero to remove that part of $G(v)$ which consists largely of noise.

The theory of filter design is quite well developed, and much has been written about it. Filters have been developed for detection of signals, prediction of signals given a previous part of the signal (extrapolation), smoothing, and many other purposes. We will say little more about the matter here, except to note that many of the operators designed in this text are stationary linear operators so their effects can be analyzed in the frequency domain. Furthermore, alternatives to these methods can be designed in the frequency domain. Examples of such operators are least-squares fitting, numerical integration, extrapolation, and numerical differentiation.

A sketch of Fourier analysis of the numerical differentiation formula,

$$f'(x) \approx g(x) = \frac{f(x+h) - f(x-h)}{2h}, \tag{526}$$

should illustrate the process. For this operator

$$\begin{aligned} G(v) &= \int_{-\infty}^{\infty} \frac{f(x+h) - f(x-h)}{2h} e^{i2\pi vx} dx \\ &= \frac{1}{2h}\left(\int_{-\infty}^{\infty} f(x+h)e^{i2\pi vx} dx - \int_{-\infty}^{\infty} f(x-h)e^{i2\pi vx} dx\right). \end{aligned} \tag{527}$$

Considering the sum of integrals above, if we change variables using $y = x + h$ in the first integral and using $y = x - h$ in the second, we produce

$$\begin{aligned} G(v) &= \frac{1}{2h}(e^{-i2\pi vh}F(v) - e^{i2\pi vh}F(v)) \\ &= \frac{sin(2\pi vh)}{ih}F(v). \end{aligned} \tag{528}$$

That is, the effect of the stationary linear operator defined by equation 526 is given by the transfer function $sin(2\pi vh)/ih$.

The desired result of differentiation is $f'(x)$, the Fourier transform of which is

$$D(v) = \int_{-\infty}^{\infty} f'(x)e^{i2\pi vx} dx. \tag{529}$$

Integrating by parts produces

$$D(v) = f(x)e^{i2\pi vx}\Big|_{-\infty}^{\infty} - i2\pi vF(v). \tag{530}$$

For the Fourier transform of $f(x)$ to exist, $f(x)$ must approach zero as $x \to \pm\infty$, so the Fourier transform of the desired result, $f'(x)$, is

$$D(v) = -i2\pi vF(v). \tag{531}$$

From equations 528 and 531 we see

$$G(v) = \frac{sin(2\pi vh)}{2\pi vh} D(v), \tag{532}$$

an equation relating in the frequency domain the desired output and the computed output. Since for $v \neq 0$ and any finite h, $|sin(2\pi vh)/(2\pi vh)| < 1$, the magnitude of the computed result is too small for all sinusoidal components with nonzero frequency, and the difficulty gets worse the higher the frequency. That is, numerical differentiation does poorly with a quickly varying (nonsmooth) input function. Furthermore, the fact that $sin(2\pi vh)/(2\pi vh)$ is negative at some frequencies means that if the input to the numerical differentiation formula is a sine wave of such a frequency (or consists predominantly of such sinusoids), the computed derivative value will have the opposite sign from the true derivative. Equation 532 also shows the result that as $h \to 0$, $G(v) \to D(v)$, so $g(x) \to d(x) = f'(x)$.

In the above we have discussed Fourier analysis of operators and functions defined on continuous variables. The analysis of operators and functions on finite discrete variables, in which category most of the operators in this text fall, is quite similar. The major difference is that integrals become sums and that both functions and their transforms are discrete; they are graphed by a series of points rather than by a continuous curve.

A finite discrete convolution is written

$$g_k = \sum_{j=-N/2+1}^{N/2} w_{k-j} f_j, \quad k = -N/2+1, -N/2+2, \ldots, -1, 0, 1, \ldots, N/2; \tag{533}$$

the finite discrete inverse Fourier transform is written

$$\hat{f}(x) = \sum_{j=-N/2+1}^{N/2} \hat{b}_j e^{-i2j\pi x/N}; \tag{534}$$

the finite discrete Fourier transform is written

$$\hat{b}_j = \frac{1}{N} \sum_{k=-N/2+1}^{N/2} f(x_k) e^{i2j\pi k/N}, \quad -N/2+1 \leq j \leq N/2, \tag{535}$$

where

$$x_k = x_0 + k\,\Delta x. \tag{536}$$

We remember that the continuous function $\hat{f}(x)$ defined by equation 534 and the discrete function $\hat{b}_j$ defined by equation 535 are periodic functions with period N. For the finite discrete convolution defined by equation 533 to fit into the structure with the finite discrete Fourier transform, the discrete impulse response function, w_k, must be cyclic with period N. If w has this property, the discrete convolution is equivalent to taking a finite discrete Fourier transform, discrete filtering (element-by-element multiplication of the finite discrete transforms of $N \cdot w$ and f, respectively), and taking the finite discrete inverse Fourier transform (see exercise 4.3.43). Also, if $f_k = f(x_k)$ is periodic with period N, then

$$\hat{f}(x_k) = f(x_k) \text{ for } k \text{ any integer,} \tag{537}$$

and the finite discrete convolution can be written

$$g_k = \sum_{j=-N/2+1}^{N/2} f_{k-j} w_j, \quad -N/2+1 \leq k \leq N/2 \tag{538}$$

(see exercise 4.3.44).

Calculating the result of a discrete convolution is a commonly required computation. Computed directly, obtaining the value of a single g_i requires N multiplications. Thus the computation of the discrete function g with N elements requires N^2 multiplications. However, we know that if f and $N \cdot w$ are real, each can be (finite discrete) transformed using the Fast Fourier transform with the requirement of only $N\ log_2(N)$ multiplications each, producing vectors $\hat{b}$ and $N \cdot W$, respectively. Then for each j, $-N/2+1 \leq j \leq N/2$, the value $\hat{b}_j$ can be multiplied by $N \cdot W_j$, requiring N multiplications of complex numbers ($4N$ real multiplications), a negligible amount compared to the number required by the FFT if N is large. Finally, we can apply the inverse FFT to the result to obtain the convolution values. Since the inverse FFT consists of the same steps as the FFT except that all exponents in the exponential coefficients appearing in the weighted sums have signs opposite those in the FFT, the final step requires $N\ log_2(N)$ real multiplications. Thus the whole operation requires approximately $3N\ log_2(N)$ multiplications. Most often, either the filter $N \cdot W$ has been synthesized directly in the frequency domain or the convolution is to be done for many input functions f, so the transform from w to $N \cdot W$ has been done once and for all. Therefore, the operations needed to transform w to $N \cdot W$ need not be counted, so the convolution executed via Fourier transform and filtering requires approximately $2N\ log_2(N)$ multiplications as compared to N^2 for the direct method. Thus, for N greater than about 10, we save some time by using the Fourier transform method; for N greater than 50, we make a significant savings.

Very often, discrete situations are analyzed by analyzing the continuous situations to which they are an analog. We analyze the function which the discrete process approaches as the difference between the discrete values goes to zero. If the discrete values were obtained by evaluating a continuous function at certain sample argument values (sampling the continuous function), we analyze the underlying continuous function. If we are synthesizing a discrete filter, we design an appropriate continuous filter and then sample the continuous function at appropriate values to produce the discrete filter we use. Conversely, as we have seen throughout this text, most numerical computer processing involves discrete approximation to continuous processes we wish to analyze. The extent to which such procedures are justified is discussed in part c of this section.

c) Continuous vs. Discrete Operators; Finite vs. Infinite Operators; Sampling. First let us investigate the relation between stationary linear operators applied in a continuous variable and the corresponding operators applied in a discrete variable. We are interested in the relationship between a continuous convolution,

$$g(x) = \int_{-\infty}^{\infty} w(x - y) f(y) dy, \tag{539}$$

and the corresponding discrete convolution,

$$g_h(x_i) = h \sum_{j=-\infty}^{\infty} w_{i-j} f(x_j), \tag{540}$$

where the x_i are evenly spaced with interval h and

$$w_{i-j} \equiv w(x_i - x_j). \tag{541}$$

In other words, we ask: How does the discrete convolution of *sampled values* of the functions w and f compare with the continuous convolution of w and f? We are also interested in the relation between the continuous and discrete Fourier transforms. In the latter part of this section, we investigate the effect of making the discrete convolution and discrete transform finite, that is, of restricting the number of samples used to a finite number.

The corresponding discrete and continuous functions that we have discussed thus far have the following property in common: The continuous function is the limiting density of the corresponding discrete function (for example, $F(v) = \lim_{\Delta v \to 0} (\hat{b}(v_j)/\Delta v)$). Here, however, we have a discrete function $f(x_j)$, $-\infty < j < \infty$, which is made up of samples of the corresponding continuous function $f(x)$ at fixed intervals, $h = x_{j+1} - x_j$ (we can without loss of generality choose our origin so that $x_0 = 0$ and thus $x_j = jh$). To compare these functions, we wish to create a continuous function $\phi_h(x)$ corresponding to the discrete function $f(x_j)$ such that two conditions hold:

1. $\phi_h(x)$ contains all of the information of the $f(x_j)$ and no more.
2. In the limit as $h \to 0$,

$$\phi_h(x) = f(x). \tag{542}$$

To satisfy condition 1, we arbitrarily define $\phi_h(x) = 0$ if $x \neq x_j$ for some j.

If condition 2 holds, in the limit as $h \to 0$, we must have

$$\begin{aligned}\int_{x_j-h/2}^{x_j+h/2} \phi_h(x)dx &= \int_{x_j-h/2}^{x_j+h/2} f(x)dx \\ &= hf(x_j).\end{aligned} \tag{543}$$

Defining $\theta(x)$ by $\phi_h(x) = \dfrac{d}{dx}\theta(x)$, from equation 543, in the limit as $h \to 0$, we must have

$$\theta(x_j + h/2) - \theta(x_j - h/2) = hf(x_j). \tag{544}$$

Since the derivative of $\theta(x) = 0$ if $x \in [x - h/2, x + h/2]$ and $x \neq x_j$, $\theta(x)$ must be constant in this interval if $x \neq x_j$, so to satisfy equation 544, $\theta(x)$ must be a step function with a discontinuity of $hf(x_j)$ at $x = x_j$, and therefore $\phi_h(x)$ is the derivative of such a step function. In section 1.4.1, we learned that the derivative of a step function with step Δy at $x = x_0$ is $\Delta y\ \delta(x - x_0)$, so

$$\phi_h(x) - hf(x_j)\delta(x - x_j), \quad \text{for } x \in [x_j - h/2, x_j + h/2], \tag{545}$$

or

$$\begin{aligned}\phi_h(x) &= \sum_{j=-\infty}^{\infty} hf(x_j)\delta(x - x_j) \\ &= hf(x) \sum_{j=-\infty}^{\infty} \delta(x - jh)\end{aligned} \tag{546}$$

satisfies our requirements.

The sum in equation 546 consists of a series of impulse functions, one at each integer multiple of h between $-\infty$ and ∞. We call this function $comb_h(x)$:

$$comb_h(x) \equiv \sum_{j=-\infty}^{\infty} \delta(x - jh). \tag{547}$$

Equation 546 can be rewritten

$$\phi_h(x) = hf(x)comb_h(x). \tag{548}$$

We shall analyze, in the frequency domain, the relation between $g(x)$ given by equation 539 and $\gamma_h(x)$, the continuous function corresponding to the discrete function $g_h(x_i)$, that is,

$$\gamma_h(x) = h \sum_{i=-\infty}^{\infty} g_h(x_i)\delta(x - x_i). \tag{549}$$

Using equation 540, equation 549 can be rewritten

$$\gamma_h(x) = \int_{-\infty}^{\infty} \omega_h(x - y)\phi_h(y)dy, \tag{550}$$

where $\omega_h(x)$ is the continuous function corresponding to the discrete function $w(x_i)$ (see exercise 4.3.45). We wish to compare

$$G(v) = W(v)F(v) \tag{551}$$

and

$$\Gamma_h(v) = \Omega_h(v)\Phi_h(v), \tag{552}$$

where Γ_h, Ω_h, and Φ_h are the continuous Fourier transforms of γ_h, ω_h, and ϕ_h, respectively. Therefore, we need an expression for $\Phi_h(v)$ and $\Omega_h(v)$.

ϕ_h is h times the product of $f(x)$ and $comb_h(x)$ (see equation 548). Therefore the Fourier transform of ϕ_h is equal to the convolution of h times $F(v)$ and $COMB_h(v)$:

$$\Phi_h(v) = h \int_{-\infty}^{\infty} COMB_h(\mu)F(v - \mu)d\mu, \tag{553}$$

where $COMB_h$ is the Fourier transform of $comb_h$.

To evaluate $COMB_h(v)$, it is useful to write $comb_h(x)$ in a way other than that given by equation 547. Remember that if $f(x)$ is defined on the interval $[-d,d]$, the function $\hat{f}$ produced by writing f as a sum of sinusoids by the finite Fourier transform (see equations 499 and 500) agrees with f on $[-d,d]$ and is cyclic with period $2d$. Thus, letting

$$d = h/2 \tag{554}$$

and

$$f(x) = \delta(x), \tag{555}$$

we see that $\hat{f}$ will be the delta function cyclically continued with period h:

$$\hat{f}(x) = comb_h(x). \tag{556}$$

From equation 499, we have

$$\hat{f}(x) = \sum_{j=-\infty}^{\infty} \hat{b}_j e^{-i2j\pi x/h}, \tag{557}$$

where equation 500 gives

$$\hat{b}_j = \frac{1}{h}\int_{-h/2}^{h/2} \delta(x)\, e^{i2j\pi x/h} dx = \frac{1}{h}. \tag{558}$$

Therefore, from equations 556, 557, and 558,

$$comb_h(x) = \frac{1}{h}\sum_{j=-\infty}^{\infty} e^{-i2j\pi x/h}. \tag{559}$$

We now take the Fourier transform of the *comb* function as expressed in equation 559:

$$COMB_h(\nu) = \int_{-\infty}^{\infty} comb_h(x) e^{i2\pi\nu x} dx = \frac{1}{h}\int_{-\infty}^{\infty}\sum_{j=-\infty}^{\infty} e^{-i2j\pi x/h} e^{i2\pi\nu x} dx. \tag{560}$$

If we switch the order of the integration and summation, an operation allowed because of the uniform convergence of Fourier approximation, we produce

$$COMB_h(\nu) = \frac{1}{h}\sum_{j=-\infty}^{\infty}\int_{-\infty}^{\infty} e^{-i2j\pi x/h} e^{i2\pi\nu x} dx. \tag{561}$$

But by equation 517 the summand in equation 561 is just $\delta(\nu - j/h)$, so

$$COMB_h(\nu) = \frac{1}{h}\sum_{j=-\infty}^{\infty} \delta(\nu - j/h) = \frac{1}{h} comb_{1/h}(\nu). \tag{562}$$

The Fourier transform of a *comb* function with interval h is a constant times the *comb* function with interval $1/h$.

Therefore, from equations 553 and 562,

$$\Phi_h(\nu) = \int_{-\infty}^{\infty} comb_{1/h}(\mu) F(\nu - \mu) d\mu, \tag{563}$$

or equivalently,

$$\Phi_h(\nu) = \int_{-\infty}^{\infty}\sum_{j=-\infty}^{\infty} \delta(\mu - j/h) F(\nu - \mu) d\mu. \tag{564}$$

Switching the integration and summation in the second form for $\Phi_h(\nu)$, we obtain

$$\begin{aligned}\Phi_h(\nu) &= \sum_{j=-\infty}^{\infty}\int_{-\infty}^{\infty} \delta(\mu - j/h) F(\nu - \mu) d\mu \\ &= \sum_{j=-\infty}^{\infty} F(\nu - j/h).\end{aligned} \tag{565}$$

Equation 565 is most usefully visualized as in figure 4-22. The transform of the continuous function corresponding to the sampled function is, at any given frequency, the sum of shifted versions of the transform of the corresponding nonsampled continuous function at that frequency. Since $\Phi_h(\nu)$ is clearly periodic with period $1/h$, we can completely characterize its behavior by finding its values on the interval $[-1/2h, 1/2h]$. The crosshatches in figure 4-22 indicate contributions to $\Phi_h(\nu)$ in $[-1/2h, 1/2h]$.

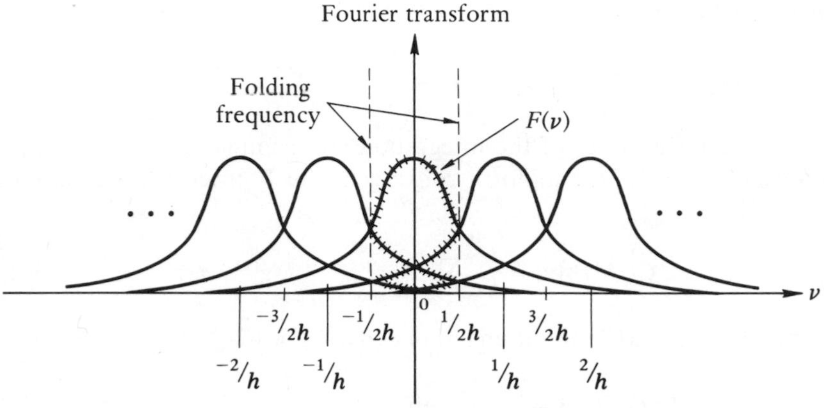

FIG. 4-22 Fourier transform of a sampled function

To obtain the value of $\Phi_h(\nu)$ on the interval $[-1/2h, 1/2h]$, we note that if $F(\nu)$ is an even function, the result of adding the shifted versions of F is equivalent to folding F at $1/2h$ and $-1/2h$, folding the folded version at $\nu = \pm 1/2h$, and so on, and adding all the folded results (if $F(\nu)$ is not even but $f(x)$ is real, at each folding the complex conjugate of the function being folded must be taken; see exercise 4.3.46). As a result of this way of looking at the composition of Φ_h, $\nu = 1/2h$ is often called the *folding frequency*. All values of F at frequencies with magnitude greater than the folding frequency are folded back into the interval $[-1/2h, 1/2h]$. That is, for any $f(x)$, components of F at frequencies greater than $1/2h$ in magnitude appear as if they were at some frequency in the interval in question. It is said that these higher frequencies are *aliased*; as far as the values at the sample points are concerned, a sinusoid at a higher frequency looks exactly like one at its aliased frequency in the interval $[-1/2h, 1/2h]$. This behavior is the same as the behavior we noted in part a of this section, where, in the finite situation, the vector obtained by sampling a sinusoid at discrete values was, for high frequencies, the same as the vector produced for some lower frequency.

If continuous analysis is to be useful in analyzing a discrete situation and if a discrete function is to be a reasonable approximation to a continuous function, F and Φ_h must agree or approximately agree on $[-1/2h, 1/2h]$. For this to be the case, F must be zero or negligible outside of this frequency range. If it is zero there, no components will be folded back into the interval of interest, and thus F will equal Φ_h on that interval. In this case, $f(x)$ can be entirely recovered from its sampled version $\phi_h(x)$ by taking the Fourier transform of ϕ_h, choosing only the cycle of this transform in $[-1/2h, 1/2h]$ while setting the remainder to zero, and taking the inverse Fourier transform of the result.

To complete the comparison between continuous and discrete convolution, we return to the comparison of $G(v)$ and $\Gamma_h(v)$, which are given by equations 551 and 552 respectively. If $\Phi_h(v) = F(v)$ and $\Omega_h(v) = W(v)$, then $G(v) = \Gamma_h(v)$. In other words, if sampling at interval h produces no aliasing for either $f(x)$ or $w(x)$, then $g(x)$ can be completely recovered from the result of the corresponding discrete convolution.

What we have analyzed above is the effect of sampling, of using a discrete approximation to a continuous function. We must also analyze the effect of using only a finite part of that discrete approximation.

Assume $f(x)$ is given only in the interval $[-p/2, p/2]$ at $x = kp/N$ for all integers such that $-N/2+1 \le k \le N/2$. Normally, we do not know f outside of this interval. Because finite Fourier analysis and resynthesis with this interval produces periodic functions with period p, it is commonly assumed that $f(x)$ is periodic with this period.

The finite discrete Fourier transform of $f(x)$ produces a discrete coefficient density function which we call $\tilde{F}(v_j)$. From equations 492, 501, 504, and 507, $v_j = j\Delta v = j(1/p)$ and

$$\tilde{F}(v_j) = \frac{\hat{b}_j}{\Delta v} = \begin{cases} \dfrac{\dfrac{1}{N}\displaystyle\sum_{k=-N/2+1}^{N/2} f(x_k)e^{i2\pi v_j x_k}}{1/p}, & -N/2+1 \le j \le N/2 \\ 0 \quad \text{otherwise} \end{cases}$$

$$= \left(\frac{p}{N}\sum_{k=-N/2+1}^{N/2} f(x_k)e^{i2\pi v_j x_k}\right) rect_{N/2p}(v_j)$$

$$= \left(\frac{p}{N}\sum_{k=-\infty}^{\infty} f(x_k) rect_{p/2}(x_k) e^{i2\pi v_j x_k}\right) rect_{N/2p}(v_j), \qquad (566)$$

where

$$rect_\mu(v) \equiv \begin{cases} 1 & \text{if } -\mu < v \le \mu \\ 0 & \text{otherwise.} \end{cases} \qquad (567)$$

We rewrite equation 566 using

$$\phi_{p/N}(x) = \frac{p}{N} f(x) \sum_{k=-\infty}^{\infty} \delta(x - kp/N) \tag{568}$$

to obtain

$$\tilde{F}(v_j) = \left(\int_{-\infty}^{\infty} \phi_{p/N}(x) rect_{p/2}(x) e^{i2\pi v_j x} dx \right) rect_{N/2p}(v_j). \tag{569}$$

The continuous function corresponding to the discrete function $\tilde{F}(v_j)$ is

$$\begin{aligned}\tilde{\Phi}(v) &= \Delta v \tilde{F}(v) comb_{\Delta v}(v) \\ &= \frac{1}{p} \left(\int_{-\infty}^{\infty} \phi_{p/N}(x) rect_{p/2}(x) e^{i2\pi vx} dx \right) rect_{N/2p}(v) comb_{1/p}(v) \\ &= \mathscr{F}(\phi_{p/N}(x) rect_{p/2}(x)) rect_{N/2p}(v)((1/p) comb_{1/p}(v)),\end{aligned} \tag{570}$$

where "$\mathscr{F}$" represents the infinite continuous Fourier transform.

Since
$$(1/p) comb_{1/p}(v) = \mathscr{F}(comb_p(x)), \tag{571}$$

$$\tilde{\Phi}(v) = \mathscr{F}(\phi_{p/N}(x) rect_{p/2}(x)) \mathscr{F}(comb_p(x)) rect_{N/2p}(v). \tag{572}$$

But the product of the Fourier transforms of two functions is equal to the Fourier transform of the convolution of the respective functions:

$$\tilde{\Phi}(v) = \mathscr{F}((\phi_{p/N}(x) rect_{p/2}(x)) * \mathscr{F}(comb_p(x)) rect_{N/2p}(v), \tag{573}$$

where "$*$" denotes convolution.

$\phi_{p/N}(x) \cdot rect_{p/2}(x)$ is zero outside $(-p/2, p/2]$. Thus $(\phi_{p/N}(x) \cdot rect_{p/2}(x)) * comb_p(x)$ is the periodic repetition with period p of $\phi_{p/N}(x) \cdot rect_{p/2}(x)$, that is, of $\phi_{p/N}(x)$ in $(-p/2, p/2]$. Since it has been assumed that $f(x)$ is periodic, and thus $\phi_{p/N}(x)$ is periodic, we have

$$(\phi_{p/N}(x) rect_{p/2}(x)) * comb_p(x) = \phi_{p/N}(x). \tag{574}$$

Using equation 574 in equation 573 produces

$$\tilde{\Phi}(v) = \Phi_{p/N}(v) rect_{N/2p}(v). \tag{575}$$

We want to compare $F(v)$ with $\tilde{\Phi}(v)$ over $(-N/2p, N/2p]$. We see that if sampling $f(x)$ produces no aliasing, that is, if

$$F(v) = \begin{cases} \Phi_{p/N}(v) & \text{for } v \in (-N/2p, N/2p] \\ 0 & \text{otherwise}, \end{cases} \tag{576}$$

then
$$\tilde{\Phi}(v) = F(v) \quad \text{for } v \in (-N/2p, N/2p]. \tag{577}$$

In summary, if $f(x)$ is periodic with period p and $f(x)$ sampled N

times in this period produces no aliasing, the infinite continuous Fourier transform of $f(x)$ and the finite discrete Fourier transform of $f(x)$ carry equivalent information. Put another way, with the above assumptions, $f(x)$ is completely defined by its finite discrete Fourier transform.

With the above results, it can be shown that given two functions, both of which are periodic with period p and neither of which produces aliasing when sampled N times (with interval p/N) in that period, finite discrete convolution of the sampled functions is equivalent to continuous convolution followed by sampling of the result as above (see exercise 4.3.48). In this case, sampled values of continuous convolution can be obtained by taking the Fast Fourier transform of each function, multiplying the results, and computing p times the inverse Fast Fourier transform of the result.

PROBLEMS (4.3.23 through 4.3.50)

For ease of reference, the exercises related to sections 4.3.1 through 4.3.5 (exercises 4.3.1 through 4.3.22) are positioned immediately following section 4.3.5 (see page 383). The exercises related to sections 4.3.6 and 4.3.7 (exercises 4.3.23 through 4.3.50) are given here at the end of section 4.3.

4.3.23. Consider the set of functions $\{f^j\}$ such that the jth member of the set is alternately $+1$ or -1, j times over $[0,2^N]$:

$$f^j(x) = (-1)^i \quad \text{for} \quad a_{ji} \le x < a_{j,i+1}, \quad i = 0,1,\ldots,j-1;$$

where $a_{j0} = 0$ and $a_{jj} = 2^N$ $(j = 1,2,\ldots)$. So $f^1(x) = 1$.

a) For $j = 2,3,4,5$ choose the a_{ji} values so that the set $\{f^j\}$ are orthogonal over the continuous interval $[0,2^N]$ according to the inner product $(g,h) = \int_0^{2^N} g(x)h(x)dx$.

These f^j are called *Walsh functions*. Normalize the f^j.

b) For $j = 2,3,4,5$ choose the a_{ji} values so that the set $\{f^j\}$ are orthogonal over a given set of not necessarily equally spaced points x_i, $i = 1,2,\ldots,2^N$. Normalize the f^j.

c) Give a method to find a least-squares approximation to $\{(x_i,y_i)|1 \le i \le 2^N\}$ for $x_i = i$ in terms of the first $m \ll 2^N$ of the f^j found in part b.

d) Discuss the suitability of the f^j for smoothing.

4.3.24. Say whether the following statements are true or false and justify your choices:

Assume we have a set of data $\{(x_i,y_i)|i = 1,2,\ldots,N\}$, where both the x_i and y_i have no error. Then a polynomial exact matching through these points always produces at least as good an interpolation as does the least-squares polynomial of degree $N/2$:

a) Near the data arguments.

b) Between data arguments but not near any data argument.

(Answer (a) and (b) separately.)

4.3.25. Let $z_i = x_i h + c$, $1 \le i \le 2n$, where $x_{i+1} - x_i = 1$. Consider the following least-squares problems:

(1) Data: $\{(x_i, y_i) | 1 \le i \le 2n\}$
Basis functions: $\{f^k(x) | 1 \le k \le 2n\}$, a set of orthogonal functions over $\{x_i | 1 \le i \le 2n\}$.

(2) Data: $\{(z_i, y_i), 1 \le i \le 2n\}$
Basis functions: $\{g^k(z) | 1 \le k \le n\}$, where $g^k(z) = f^k((z - c)/h)$.

a) Show that the g^k, $1 \le k \le 2n$, are orthogonal over $\{z_i | 1 \le i \le 2n\}$.

b) Show that the coefficient of f^k in the least-squares fit in problem 1 is equal to the coefficient of g^k in problem 2, for $1 \le k \le 2n$.

c) If the f^k are the functions defined by equation 399 and $x_i = i - M$, give an expression for g_i^k.

4.3.26. a) Using the facts that

$$\Delta sin(ax + b) = 2sin(a/2)cos(a(x + \tfrac{1}{2}) + b) \text{ and}$$
$$\Delta cos(ax + b) = -2sin(a/2)sin(a(x + \tfrac{1}{2}) + b),$$

show that

$$\sum_{x=0}^{m} sin(ax+b) = \frac{sin(a(m+1)/2)sin((am/2)+b)}{sin(a/2)},$$

and
$$\sum_{x=0}^{m} cos(ax+b) = \frac{sin(a(m+1)/2)cos((am/2)+b)}{sin(a/2)}.$$

b) Using the results of part a and the facts that

$$cos(x)cos(y) = \tfrac{1}{2}(cos(x + y) + cos(x - y)),$$
$$cos(x)sin(y) = \tfrac{1}{2}(sin(x + y) - sin(x - y)), \text{ and}$$
$$sin(x)sin(y) = \tfrac{1}{2}(cos(x - y) - cos(x + y));$$

show that if j and k are integers,

$$\sum_{x=0}^{2n-1} cos\left(\frac{k\pi x}{n}\right) sin\left(\frac{j\pi x}{n}\right) = 0 \quad \text{for all } j \text{ and } k,$$

and
$$\sum_{x=0}^{2n-1} sin\left(\frac{k\pi x}{n}\right) sin\left(\frac{j\pi x}{n}\right) = 0 \quad \text{if } 0 < j+k < 2n \text{ and } j \ne k.$$

c) Show that if k is an integer,

$$\sum_{x=0}^{2n-1} cos^2\left(\frac{k\pi x}{n}\right) = \begin{cases} 2n, & \text{if } k = 0 \text{ or } k = n \\ n, & \text{if } 0 < k < n, \end{cases}$$

and
$$\sum_{x=0}^{2n-1} sin^2\left(\frac{k\pi x}{n}\right) = n \quad \text{if } 0 < k < n.$$

4.3.27. In a way similar to that in exercise 4.3.26 show that the functions

$$f^{2k}(x) = cos(2\pi kx/(2n+1)), \quad 0 \le k \le n,$$

$$f^{2k+1}(x) = sin(2\pi kx/(2n+1)), \quad 0 \le k \le n-1,$$

are orthogonal over the arguments $\{x\} = \{0,1,\ldots,2n\}$.

4.3.28. a) Show that the function $\hat{f}(x)$ given by equation 405 (a sinusoidal series in the form of trigonometric functions) and that given by equation 420 (a sinusoidal series in the form of complex exponentials) are the same if equation 421 holds.

b) Show that the functions $\{e^{-i(j\pi x/n)} | -n+1 \le j \le n\}$ are orthogonal over $\{x_i\} = \{0,1,\ldots,2n-1\}$ according to the inner product $(f^j, f^k) = \sum_{i=1}^{2n} f^j(x_i) f^{k*}(x_i)$, where "*" indicates complex conjugate.

c) We wish to write the least-squares problem in the case where the data values y_I are real but the basis functions $f^k(x)$ are complex and orthogonal according to the inner product defined in part b. Then we wish to minimize

$$S = \sum_{I=1}^{N} \left(y_I - \sum_{j=1}^{N} (c_j + id_j) f^j(x_I) \right) \left(y_I - \sum_{j=1}^{N} (c_j + id_j) f^j(x_I) \right)^*$$

with respect to the real variables c_j and d_j, where $i = \sqrt{-1}$.
First, using $(u + v)^* = u^* + v^*$ and $(uv)^* = u^* v^*$, show

$$\left(y_I - \sum_{j=1}^{N} (c_j + id_j) f^j(x_I) \right)^* = \left(y_I - \sum_{j=1}^{N} (c_j - id_j) f^{j*}(x_I) \right).$$

Then substitute this result into S, and set to zero the derivatives of S with respect to c_k and d_k respectively to show

$$c_k = \sum_{I=1}^{N} y_I \, \mathbf{Re}(f^k(x_I)) \Big/ \sum_{I=1}^{N} (f^k(x_I) f^{k*}(x_I))$$

and

$$d_k = \sum_{I=1}^{N} y_I \, \mathbf{Im}(f^k(x_I)) \Big/ \sum_{I=1}^{N} (f^k(x_I) f^{k*}(x_I)),$$

where

$$f^k(x_I) \equiv \mathbf{Re}(f^k(x_I)) + i\,\mathbf{Im}(f^k(x_I)).$$

d) Find the least-squares coefficients of the $f^j(x)$, where $f^j(x) = e^{-i(j\pi x/n)}$, $N = 2n$, and the data is $\{(x_I, y_I) | 1 \le I \le 2n\}$. Show that these coefficients are given by equation 421.

4.3.29. Consider the function $f(x) = 1/(1 + x^2)$. Compute $f(x)$ at $x = 0,1,2,\ldots,10$.

a) Compute the orthogonal monic polynomials over these x using the recurrence relation. After each polynomial $g_k(x)$ is computed, $c_k^2 = \sum_{i=0}^{10} [g^k(i)]^2$ should be computed and new orthonormal polynomials $h^k(x) = (1/c_k) g^k(x)$ defined.

b) Least-squares fit $f(x)$ over the x above using the h^k as basis functions. Compute the $\hat{a}_j$ for $m = 0,1,\ldots,9$. Compute $|\hat{a}_9/\hat{a}_1|$, a measure of the reduction in size of the $\hat{a}_j$ with j. Also compute $|\hat{f}(9.5) - f(9.5)|$ for $m = 9$.

c) Consider the pure cosine series least-squares fit to $f(x)$ over the x above:

$$\hat{f}(x) = \sum_{k=0}^{m} \hat{a}_k \frac{cos(k\pi x/N)}{c_k},$$

where
$$c_k^2 = \begin{cases} 10 & \text{if } 1 \le k \le 9 \\ 20 & \text{if } k = 0 \text{ or } k = 10. \end{cases}$$

Note that the functions $cos(k\pi x/N)/c_k$ are orthonormal. Then

$$\hat{a}_k = \frac{y_0 + y_{10}}{\sqrt{20}} + 2 \sum_{i=1}^{9} y_i \frac{cos(k\pi i/10)}{\sqrt{10}}.$$

Compute the $\hat{a}_k$, $k = 0,1,\ldots,9$. Then compute $|\hat{a}_9/\hat{a}_1|$ and compare its value to that obtained by the polynomial fit obtained in part b. Also, compute $|\hat{f}(9.5) - f(9.5)|$ for $m = 9$ and compare its value to that obtained in part b.

d) Consider the function

$$g(x) = f(x) - \left(\frac{f(10) - f(0)}{10} x + f(0)\right),$$

which is $f(x)$ with its linear tendency subtracted. Consider the pure sine series least-squares fit to $g(x)$:

$$\hat{g}(x) = \sum_{k=1}^{m} \hat{a}_k \frac{sin(k\pi x/10)}{\sqrt{10}}.$$

The functions $sin(k\pi x/10)/\sqrt{10}$ are orthonormal. Then

$$\hat{a}_k = 2 \sum_{i=1}^{9} y_i \frac{sin(k\pi i/10)}{\sqrt{10}}.$$

Compute the $\hat{a}_k$, $k = 1,2,\ldots,9$. Then compute $|\hat{a}_9/\hat{a}_1|$ and compare its value to that obtained for the polynomial fit (part b) and the pure cosine fit (part c). Also, compute the error in the fit to f at $x = 9.5$:

$$\left|\left(\hat{g}(9.5) + \frac{f(10) - f(0)}{10} 9.5 + f(0)\right) - f(9.5)\right|$$

for $m = 9$ and compare this value to that for parts b and c.

4.3.30. Let $x_i = cos(i\pi/L)$, $i = 0,1,2,\ldots,2L-1$.

a) Show that the functions $T_0(x), T_1(x), \ldots, T_L(x)$, where $T_i(x)$ is the ith-degree Tchebycheff polynomial, are orthogonal over the x_i.

b) What is peculiar about this set of $\{x_i\}$? Interpret the effect of this peculiarity on the process of least-squares approximation using these functions and data values given at the x_i. Give in detail a method to accomplish this approximation.

c) How good are the functions $T_i(x)$ for this approximation? Discuss the rate of convergence of the coefficients multiplying the $T_i(x)$ in the least-squares fit.

4.3.31. a) Show $sin(n\theta) = sin(\theta)Q_{n-1}(cos(\theta))$, where Q_k is a kth-degree polynomial.

b) Thus show the set of functions $R_k(x) = \sqrt{1-x^2}\,Q_{k-1}(x)$ for $k = 1,2,3,\ldots,N-1$ are orthogonal over the set of points $x_i = cos(i\pi/N)$ for $i = 1,2,\ldots,N-1$.

c) What are the requirements that data must satisfy to enable one to do efficient least-squares approximation with the $R_k(x)$?

d) When these requirements are satisfied, give in detail a method for the approximation.

e) On the basis of the above and what you know about approximation in general, discuss the advantages and disadvantages of approximation using the R_k.

f) Assume we also have a data value of $y = a$ at $x = 1$ and a value of $y = b$ at $x = -1$. What can you say about the coefficients of $R_k(x)$ if we fit the set of data obtained by subtracting $\frac{b-a}{\pi}\,cos^{-1}(x_i) + a$ from each y_i? (Assume cos^{-1} has range $[0,\pi]$.) What if $\frac{a-b}{2}x_i + \frac{a+b}{2}$ is subtracted?

4.3.32. Construct a flow graph like that in table 4-14 for $N = 16$.

4.3.33. We wish to construct an FFT which produces the coefficients of the trigonometric functions in a fit through $N = 2^L$ real data points (see equation 406). In the form of the FFT developed in section 4.3.7, part a, using the complex exponential functions, the operation consists of repetitive application of the rule (see equations 460 and 461)

$$\hat{b}_j^{(k)} = \tfrac{1}{2}(\hat{b}_j^{(k-1)even} + e^{i2\pi j/M_k}\hat{b}_j^{(k-1)odd}),$$

$$\hat{b}_{j+M_k/2}^{(k)} = \tfrac{1}{2}(\hat{b}_j^{(k-1)even} - e^{i2\pi jM_k}\hat{b}_j^{(k-1)odd}), \quad 0 \le j < M_k/2,$$

where the superscript in parentheses indicates the column in the flow graph in which the coefficient in question lies, where the original data is in column 0, and where $M_k = 2^k$. By equation 421, where $n = N/2$, remembering that $\hat{b}_j \equiv \hat{b}_{N+j}$ if $j < 0$,

$$\hat{b}_j^{(k)} = \begin{cases} \hat{a}_{2j}^{(k)}, & j = 0 \text{ or } j = N/2 \\ \tfrac{1}{2}(\hat{a}_{2j}^{(k)} + i\hat{a}_{2j+1}^{(k)}), & 0 < j < N/2 \\ \tfrac{1}{2}(\hat{a}_{2(N-j)}^{(k)} - i\hat{a}_{2(N-j)+1}^{(k)}, & N/2 < j < N, \end{cases}$$

where the $\hat{a}_j^{(k)}$ are the trigonometric function coefficients in the kth column of the new FFT which are equivalent to the exponential coefficients of the old FFT.

a) Substitute the above relationships into the above rule for computing the $\hat{b}_j^{(k)}$ from the $\hat{b}_j^{(k-1)even}$ and $\hat{b}_j^{(k-1)odd}$ to produce a rule for computing the $\hat{a}_j^{(k)}$ from the $\hat{a}_j^{(k-1)even}$ and $\hat{a}_j^{(k-1)odd}$.

b) How many real multiplications are required to compute the $\hat{a}_j^{(k)}$ from the $\hat{a}_j^{(k-1)even}$ and $\hat{a}_j^{(k-1)odd}$? How many real additions?

c) Thus show that the new FFT requires $N\,log_2(N)$ real multiplications and $(3N/2)\,log_2(N)$ real additions.

4.3.34. Let

$$g(x) = \int_{-\infty}^{\infty} a(x,y)f(y)dy.$$

Show that

$$g(x-z) = \int_{-\infty}^{\infty} a(x,y)f(y-z)dy$$

for all shifts z, arguments x, and functions f iff $a(u+z,v+z) = a(u,v)$ for all arguments u and v and shifts z, that is, iff a is a function only of the difference between its first and second parameters. *Hint:* Change variables in the equation for $g(x-z)$, letting $u = x - z$ and $v = y - z$.

4.3.35. Show that

$$\int_{-\infty}^{\infty} w(x-y)f(y)dy = \int_{-\infty}^{\infty} w(y)f(x-y)dy.$$

4.3.36. Show that if

$$f^1(x) = \frac{1}{\sqrt{2\pi}\,\sigma_1} e^{-\frac{1}{2}(x/\sigma_1)^2} \quad \text{and} \quad f^2(x) = \frac{1}{\sqrt{2\pi}\,\sigma_2} e^{-\frac{1}{2}(x/\sigma_2)^2},$$

that is, both are Gaussian distribution density functions with mean zero and variance σ_1^2 and σ_2^2 respectively, then the convolution of f^1 and f^2 is also a Gaussian with mean zero and variance $\sigma_1^2 + \sigma_2^2$.

4.3.37. Let x and y be independent random variables with probability density functions $p_x(u)$ and $p_y(u)$ respectively, and let $z = x + y$. Show that the probability density function $p_z(u)$ is given by the convolution of the functions $p_x(u)$ and $p_y(u)$.

4.3.38. Let $f(x)$ be a continuous bounded function given on $[-d,d]$. Assume $f'(x)$ exists and is bounded on $[-d,d]$. Consider the approximation $\hat{f}(x)$ to $f(x)$ given by the finite discrete Fourier transform of $f(x)$ over $[-d,d]$ (equation 493 with $a = -d$). We wish to show that $\hat{f}(x)$ converges uniformly on $[-d,d]$ to $f(x)$ as the number of data points, N, approaches ∞. The following theorem, which the student may assume, will be useful:

If $h(x,y)$ is bounded and continuous on $x \in [a,b]$, $y \in [c,d]$, and $\Delta y_M = (d-c)/M$, and $y_{M,k}$ is any point in $[c + (k-1)\Delta y_M, c + k\Delta y_M]$, $\sum_{k=1}^{M} h(x,y_{M,k})\Delta y_M$ converges uniformly to $\int_c^d h(x,y)dy$ as $M \to \infty$.

a) Use the above theorem to prove that if $h(u,v)$ is bounded on $u,v \in [-\frac{1}{2},\frac{1}{2}]$, $g(u) = \int_{-1/2}^{1/2} h(u,v)\,sin(2n\pi v)dv$ approaches zero uniformly as $n \to \infty$, as follows.

(1) Define $\Delta v_M = 1/M$ and write $g(u)$ as a sum of integrals over $[(k-1)\Delta v_M, k\Delta v_M]$, where k runs between $-M/2 + 1$ and $M/2$.

(2) For the kth integral, write $h(u,v)$ as $(h(u,v) - h(u,v_{M,k})) + h(u,v_{M,k})$ where $v_{M,k}$ is a point in $[(k-1)\Delta v_M, k\Delta v_M]$ for which $h(u,v)$ is minimum over that interval. Thus show

$$|g(u)| \leq \sum_{k=-M/2+1}^{M/2} \int_{(k-1)\Delta v_M}^{k\Delta v_M} |h(u,v) - h(u,v_{M,k})|\,|sin(2n\pi v)|\,dv$$
$$+ \sum_{k=-M/2+1}^{M/2} |h(u,v_{M,k})| \left| \int_{(k-1)\Delta v_M}^{k\Delta v_M} sin(2n\pi v)dv \right|.$$

(3) In the first integral use $|sin(2n\pi v)| < 1$, integrate the second integral, and use the fact that $|cos(2n\pi v)| < 1$ to produce

$$|g(u)| \leq \left(\int_{-1/2}^{1/2} h(u,v)dv - \sum_{k=-M/2+1}^{M/2} h(u,v_{M,k})\Delta v_M \right) + FM/(n\pi),$$

where F is an upper bound for the bounded function $f(u,v)$ over $u,v \in [-\frac{1}{2},\frac{1}{2}]$.

(4) By the given theorem, the parenthesized expression in (3) above converges uniformly to zero as $M \to \infty$. For any M, there exists an n independent of u such that $FM/(n\pi)$ is as small as desired, so we have that $|g(u)|$ converges uniformly to zero as $n \to \infty$ and thus $g(u)$ converges uniformly to zero as $n \to \infty$, proving part a.

b) Define $\hat{f}_N(x) = \sum_{j=-N/2+1}^{N/2} \hat{c}_j e^{-ij\pi x/d}$, where the $\hat{c}_j$ are the coefficients of the finite continuous Fourier transform over $[-d,d]$. That is, $\hat{f}_N(x)$ is a truncated finite continuous Fourier transform over $[-d,d]$. Show that

$$|\hat{f}(x) - f(x)| \leq |\hat{f}(x) - \hat{f}_N(x)| + |\hat{f}_N(x) - f(x)|.$$

Thus, if we can show that $\hat{f}_N(x)$ converges uniformly to $f(x)$ and $\hat{f}(x)$ converges uniformly to $\hat{f}_N(x)$, we will have proved the desired result.

c) Hamming shows that

$$\hat{f}_{2n}(2\tau d) - f(2\tau d) = \int_{-1/2}^{1/2} \left[(y(\tau - \phi) - y(\tau)) \frac{cos(\pi\phi)}{sin(\pi\phi)} \right] sin(2n\pi\phi)d\phi,$$

where $y(\tau) \equiv f(2\tau d)$ for $\tau \in [-\frac{1}{2},\frac{1}{2}]$ and $y(\tau)$ is periodic with period 1.† Show that the bracketed expression in the above integral is bounded over $\tau, \phi \in [-\frac{1}{2},\frac{1}{2}]$. Thus, using the result of part a, we can conclude that $\hat{f}_N(x)$ converges uniformly to $f(x)$ as $N \to \infty$, where $N = 2n$.

d) In the same development as that mentioned above, Hamming shows that if $\hat{g}_{2n}(v) \equiv \hat{f}_{2n}(vd/n)$ and $\hat{g}(u) \equiv f(ud/n)$,

$$\hat{g}_{2n}(v) = \frac{1}{2n} \int_{-n}^{n} g(u) \frac{sin(\pi(v-u))cos\left(\frac{\pi}{2n}(v-u)\right)}{sin\left(\frac{\pi}{2n}(v-u)\right)} du.‡$$

† R. W. Hamming, *Numerical Methods for Scientists and Engineers*, section 21.4.
‡ Ibid.

By the same method it can be shown that if $\hat{g}(v) \equiv \hat{f}(vd/n)$,

$$\hat{g}(v) = \frac{1}{2n} \sum_{k=-n+1}^{n} g(k\Delta u) \frac{sin(\pi(v - k\Delta u))cos\left(\frac{\pi}{2n}(v - k\Delta u)\right)}{sin\left(\frac{\pi}{2n}(v - k\Delta u)\right)} \Delta u, \quad \text{where} \quad \Delta u = 1.$$

By the changes of variables $x = vd/n$, $z = ud/n$, obtain

$$\hat{f}_{2n}(x) = \frac{1}{2d} \int_{-d}^{d} f(z) \frac{sin\left(\frac{n\pi}{d}(x - z)\right)cos\left(\frac{\pi}{2d}(x - z)\right)}{sin\left(\frac{\pi}{2d}(x - z)\right)} dz$$

and

$$\hat{f}(x) = \frac{1}{2d} \sum_{k=-n+1}^{n} g(k\Delta x) \frac{sin\left(\frac{n\pi}{d}(x - k\Delta x)\right)cos\left(\frac{\pi}{2d}(x - k\Delta x)\right)}{sin\left(\frac{\pi}{2d}(x - k\Delta x)\right)} \Delta x,$$

where $\Delta x = d/n$. Show that the integrand in the integral just above is bounded. Thus, we can conclude by the given theorem that $\hat{f}(x)$ converges uniformly to $\hat{f}_N(x)$ as $N \to \infty$, where $N = 2n$. This completes the proof of the uniform convergence of $\hat{f}(x)$ to $f(x)$.

4.3.39. Show that the Fourier transform of a Gaussian function with mean zero and variance σ^2 is proportional to a Gaussian with mean zero and variance inversely proportional to σ^2.

4.3.40. Show that the Fourier transform of the product of two functions is the convolution of their respective Fourier transforms.

4.3.41. a) Show that the Fourier transform of the function of x, $g(x - z)$, is a constant times the Fourier transform of $g(x)$.

b) Let $g_1(x) = \int_{x-h}^{x+h} f(u)du$, and assume the Fourier transforms of $g_1(x)$ and $f(x)$ exist and are $G_1(v)$ and $F(v)$, respectively. Using the fact that the Fourier transform of any function $g(x)$ exists only if $\lim_{x \to \pm\infty} g(x) = 0$, show that

$$G_1(v) = (1/\pi v)\, sin(2\pi vh) F(v).$$

c) Let $g_2(x)$ be $g_1(x)$ approximated by Simpson's rule, that is, let

$$g_2(x) = (h/3)[f(x - h) + 4f(x) + f(x + h)].$$

Show that $G_2(v)$, the Fourier transform of $g_2(x)$, is given by

$$G_2(v) = (2h/3)(2 + cos(2\pi vh))F(v).$$

d) Write $G_2(v)$ as a multiple of $G_1(v)$. Using this result, discuss the properties of Simpson's rule integration thought of as a filter applied to the result of analytic integration.

4.3.42. Let S be the vector space of ∞-vectors,

$$y = [\ldots, y_{-2}, y_{-1}, y_0, y_1, y_2, y_3, \ldots]^T,$$

where y_j is any real number for any integer $j \in [-\infty, \infty]$.

a) We have defined stationarity of a linear operator O applied to functions of a continuous variable x. A vector in S is a function of the discrete variable j. Redefine stationarity for linear operators O applied to such functions (vectors), where O maps vectors in S to vectors in S. We have given the general form of any such operator.

b) Show that the forward-difference operator Δ is linear and stationary.

c) Show that the vectors f^ν defined by $f_j^\nu = e^{-i2\pi\nu j}$, $\nu \in [-\frac{1}{2}, \frac{1}{2}]$, are eigenvectors of any stationary linear operator from S to S.

d) Using the fact that the vectors in $F \equiv \{f^\nu \mid \nu \in [-\frac{1}{2}, \frac{1}{2}]\}$ span S and

$$\sum_{j=-\infty}^{\infty} f_j^{\nu_1} f_j^{\nu_2 *} = 0 \text{ if } \nu_1 \neq \nu_2,$$

prove that the vectors in F form a basis for S.

e) What would you mean by the transfer function corresponding to a stationary linear operator O from S to S? On what values of ν would it be defined?

4.3.43. Show that if $f(x)$ and $w(x)$ are periodic with period N, discrete convolution, $g(k) = \sum_{j=-N/2+1}^{N/2} w(k-j) f(j)$, is equivalent to taking the finite discrete Fourier transform $\hat{W}(j)$ of $w(x)$ and $\hat{F}(j)$ of $f(x)$, over the integers in $[-N/2+1, N/2]$, computing $\hat{G}(j) = N \cdot \hat{W}(j)\hat{F}(j)$, $-N/2+1 \leq j \leq N/2$, and computing the finite discrete inverse Fourier transform of $\hat{G}(j)$.

4.3.44. Show that if $f(x)$ and $w(x)$ are periodic with period N,

$$\sum_{j=-N/2+1}^{N/2} f(k-j) w(j) = \sum_{j=-N/2+1}^{N/2} w(k-j) f(j).$$

4.3.45. Show that the infinite discrete convolution given by equation 540 corresponds to a continuous function as in equation 550.

4.3.46. Let $f(x)$ be a real function with Fourier transform $F(\nu)$. Show that the Fourier transform of $\phi_h(x)$, given by equation 565, is produced by folding $F(\nu)$ at $\nu = \pm 1/2h$, while taking the complex conjugate at each folding, and summing the results.

4.3.47. Consider the Gaussian function

$$f(x) = \frac{1}{\sqrt{2\pi}\sigma} e^{-\frac{1}{2}(x/\sigma)^2}.$$

What is the relative error due to aliasing in the value at $2\pi\nu = \sqrt{2 ln(5)}/\sigma$ of the Fourier transform of $\phi_h(x)$, the sampled version of $f(x)$, if the sampling is done with interval $h = \sigma$? $\sigma/2$? 2σ? Use the definition of relative error: $(\Phi_h(\nu) - F(\nu))/F(\nu)$, and approximate the infinite sum $\Phi_h(\nu)$ by the sum of the nonnegligible summands.

4.3.48. Let $f(x)$ and $g(x)$ be periodic functions with period p which when sampled with interval p/N produce no aliasing. Show that with regard to computing the finite discrete convolution of f and g, sampling and convolution are commutative. That is, show the result of sampling each function N times with interval p/N and computing p/N times the finite discrete convolution of the results is equivalent to taking the continuous convolution of f and g and sampling the result N times with interval p/N.

4.3.49. a) Compute the inverse Fourier transform of $rect_\mu(v)$, which is defined by equation 567.

b) We have shown that if $F(v) = 0$ for $|v| \geq 1/2h$, then $F(v)$ can be retrieved from $\Phi_h(v)$, the Fourier transform of $\phi_h(x)$, which is the continuous function corresponding to the result of sampling $f(x)$ with interval h, as

$$F(v) = \begin{cases} \Phi_h(v), & -1/2h < v < 1/2h \\ 0, & |v| \geq 1/2h. \end{cases}$$

That is, $F(v) = \Phi_h(v) rect_{1/2h}(v)$.

Write the equation above in the space domain, that is, write $f(x)$ in terms of $\phi_h(x)$.

c) Verify the result of part b in both the space and frequency domains in the case $f(x) = 1$.

4.3.50. The coefficients of the finite discrete Fourier transform (equation 497 with $a = -d$) can be viewed as approximations to the coefficients of the finite continuous Fourier transform (equation 500).

a) Show that if $f(d) = f(-d) = 0$, the finite discrete Fourier transform coefficients are the approximation to the finite continuous Fourier transform coefficients produced by approximating the integration in equation 500 by using the trapezoidal rule.

b) If $f(x) = 0$ for $|x| > d$, equation 500 can be written

$$\hat{b}_j = \frac{1}{2d} \int_{-\infty}^{\infty} f(x)\, e^{ij\pi x/d} dx,$$

and equation 497 can be written

$$\hat{b}_j^{approx} = \frac{1}{2d} \sum_{k=-\infty}^{\infty} f(x_k) e^{ij\pi x_k/d} \Delta x,$$

where $x_k = k\,\Delta x$ and $\Delta x = 2d/N$ for some N. Thus

$$\hat{b}_j^{approx} = \frac{1}{2d} \int_{-\infty}^{\infty} f(x)[\Delta x\; comb_{\Delta x}(x)] e^{ij\pi x/d} dx.$$

If we define $g(x) = f(x)/2d$ and $h(x) = (f(x)/2d)[\Delta x\; comb_{\Delta x}(x)]$, then $\hat{b}_j = G(v_j)$ and $\hat{b}_j^{approx} = H(v_j)$, where $v_j = j/2d$. Write $H(v)$ as a multiple of $G(v)$.

c) Do the same as part b for Simpson's rule integration.

d) Do the same for Romberg integration.

4.4 SUMMARY

In this chapter we have discussed the general problem of approximation. We saw that the appropriate procedure for approximation depends on

1. the amount of error in the data,
2. the distribution of the error in the data, if the error is significant,
3. the function underlying the data, or at least the general shape of that function,
4. the spacing of the data arguments,
5. the number of data points given, and
6. the objective of the approximation.

Exact matching was shown to be a reasonable approximation criterion if the data has little error and the form of the underlying function is known. If the function has significant error, a method involving norm minimization should be used. Available alternatives include least-squares approximation and Tchebycheff approximation, with the choice depending on the error distribution.

If the data has little error but the underlying function is not known, we normally assume that the underlying function is smooth, and insist upon an approximation that will be smooth. We have seen that for equally spaced points, polynomial exact matching in which a single polynomial is fit to all of the data points (approximation in the large) will be smooth only if the number of data points is quite small. Otherwise, other sets of functions, like the sinusoids, will produce a better approximation in the large.

Alternatively, we can choose to do the approximation by piecing together approximations in the small, that is, approximations over a few data points. The individual approximations can be determined using exact matching or norm minimization. As described above, the amount of data error should determine which method is chosen. Furthermore, the individual approximations in the small need not approximate over intervals of the same length or the same number of data points, and they need not use the same set of basis functions. These matters should depend on the considerations listed above, for the argument range over which each individual approximation in the small is to be applied.

Spline approximation falls into the category of approximation in the small. Thus, approximation in the small can be very smooth, but it requires more storage in the computer to keep the parameters of the approximating sections than does approximation in the large with a relatively small number of basis functions.

If the data arguments are equally spaced, we have seen that sinusoidal approximation can produce a smooth fit. If the data arguments are not equally spaced, however, this is not necessarily true, and a different set of basis functions appropriate for the spacing should be found.

Many other function forms have been used for approximation. First, if the function underlying the data is known to be of a given form, it is wise to approximate the data with a function of that form unless the calculation of the approximating parameters is prohibitively slow or inaccurate. Second, some nonlinear forms have been found useful for approximation. In particular, the rational form:

$$\hat{f}(x) = \sum_{i=1}^{m_1} \hat{a}_i g^i(x) \Big/ \sum_{i=1}^{m_2} \hat{b}_i h^i(x), \tag{578}$$

where the g^i and h^i are two sets of basis functions, is commonly used.† It often produces an approximation equally good as a linear form, with fewer approximating parameters, especially if the data has areas of sudden increase or decrease.

One of the advantages of polynomial exact-matching approximation is that it is analytically simple and thus can be used as a basis for further operations, such as integration or differentiation, on the approximating function. It should be noted that least-squares approximation and approximations using other norms, as well as approximations using other sets of basis functions, can also be used for this purpose, although the process is not normally as analytically and computationally simple as with polynomial exact matching. However, in some cases, like numerical differentiation, the use of the least-squares fit as the approximation on which the operation is applied produces a much more stable result than exact matching—a result which varies less as a function of the error in the data, as long as the function underlying the data can be fit well by the basis functions involved in the least-squares approximation.

Thus the choice of approximation method depends on many factors, and no method which is best in general can be given. The user must understand both his problem and the approximation methods and choose the appropriate method for his problem accordingly.

---

† C. T. Fike, *Computer Evaluation of Mathematical Functions,* chapters 8, 9, and 10.

PROBLEMS

4.4.1. For each of the following situations, carefully specify which numerical method and parameters you would choose and why. If a standard method does not fit, say why, create a method, and justify your creation.

a) To interpolate in a table telling on the hour for each hour between opening time and one hour after closing time, the average number of shoppers in line at the checkout counters at the supermarket on Tuesdays. That is, a job scheduler may wish to know the number expected to be in line at 1:20 p.m.

b) To integrate over $[a,b]$ a function defined by data from a freshman physics experiment, where measurements are made at time $a \approx t_0, t_1, t_2, \ldots, t_n \approx b$, where each t_i is accurate but its value depends on the facility of the experimenter.

4.4.2. For each of the following problems, assume the data arguments are equally spaced and say which of the following approximation methods you would use and why.

(1) Tchebycheff approximation with polynomials
(2) Least-squares approximation with polynomials
(3) Exact-matching approximation with one polynomial
(4) Cubic-spline exact-matching approximation
(5) Fourier least-squares approximation
(6) Fourier exact-matching approximation

a) A physicist needs to find the position of a falling object at any time t, given 30 measured $(t_i, position_i)$ values. Assume the formula for a vacuum is correct:

$$\begin{aligned} \textit{position} &= \text{position at } t = 0 \\ &\quad + \text{initial velocity} \times t \\ &\quad + \text{acceleration due to gravity} \times (t^2/2). \end{aligned}$$

b) The cross-section of a sailboat hull is being designed. The designer has specified fifty (x,y) pairs, each of which represents a point on the hull cross-section. To compute drag, the positions of all points on the hull cross-section (not just of the fifty specified points) are needed.

c) A document retrieval system designer has a retrieval method with a parameter α, such that by varying α he can vary the tradeoff between the value of the measure of relevance of the set of retrieved documents and the fraction of the relevant documents which are retrieved. He has measured exactly a set of twenty-five $(\alpha_i, relevance_i)$ pairs for a sample set of documents and wishes for any given α to find the corresponding value of the relevance measure for the whole population of documents from which he has a sample.

REFERENCES

Abramson, N. *Information Theory and Coding.* New York: McGraw-Hill, 1963.

Conte, S. D. *Elementary Numerical Analysis.* New York: McGraw-Hill, 1965.

Cooley, J. W., and Tukey, J. W. "An Algorithm for the Machine Calculation of Complex Fourier Series," *Mathematics of Computation* 19 (1965): 297.

Fike, C. T. *Computer Evaluation of Mathematical Functions.* Englewood Cliffs, N.J.: Prentice-Hall, 1968.

Gentleman, W. M. "Implementing Clenshaw-Curtis Quadrature," *Communications of the ACM* 15 (1972): 337–346.

Gold, B., and Rader, C. M. *Digital Processing of Signals.* New York: McGraw-Hill, 1969.

Hamming, R. W. *Numerical Methods for Scientists and Engineers.* New York: McGraw-Hill, 1962.

Hildebrand, F. B. *Introduction to Numerical Analysis.* New York: McGraw-Hill, 1956.

Jacquez, J. A. *A First Course in Computing and Numerical Methods.* Reading, Massachusetts: Addison-Wesley, 1970.

Kaneko, T., and Liu, B. "Accumulation of Round-Off Error in Fast Fourier Transforms," *Journal of the Association for Computing Machinery* 17 (1970): 637–654.

Kunz, K. S. *Numerical Analysis.* New York: McGraw-Hill, 1957.

Lanczos, C. *Applied Analysis.* Englewood Cliffs, N.J.: Prentice-Hall.

———. *Discourse on Fourier Series.* Edinburgh: Oliver and Boyd, 1966.

McAllister, D. F. "Algorithms for Chebychev Approximation over Finite Sets," Doctoral thesis, Chapel Hill, North Carolina: University of North Carolina, 1972.

Ralston, A. *A First Course in Numerical Analysis.* New York: McGraw-Hill, 1965.

Szegö, G. *Orthogonal Polynomials.* American Mathematical Society Colloquium Publications, vol. 23, New York: American Mathematical Society, 1959.

5

NUMERICAL SOLUTION OF ORDINARY DIFFERENTIAL EQUATIONS

5.1 GOALS AND STRATEGY

Differential equations are often not solvable analytically. In such cases, a numerical solution is required. If a differential equation can be solved analytically, however, it is particularly important that it be so solved, because the numerical solution of differential equations is an error-prone process, as we shall see.

A differential equation in more than one dimension is called a *partial differential equation,* whereas a one-dimensional differential equation is called an *ordinary differential equation.* The concepts behind the numerical solution of partial differential equations are much the same as those behind the solution of ordinary differential equations, although some special problems arise and special techniques are applicable. In this book we will restrict ourselves to the solution of ordinary differential equations. The student who understands the matters involved in the solution of these equations is well prepared to understand the reasoning behind algorithms for solving partial differential equations.

A general nth-order ordinary differential equation is written

$$y^{(n)}(x) = f(x, y^{(0)}(x), y^{(1)}(x), y^{(2)}(x), \ldots, y^{(n-1)}(x)), \tag{1}$$

where $y^{(i)}(x)$ is the ith derivative of y and

$$y^{(0)}(x) = y(x). \tag{2}$$

In equation 1, f is a given function, and $y(x)$ is the function to be determined. The solution of an ordinary differential equation is a function, not a value. However, for a given value of x, $y(x)$ is a numeric value, and it is this value that we will be solving for. In particular, we will most often assume we are given a value x_{goal} at which we need the value $y(x_{goal})$.

Equation 1 can be interpreted as follows. Given the value x and the value of y and its first $n-1$ derivatives at x, equation 1 allows us to compute $y^{(n)}(x)$. Then, using the fact that as $\delta \to 0$,

$$y^{(j)}(x+\delta) = \sum_{i=j}^{n} y^{(i)}(x)\delta^{i-j}/(i-j)!, \quad 0 \le j \le n-1, \tag{3}$$

we can compute f and its first $n-1$ derivatives at $x+\delta$, where δ may be either positive or negative. Continuing this process, we can compute f and its derivatives at any point. Note that to start this process, it is necessary to have been given

$$y^{(0)}(x_0) = y_0^{(0)}, \quad y^{(1)}(x_0) = y_0^{(1)}, \ldots, \quad y^{(n-1)}(x_0) = y_0^{(n-1)}, \tag{4}$$

where the $y_0^{(i)}$ are constants for all i. Equations 1 and 4 taken together constitute a clearly specified *initial-value ordinary differential equation,* so called because the initial values (at $x = x_0$) of y and its derivatives are given.

One or more of the conditions specified in equation 4 may give values of y or its derivatives at some point or points other than x_0. For a unique solution to be assured, it is only required that (1) the total number of conditions is n, and (2) at each point at which conditions are given, called a *boundary point,* if m is the highest derivative specified at that point, all derivatives, from the zeroth, up to the mth, are specified at that point. If the conditions are not all given at the same point, the differential equation to be solved is called a *boundary-value problem.* As we will see, the numerical solution of boundary-value problems is an extension of the solution of initial-value problems. We will restrict ourselves to solving initial-value problems but will indicate at the end of the chapter how these methods can be used in solving boundary-value problems.

We will show in section 5.3 that an nth-order initial-value ordinary differential equation can be written as n simultaneous first-order initial-value ordinary differential equations, and that the solution of a set of simultaneous first-order initial-value ordinary differential equations is found by a simple extension of the methods for solving a single first-order ordinary differential equation. Thus, except at the end of this chapter, we will restrict ourselves to methods for the solution of first-order ordinary initial-value problems:

$$y'(x) = f(x,y), \tag{5}$$

$$y(x_0) = y_0. \tag{6}$$

Equation 5 gives us a direction for each point in the (x,y) plane. Starting at any point in the plane, we can move an infinitesimal distance in the specified direction, recompute the function at the new x value to get a new direction, move an infinitesimal distance in that direction, and so on, defining a function (see figure 5-1, in which a finite step width, h, is used; in the correct solution, $h \to 0$). If we start at the same x but a different y, a different function is produced. In fact, the second function cannot coalesce with or cross the first function. Assume either were possible and did occur. Let (x_c, y_c) be the point in common. Then using equation 5, we could extend y back to $x = x_0$. Its value there could not be both the y_0 given by the first condition and that given by the second condition.

What we have said is that equation 5 gives a so-called direction field on the space, defining a family of functions $y(x)$, and equation 6 picks out a particular member of the family (see figure 5-2).

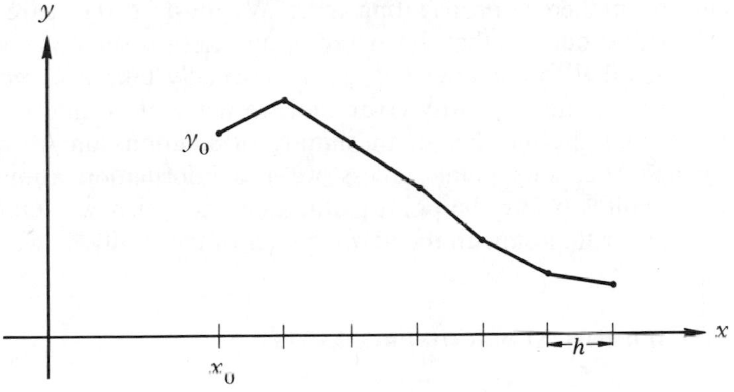

FIG. 5-1 Solution of a differential equation as extrapolation

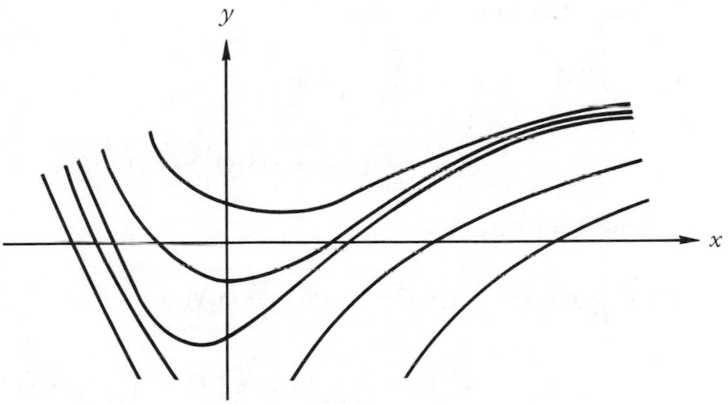

FIG. 5-2 A family of functions defined by a differential equation

The method of numerical solution of ordinary differential equations is very much like that specified above for infinitesimal distances, except that in real numerical solution we must use noninfinitesimal distances h, instead of infinitesimal distances δ. Thus, the numerical solution of first-order ordinary differential equations involves using x_i, y_i, $y_i' = f(x_i, y_i)$, and possibly x, y, and y' at previous values of x to compute $y(x_{i+1}) \equiv y_{i+1}$ (see figure 5-1). This is an extrapolation process, and we have seen before that extrapolation is very prone to error. The error problem is aggravated by the fact that y_{i+1} is used to compute y_{i+2}, y_{i+2} is used to compute y_{i+3}, and so on, thereby propagating error. We must be very careful to choose h small enough so that the extrapolation error is small.

We will see that to compute $y(x_{goal})$, we normally take a number of steps between x_0 and x_{goal}. Any error that we make at an intermediate step will change the member of the family of solutions on which we are moving. After this point, unless we use information about the function at points before the point (value of x) at which we stand, we can do no better than stay on the new member of the family.

5.2 FIRST-ORDER INITIAL-VALUE PROBLEMS

5.2.1 Rejection of Long Taylor Series

Given equation 5 with the function f given analytically, we could compute formulas for all derivatives of y by analytically differentiating equation 5, and use these derivatives in a Taylor formula to compute $y(x)$ at any desired vaue of x for which the Taylor series converges.

Differentiating equation 5, we obtain

$$\begin{aligned} y''(x) = \frac{d}{dx} y'(x) &= \frac{\partial f}{\partial x}(x,y) + \frac{\partial f}{\partial y}(x,y)\frac{dy}{dx} \\ &= \frac{\partial f}{\partial x}(x,y) + f(x,y)\frac{\partial f}{\partial y}(x,y). \end{aligned} \tag{7}$$

Differentiating again, we obtain

$$\begin{aligned} y'''(x) = \frac{d}{dx} y''(x) = {} & \frac{\partial^2 f}{\partial x^2}(x,y) + \frac{\partial f}{\partial x}(x,y)\frac{\partial f}{\partial y}(x,y) \\ & + 2f(x,y)\frac{\partial^2 f}{\partial x\,\partial y}(x,y) + [f(x,y)]^2 \frac{\partial^2 f}{\partial y^2}(x,y) \\ & + f(x,y)\left[\frac{\partial f}{\partial y}(x,y)\right]^2, \end{aligned} \tag{8}$$

and so on. In general, $y^{(n)}(x)$ depends on all partial derivatives of f of degree $n-1$ or less. Thus, evaluating a Taylor series using terms of degree up to the nth requires the evaluation of n derivatives of order $n-1$, $n-1$ derivatives of order $n-2$, and so on; that is, it requires evaluating $n(n+1)/2$ functions. If f is complicated, as it often is when we cannot solve the equation analytically, evaluation of so many derivatives is very time-consuming. (Furthermore, the human must do the analytic differentiation required.) Thus, instead of using a high-degree Taylor series to approximate over a relatively large distance, it is more efficient to divide the interval $[x_0, x_{goal}]$ into a number of pieces and successively use lower-degree Taylor series over each interval. That is, one can achieve the same overall error with less work by using low-degree Taylor series and small intervals and applying the Taylor series to move from x_0 to x_1, from x_1 to x_2, and so on, through from x_{m-1} to $x_m = x_{goal}$, rather than using one high-degree Taylor series to move from x_0 to x_m in one step. This statement is true even when one takes into account the fact that in the former method one has to be concerned about the propagation of error from step to step. Therefore, we will not consider methods that use derivatives of y higher than the first.

Because the computation of $y(x_{goal})$ involves a succession of steps, we must be particularly careful about each method's error propagation properties over many steps. In chapter 1 we learned that the word commonly used to characterize these properties is *stability*. A method is said to be stable if small errors die out after many steps, and unstable if they grow after many steps. Actually, we are interested in *relative stability*; whether the propagated error at any given point relative to the value of the function at that point grows or diminishes. That is, if error ε_i is generated up to the ith step, and if $\varepsilon_k^{prop,i}$ is the error in the kth step propagated from the error in the ith step, we are interested in the behavior of $|\varepsilon_k^{prop,i}/y_k|$ as k gets large. If this value approaches 0 as $k \to \infty$, the method is said to be relatively stable;† if the value approaches ∞, the method is said to be relatively unstable. Henceforth, when we discuss stability, we mean relative stability unless otherwise specified.

Thus, methods for solving differential equations must be evaluated with respect to three properties.

1. The truncation error generated per step of the method, that is, the error due to the fact that we are using a finite distance h rather

† This definition is not the definition of relative stability most commonly found elsewhere. However, I find the definition most commonly found elsewhere to be unsupportable. The above definition, which is used by some others (R. W. Hamming, *Numerical Methods for Scientists and Engineers*; J. M. Ortega, *Numerical Analysis*), is unlike the alternatives in that it fits into the general context of error analysis in this book.

than an infinitesimal distance δ or, equivalently, to the fact that we are truncating an infinite series.

2. The propagated error properties or stability of the method.
3. The amount of computational error generated when executing a step of the method.

5.2.2 Euler's Method

A step of a method for solving differential equations solves for y_{i+1}, given y_i and $y'_i = f(x_i, y_i)$. It is useful to write the ideal step of the method:

$$y(x_{i+1}) = y(x_i) + \int_{x_i}^{x_{i+1}} y'(x)dx$$

$$= y_i + \int_{x_i}^{x_{i+1}} f(x, y(x))dx. \tag{9}$$

By writing the problem in this form, we can solve the problem by numerical integration and thus take advantage of the favorable error properties of numerical integration. The problem could also be cast in terms of numerical differentiation, but this would be a bad choice.

We have reduced our problem to finding the best numerical integration method for numerically integrating $\int_{x_i}^{x_{i+1}} f(x, y(x))dx$, where we know f and x throughout the integration interval, but we do not know y throughout the interval. The solution of differential equations can be thought of as a combination of extrapolation, with its poor error properties, and numerical integration, with its relatively good error properties.

Since at the outset of the ith step, the only value of y we know in the interval is y_i, we can use the rectangular rule (Newton-Cotes rule for $m = 0$; see section 4.2.4, part a) for integration. In doing so, we approximate f by $f(x_i, y_i)$ throughout the whole interval. Assume that the interval width h is constant from step to step. Then the rectangular rule applied to equation 9 produces

$$\hat{y}_{i+1} = y_i + hf(x_i, y_i). \tag{10}$$

The method which moves from step to step using equation 10 to give the computed value of y_{i+1} given y_i is called *Euler's method.*

Euler's method can also be written

$$\hat{y}_{i+1} = y_i + hy'_i \tag{11}$$

(see program 5-1). In this form it is simply a Taylor series truncated after the first-degree term. It can be graphically interpreted as moving a step-length h from y_i in the direction of the tangent to $y(x)$ at x_i (see figure 5-3). Notice that h may be negative if we wish to move in a negative direction from x_0.

```
/* EULER'S METHOD FOR SOLVING FIRST-ORDER ORDINARY */
/* DIFFERENTIAL EQUATION: Y'(X) = YPRIME(X,Y); Y(XØ) = YØ */
/* SUBROUTINE YPRIME; INITIAL VALUES XØ AND YØ; */
/* INTERVAL WIDTH, H; AND NUMBER OF STEPS, N, ARE GIVEN */

   DECLARE N FIXED BINARY,  /* NUMBER OF STEPS */
           (H,              /* INTERVAL WIDTH */
           XØ, YØ,          /* INITIAL X AND Y VALUES */
           XI,              /* TABULAR ARGUMENT */
           Y(Ø:N)) FLOAT;   /* ARRAY OF SOLUTION VALUES */

/* INITIALIZE */
   Y(Ø) = YØ;
   XI = XØ;

/* RECURSIVELY COMPUTE Y(I) VALUES */
   RECURS: DO I = Ø TO N-1;
           Y(I+1) = Y(I) + H * YPRIME(XI,Y(I));
           XI = XI + H;
   END RECURS;
```

PROGRAM 5-1 Euler's method for solving first-order ordinary differential equation

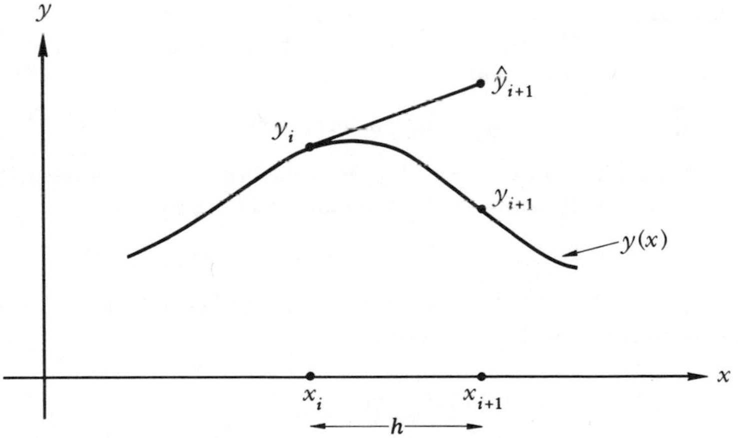

FIG. 5-3 A step of Euler's method

5.2.3 Truncation Error and Stability

In this section we will analyze the error properties of Euler's method, using a general approach to the analysis of the error of methods for the solution of differential equations.

Let us first analyze the truncation error, $\varepsilon_{i+1}^{trunc}$, of the ith step of Euler's method. This is the error in $\hat{y}_{i+1}$, assuming y_i is exact and no computational error is made in evaluating f. We know that the correct value of y at x_{i+1} is given by a Taylor series:

$$y_{i+1} = y_i + hy_i' + (h^2/2)y''(\xi_i), \quad \text{where } \xi_i \in [x_i, x_{i+1}]. \tag{12}$$

Subtracting equation 12 from equation 11, we produce

$$\varepsilon_{i+1}^{trunc} = -(h^2/2)y''(\xi_i) = -(h^2/2)y_i'' + O(h^3). \tag{13}$$

The truncation error of Euler's method is of order 2 in h. Experiments have shown that this order is too small; to keep the truncation error small, the interval width h must be so small that the number of steps required to arrive at x_{goal} becomes very large, causing much computational error, much error propagation, and much work. The methods of choice have truncation error of order 5 or 6 in h.

We will now analyze the stability of Euler's method. Assume that y_i has error ε_i. We are interested in the propagation of this error into step $i+n$ due to the recursive use of the method, neglecting truncation error or computational error generated at intermediate steps. We first analyze the effect, ε_{i+1}^{prop}, of an error ε_i in y_i on the computed value of y_{i+1}, then the effect of this error ε_{i+1}^{prop} in y_{i+1} on the computed value of y_{i+2}, and so on, finally producing the effect of $\varepsilon_{i+n-1}^{prop}$, and thus of ε_i, on ε_{i+n}^{prop}. That is, we are interested in the difference between the value of $\hat{y}_{j+1}$ as given by equation 10 and of

$$y_{j+1}^* = y_j^* + hf(x_j, y_j^*), \tag{14}$$

in which the inaccurate value of y_j is used, for $i \le j \le i+n-1$. Subtracting equation 10 with $i = j$ from equation 14, we produce

$$\varepsilon_{j+1}^{prop} = \varepsilon_j^{prop} + h[f(x_j, y_j + \varepsilon_j^{prop}) - f(x_j, y_j)]. \tag{15}$$

If we apply the mean-value theorem in the variable y to the bracketed difference in equation 15, we produce

$$f(x_j, y_j + \varepsilon_j^{prop}) - f(x_j, y_j) = \varepsilon_j^{prop} \frac{\partial f}{\partial y}(x_j, \eta_j),$$
$$\text{where} \quad \eta_j \in [y_j, y_j + \varepsilon_j^{prop}], \tag{16}$$

so

$$\varepsilon_{j+1}^{prop} = \varepsilon_j^{prop} + h\varepsilon_j^{prop} \frac{\partial f}{\partial y}(x_j, \eta_j) = \varepsilon_j^{prop}(1 + hK_j), \tag{17}$$

where
$$K_j \equiv \frac{\partial f}{\partial y}(x_j,\eta_j). \tag{18}$$

Equation 17, applied for $i \leq j \leq i+n-1$, produces

$$\varepsilon_{i+n}^{prop} = \varepsilon_i \prod_{j=i}^{i+n-1} (1 + hK_j). \tag{19}$$

The error given by equation 19 will die out if hK_j is small in magnitude and negative for all j. However, if we move in the opposite direction (change the sign of h), the error will grow, so we cannot say that the method is absolutely stable, only that for some functions and interval-width values it is absolutely stable.

But we have said that we are more interested in relative stability than absolute stability. It is difficult to say something in general for all differential equations, because the relative stability depends on the solution and thus the equation. However, if we make the approximation that the K_j are constant over j, we can draw some conclusions that apply at least over small ranges of x for which $\frac{\partial f}{\partial y}$ is approximately constant.

If K_j is constant over j, we have

$$\frac{\partial f}{\partial y} = K, \tag{20}$$

so
$$y'(x) = f(x,y) = Ky + g(x). \tag{21}$$

Equation 21 with the initial condition $y(x_0) = y_0$ has an analytic solution:

$$y(x) = y_0\, e^{K(x-x_0)} + e^{Kx} \int_{x_0}^{x} e^{-Ku} g(u)du. \tag{22}$$

By equation 19 and the fact that $K_j = K$,

$$\varepsilon_{i+n}^{prop} = \varepsilon_i(1 + hK)^n. \tag{23}$$

From equations 22 and 23, the propagated relative error r_{i+n} at step $i+n$ is

$$r_{i+n} = \frac{\varepsilon_i(1 + hK)^n}{y_0\, e^{K(x_{i+n}-x_0)} + e^{Kx_{i+n}} \int_{x_0}^{x_{i+n}} e^{-Ku} g(u)du}$$

$$= \frac{\varepsilon_i(1 + hK)^n}{y_0 e^{K(i+n)h} + e^{K(x_0+(i+n)h)} \int_{x_0}^{x_{i+n}} e^{-Ku} g(u)du}. \tag{24}$$

To simplify this equation, we note that

$$e^{hK} = 1 + hK + (h^2K^2/2)e^{\theta} > 1 + hK, \quad \text{where } \theta \in [0, hK], \tag{25}$$

so, assuming $hK > -1$, there exists a positive constant $c < 1$ such that

$$1 + hK = ce^{hK}. \tag{26}$$

Substituting equation 26 in equation 24, we have

$$\begin{aligned} r_{i+n} &= \varepsilon_i c^n e^{hKn} \Big/ \left(y_0 e^{hK(i+n)} + e^{Kx_0} e^{hK(i+n)} \int_{x_0}^{x_{i+n}} e^{-Ku} g(u)\,du \right) \\ &= \varepsilon_i c^n \Big/ \left(y_0 e^{hKi} + e^{Kx_i} \int_{x_0}^{x_{i+n}} e^{-Ku} g(u)\,du \right). \end{aligned} \tag{27}$$

Look at the last expression for r_{i+n} in equation 27. As $n \to \infty$, the numerator approaches 0, so if the denominator does not approach 0 (the usual case), the relative error approaches 0. Thus, if $\dfrac{\partial f}{\partial y}$ is constant, Euler's method is relatively stable.

We will see that a method that is relatively stable for all values of h is very unusual. Euler's method has very good stability properties. However, as we said above, its truncation error properties are so poor that we are willing to sacrifice some stability for better truncation error properties. We will develop methods with these properties in sections 5.2.4 through 5.2.8. Here, let us continue with stability analysis by taking an alternate route to the same results achieved above, namely, by applying the difference equation technique developed in section 1.3.2.

For most methods if $\dfrac{\partial f}{\partial y}$ is a constant K, the error propagation is described by an equation of the form

$$\varepsilon_{j+1} - \sum_{k=0}^{m-1} d_k \varepsilon_{j-k} = 0, \tag{28}$$

a homogeneous mth-order difference equation with constant coefficients, where the coefficients d_k are functions of the interval width h and K. We know (see appendix A) that the solution of this equation is

$$\varepsilon_n = \sum_{k=1}^{m} c_k \rho_k^n, \tag{29}$$

where the c_k depend on initial conditions and the ρ_k are functions of the d_k, that is, of h and K. We are interested in the behavior of $\varepsilon_{i+n}/y_{i+n}$ for an initial error ε_i made at a fixed i, as n becomes large. From equations 22 and 29, we can write

$$r_{i+n} = \sum_{j=1}^{m} c_j\rho_j^n \Big/ \left(y_0 e^{Khi} e^{Khn} + e^{Kx_i} e^{Khn} \int_{x_0}^{x_{i+n}} e^{-Ku} g(u)du\right)$$

$$= \sum_{j=1}^{m} c_j(\rho_j/e^{Kh})^n \Big/ \left(y_0 e^{Khi} + e^{Kx_i} \int_{x_0}^{x_{i+n}} e^{-Ku} g(u)du\right). \tag{30}$$

Assume that the magnitude of the denominator of equation 30 does not approach either 0 or ∞ as n gets large. That it should approach 0 is a very unusual circumstance. Moreover, if x_{i+n} is bounded as $n \to \infty$, the denominator of equation 30 will be bounded if $g(u)$ is reasonably well behaved. Since we are most often interested in finding solutions over a finite range of x, x_{i+n} is normally bounded; n getting large implies h becoming small, not x_{i+n} approaching ∞.

If the above assumption holds, whether the relative error grows or diminishes with n depends upon the numerator of the right side of equation 30. In particular, if there exists any j such that

$$|\rho_j(h,K)| > e^{Kh}, \tag{31}$$

the relative error will grow with n, and the method will be relatively unstable. If for all j,

$$|\rho_j(h,K)| < e^{hK}, \tag{32}$$

it will be relatively stable.

In the case of Euler's method, the difference equation is given by equation 17, which we can rewrite as

$$\varepsilon_{j+1} - (1 + hK)\varepsilon_j = 0. \tag{33}$$

This is a first-order difference equation with

$$\rho_1 = 1 + hK. \tag{34}$$

Since we have shown that

$$|1 + hK| \le e^{hK}, \tag{35}$$

Euler's method is relatively stable.

It should be repeated that the fact that a method is relatively stable does not mean that the error in y_{i+n} becomes zero (relatively) as $n \to \infty$, only that the error *propagated from error generated in step i* becomes zero. The overall error ε_k in $\hat{y}_k$ is the sum of the propagation of errors in all previous steps. For example, in Euler's method,

$$\varepsilon_k = \sum_{i=0}^{k-1} \varepsilon_i^{gen} \prod_{j=i}^{k-1} (1 + hK_j), \tag{36}$$

where ε_i^{gen} is the sum of the computational error and truncation error made at step i.

5.2.4 Heun Corrector

We will now attempt to improve Euler's method so that its truncation error is of higher order in h than 2. Let us consider figure 5-3. What is the difficulty with Euler's method? It is that the value of the derivative used to determine the direction in which we move a step ahead depends only on the beginning of the interval. Put another way, the difficulty is that we are only using the rectangular rule (Newton-Cotes method for $m = 0$) for our numerical integration. This rule gives a biased estimate of the average value of $y' = f$ over the interval—an estimate at only one end of the interval. We would do better to use the average of the f values at each end of the interval, that is, use the trapezoidal rule (see section 4.2.4, part a):

$$\hat{y}_{i+1} = y_i + h\frac{f(x_i, y_i) + f(x_{i+1}, y_{i+1})}{2}. \tag{37}$$

Equation 35 is called *Heun's method.* The difficulty is that we do not know y_{i+1} so we cannot compute $f(x_{i+1}, y_{i+1})$. But we can approximate y_{i+1} by using Euler's method. If we use this approximate value of y_{i+1} in equation 37, we might produce a better value for y_{i+1}:

$$y_{i+1(c)} = y_i + h((f(x_i, y_i) + f(x_{i+1}, y_{i+1(p)}))/2), \tag{38}$$

where $y_{i+1(p)}$ is the value of y_{i+1} predicted by Euler's method. We say we use Euler's method as a *predictor* for y_{i+1} and Heun's method as a *corrector* for y_{i+1} (see program 5-2).

```
/* EULER-HEUN METHOD FOR SOLVING FIRST-ORDER ORDINARY */
/* DIFFERENTIAL EQUATION: Y'(X) = YPRIME(X,Y); Y(XØ) = YØ */
/* SUBROUTINE YPRIME; INITIAL VALUES XØ AND YØ; */
/* INTERVAL WIDTH, H; AND NUMBER OF STEPS, N, ARE GIVEN */

    DECLARE N FIXED BINARY, /* NUMBER OF STEPS */
            (H,             /* INTERVAL WIDTH */
            XØ, YØ,         /* INITIAL X AND Y VALUES */
            XI,             /* TABULAR ARGUMENT */
            Y(Ø:N),         /* ARRAY OF SOLUTION VALUES */
            LFDERIV,  /* DERIVATIVE AT LEFT END OF INTERVAL */
            RTDERIV,  /* DERIVATIVE AT RIGHT END OF INTERVAL */
            PRED)     /* APPROXIMATE VALUE AT RIGHT END */
            FLOAT;    /* OF INTERVAL */
```

PROGRAM 5-2 Euler-Heun predictor-corrector method for solving first-order ordinary differential equation (*continued on next page*)

```
/* INITIALIZE */
   Y(Ø) = YØ;
   XI = XØ;

/* RECURSIVELY COMPUTE Y(I) VALUES */
   RECURS: DO I = Ø TO N-1;
                 LFDERIV = YPRIME(XI,Y(I));
                 PRED = Y(I) + H * LFDERIV;
                 RTDERIV = YPRIME(XI + H,PRED);
                 Y(I+1) = Y(I) + H * (LFDERIV + RTDERIV) / 2;
                 XI = XI + H;
   END RECURS;
```

PROGRAM 5-2 (*continued*)

Let us analyze the truncation error in Heun's method. Expanding the second term in the numerator of equation 38 in a Taylor series in y, we produce

$$f(x_{i+1},y_{i+1(p)}) = f(x_{i+1},y_{i+1}) + \varepsilon_{i+1(p)}\frac{\partial f}{\partial y}(x_{i+1},\zeta_{i+1}), \tag{39}$$

where the truncation error in Euler's method,

$$\varepsilon_{i+1(p)} \equiv y_{i+1(p)} - y_{i+1}, \tag{40}$$

and $\zeta_{i+1} \in [y_{i+1(p)}, y_{i+1}]$. If we define

$$K_{i+1} \equiv \frac{\partial f}{\partial y}(x_{i+1},\zeta_{i+1}), \tag{41}$$

equation 38 becomes

$$y_{i+1(c)} = y_i + (h/2)(y_i' + y_{i+1}' + K_{i+1}\varepsilon_{i+1(p)}). \tag{42}$$

Expanding y_{i+1}' in a Taylor series about x_i, we produce

$$y_{i+1}' = y_i' + hy_i'' + (h^2/2)y_i''' + O(h^3), \tag{43}$$

so
$$\begin{aligned} y_{i+1(c)} &= y_i + (h/2)(2y_i' + hy_i'' + (h^2/2)y_i''' + O(h^3) + K_{i+1}\varepsilon_{i+1(p)}) \\ &= y_i + hy_i' + (h^2/2)y_i'' + (h^3/4)y_i''' + O(h^4) + (h/2)K_{i+1}\varepsilon_{i+1(p)}. \end{aligned} \tag{44}$$

By Taylor's theorem, we know that

$$y_{i+1} = y_i + hy_i' + (h^2/2)y_i'' + (h^3/6)y_i''' + O(h^4). \tag{45}$$

Subtracting this equation from equation 44, we find that the truncation error in the corrector is

$$\varepsilon_{i+1(c)} = (h^3/12)y_i''' + (h/2)K_{i+1}\varepsilon_{i+1(p)} + O(h^4). \tag{46}$$

Thus if the error in the predictor is $O(h^2)$ as it is in Euler's method, the truncation error in the Heun corrector is of order 3 in h, an improvement for small h.

If Euler's method is used to compute the predictor, the truncation error in the Heun corrector is

$$\varepsilon_{i+1(c)} = (h^3/4)((y_i'''/3) - K_{i+1}y_i'') + O(h^4). \tag{47}$$

The term $K_{i+1}y_i''$, that is, the term due to the predictor error, may be a significant contributor to the truncation error in the corrector. On the other hand, we can see from equation 46 that if the predictor error were $O(h^3)$, the error in the Heun corrector due to the predictor error would be $O(h^4)$ compared to the inherent $O(h^3)$ truncation error of Heun's method. How can we produce a predictor with $O(h^3)$ error? First, we can reapply Heun's method (equation 38) using, as a predictor, the result of Heun's method with the Euler predictor. In fact, this process can be iterated, but with little gain, since after the second application of Heun's method, the principal source of the error of order 3 in h is not diminished.

5.2.5 Modified Euler Predictor; with Heun Corrector to Form Predictor-Corrector

The difficulty with the method involving two applications of Heun's method per step is that it requires three evaluations of f per step—one for the Euler predictor and one for each of the two applications of the Heun corrector. We would like to find an initial predictor requiring only one computation of f and having $O(h^3)$ error. Such a method can be developed. We rewrite equation 9 as

$$y_{i+1} = y_i + hy'_{avg,i}. \tag{48}$$

We have said that it is better for the estimate of $y'_{avg,i}$ to depend on y' values symmetrically spaced in $[x_i, x_{i+1}]$ than on values nonsymmetrically spaced in that interval (for example, one at one end). This implies that we are better off using a rectangular rule with y' evaluated at the middle of the interval than with y' evaluated at one end of the interval:

$$\hat{y}_{i+1} = y_i + hf(x_{i+1/2}, y_{i+1/2}). \tag{49}$$

We have $y_{i+1/2}$ no more than we have y_{i+1}, but if we consider the interval of width $2h$ starting at x_{i-1}, we see that y' evaluated at the middle of the interval is y_i', which we do have available. Therefore we can produce a *modified Euler's method,*

$$y_{i+1(p)} = y_{i-1} + 2hy_i', \tag{50}$$

for which we start at x_{i-1} and move forward for a distance $2h$ with slope defined by the derivative at the middle of the interval. Does this indeed provide a better estimate of y_{i+1} than Euler's method? We prove that it does by the common method of expanding everything in a Taylor series about x_i, producing

$$\begin{aligned}\varepsilon_{i+1(p)} &\equiv y_{i+1(p)} - y_{i+1} = y_{i-1} + 2hy_i' - y_{i+1} \\ &= y_i - hy_i' + (h^2/2)y_i'' - (h^3/6)y_i''' + O(h^4) \\ &\quad + 2hy_i' - (y_i + hy_i' + (h^2/2)y_i'' + (h^3/6)y_i''' + O(h^4)) \\ &= -(h^3/3)y_i''' + O(h^4).\end{aligned} \tag{51}$$

The error is of order 3 in h as we suspected. Thus, if we use this predictor with the Heun corrector, we will produce a result for which

$$\varepsilon_{i+1(c)} = (h^3/12)y_i''' + O(h^4), \tag{52}$$

and only two evaluations of f are required per step.

Let us do a stability analysis of the methods we have developed above. First let us do an analysis of the modified Euler's method used by itself, with no corrector. Then we have

$$\hat{y}_{i+1} = y_{i-1} + 2hf(x_i, y_i). \tag{53}$$

If we use inaccurate values for y_i and y_{i-1}, we have

$$y_{i+1}^* = y_{i-1}^* + 2hf(x_i, y_i^*). \tag{54}$$

The propagated error is obtained by subtracting equation 53 from equation 54, producing

$$\varepsilon_{i+1} = \varepsilon_{i-1} + 2hK_i\varepsilon_i. \tag{55}$$

Assuming $\dfrac{\partial f}{\partial y} = K$, the roots for the quadratic equation corresponding to the difference equation 55 are

$$\begin{aligned}\rho_1, \rho_2 &= hK \pm \sqrt{1 + (hK)^2} \\ &\approx 1 + hK + \frac{(hK)^2}{2} - \frac{(hK)^4}{4}, -\left(1 - hK + \frac{(hK)^2}{2} - \frac{(hK)^4}{4}\right).\end{aligned} \tag{56}$$

$|\rho_1| \le e^{hK}$ except when hK is negative and small in magnitude, in which case it is at most .3 percent greater than e^{hK}. $|\rho_2| < e^{hK}$ for $hK > 0$, but it can be considerably greater than e^{hK} for $hK < 0$. For example, for $hK = -1$, $|\rho_2|/e^{hK} = 2.4e$—the relative error at one step can magnify by a factor that is greater than 6 over that of the preceding step. More precisely, because for all $hK < 0$,

$$\sqrt{1 + (hK)^2} - hK > e^{hK} \tag{57}$$

(the left side of this inequality is an increasing function of $|hK|$ and the right side is a decreasing function of $|hK|$, with both sides remaining positive), the larger the magnitude of hK for hK negative, the worse the stability is. From these relations, taken together, we conclude that the modified Euler's method is relatively stable for $hK > 0$, but seriously unstable for $hK < 0$ and not tiny in magnitude.

Intuitively, what causes this difficulty? Why is the modified Euler's method unstable for negative hK when the simple Euler's method is stable? The problem is one of feedback. The difference between Euler's method and the modified Euler's method is that the modified Euler's method uses points previous to x_i, whereas Euler's method does not. The error in these previous points can propagate in an undesirable way into $\hat{y}_i$, so that when the values $\hat{y}_i$ and $\hat{y}_{i-1}$ are used in the same formula, the errors in these values combine in a particularly bad manner. It is not the case that whenever previous points are used, the error propagation is bad, but in this situation we have to be concerned about such a possibility.

But we do not propose to use the modified Euler's method by itself; we intend to use it only as a predictor. Since its stability problems have only to do with its iterative use with no intervening methods applied, the combination of it and the Heun corrector may not be so unstable. Let us analyze this question.

The result of one step of the Euler-Heun predictor-corrector method is written as

$$\hat{y}_{i+1} = y_i + (h/2)[f(x_i,y_i) + f(x_{i+1},y_{i-1} + 2hf(x_i,y_i))]. \tag{58}$$

If y_i values with error are used, we have

$$y^*_{i+1} = y^*_i + (h/2)[f(x_i,y^*_i) + f(x_{i+1},y^*_{i-1} + 2hf(x_i,y^*_i))]. \tag{59}$$

Subtracting equation 58 from equation 59 and assuming that $\dfrac{\partial f}{\partial y} = K$, we arrive at the difference equation

$$\varepsilon_{i+1} = \varepsilon_i + (h/2)[K\varepsilon_i + K(\varepsilon_{i-1} + 2hK\varepsilon_i)] \tag{60}$$

or equivalently,

$$\varepsilon_{i+1} - (1 + (hK/2) + (hK)^2)\varepsilon_i - (hK/2)\varepsilon_{i-1} = 0. \tag{61}$$

We can show that for $hK < 0$ and $> -.601$, both $|\rho_1|$ and $|\rho_2|$ are less than e^{hK} (see exercise 5.2.2). At $hK = -.601$, $|\rho_2|$ becomes greater than e^{hK}. Therefore for negative hK, this method is stable as long as the magnitude of hK is less than .601. For $hK > 0$, $|\rho_2| < e^{hK}$, but $|\rho_1| > e^{hK}$ for $hK < .701$. But the maximum ratio by which $|\rho_1|$ is greater than e^{hK} is only .16 percent, so the instability is very slight.

We have shown that the Heun corrector used with the modified Euler predictor is at worst slightly unstable if hK is small in magnitude. It is a common property of higher-order methods that the stability properties are acceptable if h is small enough (in magnitude) but the method becomes unstable if h is too large.

5.2.6 Automatic Error Estimation and Mop-Up

Comparing the truncation error in the modified Euler predictor (equation 51) and the Heun corrector using the modified Euler predictor (equation 52), we see that for h small enough that the $O(h^4)$ terms are negligible, the errors of the predictor and of the corrector have opposite signs. That is, for small h, the predictor and corrector values straddle the correct value. Thus the difference between the predictor value and the corrector value can be used as a bound on the error in the corrector value. If the true value is between the predictor and corrector values, the worst that can happen is that the true value is equal to the predictor, in which case

$$|\varepsilon_{i+1(c)}| \leq |y_{i+1(c)} - y_{i+1(p)}|. \tag{62}$$

Equations 51 and 52 can be used to give a more accurate estimate of the error in the corrector:

$$\varepsilon_{i+1(c)} = (y_{i+1(c)} - y_{i+1(p)})/5 + O(h^4). \tag{63}$$

This value can be used as an estimate of the error, or it can be subtracted from the corrected value to produce an improved value for y_{i+1}:

$$y_{i+1(m)} = y_{i+1(c)} - (y_{i+1(c)} - y_{i+1(p)})/5 = (4y_{i+1(c)} + y_{i+1(p)})/5. \tag{64}$$

Such an approach to improving $\hat{y}_{i+1}$ is sometimes known as *mopping up*. This value of y_{i+1} after mop-up can also be obtained by noting that from equation 51,

$$y_{i+1(p)} = y_{i+1} - (h^3/3)y_i''' + O(h^4), \tag{65}$$

and from equation 52,

$$y_{i+1(c)} = y_{i+1} + (h^3/12)y_i''' + O(h^4). \tag{66}$$

Since we have two estimates of the same value, each with an error estimate, we can produce a better estimate: an appropriately weighted average of the two. To zero the coefficient of h^3, we should weight equation 66 by $\frac{4}{5}$ and equation 65 by $\frac{1}{5}$, producing

$$y_{i+1(m)} = (4y_{i+1(c)} + y_{i+1(p)})/5 + O(h^4). \tag{67}$$

Thus using the modified Euler predictor, the Heun corrector, and mop-up, we can produce a method with truncation error of order 4 in h. We do not have an error estimate for the value $y_{i+1(m)}$ thus produced, but its error is normally less than the error in the corrector, so we can use as a conservative estimate of the error in $y_{i+1(m)}$ the value given by equation 63, neglecting the term $O(h^4)$. This procedure of computing an error estimate from values computed in the method, rather than by evaluating a formula not closely related computationally to the solution process, is called *automatic error estimation.*

Mopping up changes the stability properties of a solution method, often for the worse. The stability of our particular mopped-up formula can be obtained by analyzing the equation produced by substituting equation 58 for the corrector and equation 50 for the predictor in equation 64. Interestingly enough, for this formula the stability is improved by mopping up. It can be shown that the mopped-up formula is stable for all positive hK and for negative hK if $|hK| < 1.28$ (see exercise 5.2.3).

Notice that mop-up is possible because the order of h in the error in the predictor and in the corrector is the same. The ability to mop up is another important gain of using the modified Euler's method rather than the simple Euler's method for the predictor. All of the predictor-corrector methods that we will see have the property that the predictor and corrector have errors which are of the same order in h and for which the respective low-order terms of the error have opposite signs, allowing automatic error estimation.

The importance of automatic error estimation cannot be overemphasized. For any method, h must be chosen small enough not only to give the method stability but also to keep the truncation error appropriately small. On the other hand, choosing h too small leads to too many steps in moving from x_0 to x_{goal}, thereby making the work and the generated and propagated error greater than necessary. We see that choosing h appropriately is critical to the success of the solution process. But with differential equations of any complexity, it is impossible to accurately predict, before the solution process begins, the value of h needed to obtain the error properties desired, because that value depends on the solution. Thus, what we wish to do is to estimate the error as we go along. If it is too large at a particular step, we can use a smaller h from that step on. If it is too small, we can increase h. To make this evaluation, we need a way to estimate the error at a step. An estimate could be produced by evaluating an error formula involving higher derivatives of $y(x)$. But we know that it is inefficient to compute higher-order derivatives of $y(x)$, so we do not wish to evaluate the error formula directly. Fortunately, automatic error estimation makes it unnecessary for us to use this formula. We simply compute the error estimate as a function of the predictor and corrector values, which we compute in any case. Very little extra computation is required for the error analysis.

To start, we choose a tolerance τ within which we want to compute $y(x_{goal})$. Assuming interval width h, the method will require $(x_{goal} - x_0)/h$ steps. Assuming that the error in each step neither diminishes nor grows (a reasonable and somewhat conservative assumption if the process is stable), the error in $y(x_{goal})$ will be less than or equal to the sum of the errors made in each step. Assuming that the error made in each step is less than ε, where this error includes both truncation error and computational error, we wish that

$$\left|\frac{x_{goal} - x_0}{h}\varepsilon\right| \leq \tau. \tag{68}$$

In other words, we want the error in each step to satisfy the relation

$$|\varepsilon/h| \leq \tau/|x_{goal} - x_0|. \tag{69}$$

At any step, we can bound the truncation-error part of ε by using automatic error estimation. Normally, this truncation error is the major part of the error. If $|\varepsilon/h|$ is larger than the right side of relation 69, we should halve h for succeeding steps. If ε is much too small, we can double the value of h. Notice that, for a method which has truncation error of order 4 in h, doubling h increases $|\varepsilon/h|$ by a factor of approximately 8. Therefore, if $|\varepsilon/h|$ computed by automatic error estimation is less than one-eighth of the value of the right side of relation 69, we can double h.

With the modified Euler-Heun method described above, if h is doubled and x_i is the last point at which $y(x)$ has been computed, the point x_{i-1} used in the modified Euler predictor is the double interval back from x_i; that is, the y_{i-1} used in the formula is $y(x_{i-2})$ in terms of the old h. Thus, to allow ourselves to double h, we must save both y_{i-1} and y_{i-2}.

Halving the interval width h is not so simple. We need the value of y at a point that is one-half of the old interval back from x_i. This value can be produced using Hermite interpolation with the arguments x_{i-1} and x_i (see section 4.2.4, part c), or by some other interpolation method.

A difficulty which arises when the modified Euler's method is used as a predictor is that we do not have the value of y_{-1} required to apply the method for $i = 0$, that is, when we start. The method is not self-starting. We can get around this difficulty by using at the first step the simple Euler's method and then correcting with Heun's method to produce a prediction with error of order 3 in h. We use this prediction in a reiteration of the Heun corrector to produce the desired result. Mop-up and automatic error estimation are not applicable at this first step, but they can be applied at all successive steps.

5.2.7 Multistep Predictor-Corrector Methods

Studies have shown that methods with error of order 5 or 6 in h provide the best compromise between method complexity and truncation-error size. To produce predictor-corrector methods of this order without using derivatives higher than the first, we must use points previous to x_{i-1} in the extrapolation formula. The predictor will be of the form

$$y_{i+1(p)} = A_0 y_i + hB_0 y'_i + A_1 y_{i-1} + hB_1 y'_{i-1} + A_2 y_{i-2} + hB_2 y'_{i-2} + \cdots, \tag{70}$$

and the corrector will be of the form

$$y_{i+1(c)} = hD_{-1} y'_{i+1} + C_0 y_i + hD_0 y'_i + C_1 y_{i-1} + hD_1 y'_{i-1} + C_2 y_{i-2} + hD_2 y'_{i-2} + \cdots. \tag{71}$$

The predictor involves the point x_i and previous points only. The corrector involves these points and the derivative of y at x_{i+1}. That the derivative terms are multiplied by h comes from the fact that y' is the integrand in equation 9; when the integration is made into a quadrature, the multiplication by h is produced.

The problem is to find the constants A_j, B_j, C_j, and D_j in equations 70 and 71 so that these formulas have appropriate truncation-error and stability properties. Experience shows that at least two coefficients in each formula must be determined by stability considerations. The remaining coefficients can be determined using a *method of undetermined coefficients* to ensure that the truncation error has the proper order in h.

For example, assume we wish to produce a method with error of order 5 in h. Because everything is developed from a Taylor series, we expect the h^5 to be multiplied by $y^{(5)}(\theta)$. That is, the method should be exact for $y(x)$ equal to any polynomial of degree 4 or lower, since, for such a polynomial, $y^{(5)}(x) \equiv 0$. Since all operators in the problem are linear, this statement is equivalent to saying that the method must be exact for the basis polynomials, $y = 1$, $y = x - x_i$, $y = (x - x_i)^2$, $y = (x - x_i)^3$, and $y = (x - x_i)^4$. These five conditions will determine five of the constants in equation 70 and in 71. Since at least two more parameters are needed to control stability, at least seven terms are required in equation 70 and in 71. Thus we must use values at x_{i-3} through x_i in the predictor and at x_{i-2} through x_{i+1} in the corrector. So the predictor is of the form

$$y_{i+1(p)} = A_0 y_i + hB_0 y_i' + A_1 y_{i-1} + hB_1 y_{i-1}' + A_2 y_{i-2} + hB_2 y_{i-2}' + A_3 y_{i-3} + hB_3 y_{i-3}', \tag{72}$$

and the corrector is of the form

$$y_{i+1(c)} = hD_{-1} y_{i+1}' + C_0 y_i + hD_0 y_i' + C_1 y_{i-1} + hD_1 y_{i-1}' + C_2 y_{i-2} + hD_2 y_{i-2}'. \tag{73}$$

Let the parameters left to control stability be A_1, A_2, and A_3 in equation 72 and C_1 and C_2 in equation 73. If our method of undetermined coefficients is applied to equation 72, the following results are produced.

The value of $y_{i+1(p)}$ must be equal to y_{i+1} (that is, $y(x_{i+1})$) if $y(x)$ is any polynomial of degree 4 or less. Thus, for such $y(x)$,

$$y_{i+1} = A_0 y_i + hB_0 y_i' + A_1 y_{i-1} + hB_3 y_{i-3}'. + A_2 y_{i-2} + hB_2 y_{i-2}' + A_3 y_{i-3} + hB_3 y_{i-3}'. \tag{74}$$

For

$$y(x) = 1, \tag{75}$$

we have

$$y'(x) = 0, \tag{76}$$

which, with equation 74, implies that

$$1 = A_0 \cdot 1 + hB_0 \cdot 0 + A_1 \cdot 1 + hB_1 \cdot 0 + A_2 \cdot 1 + hB_2 \cdot 0 + A_3 \cdot 1 + hB_3 \cdot 0 = A_0 + A_1 + A_2 + A_3. \tag{77}$$

Equation 74 must hold for any first-degree polynomial, in particular for

$$y(x) = x - x_i, \tag{78}$$

for which

$$y'(x) - 1. \tag{79}$$

Substituting equations 78 and 79 in equation 74, we arrive at

$$h = A_0 \cdot 0 + hB_0 \cdot 1 + A_1 \cdot (-h) + hB_1 \cdot 1 + A_2 \cdot (-2h) + hB_2 \cdot 1 + A_3 \cdot (-3h) + hB_3 \cdot 1. \tag{80}$$

Dividing through by h, we produce

$$1 = -A_1 - 2A_2 - 3A_3 + B_0 + B_1 + B_2 + B_3. \tag{81}$$

Similarly, equation 74 is true for any quadratic $y(x)$, in particular for

$$y(x) = (x - x_i)^2, \tag{82}$$

for which

$$y'(x) = 2(x - x_i). \tag{83}$$

Substituting equations 82 and 83 in equation 74 and dividing through by h^2 produces

$$1 = A_1 + 4A_2 + 9A_3 - 2B_1 - 4B_2 - 6B_3. \tag{84}$$

Similarly, letting $y(x)$ be a cubic produces

$$1 = -A_1 - 8A_2 - 27A_3 + 3B_1 + 12B_2 + 27B_3, \tag{85}$$

and letting it be a quartic produces

$$1 = A_1 + 16A_2 + 81A_3 - 4B_1 - 32B_2 - 108B_3. \tag{86}$$

If we let A_1, A_2, and A_3 be parameters, equations 77, 81, 84, 85, and 86 are five equations in five unknowns: A_0, B_0, B_1, B_2, and B_3. These equations can be solved for the five unknowns in terms of A_1, A_2, and A_3. A similar analysis can be applied to equation 73 to produce an equation for C_0, D_{-1}, D_0, D_1, and D_2 in terms of C_1 and C_2.

Substituting

$$y'_{i+1} = f(x_{i+1}, y_{i+1(p)}) \tag{87}$$

in equation 73, where $y_{i+1(p)}$ is given by equation 72, we produce an expression for $y_{i+1(c)}$ that gives a correct value of y_{i+1} if $y(x)$ is any polynomial of degree 4 or lower, and is a function of the parameters A_1, A_2, A_3, C_1, and C_2. Assuming we use inaccurate y_i values rather than correct ones, we produce an equation from which we can subtract the same equation with no input error, producing a 4th-order difference equation for ε_i. The roots ρ_j of the polynomial corresponding to this difference equation will be functions of the five parameters A_1, A_2, A_3, C_1, and C_2. We can choose these parameters to make the stability as good as possible. We should also choose them to make the predictor and corrector formulas as simple as possible.

A choice of $A_1 = A_2 = A_3 = C_1 = C_2 = 0$ for parameter values does a very good job. This choice produces the so-called *Adams-Bashforth predictor,*

$$y_{i+1(p)} = y_i + (h/24)(55y'_i - 59y'_{i-1} + 37y'_{i-2} - 9y'_{i-3}), \tag{88}$$

and the so-called *Adams-Moulton corrector,*

$$y_{i+1(c)} = y_i + (h/24)(9y'_{i+1} + 19y'_i - 5y'_{i-1} + y'_{i-2}) \tag{89}$$

(see exercise 5.2.7). We must analyze the truncation error of this predictor and corrector.

For both the Adams-Bashforth predictor and the Adams-Moulton corrector (assuming the correct value of y'_{i+1} is used, not an approximate one), the truncation error is of the form $Ch^5y^{(5)}(\theta)/5!$, where θ is in the interval covered by the points used in the formula, that is, for the predictor $\theta \in [x_{i-3}, x_{i+1}]$ and for the corrector $\theta \in [x_{i-2}, x_{i+1}]$. Equivalently, the error can be written in the form $Ch^5 y_i^{(5)}/5! + O(h^6)$. We must find the constants C_p and C_c for the predictor and corrector errors, respectively. These can be found by letting $y(x)$ be a function such that $y^{(5)}(\theta)$ is a constant, that is, letting $y(x)$ be a quintic:

$$y(x) = (x - x_i)^5. \tag{90}$$

Then, in general,

$$y_{i+1(p)} = y_{i+1} + \varepsilon_{i+1(p)}, \tag{91}$$

so for the special $y(x)$ given by equation 90,

$$\varepsilon_{i+1(p)} = C_p h^5 y^5(\theta)/5! = C_p h^5. \tag{92}$$

Substituting equations 88 and 92 in equation 91, we have

$$y_i + (h/24)(55y_i' - 59y_{i-1}' + 37y_{i-2}' - 9y_{i-3}') = y_{i+1} + C_p h^5. \tag{93}$$

Using the values derived from equation 90 in equation 93, we arrive at

$$0 + (h/24)(0 - 59 \cdot 5 \cdot (-h)^4 + 37 \cdot 5 \cdot (-2h)^4 - 9 \cdot 5 \cdot (-3h)^4) = (h)^5 + C_p h^5. \tag{94}$$

We can solve this equation for C_p, producing

$$C_p = -251/6. \tag{95}$$

So the truncation error of the Adams-Bashforth predictor is

$$\varepsilon_{i+1(p)} = -(251/720)h^5 y^{(5)}(\theta) = -(251/720)h^5 y_i^{(5)} + O(h^6), \tag{96}$$

where $\theta \in [x_{i-3}, x_{i+1}]$. Similarly, we can show that the truncation error of the Adams-Moulton corrector is

$$\varepsilon_{i+1(c)} = (19/720)h^5 y_i^{(5)} + O(h^6). \tag{97}$$

Equation 97 holds even if the predictor rather than the correct value of y_{i+1} is used in the computation of y_{i+1}'. As was the case with the modified Euler-Heun method, since the truncation error of the predictor and corrector are of the same order in h, the error in the corrector due to predictor error is in the $O(h^6)$ term (see exercise 5.2.8).

As promised, assuming the $O(h^6)$ terms are negligible, the predictor and corrector errors have opposite signs, so the predictor and corrector straddle the correct answer. Thus we can produce an automatic error estimate,

$$\varepsilon_{i+1(c)} \approx (19/270)(y_{i+1(c)} - y_{i+1(p)}). \tag{98}$$

Furthermore, we can mop up to make the truncation error $O(h^6)$ by the formula,

$$y_{i+1(m)} = (251y_{i+1(c)} + 19y_{i+1(p)})/270. \tag{99}$$

```
/* ADAMS METHOD FOR SOLVING FIRST-ORDER ORDINARY DIFFERENTIAL EQUATION: */
/* Y'(X) = YPRIME(X,Y); Y(XØ) = YØ */
/* SUBROUTINE YPRIME; INTERVAL WIDTH, H; AND NUMBER OF STEPS, N, ARE GIVEN */
/* Y(I), Ø <= I <= 3, ARE GIVEN, PRESUMABLY COMPUTED BY ANOTHER METHOD */
   DECLARE (XØ, XI,           /* INITIAL ARGUMENT AND PRESENT ARGUMENT */
            Y(Ø:N),           /* SOLUTION VALUES */
            H, PRED,          /* INTERVAL WIDTH AND PREDICTOR VALUE */
            FI, FI1, FI2, FI3)  /* PREVIOUS DERIVATIVE VALUES */
            FLOAT,
            N FIXED BINARY; /* NUMBER OF STEPS */
/* COMPUTE INITIAL DERIVATIVES AND ARGUMENT VALUE */
   FI3 = YPRIME(XØ,Y(Ø));
   XI = XØ + H;
   FI2 = YPRIME(XI,Y(1));
   XI = XI + H;
   FI1 = YPRIME(XI,Y(2));
   XI = XI + H;
/* RECURSIVELY COMPUTE REMAINING SOLUTION VALUES */
   RECURS: DO I = 3 TO N-1;
            FI = YPRIME(XI,Y(I));  /* NEW DERIVATIVE */
   /* PREDICTOR */
            PRED = Y(I) + (H/24) * (55*FI - 59*FI1 + 37*FI2 - 9*FI3);
            FI3 = FI2;  /* UPDATE DERIVATIVES AND ARGUMENT */
            FI2 = FI1;
            FI1 = FI;
            XI = XI + H;
   /* CORRECTOR */
            Y(I+1) = Y(I) + (H/24) * (9*YPRIME(XI,PRED) + 19*FI1 - 5*FI2 + FI3);
   /* MOP-UP */
            Y(I+1) = (251*Y(I+1) + 19*PRED) / 270;
   END RECURS;
```

PROGRAM 5-3 Adams predictor, corrector, mop-up method for solving first-order ordinary differential equation

The *Adams method,* made up of the above predictor, corrector, and mop-up (see program 5-3), is relatively stable for $hK > -0.69$. Lapidus and Seinfeld compare this method to other predictor-corrector methods with fifth-order error initially and sixth-order error after mop-up, and conclude that the Adams method is the best general-purpose method.† There are methods that have better truncation-error properties, for example, Milne's method,‡ but stability problems make these methods less desirable. The form of the Adams method recommended by Lapidus and Seinfeld involves a mop-up on the predictor, based on the difference between the previous predictor and corrector, followed by a corrector using this improved predictor, followed by a mop-up on the corrector, based on the difference between the corrector and the unimproved predictor (see exercise 5.2.9).

The Adams method requires $x_0, x_1, x_2,$ and x_3 to get started. To compute x_1, x_2, and x_3, we need a method which has fifth-order or sixth-order error and which does not use any points preceding the current point (x_i). As shown above, no predictor-corrector method can have this characteristic. In the following section, we develop a method such as we require.

5.2.8 Runge-Kutta Methods

We require a self-starting method that has error of order 5 in h and does not use derivatives higher than the first. To achieve these goals, we must use more points than just x_i, but we are not allowed to use preceding points. Since we may not look back, we must look ahead. We are trying to approximate the integral $\int_{x_i}^{x_{i+1}} y'(x)dx = hy'_{avg,i}$. To compute $y'_{avg,i}$, we should have values of y' at values of x symmetrically distributed across the interval $[x_i, x_{i+1}]$. We start out with only $y(x_i)$. To obtain a better estimate of the average slope, our first step should be to predict the slope at the single point symmetrically spaced in the interval, namely, the midpoint of the interval. We can approximate the y value at this point using the half-interval form of Euler's method:

$$\hat{y}_{i+\frac{1}{2}(1)} = y_i + (h/2)y'_i = y_i + (h/2)f(x_i, y_i). \qquad (100)$$

Given this estimate of $y_{i+\frac{1}{2}}$, $f(x_{i+\frac{1}{2}}, \hat{y}_{i+\frac{1}{2}(1)})$ is a better estimate than y'_i of the average slope in the interval, so this average slope estimate can be used to obtain an even better estimate of the slope at $x_{i+\frac{1}{2}}$. That is, we can

† L. Lapidus and J. H. Seinfeld, *Numerical Solution of Ordinary Differential Equations,* section 4.16.

‡ Ibid, section 4.5.

use Euler's method with the better slope estimate to compute

$$\hat{y}_{i+\frac{1}{2}(2)} = y_i + (h/2)f(x_{i+\frac{1}{2}}, \hat{y}_{i+\frac{1}{2}(1)}). \tag{101}$$

Using this value, we can compute an improved estimate of the average slope, with which we can move a whole interval to compute an estimate for y_{i+1}:

$$\hat{y}_{i+1(1)} = y_i + hf(x_{i+\frac{1}{2}}, \hat{y}_{i+\frac{1}{2}(2)}). \tag{102}$$

We could accept this estimate of y_{i+1} as our answer, but it has truncation error of order only 3 in h (see exercise 5.2.11). So we do some mopping up with our four estimated values of y' (y'_i, $\hat{y}'_{i+\frac{1}{2}(1)}$, $\hat{y}'_{i+\frac{1}{2}(2)}$, and $\hat{y}'_{i+1(1)}$), thereby computing a final estimate for the average slope,

$$\begin{aligned} \hat{y}'_{avg,i} = {} & k_1 f(x_i, y_i) + k_2 f(x_{i+\frac{1}{2}}, \hat{y}_{i+\frac{1}{2}(1)}) \\ & + k_3 f(x_{i+\frac{1}{2}}, \hat{y}_{i+\frac{1}{2}(2)}) + k_1 f(x_{i+1}, \hat{y}_{i+1(1)}), \end{aligned} \tag{103}$$

and thus a final estimate for y_{i+1},

$$\hat{y}_{i+1} = y_i + hy'_{avg,i}. \tag{104}$$

This complicated process to compute $\hat{y}_{i+1}$, the *Runge-Kutta method of fourth order,* is illustrated in figure 5-4 and fully specified in program 5-4.

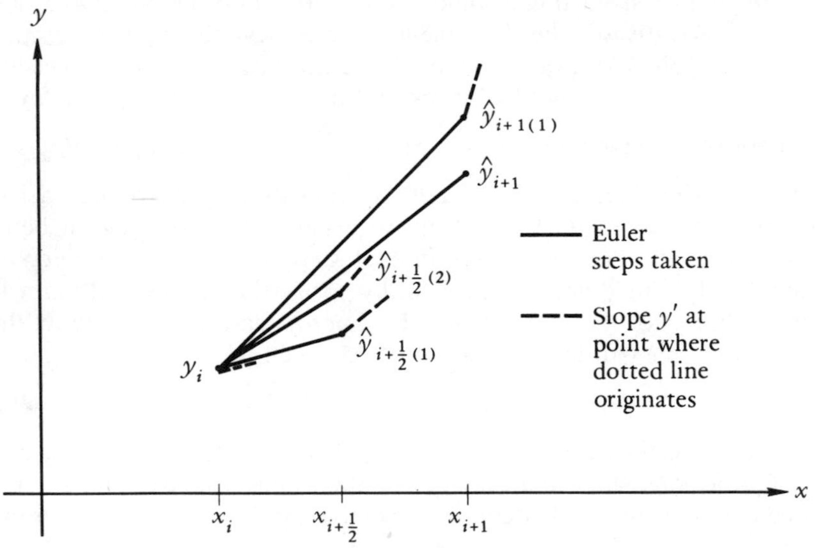

FIG. 5-4 Runge-Kutta method

```
/* RUNGE-KUTTA FOURTH-ORDER METHOD FOR SOLVING FIRST-ORDER */
/* ORDINARY DIFFERENTIAL EQUATION: Y'(X) = YPRIME(X,Y); Y(XØ) = YØ */
/* SUBROUTINE YPRIME; INITIAL VALUES XØ AND YØ; */
/* INTERVAL WIDTH, H; AND NUMBER OF STEPS, N, ARE GIVEN */

   DECLARE (XØ, YØ,              /* INITIAL VALUES */
            Y(Ø:N),              /* SOLUTION VALUES */
            XI,                  /* TABULAR ARGUMENT */
            YI,                  /* TRIAL SOLUTION VALUE */
            H,                   /* INTERVAL WIDTH */
            HALFINT,             /* HALF-INTERVAL WIDTH */
            F1, F2, F3, F4)      /* DERIVATIVES */
            FLOAT,
            N FIXED BINARY;      /* NUMBER OF STEPS */

/* INITIALIZE */
   XI = XØ;
   Y(Ø) = YØ;

/* RECURSIVELY COMPUTE SOLUTION VALUES */
   RECURS: DO I = Ø TO N-1;
            F1 = YPRIME(XI,Y(I));
            YI = Y(I) + HALFINT * F1;
            F2 = YPRIME(XI+HALFINT,YI);
            YI = Y(I) + HALFINT * F2;
            F3 = YPRIME(XI+HALFINT,YI);
            YI = Y(I) + H * F3;
            F4 = YPRIME(XI+H,YI);
            Y(I+1) = Y(I) + H * (F1 + 2*(F2+F3) + F4) / 6;
            XI = XI + H;
   END RECURS;
```

PROGRAM 5-4 Runge-Kutta fourth-order method for solving first-order ordinary differential equation

Note that the weights k_1, k_2, and k_3 in equation 103 are symmetric across the interval. These weights are chosen to maximize the order of h in the truncation error of the method (see exercise 5.2.12). The result is $k_1 = \frac{1}{6}$, and $k_2 = k_3 = \frac{1}{3}$. The resulting truncation error is $O(h^5)$. The precise expression for the error—a very complicated function of f and all of its partial derivatives—will not be given here.

Runge-Kutta methods of the above form, namely,

$$\hat{y}_{i+1} = y_i + h\sum_j k_j y'_{i+r_j} = y_i + h\sum_j k_j f(x_{i+r_j}, \hat{y}_{i+r_j}), \tag{105}$$

where the $\hat{y}$ values used as arguments of f are computed by look-ahead, can be developed with error of any order desired. The most popular are the fourth-order method (the one with error $O(h^5)$) given above and some fifth-order methods.† Any of these methods can be used either to compute starting values for a predictor-corrector method or as a method in itself. Generally these Runge-Kutta methods have the following properties:

1. When used by themselves, they are quite stable. For a small enough h, all are stable.
2. They are quite time-consuming. For example, the fourth-order method given above requires four evaluations of the function f per step as compared to two evaluations of f per step when the Adams method is used.
3. They are self-starting. This feature provides a storage advantage in that previous values need not be saved.
4. They do not provide automatic error estimation. Automatic error estimation can be provided by computing two Runge-Kutta formulas having the following property: the difference in their values has the same order of magnitude as the error in one of them. However, this approach requires even more computation than the single Runge-Kutta method.

5.2.9 Overall Solution Algorithm

The Runge-Kutta methods are not uncommonly used, especially for computing starting values and required mid-interval values for predictor-corrector methods and in cases where stability or storage requirements are the dominant issues. However, because of the inefficiency of the Runge-Kutta methods, the high quality of the Adams predictor-corrector method (even though it is not quite as stable as the Runge-Kutta methods), and the automatic error estimation capability of the Adams method, the latter is more popular. In fact, the Adams method, augmented by a Runge-Kutta method to compute starting values and required mid-interval values, is the most commonly used method for solving first-order ordinary differential equations in one unknown. This algorithm, including recommended details, is given below. In developing these details, we assume that the computational error at each step is negligible compared to the truncation error. If this assumption does not hold, the method for estimating a step's generated error, ε_{step}, and thus that for choosing the value of h, can be modified to take this fact into account (see exercise 5.2.14).

1. Using equation 69, compute the upper bound on $1/h$ times the generated error per step:

† Lapidus and Seinfeld, *Numerical Solution of Ordinary Differential Equations*, Chap. 2.

$$|\varepsilon_{step}/h| < \tau_{goal}/|x_{goal} - x_0| = C, \tag{106}$$

where τ_{goal} is the tolerance on the error in $y(x_{goal})$.

2. Find the correct interval width h by either of two approaches:
   a) (1) Choose an interval-width guess h_1.
   (2) Do three steps of the Runge-Kutta method with $h = h_1$.
   (3) Apply the Adams method for one step with $h = h_1$, and compute the automatic error estimate E (that is, the estimate of ε_{step}). Since the actual error is $O(h_1^6)$, $|\varepsilon_{step}/h_1| = O(h_1^5)$. Therefore,

   $$E/h_1 = Kh_1^5 \text{ for some constant } K, \tag{107}$$

   and we desire the interval width h such that

   $$Kh^5 = C. \tag{108}$$

   Therefore,

   $$h = (C/K)^{1/5} = (Ch_1^6/E)^{1/5}. \tag{109}$$

   b) Compute $y^{(5)}$ as a function of f and its partial derivatives, and evaluate this fifth derivative at (x_0,y_0). Choose h such that

   $$|(19/270)y^{(5)}(x_0,y_0)|h^4 = C, \tag{110}$$

   that is, let

   $$h = \left(\frac{270C}{19|y^{(5)}(x_0,y_0)|}\right)^{1/4}. \tag{111}$$

3. Using the h determined in part 2, do three steps of the Runge-Kutta method to compute $\hat{y}_1$, $\hat{y}_2$, and $\hat{y}_3$. Let $i = 3$.
4. Using the present value of h, do one step of the Adams method to compute $\hat{y}_{i+1}$, and compute the automatic error estimate E.
5. If $|E| > C|h|$, go to step 6. If $E < C|h|/32$, go to step 7. Otherwise go on to the next interval with the same h, that is, increment i and go to step 4.
6. Halve h. Assuming $\hat{y}_{i+1}$ is the value just computed and subscripts are in terms of the old h value, use the Runge-Kutta method with the new h value starting at x_i to compute $\hat{y}_{i+1/2}$ and the Runge-Kutta method starting at x_{i-1} to compute $\hat{y}_{i-1/2}$. Increment i and go to step 4 with the new h value, that is, start from x_{i+1} to compute the next $\hat{y}$ value.
7. Double h, increment i, and go to step 4. To allow this interval doubling, at all steps we must save not only $\hat{y}_i$, $\hat{y}'_i$, $\hat{y}_{i-1}$, $\hat{y}'_{i-1}$, $\hat{y}'_{i-2}$, and the new value $\hat{y}_{i+1}$, but also $\hat{y}'_{i-3}$ through $\hat{y}'_{i-5}$.

We may fail to arrive exactly at x_{goal} by this method. In this case, we can use the Runge-Kutta method for the last fractional step. Another situation which may arise is that we have many goal values for x. In this case, the above method should be applied for the farthest x values from x_0 in each direction; other x values should be computed using the Runge-Kutta method starting at the computed x value closest to the goal value in question.

Finally, we may wish to be able to compute the solution for any given x value in some range. That is, we want a subroutine which, for any argument x, computes the value $y(x)$, where y is the solution of the given differential equation. We do not wish to re-solve the differential equation each time we need a new value. Rather we should tabulate the solution at appropriate x values as determined by the material in chapter 4. Then we should interpolate in this table whenever we need a $y(x)$ value. In designing our interpolation method, we must remember that the tabular values which come from the solution of the differential equations will have significant error as determined by the error properties of our solution method (see exercise 5.2.15).

PROBLEMS

5.2.1 Assume a method for the numerical solution of differential equations has truncation error ε_i^{trunc} at step i, computational error ε_i^{comp} at step i, and propagation factor C_i in step i (if ε_i^{inp} is the error in y_i, the propagated error will be $C_i\varepsilon_i^{inp}$).

a) What will be the overall error in $\hat{y}_{i+1}$, the result of step i, given ε_i^{trunc}, ε_i^{comp}, ε_i^{inp}, and C_i?

b) Note that the result of part a is ε_{i+1}^{inp}. Give an expression for the overall error in the result of the nth step, given ε_0^{inp}; ε_i^{trunc}, $1 \le i \le n$; ε_i^{comp}, $1 \le i \le n$; and C_i, $1 \le i \le n$.

c) Make the following assumptions: $|\varepsilon_0^{inp}| < b_0$; the generated relative error in computing $f(x_i,y_i)$ is bounded by $r_{i(f)}$; the generated relative error in multiplication is bounded by $r_\times$; the generated relative error in addition is bounded by r_+; $\frac{\partial f}{\partial y}(x_i,\eta_i) < K_i$ if $\eta_i \in [y_i, y_i + \varepsilon_i^{inp}]$; and $y''(\xi_i) < D_i$ if $\xi_i \in [x_i, x_{i+1}]$. Further assume that h has no error and $\left|h\frac{\partial f}{\partial y}(x_i,\eta_i)\right| \ll 1$. Bound the error in the result of the nth step of applying Euler's method with interval h to $y'(x) = f(x,y)$, $y(x_0) = y_0$, in terms of b_0, h, r_+, $r_\times$, and $\{f(x_i,y_i), r_{i(f)}, K_i, D_i | 1 \le i \le n\}$.

5.2.2. a) Solve the difference equation given by equation 61.

b) Show that $\rho_1(hK)$ and $\rho_2(hK)$, the two roots of the characteristic equation corresponding to equation 61, have the following properties:

(1) If $-.601 < hK < 0$, $|\rho_1(hK)| < e^{hK}$ and $|\rho_2(hK)| < e^{hK}$.
(2) If $hK > -.601$, $|\rho_2(hK)| > e^{hK}$.
(3) *If* $0 < hK < .701$, $|\rho_1(hK)| > e^{hK}$ and $|\rho_2(hK)| < e^{hK}$.
(4) $\underset{0<hK<.701}{Max} (|\rho_1(hK)|/e^{hK}) = 1.0016$.

5.2.3. Show that the Euler-Heun predictor-corrector method with mop-up is relatively stable if $hK > -1.28$.

5.2.4. Determine the interval width h so that four decimal places of accuracy are obtained for $y(.5)$ when applying Euler's method to the differential equation $y' = x^2 + y$, $y(0) = 1$.

5.2.5. Given the differential equation $y' = f(x,y)$, $y(x_0) = y_0$,

a) Consider a predictor of the form

$$y_{i+1(p)} = Ay_i + By_{i-1} + h(Cy'_i + Dy'_{i-1}) \text{ with truncation error } O(h^4).$$

(1) Find A, B, C, and D.
(2) Find the truncation error of this predictor.
(3) Compare this predictor with the formula produced by extrapolation using the Hermite interpolation formula (see chapter 4, equations 234–236).

b) Consider a corrector of the form

$$y_{i+1(c)} = Fy_i + Gy_{i-1} + h(Hy'_i + Iy'_{i+1}) \text{ with truncation error } O(h^4).$$

(1) Find F, G, H, and I.
(2) Find the truncation error of this corrector (assuming no error in y'_{i+1}).

c) Give a mop-up formula for this predictor-corrector pair.

d) For the special equation $y' = Ky$, $y(x_0) = y_0$, analyze the relative stability of this predictor, corrector, mop-up operation for small h. *Hint:* Show $\max_i |\rho_i(hK)|/e^{hK} = 1$ for $h = 0$. Then analyze derivatives of this ratio.

5.2.6. a) Consider a predictor of the form

$$\hat{y}_{i+1} = A_0y_i + A_1y_{i-1} + h(B_0y'_i + B_1y'_{i-1} + B_2y'_{i-2}) + Eh^4y^{(4)}(\theta)/4!.$$

Let A_1 be a parameter and find A_0, B_0, B_1, B_2, and E in terms of A_1.

b) Consider a corrector of the form

$$\hat{y}_{i+1} = C_0y_i + C_1y_{i-1} + h(D_{-1}y'_{i+1} + D_0y'_i + D_1y'_{i-1}) + Fh^4y^{(4)}(\theta)/4!.$$

Let C_1 be a parameter and find C_0, D_{-1}, D_0, D_1, and F in terms of C_1.

c) Find a mop-up for this predictor-corrector pair in terms of A_1 and C_1.

d) Analyze the stability of this predictor, corrector, mop-up operation.

5.2.7. a) Show that letting $A_1 = A_2 = A_3 = C_1 = C_2 = 0$ in equations 72 and 73 produces the Adams-Bashforth predictor (equation 88) and the Adams-Moulton corrector (equation 89).

b) Show that the Adams-Bashforth predictor can be produced by

$$\hat{y}_{i+1} = y_i + \int_{x_i}^{x_{i+1}} \hat{y}'(x)dx,$$

where $\hat{y}'(x)$ is the approximation to $\hat{y}'(x)$ produced by the Newton backward interpolation formula through third differences (see exercise 4.2.19):

$$\hat{f}(x_0 + sh) = f_0 + \binom{s}{1}\nabla f_0 + \binom{s+1}{2}\nabla^2 f_0 + \binom{s+2}{3}\nabla^3 f_0,$$

where $\nabla f(x) \equiv f(x) - f(x-h)$.

5.2.8. Show that mop-up applied to the Adams predictor-corrector method produces a truncation error of order 6 in h.

5.2.9. We know that a constant, C, times the difference between the values, at a given step, of the corrector and the predictor of a predictor-corrector method, gives the low-order term (in h) of the truncation error in the corrector at that step in cases where the predictor and corrector have truncation errors of the same order in h.

a) Show that a constant (say, D) times this same difference gives the low-order term of the truncation error in the predictor of the *next* step.

b) What is the value of D for the Adams predictor and corrector?

c) We can produce a doubly mopped-up predictor-corrector sequence.

(1) Compute $y_{i+1(p)}$ from the Adams-Bashforth predictor.

(2) Compute $y_{i+1(pm)} = y_{i+1(p)} - D(y_{i(c)} - y_{i(p)})$.

(3) Compute $y_{i+1(c)}$ from the Adams-Moulton corrector using $y_{i+1(pm)}$ as the estimate of y_{i+1}.

(4) Compute $\hat{y}_{i+1} = y_{i+1(cm)} = y_{i+1(c)} - C(y_{i+1(c)} - y_{i+1(p)})$.

What is the order of h in the truncation error in $y_{i+1(pm)}$? In $y_{i+1(c)}$? In $\hat{y}_{i+1}$? Why is the above method better than the singly mopped-up Adams method? Why cannot $y_{i+1(pm)}$ appear in the corrector mop-up (4 above) in place of $y_{i+1(p)}$?

5.2.10. Use the Adams method to compute the solution of $y' = -2xy^2$, $y(0) = 1$ (which has the solution $y(x) = 1/(1 + x^2)$) on $[0,4]$ with $h = .4$ as follows:

a) Using the predictor only.

b) Using the predictor and corrector with no mop-up.

c) As in part b but with single mop-up.

d) As in part b but with double mop-up (see exercise 5.2.9).

5.2.11. Show the estimate of y_{i+1} given by equation 102 has order 3 in h.

5.2.12. Show that the values of the weights k_1, k_2, and k_3 given before equation 105 maximize the order of h in the truncation error of the estimate of y_{i+1} given by equation 104, where $\hat{y}'_{avg,i}$ is given by equation 103.

5.2.13. Show how Richardson extrapolation (see section 4.2.4, part b) can be applied to the Runge-Kutta method.

5.2.14. Modify the method for choosing h given in section 5.2.9 to reflect the assumption that the standard deviation of the computational error at each step has the value d, which is not very much less than the truncation error at each step.

5.2.15. Assume $y(x)$ is the solution of the differential equation $y'(x) = f(x,y)$, $y(0) = y_0$, where $f(x,y)$ is given analytically and y_0 is a known constant. Further assume we cannot solve for y analytically.

Consider the problem of designing a library subroutine to give the solution to this differential equation at any value in $[-1,1]$ of the input parameter x. We would be willing to use much computer time in preparing the subroutine so that at each application of the subroutine the computer time used would be minimal. We would not want to apply a numerical method for the solution of differential equations for each execution of the subroutine because such methods are time-consuming.

One possible method for resolving this problem would be to use a numerical method for the solution of differential equations to solve for y once at a set of points to be tabulated, and have the subroutine interpolate in the table. Ignoring roundoff, discuss in detail how you would go about implementing this method so that for every $x \in [-1,1]$, $y(x)$ could be found to within a constant given error ε. Justify your suggestions. Show your error analysis. Assume $|y^{(n)}(x)| < M_n$ for all n, where the values of M_n are given. Also assume the numerical method you choose for solving the differential equation has propagation factor 1, that is, the overall error at any step is the sum of the errors generated at all steps up to and including that step.

Be careful to address the question of the spacing of the tabular arguments and the interpolation method to be used with the table.

5.2.16. Say whether the following statement is true or false and justify your choice: Decreasing the step size at step i of the application of a predictor-corrector method always decreases the error in y_n where $n > i$.

5.2.17. We wish to find numerically the zero of the solution of the differential equation $\dfrac{dy}{dx} = -x^2y - x^3 - 1$, $y(0) = 1$, with an error of less than 10^{-2}.

a) What combination of standard numerical methods would you use to solve this problem? Justify your choice.

b) Solve the problem.

5.3 HIGHER-ORDER INITIAL-VALUE PROBLEMS

The general nth-order initial-value differential equation is given by

$$y^{(n)}(x) = f(x, y^{(0)}, y^{(1)}, y^{(2)}, \ldots, y^{(n-1)}), \tag{112}$$

$$y^{(0)}(x_0) = y_0^{(0)}, \quad y^{(1)}(x_0) = y_0^{(1)}, \ldots, \quad y^{(n-1)}(x_0) = y_0^{(n-1)}, \tag{113}$$

where the $y_0^{(j)}$ are constants. This system can be converted to n simultaneous first-order differential equations by defining n variables, $y_{(0)}, y_{(1)}, \ldots, y_{(n-1)}$, as follows:

$$y_{(0)}(x) \equiv y^{(0)}(x), \quad y_{(1)}(x) \equiv y^{(1)}(x), \ldots, y_{(n-1)}(x) \equiv y^{(n-1)}(x). \tag{114}$$

These variables satisfy the relations

$$y'_{(0)}(x) = y_{(1)}(x), \quad y'_{(1)}(x) = y_{(2)}(x), \ldots, \quad y'_{(n-2)}(x) = y_{(n-1)}(x), \tag{115}$$

a series of first-order differential equations with corresponding initial values given by

$$y_{(0)}(x_0) = y_0^{(0)}, \quad y_{(1)}(x_0) = y_0^{(1)}, \ldots, \quad y_{(n-2)}(x_0) = y_0^{(n-2)}. \tag{116}$$

Equations 112 and 113 become

$$y'_{(n-1)}(x) = f(x, y_{(0)}, y_{(1)}, \ldots, \quad y_{(n-1)}), \tag{117}$$

$$y_{(n-1)}(x_0) = y_0^{(n-1)}, \tag{118}$$

another first-order differential equation, but in many unknowns. Equations 115 and 117, together with the initial conditions 116 and 118, are a set of n simultaneous first-order differential equations in n unknowns with initial conditions.

The system of equations 115 through 118 is a special case of the general initial-value problem of n simultaneous first-order ordinary differential equations in n variables:

$$\begin{aligned}
y'_{(n-1)}(x) &= f_{(n-1)}(x, y_{(0)}, y_{(1)}, \ldots, y_{(n-1)}) \\
y'_{(n-2)}(x) &= f_{(n-2)}(x, y_{(0)}, y_{(1)}, \ldots, y_{(n-1)}) \\
&\vdots \\
y'_{(0)}(x) &= f_{(0)}(x, y_{(0)}, y_{(1)}, \ldots, y_{(n-1)}),
\end{aligned} \tag{119}$$

$$\begin{aligned}
y_{(n-1)}(x_0) &= y_0^{(n-1)} \\
y_{(n-2)}(x_0) &= y_0^{(n-2)} \\
&\vdots \\
y_{(0)}(x_0) &= y_0^{(0)}.
\end{aligned} \tag{120}$$

If we can solve these equations, we can solve the special case given by equations 115 through 118 and therefore the nth-order initial-value differential equation given by equations 112 and 113.

Solving the general case given by equations 119 and 120 is accomplished by a straightforward sequential application of the method of choice for one unknown. For example, consider the Adams method.

Assume that at the present step we have values at x_i, x_{i-1}, x_{i-2}, and x_{i-3} of each of the n variables $y_{(n-1)}, y_{(n-2)}, \ldots, y_{(0)}$. Initially, we do not have these values, but we can compute them by application of the extension of the Runge-Kutta method (see exercise 5.3.1). Then using each of equations 119 in equation 88 (the Adams-Bashforth predictor), we produce

$$\begin{aligned} y_{(j)i+1(p)} = y_{(j)i} + (h/24)\big[&55f_{(j)}(x_i, y_{(0)i}, y_{(1)i}, \ldots, y_{(n-1)i}) \\ &- 59f_{(j)}(x_{i-1}, y_{(0)i-1}, y_{(1)i-1}, \ldots, y_{(n-1)i-1}) \\ &+ 37f_{(j)}(x_{i-2}, y_{(0)i-2}, y_{(1)i-2}, \ldots, y_{(n-1)i-2}) \\ &- 9f_{(j)}(x_{i-3}, y_{(0)i-3}, y_{(1)i-3}, \ldots, y_{(n-1)i-3})\big], \quad 0 \le j \le n-1. \end{aligned} \tag{121}$$

Thus we can compute a predictor for each of the n variables $y_{(j)}$. These predictors and the values of the $y_{(j)}$ at x_i, x_{i-1}, and x_{i-2} can be used to compute corrected values of the variables using the Adams-Moulton formula (equation 89), where the equations 119 are used to compute the derivatives in the formula. The predicted values of each variable can be combined with the corrected values to produce a mopped-up value, finishing this step. Furthermore, we can produce an automatic error estimate for each variable.

Thus we can solve numerically any set of n simultaneous ordinary differential equations with initial values given, and, in particular, we can solve any nth-order initial-value ordinary differential equation, because it can be cast as such a set of simultaneous differential equations.

PROBLEMS

5.3.1. Discuss how the Runge-Kutta method described in section 5.2.8 can be applied to the solution of n simultaneous ordinary differential equations in n variables, with initial conditions.

5.4 BOUNDARY-VALUE PROBLEMS

If the boundary conditions given for the differential equation are not all at the same point, the above method is not directly applicable, because we have no initial values from which to start. However, we can reduce this so-called boundary-value problem to an iterated initial-value problem as follows, producing one technique for solving the former.

Assume there are just two points at which boundary conditions are given. If there are more, the problem becomes more complicated, but the basic approach is the same. Furthermore, let us assume that we have a second-order differential equation

$$y''(x) = f(x,y,y') \tag{122}$$

with two boundary conditions, one at $x = a$ and one at $x = b$:

$$y(a) = y_a, \qquad y(b) = y_b. \tag{123}$$

Again, solving higher-order equations with more boundary conditions at a and b is conceptually an extension of the following method applied to the above second-order equation.

We guess at the initial value y'_a to make an initial-value problem: equation 122 with the first boundary condition and the guessed condition. Then we solve this initial-value problem, computing a value $y(b)$. In general, the computed value will not be equal to the boundary value given. However, since the equations in a family of solutions do not cross, the direction by which the computed value is off from the correct value of $y(b)$ is the same direction by which y'_a is off from what it should have been. We can thus change the guess y'_a in the correct direction and repeat the initial-value problem solution, obtaining a new value $y(b)$. We are by this process trying to solve the nonlinear equation in one unknown, y'_a,

$$g(y'_a) - y_b = 0, \tag{124}$$

where $g(y'_a)$ is the value $y(b)$, where $y(x)$ is the solution of the initial-value problem

$$y''(x) = f(x,y,y'), \tag{125}$$

$$y(x_a) = y_a, \quad y'(x_a) = y'_a. \tag{126}$$

We can apply our iterative methods for solving nonlinear equations, as described in chapter 3, to this problem. There is no conceptual difference between this problem and the problems we solved in chapter 3. The only difficulty is that computing the function g is particularly hard because it involves the numerical solution of an initial-value differential equation. When we find the value y'_a which solves equation 124, we have an initial-value problem by which we can evaluate any value of y desired. Thus, we have solved the boundary-value problem.

5.5 SUMMARY

The numerical solution of differential equations is a particularly appropriate subject with which to end this text. It is a problem which involves computational error, approximational error, and, in a particularly virulent form, propagated error. It involves the solution of equations and it especially involves methods of approximation. In the above solution methods, we have used polynomial exact-matching approximation methods, but these are not the only approximation methods that can be used. We can use basis functions other than the polynomials in this approximation and criteria other than exact matching for closeness of approximation.

This text was designed with the idea in mind that the student who learns the basic notions in this book should have the mathematical and numerical analytic concepts necessary to solve most numerical problems. We have not covered all useful numerical methods. In some cases, we have not covered even the method of choice for a given problem. But the basic concepts presented in the book should enable the student to understand the majority of these methods and to develop methods of his own.

REFERENCES

Hamming, R. W. *Numerical Methods for Scientists and Engineers.* New York: McGraw-Hill, 1962.

Lapidus, Leon, and Seinfeld, John H. *Numerical Solution of Ordinary Differential Equations.* New York: Academic Press, 1971.

Ortega, James M. *Numerical Analysis.* New York: Academic Press, 1972.

Ralston, Anthony. *A First Course in Numerical Analysis.* New York: McGraw-Hill, 1965.

APPENDIXES

APPENDIX A. SOLUTION OF HOMOGENEOUS DIFFERENCE EQUATIONS

Consider a sequence of variables $y_0, y_1, y_2, \ldots$. A *difference* of y_i is defined as $y_{i+1} - y_i$. Any linear combination of $y_i, y_{i+1}, y_{i+2}, \ldots, y_{i+n}$ can be rewritten in terms of y_i, the difference of y_i, the difference of y_{i+1}, and so on, through the difference of y_{i+n-1}. If we define a second difference as the difference of two first differences as previously defined, and similarly define higher-order differences, we see that the original equation can be written in terms of the zeroth difference of y_i (y_i itself), the first difference of y_i, the second difference of y_i, and so on, through the nth difference of y_i. Thus, an equation of the form

$$\sum_{k=0}^{n} a_k y_{i+k} = f_i \tag{1}$$

is called a *linear nth-order difference equation* with constant coefficients. If the function on the right side of the equation is zero, the difference equation is said to be *homogeneous.*

Difference equations have a great deal in common with differential equations. In fact, they become differential equations in the limit as the difference between the arguments corresponding to the y_i approaches zero. As with differential equations, nth-order homogeneous difference equations with constant coefficients have n distinct solutions; that is, any solution is a linear combination of these n basic solutions. Commonly, these basic solutions are of the form ρ^i (compare solutions of homogenous linear differential equations, which are of the form ρ^x, where $\rho = e^\alpha$ for some α). As with initial-value differential equations, the constants in the linear combination of these basic solutions are determined by initial values $y_0, y_1, \ldots, y_{n-1}$.

To find the values of ρ for the difference equation

$$\sum_{k=0}^{n} a_k y_{i+k} = 0, \tag{2}$$

we substitute $y_j = \rho^j$ in this equation, producing

$$\sum_{k=0}^{n} a_k \rho^{i+k} = 0. \tag{3}$$

This implies either $\rho = 0$ (the trivial solution) or

$$\sum_{k=0}^{n} a_k \rho^k = 0, \tag{4}$$

a polynomial equation. If this polynomial has n distinct roots $\rho_1, \rho_2, \ldots, \rho_n$, the basic solutions are

$$y_i = \rho_j^i, \; j = 1,2,\ldots,n. \tag{5}$$

If any root ρ_j has multiplicity $m > 1$, the basic solutions associated with that root are ρ_j^i, $i\rho_j^i$, $i^2\rho_j^i$, $\ldots$, $i^{m-1}\rho_j^i$.

APPENDIX B.
NOTATION

1. SPECIAL SYMBOLS AND NOTATIONAL CONVENTIONS

$a \approx b$	a is approximately equal to b
$a \nless b$	a is not less than b
$a \lesssim b$	a is less than a number approximately equal to b
$a \equiv b$	a is defined as b, a is equivalent to b
$a \ll b$	a is very much less than b
$a \gg b$	a is very much greater than b
$a \cdot b$	a times b
$a \times b$	a times b
$a \pm b$	a plus or minus b
$+, -, \times, \div$ as subscripts to error	error in corresponding arithmetic operation
$x \to b$	x approaches b
$x \downarrow b$	x approaches b from above
$\forall x$	for all x
$\ni:$	such that
$\Rightarrow$	implies
$\Leftarrow$	is implied by
$\Leftrightarrow$	is true if and only if
$x \in I$	x is in the interval I, x is a member of the set I
$x \notin I$	x is not in the interval I, x is not a member of the set I
$+*$	addition with error as implemented on a computer
$x*$	x with error
$f * g$	convolution between functions f and g
$\|x\|$	magnitude of x
(a,b)	open interval with endpoints a and b
$[a,b]$	closed interval with endpoints a and b
$[a,b)$	interval closed at endpoint a and open at endpoint b
$[x_1,x_2,x_3,x_4]$	smallest closed interval including x_1, x_2, x_3, and x_4
$\{x_i\}$	sequence of x_i, set of x_i
$\{x_i \mid x_i < 0\}$	set of x_i such that $x_i < 0$

$\sum_{\substack{i=1 \\ i \neq j}}^{n} x_i$	sum of x_i over all i between $i = 1$ and $i = n$ such that $i \neq j$
$\prod_{\substack{i=1 \\ i \in S}}^{n} x_i$	product of x_i over all i between $i = 1$ and $i = n$ such that $i \in S$
$\lfloor x \rfloor$	the greatest integer less than or equal to x
$k!$	k factorial $\equiv k(k-1)(k-2)\cdots(2)(1)$
$\binom{n}{i}$	$n(n-1)(n-2)\cdots(n-k+1)/k!$
$ln(x)$	natural logarithm of x
$\delta(x-a)$	Dirac delta function (impulse) at $x = a$
$comb_h(x)$	function made of an impulse at $x = kh$ for all integers k
$rect_u(x)$	function which is 1 for $x \in (-u,u]$ and zero elsewhere
$f(x)$	function f with argument variable x
f	function f, or vector f with elements $f_i \equiv f(x_i)$
$f^i(x), f_i(x)$	ith function
$f'(x)$	first derivative of function $f(x)$
$f''(x)$	second derivative of function $f(x)$
$f^{(m)}(x)$	mth derivative of function $f(x)$
$J(f)$	Jacobian of the function f
$f^{-1}(x)$	the function inverse to f
$f[x_0,x_1,x_2,x_3]$	divided difference of f using arguments x_0, x_1, x_2, and x_3
$\Delta f(x), \Delta f_i$	forward difference of f
$\nabla f(x), \nabla f_i$	backward difference of f
$\delta f(x), \delta f_i$	central difference of f
$\mathscr{F}(f(x))$	infinite continuous Fourier transform of $f(x)$
$F(v)$	infinite continuous Fourier transform of $f(x)$ with argument frequency v (and similarly with the capital of any letter)
$\tilde{F}(v_j)$	discrete coefficient density of $f(x)$ at frequency v_j
$\phi_h(x)$	the discrete function with interval h corresponding to the continuous function $f(x)$ (and similarly with the Greek letter corresponding to any letter)
$[s(f)](x)$	the function obtained by applying the operator s to the function f, with argument variable x
$O(x^n)$	terms which when divided by x^n approach zero in the limit (as $x \to 0$ or $x \to \infty$, depending on the context)
$Pr(\ \ )$	the probability that the event inside the parentheses occurs

$P_x(i)$ the probability that $x = i$ (discrete probability)
$P_{x,y}(i,j)$ the probability that both $x = i$ and $y = j$ (joint probability)
$P_{x|y}(i|j)$ the probability that $x = i$ given that $y = j$ (conditional probability)
$p_x(u)$ the probability density at $x = u$
$p_{x,y}(u,v)$ the joint probability density at $x = u$ and $y = v$
$p_{x|y}(u|v)$ the conditional probability density at $x = u$ given $y = v$
$var(x)$ the variance of the random variable x
$cov(x,y)$ the covariance of the random variables x and y
$\hat{x}$ estimator of x
$\underline{x}$ vector x
x_i the ith element of the n-vector x, or
the ith element of a sequence $\{x_i\}$, or
the ith tabular argument
$\overline{x}$ sample mean of the random variable x, or
the complex conjugate of x
x^i x to the ith power, or
the ith vector
$x^{(i)}$ the ith vector in an iterative sequence
Ø zero, in a program

In the context of vectors and matrices:

A (any capital Roman letter) a matrix
a (any lowercase Roman letter) a vector
α (any lowercase Greek letter) a scalar
A^T the transpose of A
A^{-1} the inverse of A
A_{ij} the ijth element of A
a_i the ith element of a
$A^{(i)}$ the ith matrix
A' a modified version of A
a^i the ith column vector of A
$a^{(i)}$ the ith row vector of A
$\begin{bmatrix} & \end{bmatrix}$ a matrix, or a vector (if only one column)
$det(A)$ the determinant of A
$\|A\|, \|a\|$ the norm of A, of a
$\|A\|_p, \|a\|_p$ the $\mathscr{L}_p$ norm of A, of a

$\|A\|$, $\|a\|$	the $\mathscr{L}_2$ norm of A, of a
$cond(A)$	the condition number of A
$cond_p(A)$	the $\mathscr{L}_p$ condition number of A
(a,b)	the inner product between a and b
D_A	the diagonal matrix agreeing with A on the diagonal
L_A	the strictly lower triangular matrix agreeing with A in its strictly lower triangular part

2. COMMON USAGE OF ALPHABETICAL SYMBOLS

(See also part 1.)

$[a,b]$	interval for approximation, or integration limits
a, b, c	constants
$\hat{a}, \hat{b}$	least-squares approximation coefficients
a_i, b_i	coefficients in a linear approximation
a_i, b_i, c_i, d_i	polynomial coefficients
b	absolute error bound, or base of computer arithmetic, or n-vector on right side of linear equation
B	absolute error bound, or matrix operating on error vector in iteration
c_i	curvature at $x = x_i$, or approximate convergence factor at step i of an iteration
d	interval endpoint in finite Fourier transform, or vector of inner products on right side of normal equations for least-squares approximation
d^i	principal vector
e	the base of the natural logarithm $= 2.718\ldots$
e^i	the unit n-vector with a single nonzero element in position i
E	expected value, or shifting operator, or matrix approximation to A in iterative method for solution of linear equations
f, g, h	functions
$f(x)$	input function to operator, or function whose root is to be found, or underlying function in approximation problem

$f(x,y)$	right side of first-order ordinary differential equation
$\hat{f}(x)$	approximating function
$f_i(x)$	ith function in function vector
$f_z(x)$	$f(x)$ shifted right by z
$f^j(x)$	jth function in a set or sequence of functions, or jth basis function for linear approximation
f^j	jth basis vector for linear approximation—with elements $f^j(x_i)$
F	matrix of inner products on left side of normal equations for least-squares approximation
$g(x)$	output function from operator, or function in relaxation, $x = g(x)$, producing an iteration
$g_i(x)$	ith function in multidimensional relaxation, producing an iteration
$g_h(x_i)$	result of discrete convolution with interval h
h	interval width
i	$\sqrt{-1}$
i, j, k, l, m, n	integer variables or constants
I	identity matrix, or interval, or integral
k	constant, or convergence factor
k_i	ratio of standard deviation of data value y_i to that of y_1
K	constant, or value of $\frac{\partial f}{\partial y}$ in first-order ordinary differential equation
K_j	value of $\frac{\partial f}{\partial y}(x_j, y_j)$ in first-order ordinary differential equation
L	vector space, or lower triangular matrix
L_n	vector space of all n-vectors
L_A	vector space containing all images under transformation by A
$L_j(x)$	Lagrange polynomial corresponding to x_j
$\mathscr{L}_p$	the norm of an n-vector defined by $\left(\sum_{i=1}^{n} \lvert x_i \rvert^p\right)^{1/p}$

m	slope, or multiplicity of a root, or number of basis functions used in a linear approximation
m,n	matrix dimensions
M	a nonnegative bound < 1
$n(x)$	noise function
N	number of data points used in an approximation, or that number minus 1
$p(x), q(x), r(x), s(x)$	polynomials
$\hat{p}(x)$	best approximating polynomial
$\hat{p}_{(m)}(x)$	best approximating polynomial of degree m
$p_n(x)$	polynomial of degree n
$p_{i_1 i_2 i_3 \ldots i_n}(x)$	approximating polynomial through data points $i_1, i_2, i_3, \ldots, i_n$
P	permutation matrix
$P_i(x)$	Legendre polynomial of degree i
r	remainder, or relative error, or relative error bound
r_{xy}	estimator of correlation between random variables x and y
R	relative error bound, or covariance matrix
s	scaling vector, or interval counter, or operator
$s(x)$	signal function
s_i	slope at x_i
s_x^2	estimator of variance of random variable x
s_{xy}	estimator of covariance of random variables x and y
S	a sum
t	time
$T_k(x)$	Tchebycheff polynomial of degree k
$T^i(h)$	result of i steps of Romberg integration with interval h
u,v,w	arguments of probability density functions
U	upper triangular matrix, or matrix of eigenvectors of A^T
v^i	eigenvector
V	matrix of eigenvectors
$w(x)$	impulse response function, or weighting function

w_i	ith weight, or
	ith modified data value
W	exponential multiplier in Fast Fourier transform
$W(v)$	transfer function
x	argument
x_i	argument point
x_{goal}	most distant argument from initial argument at which solution to ordinary differential equation is required
x, y, z	random variables
x^+, x^-	interval limits for root
$y(x)$	data function, or
	unknown function in ordinary differential equation
y_i	data value at x_i, or
	solution of ordinary differential equation at x_i
$\hat{y}_i$	computed solution to ordinary differential equation at x_i
$y_{(p)}$	predictor solution to ordinary differential equation
$y_{(c)}$	corrector solution to ordinary differential equation
$y_{(m)}$	mopped-up solution to ordinary differential equation
$y_{(pm)}$	mopped-up predictor solution to ordinary differential equation
$y_{(cm)}$	mopped-up corrector solution to ordinary differential equation
z	root, or
	original argument
z_i	original argument point
α	order of convergence
α, β	coefficients in the power method and Newton-Bairstow method
δ	small change, or
	tolerance
δ_i	error in approximation at x_i
δ^i	error vector at ith step of iterative method
δx	small change in x
Δx	small change in x, or
	interval in x
$\Delta(x)$	error polynomial
Δ^i	change in iterate at ith iteration
ε	error, or
	small number
$\varepsilon(x)$	error in x
ε_x	error in x
ε_i	error in ith iterate, or

	error in data at x_i
ε^i	error in ith iterate
ε^{comp}	computational error
ε^{gen}	generated error
ε^{inp}	input error
$\varepsilon^{overall}$	overall error
ε^{prop}	propagated error
ε^{trunc}	truncation error
$\zeta, \eta, \theta, \phi$	parameters in error term of approximation
θ, ϕ	angle
λ	eigenvalue
Λ	diagonal matrix of eigenvalues
μ	mean, or averaging operator
μ_x	mean of random variable x
ν	frequency
ξ	root, or limit of sequence, or parameter in error term of polynomial exact-matching approximation
ξ_i	coefficient of ith eigenvector in expansion of a vector
π	3.141592...
ρ_{xy}	correlation between random variables x and y
ρ_i	basic solution of a difference equation
$\rho_i(\alpha)$	element of sequence with convergence factor as limit point if α is order of convergence
σ	standard deviation
σ_i	standard deviation of data value y_i
σ_x^2	variance of random variable x
σ_{xy}	covariance of random variables x and y
$\hat{\sigma}_{(m)}$	estimator of σ using m-function approximation
σ^{gen}	standard deviation of generated error
$\sigma^{overall}$	standard deviation of overall error
σ^{prop}	standard deviation of propagated error
τ	tolerance
ϕ^x	vector of approximation basis functions evaluated at x
ϕ^i	vector of approximation basis functions evaluated at x_i
$\psi(x)$	$\prod_{i=0}^{N}(x - x_i)$, where the x_i are the argument points used in a polynomial exact-matching approximation

APPENDIX C. SPECIFICATIONS AND CONVENTIONS OF THE PROGRAMMING LANGUAGE USED

The algorithms discussed in this book can be expressed in any of several programming languages. PL/I has been chosen because it produces easily readable programs. The reader who is not familiar with PL/I may benefit from the following summary of the parts of PL/I used and of the programming conventions adhered to.

A PL/I program consists of a series of statements, which are executed in sequence, except as indicated below. Each statement may be preceded by an alphanumeric label followed by a colon, and it ends with a semicolon. Statements are not affected by alignment or line changes, but indentation is used to indicate to the reader the extent of loops and other coherent program sections.

Comments are enclosed by /* and */ and appear wherever they are needed for documentational purposes. Comments on lines by themselves describe the effects of the program sections they precede. Comments on lines with PL/I statements indicate the effects of those statements.

The PL/I statements used can be classified as shown in the following table. In this table, uppercase letters are used for PL/I keywords and variables, and lowercase letters designate general constructs in PL/I.

Statement Type	Specification	Examples
Assignment	A statement with a variable name followed by the assignment symbol, =, followed by an arithmetic expression assigns the result of evaluation of the expression to the variable represented by the name to the left of the assignment symbol.	A = B + C*D;
Branch	This type of statement consists of the words GO TO followed by a statement label. It modifies the sequence of execution of statements. A new sequence is begun at the statement preceded by the label specified.	GO TO PLACE;

(*table continued*)

<table>
<tr><th>Statement Type</th><th>Specification</th><th>Examples</th></tr>
<tr><td>Conditional</td><td>Choosing one of two groups of statements to be executed, depending on the truth or falsehood of a condition, is the effect of the statement of the form
IF condition
THEN then-clause
ELSE else-clause
where “condition” is a condition which is either true or false (for example, $A < B$); “then-clause” is the program section to be executed if the condition is true; and “else-clause” is the program section to be executed if the condition is false. Each clause is either a PL/I statement or a group of statements which starts with a statement consisting of a label followed by the word DO (for example, NAME: DO;) and ends with the word END followed by the label of that DO statement (for example, END NAME;). The word ELSE followed by the else-clause is omitted if nothing particular is to be done only if the condition is false.
After either the then-clause or the else-clause (if specified) is executed, the next statement to be executed (assuming no GO TO statement referring to a statement outside the clause is first executed) is the one following the else-clause.</td><td>IF A < B
THEN C = D;
ELSE NAME: DO;
C = F;
D = E;
END NAME;

IF C = E
THEN GROUP: DO;
A = B + C;
B = D;
C = D;
END GROUP;</td></tr>
<tr><td>Loop</td><td>A loop is initiated by a labeled statement of the form
label: DO index-variable =
starting-value TO ending-
value BY increment;
It is ended by a statement of the form
END label;
where “label” is the label of the corresponding DO statement. The END statement is aligned with the label of the DO statement, and all statements</td><td>LOOP: DO I = 1 TO N BY 2;
SUM = SUM + I;
END LOOP;</td></tr>
</table>

(*table continued*)

Statement Type	Specification	Examples
	between these two (the interior statements of the loop) are indented for readability. This construction has the following effect: First, the index variable is given the starting value in an initialization step. Next, a test is made to see whether the index variable has a value beyond (greater than if the increment is positive, less than if the increment is negative) the ending value. If so, the loop is terminated, and execution continues with the statement following the END statement of the loop. If not, the interior statements of the loop are executed until the END statement of the loop is reached, either in sequence or as the result of a GO TO statement. Then, the increment is added to the value of the index variable, and the sequence described by this paragraph is repeated until the test is satisfied. If the "BY increment" part of the DO statement is omitted, the increment is assumed to be 1.	SEQ: DO I = −3 TO 17; A(I) = 1; IF C < D(I) THEN GO TO SEQEND; A(I) = A(I) + 1; SEQEND: END SEQ;
Declaration of variables	The DECLARE statement is used to specify whether variables are fixed-point or floating-point, whether they are simple or array variables, and if the latter, the number of indices (subscripts) and the range of each. The statement is of the form DECLARE variable-list property-list, variable-list property-list, ..., variable-list property-list; where each variable-list is either a single variable name or an array name, or a parenthesized list of variable names and array names, with the members of the list separated by commas. Each	DECLARE (A, B(N), C, KK, D(0:3)) FLOAT, G(7,2,3) FIXED BINARY;

(*table continued*)

Statement Type	Specification	Examples
	property-list is made up of one or more properties (attributes) separated by blanks. The statement specifies that the variables in each variable-list have the properties specified in the associated property-list. In this book, properties are restricted to FLOAT, indicating floating-point variables, and FIXED BINARY, indicating fixed-point (binary integer) variables. Next to each variable is a comment indicating the use of the variable. The only variables which do not appear in a DECLARE statement are those used to index DO loops (which are always fixed-point variables in this book). An array comprises a group of data items with identical properties. A variable name that refers to an array is specified by following the name by a parenthesized list of index ranges separated by commas. A range is indicated for each index in the array. (For example, G(7,2,3) represents a triply indexed array.) Each index range is a single number or variable name, indicating that the range of the associated index is from 1 to the value of the number or variable, or it is a pair of such numbers or variable names separated by a colon, indicating that the range of the index is between the values of the two numbers or variables.	

In executable statements (all of the above except the DECLARE statement), array-variable names are followed by a parenthesized list of numbers, variable names, and arithmetic expressions. The value of each index is that of the associated number, variable, or expression at the time of execution of the statement. Any name immediately followed by a parenthesized list that has not been specified as an array in a DECLARE statement is the name of a function subroutine which is being invoked (for example, A = ABS(B); causes the PL/I absolute-value function to be invoked to determine the absolute value of the variable represented by B and assigns that value to the variable A).

INDEX

NOTE: Numbers in **boldface** refer to relevant sections of the book; individual page references within such a section are not given for the entry for which the section reference appears.